Infectious Disease Informatics

Vitali Sintchenko
Editor

Infectious Disease Informatics

Springer

Editor
Vitali Sintchenko
Centre for Infectious Diseases and Microbiology
Sydney Medical School
The University of Sydney
Sydney, NSW 2006
Australia
vsintchenko@usyd.edu.au

ISBN 978-1-4419-1326-5 e-ISBN 978-1-4419-1327-2
DOI 10.1007/978-1-4419-1327-2
Springer New York Dordrecht Heidelberg London

Library of Congress Control Number: 2009933097

Printed on acid-free paper

Springer is part of Springer Science+Business Media (www.springer.com)

Preface

There are several reasons to be interested in infectious disease informatics. First, it is of practical significance to understand how the technology revolution has been reshaping infectious disease research and management, as rapid advances in genome-associated technologies have changed the very nature of the questions we can ask. Second, the emerging evidence has confirmed that the application of information technologies in healthcare enhances our ability to deal with infectious diseases. Finally, the implementation of electronic health records has created new and exciting opportunities for secure, reliable and ethically sound clinical decision support and biosurveillance guided by the genomics of pathogens with epidemic potential.

This volume addresses the growing need for the critical overview of recent developments in microbial genomics and biomedical informatics relevant to the control of infectious diseases. This field is rapidly expanding, and attracts a wide audience of clinicians, public health professionals, biomedical researchers and computer scientists who are fascinated by the complex puzzle of infectious disease. This book takes a multidisciplinary approach with a calculated move away from the traditional health informatics topics of computerized protocols for antibiotic prescribing and pathology testing. Instead authors invite you to explore the emerging frontiers of bioinformatics-guided pathogen profiling, the system microbiology-enabled intelligent design of new drugs and vaccines, and new ways of real-time biosurveillance and hospital infection control. Throughout the book, references are made to different products supplied by public sources and commercial vendors, but this is not an endorsement of these products or vendors.

I am deeply grateful to all of the contributors for the generous sharing of their knowledge and expertise. Special thanks go to my good friends and colleagues at the Centre for Infectious Diseases and Microbiology at Westmead Hospital and The University of Sydney, and at the Centre for Health Informatics at the University of New South Wales for their encouragement and support. I would also like to thank Jeffrey Ciprioni, Jenny Wolkowicki and Rajesh Harini from Springer Life Science for their continual help during this book's production.

I hope that this volume will be useful to those already working in the field but seeking to broaden their horizons. I also hope that it will encourage interest in infectious disease informatics among readers in general.

Sydney — Vitali Sintchenko

Contents

Contributors

Samia Aci
Institut de Recherches en Technologies et Sciences pour le Vivant, Grenoble, France

Stephen Anthony
Centre for Health Informatics, University of New South Wales, Sydney, NSW, Australia

Lyn-Marie Birkholtz
Department of Biochemistry, School of Biological Sciences, and African Centre for Gene Technologies, Pretoria, South Africa

Vincent Breton
IN2P3, Clermont-Ferrand, France

Loraine Brillet
Institut de Recherches en Technologies et Sciences pour le Vivant, Grenoble, France

Caroline Buckee
Department of Zoology, University of Oxford, Oxford, UK

Peter Burger
Bioinformatics and Computational Biology Unit, University of Pretoria, Pretoria, South Africa

Richard Christen
University of Nice Sophia-Antipolis, and Institute of Developmental Biology and Cancer, Parc Valrose, Centre de Biochimie, Nice, France

Tobias Cohen
Brown University School of Medicine, Providence, RI, USA

Enrico Coiera
Centre for Health Informatics, University of New South Wales, Sydney, NSW, Australia

Lindsay G. Cowell
Department of Biostatistics and Bioinformatics, Duke University Medical Center, Durham, NC, USA

Russell J. Crawford
Swinburne University of Technology, Faculty of Life and Social Sciences, Melbourne, VIC, Australia

Ana Lucia da Costa
IN2P3, Clermont-Ferrand, France

Matthew N. Davies
SGDP Centre, Institute of Psychiatry, London, UK

Tjaart de Beer
Bioinformatics and Computational Biology Unit, University of Pretoria, Pretoria, South Africa

Anne De Groot
Institute for Immunology and Informatics, University of Rhode Island, Providence, RI, USA

Riet De Smet
Department of Microbial and Molecular Systems, K.U. Leuven, Leuven, Belgium

Claudio Donati
Novartis Vaccines and Diagnostics, Siena, Italy

Ana Carolina Fierro
Department of Microbial and Molecular Systems, K.U. Leuven, Leuven, Belgium

Darren R. Flower
The Jenner Institute, University of Oxford, Oxford, UK

Christian Forst
UT Southwestern Medical Center in Dallas, TX, USA

Blanca Gallego
Centre for Health Informatics, University of New South Wales, Sydney, NSW, Australia

Phillip Giffard
Professor, Menzies School of Medical Research, Darwin, NT, Australia

Gwendolyn L. Gilbert
Centre for Infectious Diseases and Microbiology, Institute of Clinical Pathology and Medical Research, Westmead Hospital, The University of Sydney, Sydney, NSW, Australia

Sunetra Gupta
Department of Zoology, University of Oxford, Oxford, UK

Afra Held
Centre for Health Informatics, University of New South Wales, Sydney, NSW, Australia

Bala Hota
Rush University Medical Center, Chicago, IL, USA

Jonathan Iredell
Centre for Infectious Diseases and Microbiology, Westmead Hospital, Sydney West Area Health Service, Sydney, NSW, Australia

Elena Ivanova
Faculty of Life and Social Sciences, Swinburne University of Technology, Melbourne, VIC, Australia

Fourie Joubert
Bioinformatics and Computational Biology Unit, University of Pretoria, Pretoria, South Africa

Arkadiy Kurilenko
Faculty of Life and Social Sciences, Swinburne University of Technology, Melbourne, VIC, Australia

Ross Lazarus
Channing Laboratory, Harvard Medical School, Boston, MA, USA

Karen Lemmens
Department of Microbial and Molecular Systems, K.U. Leuven, Leuven, Belgium

Braam Louw
Department of Biochemistry, School of Biological Sciences, and African Centre for Gene Technologies, Pretoria, South Africa

Kathleen Marchal
Department of Microbial and Molecular Systems, K.U. Leuven, Leuven, Belgium

Eric Maréchal
Institut de Recherches en Technologies et Sciences pour le Vivant, CEA Grenoble, France

William Martin
EpiVax, Inc., Providence, RI, USA

Duccio Medini
Novartis Vaccines and Diagnostics, Siena, Italy

Lenny Moise
EpiVax, Inc. and University of Rhode Island and Brown University Medical School, Clinica Esperanza/Hope Clinic GAIA Vaccine Foundation, Providence, RI, USA

Matthew O'Sullivan
Centre for Infectious Diseases and Microbiology, University of Sydney, Sydney, NSW, Australia

Sally Partridge
Centre for Infectious Diseases and Microbiology, Westmead Hospital, Sydney West Area Health Service, Sydney, NSW, Australia

Rino Rappuoli
Novartis Vaccines and Diagnostics, Siena, Italy

Sylvaine Roy
Institut de Recherches en Technologies et Sciences pour le Vivant, Grenoble, France

Jaron Schaeffer
Centre for Health Informatics, University of New South Wales, Sydney, NWS, Australia

Michael Selgelid
Centre for Applied Philosophy and Public Ethics, Australian National University, Canberra, ACT, Australia

Vitali Sintchenko
Centre for Infectious Diseases and Microbiology, Sydney Medical School, The University of Sydney, Sydney, NSW 2006, Australia

Barry Smith
Department of Philosophy, Center of Excellence in Bioinformatics and Life Sciences and National Center for Biomedical Ontology, University of Buffalo, Buffalo, NY, USA

Guy Tsafnat
Centre for Health Informatics, University of New South Wales, Sydney, NSW, Australia

Hélène Valadié
Institut de Recherches en Technologies et Sciences pour le Vivant, Grenoble, France

Feng Wang
Faculty of Life and Social Sciences, Swinburne University of Technology, Melbourne, Victoria, Australia

Gordon Wells
Bioinformatics and Computational Biology Unit and University of Pretoria, Pretoria, South Africa

Chapter 1
Informatics for Infectious Disease Research and Control

Vitali Sintchenko

1.1 Introduction

Infectious disease informatics has been defined as a new field that studies knowledge creation, sharing, modeling and management in the domain of infectious diseases (Zeng et al. 2005). Its emergence has been fueled by rapid increases in the amount of biomedical and clinical data, and demands for data analyses. The resulting combinations of experimental and informatics evidence have reshaped the ways of conducting infectious disease research, raising the expectation of better control of infectious diseases. The authors of this book argue that informatics has not only changed the scale on which the infectious disease research is being done but has also conceptually opened up different ways of managing patients and making discoveries in the field of infectious diseases.

The goals of infectious disease informatics are lofty and include the optimization of the development of antimicrobials, the improved design of more effective vaccines, the identification of biomarkers for transmissibility and clinical outcomes of infectious diseases, and a better understanding of host-pathogen interactions. In the last two decades, the emergence of new informatics methods and integrated databases has facilitated the realization of these goals. This chapter outlines the major challenges and opportunities that infectious disease informatics faces in the twenty-first century.

V. Sintchenko
Centre for Infectious Diseases and Microbiology, Syndey Medical School,
The University of Sydney, Sydney, NSW 2006, Australia

V. Sintchenko (ed.), *Infectious Disease Informatics*,
DOI 10.1007/978-1-4419-1327-2_1, © Springer Science+Business Media, LLC 2010

1.2 Handling New Data Types

1.2.1 Microbial Genome Assembly and Annotation

"New Age" infectious disease informatics rests on advances in microbial genomics, the sequencing and comparative study of the genomes of pathogens, and proteomics or the identification and characterization of their protein related properties and reconstruction of metabolic and regulatory pathways (Bansal 2005). The speed of microbial genome sequencing has been steadily accelerating since the introduction of modern DNA sequencing methods more than thirty years ago (Sanger et al. 1977). The accumulation of sequenced genomes of bacteria shows a good fit to exponential functions with a doubling time of approximately 20 months (Koonin and Wolf 2008). Despite the historical bias towards the "working horses" of bacterial genomics, such as commensals *E. coli* and *B. subtilis* (Collado-Vides et al. 2008), the depth and breadth of the coverage of sequences belonging to different species of viral, bacterial, fungal and protozoan pathogens has been rapidly expanding.

Microbial genomes are thousands or millions of base pairs in length, requiring both a global view of the genome and the ability to zoom in on details for the purpose of analysis and annotation. Annotation is the extraction of biological knowledge from raw nucleotide sequences (Médigue and Moszer 2007). Such decoding of the genomes allows the prediction of protein-coding genes and therefore, the proteins the organism is able to produce. Desktop computer sequence editors such as Chromas Lite (http://chromas-lite.software.informer.com/), Trace Edit (http://www.ridom.de/traceedit/) or commercial products like LaserGene (http://www.dnastar.com/products/lasergene.php) or Sequencher (http://www.sequencher.com/) are helpful in the initial sequence assessment. The task of assembling of sequences from re-sequencing experiments, when a reference sequence is available, can be supported by tools like TraceEditpro (http://www3.ridom.de/traceeditpro/) or SeqScape.

Different software pipelines have been developed to automate microbial genome annotation and assembly (Table 1.1). The Integrated Microbial Genome (IMG) system, hosted by the Joint Genome Institute (JGI), and the RAST (Rapid Annotation using Subsystem Technology) server are examples of open resources. Major sequencing centers offer genome viewers and browsers through their websites (McNeil et al. 2007). For example, Manatee (J. Craig Venter Institute (JCVI)) has been developed to view and to alter initial automatic annotations of prokaryotic genomes. The Sanger Institute's Pathogen Sequencing Unit has been maintaining freeware for sequence analysis, viewing and annotation, such as Artemis and the Artemis Comparison Tool (ACT) (Carver et al. 2008). The alignment of genomes of three strains of *Staphylococcus aureus* using ACT is shown in Fig. 1.1. Alternatively, multiple genome alignments in the presence of large-scale evolutionary events, such as rearrangement and inversion, can be efficiently constructed and visualized using the Mauve program (http://gel.ahabs.wisc.edu/mauve/download.php) (Darling et al. 2004). These tools assist in the rapid identification of protein-coding

Table 1.1 Bioinformatics analysis tools

Analysis tasks	Tools	URL
ORF or gene identification	ORF Finder	http://www.ncbi.nlm.nih.gov/gorf.html
	GeneMark	http://opal.biology.gatech.edu/GeneMark/genemarks.cgi
	GLIMMER	http://www.cbcb.umd.edu/software/glimmer/
Sequence alignment	ClustalW	http://www.ebi.ac.uk/clustalw/
	Tcoffee	http://www.tcoffee.org/Projects_home_page/
	MUSCLE	http://www.drive5.com/muscle/
Genome annotation	RAST	http://rast.nmpdr.org/
	Artemis and ACT	http://www.sanger.ac.uk/Software/
	IMG	http://rast.nmpdr.org/
	MAUVE	http://genome-alignment.org/mauve/
Phylogenetic analysis	Phylogeny programs	http://evolution.genetics.washington.edu/phylis/software.html
	SplitsTree	http://www.splitstree.org
	MEGA	http://www.megasoftware.net
Microarray analysis	Gene Expression Omnibus	http://www.ncbi.nih.gov/geo/
	Microarray informatics EBI	http://www.ebi.ac.uk/microarray
Metabolic pathway analysis	KEGG	http://www.genome.ad.jp/kegg/kegg2.html
	UniPathway	http://www.grenoble.prabi.fr/obiwarehouse/unipathway
Whole genome visualization	BacMap	http://wishart.biology.ualberta.ca/BacMap/index_2.html
	GenomeAtlas	http://www.cbs.dtu.dk/services/GenomeAtlas/

genes, as well as other features like non-coding RNA genes, repetitive sequences or recently acquired DNA.

Web servers like Integrated Microbial Genomes (Joint Genome Institute; http://img.jgi.doe.gov) or the Bacterial Annotation System (BASys, http://wishart.biology.ualberta.ca/basys/cgi/submit.pl) also support comparative analysis and the automated annotation of bacterial genomic (chromosomal and plasmid) sequences (Van Domselaar et al. 2005). They accept raw sequence data and gene identification information, and provide textual annotation and hyperlinked image output.

Strings of nucleotides are assembled into draft sequences that can be characterized by the following: (1) > 90% of genome in contigs, (2) average contig length > 5 kb, (3) >90% of a set of conserved genes present, (4) contig N90 length > 5 kb, (5) >90% of bases > 5× read coverage, (6) scaffold N90 length > 20 kb. The information used to annotate genomes comes from three types of analysis: (1) ab initio gene finding programs, which are run on the DNA sequence to predict protein coding genes; (2)

Fig. 1.1 Alignment of genomes of three strains of *Staphylococcus aureus*. DNA sequences that find a perfect match are connected with red lines or blocks. *Blue areas* are inversions or transitions and *white areas* represent indels. The figure was produced using Artemis software (The Wellcome Trust Sanger Institute, UK)

evidence-based gene calling or translating alignments of the DNA sequence to known proteins; and (3) aligning cDNAs from the same or related species. Gene finding has progressed far beyond the simple identification of open reading frames. The programs aligning cDNA and protein sequences to genomic DNA can locate the protein coding regions by searching the publicly available databases or by applying machine learning algorithms such as Hidden Markov Models (HMM). There is a long list of such programs including GeneMark, mORFind, PRODIGAL (Prokaryotic Dynamic programming Genefinding Algorithm), Argon and GLIMMER (Gene Locator and Interpolated Markov Modeller) (Delcher et al. 1999; Suzek et al. 2001; Majoros 2007). They differ in the time required for automated annotation as well as the quality of gene calling (Guigo et al. 2006). Problems with the accuracy of current gene finders reflect not only the performance of their algorithms but also the quality of the primary resources and the abundance of non-coding DNA regions in microbial genomes. Genome assembly annotation methods and tools including new applications for RNA genes, were reviewed in detail elsewhere (Stothard and Wishart 2006; Médigue and Moszer 2007; Brent 2008; Pop and Salzberg 2008).

Recent breakthroughs in high-throughput sequencing technologies have posed new challenges for genome assembly, annotation and analysis. These technologies make it feasible to sequence not only static genomes but also entire transcriptomes expressed under different conditions (Shendure and Ji 2008). However, they can produce read lengths as short as 35–40 nucleotides, which cannot be analyzed with software developed for Sanger data as they are often non-unique, lack neighborhood context and have a different distribution of errors. The task of linking such short-reads may be accomplished using a comparative assembly algorithm, in which new sequences are put together by mapping them onto close relatives or the "reference genomes." Not surprisingly, the comparative assembly strategy works best when the two species are more than 90% identical. Alternatively, when no "reference genome" is available, the new cohort of assembly algorithms based on de Bruijn graphs – a way to transform sequence data into a network structure – has risen to the task (Chaisson and Pevzner 2008; MacLean et al. 2009). Strategies and systems that address these new challenges have recently been reviewed elsewhere (Pop and Salzberg 2008; MacLean et al. 2009; Ussery et al. 2009). Tables 1.1 and 1.2 provide examples of informatics tools for pathogen annotation and analysis.

1.2.2 Meta-Omics: Metagenomics and Metaproteomics

The metagenomics or the sequencing of genomes of complex mixed communities has emerged at the interface of genomics, microbiology and information technology. This field examines the interplay of hundreds of microbial species present at specific sites of potential infections in space and time (Hutchinson 2007; Smarr et al. 2009). Significantly, metagenomics has extended its focus from environmental microorganisms to microbial communities or "community whole genome sequences" of the human host (Field et al. 2006; Verberkmoes et al. 2009).

Table 1.2 Examples of bioinformatics resources for pathogens with epidemic potential

Analysis	Tools	URL
Sequence databases and tools	GenBank	http://www.ncbi.nlm.nih.gov/sites/entrez
	Protein Data Bank	http://www.rcsb.org/pdb/
	Microbial Genome Database	http://mbgd.genome.ad.jp/
Workbenches	Virology on the WWW	http://www.virology.net
	Viral Bioinformatics	http://www.biovirus.org
	Research	http://www.microbase.gr
	Microbase	http://xbase.bham.ac.uk/
	xBASE	
	SEED	http://www.theseed.org
	Influenza Virus Resources	http://www.ncbi.nih.gov/genomes/FLU/FLU.html
		http://www.biohealthbase.org
		http://www.flu.lanl.gov/
Pathogen specific datasets	European Hepatitis C database	http://euhcvdb.ibcp.fr/euHCVdb/
		http://hcv.lanl.gov/content/hcv-db/index
	Hepatitis C database	http://www.hiv.lanl.gov/content/index
	HIV databases	
	Poxvirus Resource	http://www.poxvirus.org
	SARS Bioinformatics Suite	http://athena.bioc.uvic.ca/database.php?db = cooronaviridae
	DengueInfo	
		http://www.dengueinfo.org
	Neisseria.org	http://neisseria.org/
	TB Database	http://www.tbdb.org/
	Plasmodium Genome Resource	http://plasmodb.org/plasmo/
Antimicrobial resistance	ARDB	http://ardb.cbcb.umd.edu
	ARGO	http://www.argodb.org/
	Compendium of TEM genes	http://www.lahey.org/studies

Most of the 10–100 trillion microorganisms in the human gastrointestinal tract live in the colon (Turnbaigh et al. 2007). The genomes of these microbial symbionts have been collectively defined as the microbiome or ecosystem in which the number of microbial genes is estimated to be many folds higher than those present in the human genome. The Human Gut Microbiome Initiative, a logical conceptual extension of the Human Genome Project, aims to discover genomes of at least 100 new intestinal species. This approach has targeted the totality of genes involved in the gut biofilms, the mechanisms of horizontal gene transfer, and the role of the microbial pan-genome (Field et al. 2006). The Microbiome project aims to address some of the most inspiring and fundamental scientific questions today in order to identify new ways to determine health and predisposition to diseases and define parameters needed to design, implement and monitor strategies for intentionally manipulating the human microflora (Turnbaigh et al. 2007).

1.2.3 Global Genome Analysis

In addition to conventional strings of nucleotides, large-scale sequencing can provide new types of data reflecting global genome architecture and the properties of pathogens. These data include the size of a genome and its nucleotide composition, the locations of genes and intergenic regions, GC percentage and gene density. Microbial genomes are compared by the number of particular sets of genes, gene order (synteny) and the presence or absence of important genes. Other metrics include gene set properties (the number of two component system regulatory genes) and nucleotide sequence-based measures (distance between paired two-component system genes and consensus sequence) (Whitworth 2008; Ussery et al. 2009). These metrics represent a global view of genomes but often have limited biological meaning. Thus, "signature" sequences have been suggested as a means of identifying organisms or genes with sequence profiles correlating with the pathogen phenotype or disease outcomes. Examples of genome characteristics that are more directly related to biologically important behavior are bacterial IQ (a measure of the number of signal transduction proteins as a function of genome size) and extrovertedness (the proportion of signaling proteins predicted to sense external stimuli) (Galperin 2005).

Analyses of genomics data challenge the traditional taxonomy of microbial species. Recent projects have focused on producing simple analytical diagnostic tools based on strong taxonomic knowledge collated in the DNA reference libraries such as the DNA Barcode of Life Data System (BOLD; http://www.boldsystems.org). These types of data enable the acquisition, storage, analysis and publication of DNA barcode results, and provide clues about the global distribution of species. Their genetic diversity and structure is based on two postulates: first, that every species is represented by a unique DNA barcode (indeed there are 4^{650} possible ATGC combinations compared to an estimated 10 million species remaining to be discovered (Frézal and Leblois 2008)), and second, that the genetic variation between species exceeds the variation within species. DNA barcoding requires a minimum sequence length of 500 bp and more than three individual sequences per species. The initial Barcode of Life framework was based on the sequence of a single universal marker – the cytochrome c oxidase gene – but has evolved since then, giving rise to a flexible description of DNA barcoding, a larger range of applications and the broader use of the term "barcode" (Frézal and Leblois 2008). For example, the whole microbial genome's barcodes were defined as frequency distributions of periodic DNA sequences or *k*-mers across the whole genome (Zhou et al. 2008). It has been postulated that such barcode similarities are proportional to the genomes' phylogenetic closeness and could be utilized in metagenome analyses (Zhou et al. 2008).

Microbial species diversity can be also estimated by the average nucleotide identity (ANI) using the list of orthologs and deriving the overall divergence of the core genome by averaging the percentages of identity at the nucleotide level (Konstantinidis and Tiedje 2005). Another approach to measure distances between genomes is based on estimating the proportion of common genes by calculating the ratio of orthologs to the total number of genes of the reference genome. More recently, similar methods such as DNA content, BLAST distance phylogeny and the MUM (maximal unique and exact matches) index have been suggested as more sensitive measures for intra-species comparisons (Deloger et al. 2009).

1.3 Changing the Way Discoveries Are Made

1.3.1 Knowledge Discovery from Comparative Genomics

The true power of large-scale comparative genomic studies lies in their ability to identify and characterize biological trends and rules that explain particular phenomena (Field et al. 2006). Computational methods have become essential steps in formulating hypotheses about gene functions. The comparative approach has not only yielded fundamental insights into the function and evolution of microbial genomes, but has also led to practical results. Comparative genomics has allowed the accurate estimation of the structure of genomes and the speed of gene movements, including the role of natural selection versus genetic drift, the origin of the pandemic strains, and the ecology of a pathogen in its natural reservoir (Chen et al. 2005; Yang et al. 2008a). Computational studies identified unexpected relationships between genomic features and ecological niches, demonstrated diversity in the microbial world and helped to reconstruct evolutionary relationships among genomes (Binnewies et al. 2006; Field et al. 2006).

Comparisons made between different genomes can also generate new hypotheses for testing, usually relating to the unexpected presence or absence of particular genes with respect to other genomes (Whitworth 2008). The studies of three main forces shaping genome evolution – gene loss, gain and change – have been especially fruitful in this respect (Burrack et al. 2007; Whitworth 2008). Discoveries of gene duplication in many bacterial pathogens, resulting in increased numbers of key gene clusters or the expansion of important protein families have led to the development of new diagnostic methods. For example, the gene clusters encode a secreted protein called the early secretory antigenic target 6 or ESAT6, which was identified as one of the key virulence factors in *Mycobacterium tuberculosis* and was subsequently used in the interferon-gamma release assays for the diagnosis of tuberculosis (Pallen and Wren 2007; Behr 2008).

Comparative genomics has also revealed that pathogens undergo a process of genome decay or a reduction in the number of biosynthetic pathways, resulting in a dependence on the infected host for certain essential functions. The most surprising snapshots of genome decay have come from relatively recently emerged pathogens that have changed their lifestyles by adopting a simpler host-associated niche. For example, the genomes of *Yersinia pestis* (Parkhill et al. 2001b) and *Salmonella enterica* serovar Typhi (Parkhill et al. 2001a) contain hundreds of pseudogenes. These findings challenge the traditional view that bacterial genomes never contain "junk" DNA and that every gene in a bacterial genome must have a function. Instead, every genome should be viewed as a work in progress, burdened with some non-functional "baggage of history" (Pallen and Wren 2007).

As the smallest-scale variation in microbial genomes occurs at the level of single-nucleotide polymorphisms (SNPs), SNP detection has been applied extensively to many pathogens (Yao et al. 2008). While SNPs are generally considered rare, at one per several thousand base pairs, two genomes of *M.tuberculosis* of 4 Mb each may have some 1,0002008 SNPs between two isolates (Behr). Whole-genome sequencing has been proven as an even more powerful tool to detect SNPs. It enabled the differentiation of *Escherichia coli* strains that had diverged for as few as 200 generations (Shendure and Ji 2005) and revealed genomic changes in pathogens in the process of human infection (Chen et al. 2006; Forst 2006; Pallen and Wren 2007).

1.3.2 Automatic Recognition of Functional Regions

In the pre-informatics era, virulence factors were typically identified either by biochemical studies or through genetic screens. Informatics has enabled innovative strategies for the recognition of virulence gene recognition through the analysis of genetic signatures (Pallen and Wren 2007). Despite the variety of microbial life styles and associated genomic and metabolic complexity, pathogen genomes share common architectural principles. As a result, computational techniques assist in exploring similarities between virulence factors and other genes with known functions. This association can then be tested using targeted genetic methods such as the inactivation of the putative virulence gene followed by the comparison of phenotypes of the original and modified microorganisms (Chen et al. 2005; Raskin et al. 2006). A strategy that does not rely on sequence similarity for identifying potential genes is the detection of coding sequences, which is based the gene context "grammars" supplemented with machine learning models (Garrido et al. 2008). For example, functional gene recognition tools GeneMark and GLIMMER employ Hidden Markov models, in which the preceding nucleotide bases are used to predict the next base in a coding region, and the algorithm is trained on a trusted set of sequences. Gene coding regions are then identified using probability estimates of the correct coding "grammar" in a region (Dougherty et al. 2002). Different statistical and machine learning methods for gene prediction have been reviewed elsewhere (Majoros 2007).

Gene-gene interactions specifically associated with a phenotype or a particular disease can be explored with or without a prior biological knowledge. Several techniques utilizing Bayesian networks, pair-wise mutual information and graphical

Gaussian models have been proposed for this purpose. Coupled with biological knowledge, the identification of such phenotype-specific interactions can shed light on the responsible pathways. The complexity of data handling and visualization has led to efforts to develop dedicated comparative genomics resources such as GenDB (Meyer et al. 2003), CMR, ACT, (Table 1.1) xBASE and Microbes OnLine as well as data management systems such as SEED (Table 1.2) (Chaudhuri et al. 2008).

1.3.3 Enabling the Dynamic View of Infectious Diseases

Informatics has been instrumental in the change from static to a dynamic view of the microbial world. In contrast to the static view of genome annotations focused on the gene or protein prediction, the dynamic view places information obtained into a biological context to identify interactions between the genomic components and the reconstruction of regulatory networks (Médigue and Moszer 2007; Sakata and Winzeler 2007). Under the network vision of the microbial world, microbial chromosomes are not envisaged as strictly defined genotypes gradually changing in time but rather as islands of temporary, relative dynamic stability that form tightly connected (vertically and horizontally) areas of the network (Koonin and Wolf 2008). The infection cycle should be considered as a whole and the links between growth, virulence, immune evasion and transmission should be assessed (Restif 2009).

Biological interactions vary in their nature and are spatially and temporally heterogeneous. One can abstract the actions of proteins and metabolites by representing genes acting on other genes as a gene network or as genetic regulatory, transcription or expression networks. Such networks can be constructed using computationally assigned functional linkages inferred by Rosetta Stone, Operon or similar methods (Rachman and Kaufmann 2007; Harrington et al. 2008), and often point to highly connected and central proteins frequently referred to as "hubs" (Wu et al. 2008). Biological interaction and communication networks share several commonalities: they are scale free (only a few nodes are highly connected) and are small world networks (highly clustered with short distances between any two nodes) (Kann 2008). Increasingly, disease pathogenesis and the mechanisms of drug action are viewed from a biological systems perspective (Wu et al. 2008). From this perspective, a deeper understanding of infectious diseases may rely on an exhaustive characterization of all potential interactions occurring between proteins encoded by viruses and those expressed in infected cells. Thus, the integration of all protein-protein interactions into an infected cellular network, or "*infectome*," offers a powerful framework for the virtual modeling and analysis of infections (Navrati et al. 2009). The terms "*interactome*" and "*phenomics*" have been coined in this context (Lussier and Liu 2007).

Numerous resources have been developed to explore host-pathogen interactions (PHI) (Table 1.3). Specifically, PHI-base (Winnenburg et al. 2006), PHIDIAS (Xiang et al. 2007), BioHealthBase (Squires et al. 2008), PIG (Driscoll et al. 2009) VirusMINT (Chatr-aryamontri et al. 2009) and VirHostNet (Navrati et al. 2009) have been

Table 1.3 Knowledge discovery tasks from the host-pathogen interactions

Levels	Microbial genomes	Microbial proteins	Microbial metabolome	References
Human proteins	Gene-protein interactions, networks, defining protein functions	Protein-protein interactions, protein structure prediction, epitope mapping		An and Faeder 2009 Chatr-aryamontri et al. 2009 Driscoll et al. 2009 Garrido et al. 2008 Kann 2008 Xiang et al. 2007
Human metabolites	Pathway mapping and reconstruction	Protein function prediction	Pathway comparison	Lisacek et al. 2006 Winnenburg et al. 2006
Human phenome	Genotype-patient outcome mapping	Disorder prediction, virulence prediction	Biomarker discovery	Burrack and Higgins 2007 Forst 2006
			Virulence prediction	Lengauer et al. 2007
			Drug resistance prediction	Navrati et al. 2009 Raman et al. 2008
	Effect of diseases on gene expression	Drug target identification		Reddy et al. 2009 Squires et al. 2008
	Disease reclassification	Drug resistance prediction		Stavrinides et al. 2008
	Disorder prediction, virulence prediction			
	Drug resistance prediction			

suggested to study and visualize pathogen-related pathways. For example, the VirHostNet is a knowledge base for the management and analysis of proteome-wide virus-host interaction networks and a resource of manually curated interactions defined for a wide range of viral species (Navrati et al. 2009). Genomic and proteomic data is often informationally synergistic, allowing for the reconstruction of known pathways from the first principles. The combination of these forms of data have been used to identify libraries of recurring motifs, where the mixed semantics of the pattern promises to be more informative than any single data source taken in isolation in building biological networks (Michael et al. 2008; Stavrinides et al. 2008).

Systems biology has arisen from various attempts to move away from the reductionist approach, which is hindered by the difficulty of breaking a system into separable and meaningful parts. It encompasses several high-throughput analytic technologies, including genomics, transcriptomics to measure gene expression and its regulation at the level of messenger RNA and microRNA production, proteomics to measure changes in protein production, and computational biology, which depends on analytic software packages for analyzing, organizing, and interpreting those data (Sakata and Winzeler 2007). Such an approach treats pathogens and their environments as a series of hierarchical levels or networks from gene products to whole organisms and integrates the time dimension in order to structure knowledge and to determine rules that would allow navigation between levels (Lisacek et al. 2006). This approach demands new tools for data management, the integration of which offers the opportunity to correlate multiple lines of evidence and to reduce uncorrelated noise.

1.3.4 Cross-Validating the Knowledge Sources

The major difference between the pre- and post-genomics eras is that one can now potentially account for and keep track of all components at once. However, the gathering of a large collection of data does not guarantee that we can make sense of it or that new knowledge will emerge (Collado-Vides et al. 2009). The chance for enriching biomedical knowledge can be increased by mixing various streams of data and gaining robustness from the "cross-validation" of the knowledge sources (Guyet et al. 2007). Public websites like Galaxy (http://galaxy.psu.edu) and InterPro (http://www.ebi.ac.uk/interpro/) offer integration toolsets for genomics and proteomics analyses.

As generating data remains a costly undertaking, computational models have a pivotal role to play in the integrative science. They help researchers to illuminate the underlying processes and identify the key questions that need to be addressed experimentally (Restif 2009). Compared to conventional, small-scale experimental approaches, they give a wider, often more relevant view of host responses to infections or other health insults. These computational models have the capacity to guide and direct wet lab experimental efforts complimenting traditional in vivo, in situ, and in vitro testing with the emerging in silico approach (Lengauer et al. 2007;

Raman et al. 2008). Some impressive starts have been made on bacterial models in the form of simulation tools. For example, the reconstruction of metabolic networks gave birth to the first examples of in silico strains that can be utilized to explore alternative ways of identifying new drug targets (Jamshidi and Palsson 2007). The end result of these simulations may be the genomic bioengineering of microorganisms based on knowledge of interacting systems and networks of genes and gene products.

Text mining tools are being created to query the PubMed literature database and to integrate the available genomic and proteomic information to map the genes and their interrelationship with particular networks of a disease (Korbel et al. 2005; Jelier et al. 2008; Rzhetsky et al. 2008; Zaremba et al. 2009). An unsupervised, systematic approach for associating genes and phenotypic characteristics (G2P) that combines literature mining with comparative genome analysis has been successfully applied and has uncovered clusters of unsuspected G2P associations (Korbel et al. 2005).

1.4 Enabling Knowledge Communities: eScience

The phase of history in which biomedical science could be significantly advanced by individual researchers without data sharing has come to a close. The global, collaborative analyses of data and the exchange of the results across social, political and technological boundaries have created the demand for new cyber-infrastructures for research. There has been a major effort, in the form of e-Science, to develop technologies to fulfill these demands (Craddock et al. 2008).

1.4.1 Novel Infrastructures Support Knowledge Communities

The chance of making a discovery or replicating the finding is greatly increased if there are effective mechanisms for different groups to share data and thereby enlarge the number of samples that are studied. This paradigm has been successful in both human genomics and infectious disease research (e.g., including the rapid discovery and identification of emerged pathogens such as the Nipah virus and the novel coronavirus that caused the SARS epidemic). Post-genomic era solutions such as federated databases and other technologies that enhance connectivity and data retrieval have created a new knowledge environment (Birkholtz et al. 2006; Thorisson et al. 2009). The level of technical competence required of the users is being reduced by the provision of "off-the-shelf" solutions. For example, the GEN2PHEN project offers "database-in-a-box" installation packages, which include an open-source complete genetic association database system with the option for federation (Thorisson et al. 2009).

Alternative infrastructures for e-Science with significant advantages over conventional Internet technologies are offered by grid and cloud computing and the Semantic Web (Numann and Prusak 2007; Craddock et al. 2008). First, grids provide unique access to high performance computing power, distributed applications and sources (see Chap. 14 for examples). Second, grids increase data storage spaces, and allow data and tools to be shared by geographically dispersed users. However, developing and maintaining grid or cloud architectures remains a complex task and requires further advances in security and privacy models before they can be embraced by diagnostic laboratories (Lisacek et al. 2006).

1.4.2 Data Aggregation

Tasks that require an e-Science approach or global science that is performed in silico are typically computationally intensive and use heterogeneous resources that must be integrated across distributed networks (Craddock et al. 2008). Increasingly, the genomic, proteomic and metabolomic data have to be integrated with traditional literature in a machine-readable way. Typical sets of experimental data yield component lists with quantitative content data and a catalog of interactions and networks. This requires the establishment of a middleware to convert experimental data into a format suitable for manipulation and viewing by end-users. For example, the Generic Model Organism Database project (GMOD; http://gmod.org) aims to link experimental data with corresponding contextual meta-data about experimental conditions and protocols in a multi-user, multi-center environment. It offers a collection of open source tools for creating and managing genome-scale biological databases ranging from a small database of genome annotations to a large web-accessible community database. Another approach is to trade off the width of integration for more depth with regard to a particular analysis task, and to employ workflow systems such as InforSense (http://www.inforsense.com) or Taverna (http://taverna.sf.net). These act as glue layers between various data sources and analysis packages and are also often referred to as pipelines, in silico protocols or *e*-experiments (Turnbaigh et al. 2007). "Pipeline" is mostly used to describe executable workflows, while the other terms are dedicated to abstract workflows (Lisacek et al. 2006).

Many innovative solutions for the multi-dimensional integration of data produced by experimental laboratories have been introduced by Bioinformatics Resource Centers for Biodefense and Emerging/Re-Emerging Infectious Diseases through regional Biodefense Centers of Excellence (McNeal et al 2007; Greene et al. 2007). Sets of task- and domain-specific online query and display tools are being developed to allow the end-user to view data in a number of different formats and to run informative comparisons of data with existing libraries (Louie et al. 2007; Glassner et al. 2008). The most striking change in data collection and representation is expressed by the move from flat databases to atlases or collections of interconnected maps (Lisacek et al. 2006).

The uneven content and quality of data and the constant evolution of biomedical knowledge remain the main obstacles to data integration (Lisacek et al. 2006). The quality of data is affected by a number of factors including the accuracy of the mapping algorithms and reference datasets, the standardization of data formats and the level of detail of the experiment description (Stead et al. 2008). In addition, an increasing number of genomes are being released in "draft" form, before the finishing stage of a sequencing project, with high sequencing error rates (De Keersmaesher et al. 2006; Médigue and Moszer 2007). Recent developments in databases and browsers for genomics have been summarized by Schattner (2008).

There is an urgent need for data structures suitable for infectious disease space that can be applied to emerging "omics" data sets. The *Pathogen Information Markup Language* (PIML) has also recently been introduced to enhance the interoperability of microbiology datasets for pathogens with epidemic potential (He et al. 2005) by capturing the data elements that describe determinants of pathogen profiles. However, the jury is still out on the question of which data integration architectures are best suited to assembling large scale and highly diverse genomic data.

Integrating high-throughput techniques with other analytic tools brings a new understanding of infectious processes and introduces an era of personalized strategies for managing infectious diseases. In this way, informatics becomes an irreplaceable platform for the constant cross-fertilization and interplay between focused and genome-wide studies.

1.5 Translating "Omics" into Clinical Practice

1.5.1 Rapid Identification of Pathogens

Rapid and standardizable molecular identification systems have emerged during the last decade, with the development of sequence based species identification and sub-typing as the alternative to slow, labor-intensive and underpowered phenotypic techniques. Molecular identification usually relies on the detection of a single gene or multiple gene targets, or requires the comparison of whole microbial genomes. For example, in the pragmatic world of diagnostic bacteriology, conserved housekeeping genes such as the 16S rRNA gene, *rpoB* gene and others have been accepted as reliable targets. They are found in all microorganisms and show enough sequence conservation for accurate alignment as well as enough variation for phylogenetic analyses (Christen 2008). Furthermore, the 16S rRNA gene based phylogeny is sufficiently congruent with those based on whole genome approaches. Sequencing of six to eight genes or loci, as it typically done in multilocus sequence typing analysis, may constitute a reasonable compromise between single gene-based and whole genome-based methods for species diversity studies.

To streamline the process of the translation of sequencing-based identification into clinical practice, the concept of the pathogen profile has been introduced (Sintchenko et al. 2007). A pathogen profile is a single, multivariate observation or set of observations, comprised of classes of specific attributes (e.g., genome, transcriptome, proteome or metabolome data), which are designed to allow the interrogation of existing or future databases, and the integration of genomics and post-genomics data with clinical observations and patient outcomes. The profile may indicate the probability that a specific marker is associated with a clinically relevant phenotype such as in vivo antimicrobial resistance or high transmissibility. This information allows the classification of strains into "risk groups" for treatment failure or a propensity to cause outbreaks of infections. It is often important to capture the quantitative information about a pathogen, in vivo, i.e. viral or bacterial loads and their units of measurement. In contrast to traditional subtyping, which is based on phenotypic characteristics such as serotype, biotype, phage type or antimicrobial susceptibility, genetic profiling describes the phenotypic potential in the nucleic acid sequence. A pathogen profile is a synthesis of various markers and clinical end-points, which can be extracted from medical charts that characterize an individual patient's clinical and public health outcomes. The profile may be heuristic, when only a single genetic marker is associated with a specific patient outcome, while more insights can be achieved when attributes from different levels of the biological hierarchy (i.e. gene detection, gene expression, metabolite profiles etc) corroborate and complement each other. Machine learning algorithms, such as E-Predict (Urisman et al. 2005), are being developed to identify viruses and bacteria present in clinical samples. These profiles are based on the microarray hybridization patterns or DNA sequences of pathogens.

1.5.2 Guiding Antibiotic Prescribing Decisions

Many computerized evidence-based guidelines and decision support systems (DSS) have been designed to improve the effectiveness and efficiency of antibiotic prescribing (Samore et al. 2005; Buising et al. 2008). The most frequently utilized are electronic guidelines and protocols, especially for the empirical selection of antibiotics. The majority of DSS result in improvement in clinical performance and, in at least half of the published trials, in improved patient outcomes (Finch and Low 2002; Sintchenko et al. 2007; Sintchenko et al. 2008a). The revival of interest in prescribing-decision-support reflects the recent change in emphasis from support for diagnostic decisions towards support for patient management, and the changing focus from systems targeting a broad range of clinical diagnoses to task- and condition-specific decision aids. Despite reported successes of individual applications, the safety of electronic prescribing systems in routine practice has recently been identified as an issue of potential concern.

Bioinformatics assisted prescribing has become a new frontier in reducing the complexities of prescribing combinations of antimicrobials in the era of multidrug resistance. The great diversity of mutational patterns contributing to antimicrobial resistance complicates the choice of optimal therapies. A range of bioinformatics tools to predict drug resistance or response to therapy from a genotype, have been developed to support clinical decision-making (Beerenwinkel et al. 2003; Lengauer and Singh 2006). These tools use either a statistical approach, in which the inferred model and prediction are treated as regression problems, or machine learning algorithms, in which the model is addressed as a classification problem (Sintchenko et al. 2008a). A statistical learning approach to the ranking of therapeutic choices often relies on a direct correlation between the baseline microbial profiles, the therapeutic decision and the patient's response to treatment (e.g., expected reduction in viral load resulting from anti-HIV combination therapy). For example, several susceptibility scores have been used for combination antiretroviral therapy. These take into account specific resistance mutations and add up the activities of individual drugs in the regimen (Lengauer and Singh 2006). Computer-assisted therapy depends on the availability of widely shared databases that can correlate quality-controlled data from genotypic resistance assays and treatment regimens with short- and long-term clinical outcomes. Databases such ARDB (Liu and Pop 2009) capture differences in antimicrobial sensitivities and reflect variation in the amino acid composition of resistant microbes, but simply counting mutations may not be enough to predict functional differences, which affect treatment outcomes.

1.5.3 Linking Genomics to Clinical Outcomes

The molecular profiling of pathogens is based on the concept that various pathogens can be associated with different clinical outcomes. It brings together the pathogen and host factors as the pathogenesis and natural history of infection are determined by both the pathogen and human genetic susceptibility. The effectiveness of combining host and pathogen genetics in a single system or "genetics-squared" has been proven in studies of viral infections (Persson and Vance 2007). Investigations of the impact of host genetics on the susceptibility to HIV infection and the rate of disease progression have mainly used a candidate gene approach to reveal associations with a number of different genes. The genome-wide association studies look at the genetic variation across the human genome in order to uncover factors not previously suspected of influencing infection outcomes. For example, this strategy identified variants of the HIV virus associated with differences in the control of viral load at set points and in disease progression. However, unraveling the interaction between the host and microbial genetic factors requires large clinical trials, reinforcing the role of collaborative networks and data repositories.

Informatics methods have become critical for data mining to decipher links between genetic variation and disease pathogenesis in order to define markers of disease progression, to guide the optimum use of therapeutics and to refine the drug and vaccine development (Mansmann 2005). A better understanding of the function of genes and other parts of the genome has enabled the reverse engineering approach, which may lead to the characterization and discovery of potential drug targets, vaccine candidates and diagnostic or prognostic markers (Davies and Flower 2007; Yang et al. 2008b). Proteins with essential biological functions present in multiple pathogens could be the best drug targets. Once the target genes essential for pathogen survival are identified, their susceptibility to specific compounds derived from large chemical libraries is examined in silico and in vitro (Muzzi et al. 2007; Biswas et al. 2008).

1.5.4 Tracing Pathogens with Epidemic Potential

Increases in the use of electronic medical records and the availability of information technology tools have created opportunities for the automation of surveillance and facilitation of surveillance based on either syndromic or disease-specific signals (Amadoz and Gonzales-Candelas 2007; M'ikanatha et al. 2007). The automation of data collection improves the time and completeness of surveillance and allows infection control professionals to focus on interventions (Hota et al. 2008; Young and Stevenson 2008).

The comparison of chromosomal sequences allows the identification of the unique genomic signatures of pathogens for the purposes of infection control and "microbial forensics." Molecular typing methodologies, in contrast to classical phenotypic methods, allow the discrimination of variations among strains within a species, the elucidation of the route of contamination, the identification of the source of infection as well as the analysis of epidemics. The identification of the natural reservoir and any possible intermediate hosts of pathogens is critical for understanding the transmission modes, designing a long-term disease control strategy, and preventing future reintroduction (Sintchenko and Gallego 2009). Bioinformatics assisted biosurveillance addresses the inefficiencies of traditional surveillance, as well as the need for a more timely and comprehensive infectious disease monitoring and control. It leverages on recent breakthroughs in the rapid, high-throughput molecular profiling of microorganisms and text mining, as well as on the growing electronic body of knowledge about the molecular epidemiology of pathogens with epidemic potential. Such a framework combines the genetic and geographic data of a pathogen to reconstruct its history and to identify the migration routes through which the strains spread regionally and internationally (Cantón 2005; Sintchenko et al. 2008b). Computer-based geographic information systems (GIS) have offered an efficient way to visualize the dynamics of the transmission of infections, especially in the setting of a community outbreak (McKee et al. 2000; Schreiber et al. 2007).

Another way to track infectious diseases of public health concern is to monitor health-seeking behavior in the form of queries to online search engines used by the general public or health professionals. Epidemics of seasonal influenza in areas with a large population of Internet users have been successfully detected using Google search data and then correlated with visits to a doctor (Ginsberg et al. 2009; Brownstein et al. 2009). The advent of news aggregators has led to the development of new disease surveillance tools that can continuously mine, categorize, filter, and visualize multilingual online information about epidemics. The Global Public Health Intelligence Network (GPHIN), developed almost a decade ago by Health Canada in collaboration with WHO, HealthMap (http://www.healthmap.org/en) (Fig. 1.2) or Geosentinel (http://www.istm.org/geosentinel/main.html) among many others are examples of such early warning systems. Resources for infection prevention and control on the World Wide Web have been recently reviewed elsewhere (Brownstein et al. 2009; Johnson et al. 2009)

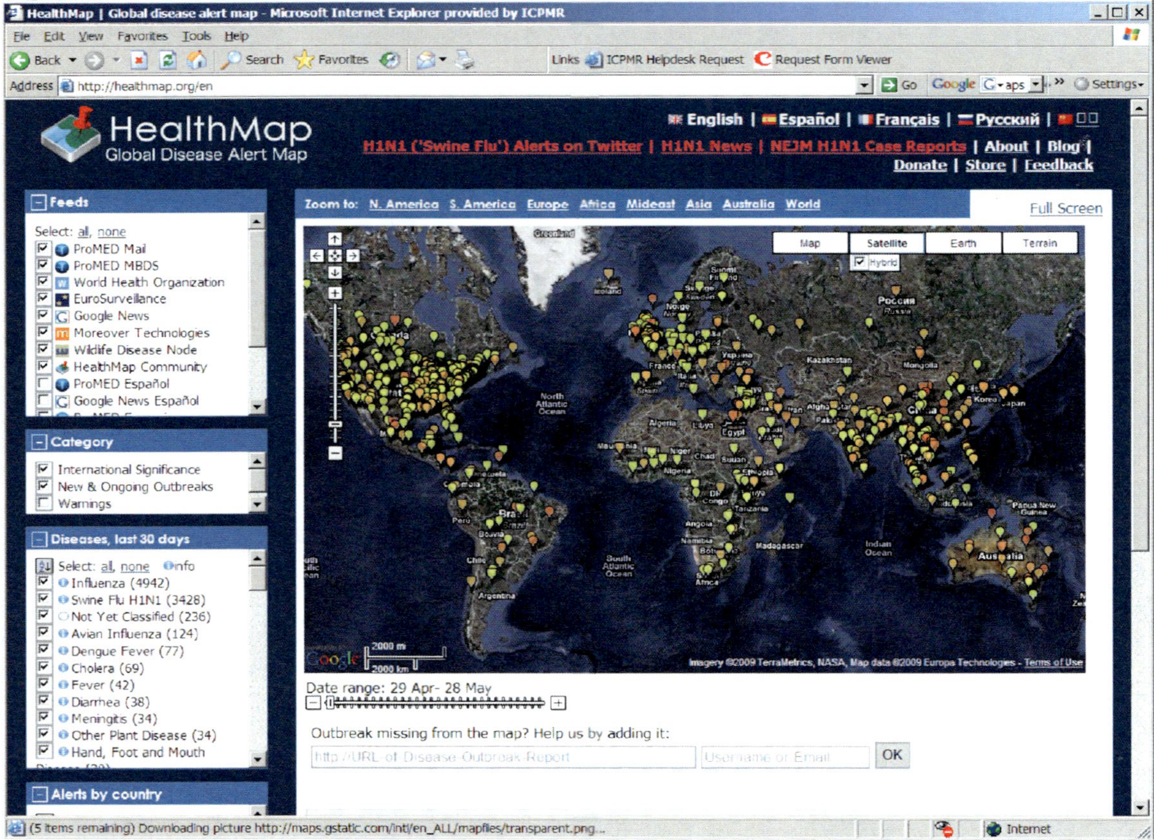

Fig. 1.2 Screen shot of HealthMap (Global Disease Alert Map) displaying reports about recent outbreaks in English language sources (with permission, http://www.healthmap.org/en)

1.6 Conclusions

The reductionist approach to biomedical research focusing on the study of cells and molecules has peaked with the sequencing of the human genome. However, it is becoming increasingly clear that "taking apart" analyses have reached their limit, and the time has perhaps come for integrative science (An and Faeder 2009). Developments in informatics have been critical in supporting and engaging with both reductionist and integrative paradigms. On one hand, informatics has equipped comparative genomics with tools to scrutinize genes and explore genetic polymorphisms. On the other hand, informatics has enabled the generation of integrative and testable hypotheses through the discovery of knowledge in databases and through the study of gene-phenotype connections between a pathogen and its host environment. A variety of data sets can be integrated, including the patient's demographic and clinical presentation, the laboratory results, the pathogen's gene regulation and expression, and metabolic maps with different parameters reflecting the phenotypic behavior of a pathogen and host factors. In early years some skeptics saw informatics-assisted research as a distraction of effort and funding away from traditional hypothesis-driven inquiry. Since then, infectious disease informatics has verified its status as a platform for hypothesis generation and testing (Sintchenko et al. 2007).

New breakthroughs in infectious disease informatics (IDI) are the result of cross-pollination between different disciplines that use technologies to gather and disseminate knowledge (Fig. 1.3). Microbial genome sequence analysis and metagenomics have contributed intriguing new data types and data sources to IDI. Bioinformatics has brought to the IDI a range of analytic tools, databases and data standards. Conventional health informatics and computer science has provided high performance solutions for the data storage, sharing, analysis and visualization as well as clinical terminology libraries, data standards, decision support and technology evaluation frameworks. Importantly, the infectious disease informatics community has fed the lessons learnt from the implementation of clinical and public health systems back to the broader audience.

As the subsequent chapters of this volume testify, infectious disease informatics is set to lead to the more targeted and effective prevention, diagnosis and treatment of infections through a comprehensive review of the genetic repertoire and metabolic profiles of pathogens. The post-genomic era offers new opportunities for the efficient discovery of safe and efficacious subunit vaccines by shortcutting the enormous economic burden of the experimental process. Our analytical capacity has already become the rate-limiting step in biomedical research. At the same time, it provides an opportunity to apply the engineering paradigm to biomedical research, thereby mandating the development of tools that can dynamically represent a body of current knowledge. However, the simplistic application of brute force computational power to massive reams of biomedical data is unlikely to result in meaningful mechanistic insight. It cannot be overstressed that informatics initiatives should compliment "wet laboratory" practices. An iterative loop of discovery and validation between the two methodologies remains the best way forward.

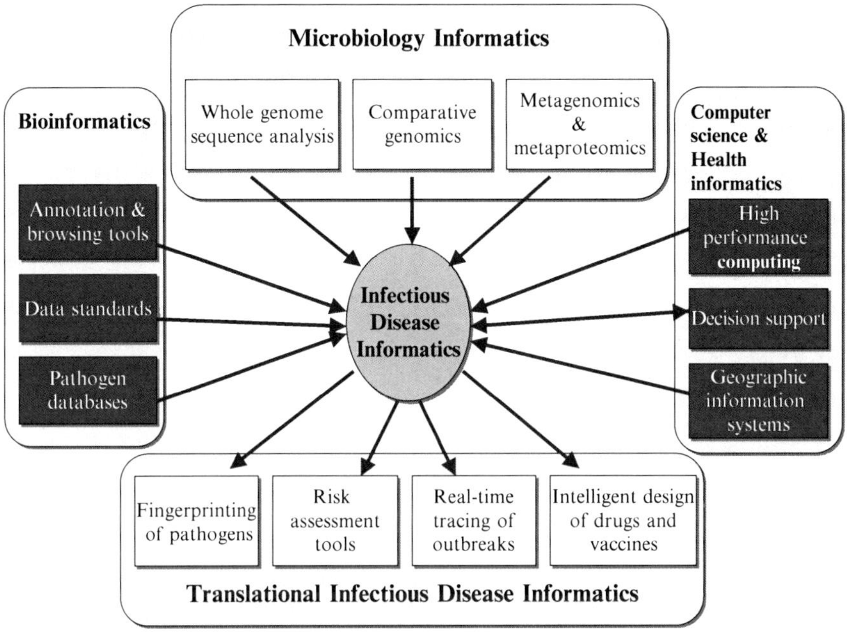

Fig. 1.3 Inter-relations between branches of informatics and bioinformatics domains

References

Amadoz A, Gonzales-Candelas F (2007) epiPATH: an information system for the storage and management of molecular epidemiology data from infectious pathogens. BMC Infect Dis 7:32

An GC, Faeder JR (2009) Detailed qualitative dynamic knowledge representation using a BioNet Gen model of TLR-4 signaling and preconditioning. Math Biosc 217:53–63

Bansal AK (2005) Bioinformatics in microbial biotechnology – a mini review. BMC Microb Cell Factor 4:19

Beerenwinkel N et al (2003) Geno2Pheno: Estimating phenotypic drug resistance from HIV-1 genotypes. Nucleic Acids Res 31:3850–3855

Behr MA (2008) Mycobacterium du jour: what's on tomorrow's menu? Microb Infect 10:968–972

Binnewies TT, Motro Y, Hallin PF et al (2006) Ten years of bacterial genome sequencing: comparative-genomics-based discoveries. Funct Integr Genomics 6:165–185

Birkholtz L-M et al (2006) Integration and mining of malaria molecular, functional and pharamacological data: how far are we from a chemogenomic knowledge space? Malaria J 5:110

Biswas S, Raoult D, Rolain J-M (2008) A bioinformatic approach to understanding antibiotic resistance in intracellular bacteria through whole genome analysis. Int J Antimicrob Agents 32:207–220

Brent MR (2008) Steady progress and recent breakthroughs in the accuracy of automated genome annotation. Nat Rev Genet 9:62–73

Brownstein JS, Freifeld CC, Madoff LC (2009) Digital disease detection - harnessing the Web for public health surveillance. N Engl J Med 360:2153–2157

Buising KL, Thursky KA, Black JF (2008) Improving antibiotic prescribing for adults with community acquired pneumonia: does a computerised decision support system achieve more than academic detailing alone?-A time series analysis. BMC Med Inform Dec Mak 8:35

Burrack LS, Higgins DE (2007) Genomic approaches to understanding bacterial virulence. Curr Opin Microbiol 10:4–9

Cantón R (2005) Role of the microbiology laboratory in infectious disease surveillance, alert and response. Clin Microbiol Infect 11(Suppl 1):S3–S8

Carver T, Berriman M, Tivey A et al (2008) Artemis and ACT: viewing, annotating and comparing sequences stored in a relational database. Bioinform 24:2672–2676

Chaisson MJ, Pevzner PA (2008) Short read fragment assembly of bacterial genomes. Genome Res 18(2):324–330

Chatr-aryamontri A, Ceol A, Peluso D, Nardozza A, Panni S et al (2009) VirusMINT: a viral protein interaction database. Nucleic Acids Res 37:D669–D673

Chaudhuri RR et al (2008) xBASE2: a comprehensive resource for comparative bacterial genomics. Nucleic Acids Res 36:D543–D546

Chen L et al (2005) VFDB: a reference database for bacterial virulence factors. Nucleic Acids Res 33:D325.

Chen SL et al (2006) Identification of genes subject to positive selection in uropathogenic strains of *Escherichia coli*: a comparative genomics approach. Proc Natl Acad Sci USA 103:5977–5982

Christen R (2008) Identification of pathogens – a bioinformatic point of view. Curr Opin Bitech 19:266–273

Collado-Vides J, Salgado H, Morett E et al (2008) Bioinformatics resources for the study of gene regulation in bacteria. J Bacteriol 191:23–31

Craddock T, Harwood CR, Hallinan J, Wipat A (2008) e-Science: relieving bottlenecks in large-scale genome analyses. Nat Rev Microbiol 6:948–954

Darling ACE, Mau B, Blatter FR, Perna NT (2004) Mauve: multiple alignment of conserved genomic sequence with rearrangements. Genome Res 14(7):1394–1403

Davies MN, Flower DR (2007) Harnessing bioinformatics to discover new vaccines. Drug Discov Today 12:389–395

De Keersmaecker SCJ, Thijs IMV, Vanderleyden J, Marchal K (2006) Integration of omics data: how well does it work for bacteria? Mol Microbiol 62:1239–1250

Delcher AL, Harmon D, Kasif S et al (1999) Improved microbial gene identification with GLIMMER. Nucl Acids Res 27:4636–4641

Deloger M, El Karoui M, Petit M-A (2009) A genomic distance based on MUM indicates discontinuity between most bacterial species and genera. J Bacteriol 191:91–99

Dougherty TJ, Barrett JF, Pucci MJ (2002) Microbial genomics and novel antibiotic discovery: new technology to search for new drugs. Curr Pharmac Design 8:1119–1135

Driscoll T, Dyer MD, Murali TM, Sobral BW (2009) PIG - the pathogen interaction gateway. Nucleic Acids Res 37 (Database Issue):D647–D650

Field D, Wilson G, van der Gast C (2006) How do we compare hundreds of bacterial genomes? Curr Opin Microbiol 9:499–504

Finch RG, Low DE (2002) A critical assessment of published guidelines and other decision-support systems for the antibiotic treatment of community-acquired respiratory tract infections. Clin Microbiol Infect 8(Suppl 2):69–91

Forst CV (2006) Host-pathogen systems biology. Drug Discov Today 11:220–227

Frézal L, Leblois R (2008) Four years of DNA barcoding: current advances and prospects. Infect Genet Evol 8:727–736

Gallego B, Sintchenko V, Wang Q et al (2009) Biosurveillance of emerging biothreats using scalable genotype clustering. J Biomed Inform 42:66–73

Galperin MY (2005) A census of membrane-bound and intracellular signal transduction proteins in bacteria: bacterial IQ, extroverts and introverts. BMC Microbiol 5:35

Garrido C, Roulet V, Chueca N et al (2008) Evaluation of eight different bioinformatics tools to predict viral tropism in different human immunodeficiency virus type 1 subtypes. J Clin Microbiol 46:887–891

Ginsberg J, Mohebbi MH, Patel RS, Brammer L et al (2009) Detecting influenza epidemics using search engine query data. Nature 457:1012–1014

Glasner JD et al (2008) Enteropathogen Resource Integration Center (ERIC): bioinformatics support for research on biodefense-relevant enterobacteria. Nucleic Acids Res 36:D519–D523

Greene JM, Collins F, Lefkowitz et al (2007) National Institute of Allergy and Infectious Diseases Bioinformatics Resource Centers: new assets for pathogen informatics. Infect Immun 75:3212–3219

Guigó R, Flicek P, Abril JF et al (2007) EGASP: the human ENCODE Genome Annotation Assessment Project. Genome Biol 7(Suppl 1):S21–S31

Guyet T, Garbay C, Dojat M (2007) Knowledge construction from time series data using a collaborative exploration system. J Biomed Inform 40:672–687

Harrington ED, Jensen LJ, Bork P (2008) Predicting biological networks from genomic data. FEBS Lett 582:1251–1258

He Y, Vines RR, Wattam AR, Abramochkin GV et al (2005) PIML: the Pathogen Information Markup Language. Bioinform 21:116–121

Hota B, Jones RC, Schwartz DN (2008) Informatics and infectious diseases: what is the connection and efficacy of information technology tools for therapy and health care epidemiology. Am J Infect Control 36:S47–S56

Hutchinson CA (2007) DNA sequencing: bench to bedside and beyond. Nucleic Acids Res 35:6227–6237

Jamshidi N, Palsson BO (2007) Investigating the metabolic capabilities of *Mycobacterium tuberculosis* H37Rv using the *in silico* strain *iNJ661* and proposing alternative drug targets. BMC Syst Biol 1:26

Jelier R, Schuemie MJ, Veldhoven A et al (2008) Anni 2.0: a multipurpose text-mining tool for the life sciences. Genome Biol 9(6):R96

Johnson LE, Reyes K, Zervos MJ (2009) Resources for infection prevention and control on the World Wide Web. Clin Infect Dis 48:1585–1595

Kahveijian A, Quackenbush J, Thompson JF (2008) What would you do if you could sequence everything? Nat Biotech 26:1125–1133

Kann MG (2008) Protein interactions and disease: computational approaches to uncover the etiology of diseases. Brief Bioinform 8:333–346

Kommedal Ø, Karlsen B, Sæbø Ø (2008) Analysis of mixed sequencing chromatograms and its application in direct 16S rRNA gene sequencing of polymicrobial samples. J Clin Microbiol 46:3766–3771

Konstantinidis KT, Tiedje JM (2005) Genomic insights that advance the species definition for prokaryotes. Proc Natl Acad Sci USA 102:2567–2572

Koonin EV, Wolf YI (2008) Genomics of bacteria and archaea: the emerging dynamic view of the prokaryotic world. Nucleic Acids Res 36:6688–6719

Korbel JO, Doerks T, Jensen LJ, Perez-Iratxeta C et al (2005) Systematic association of genes to phenotypes by genome and literature mining. PloS Biology 3:e134

Kumar S, Nei M, Dudley J, Tamura K (2008) MEGA: a biologist-centric software for evolutionary analysis of DNA and protein sequences. Brief Bioinform 9:299–306

Lengauer T, Sing T (2006) Bioinformatics-assisted anti-HIV therapy. Nat Rev Microbiol 4:790–797

Lengauer T, Sander O, Sierra S et al (2007) Bioinformatics prediction of HIV coreceptor usage. Nat Biotech 25:1407–1410

Lisacek F, Cohen-Boulakia S, Appel RD (2006) Proteome informatics II: bioinformatics for comparative proteomics. Proteom 6:5445–5466

Liu B, Pop M (2009) ARDB – Antibiotic Resistance Genes Database. Nucleic Acids Res 37:D443–447

Louie B et al (2007) Data integration and genomic medicine. J Biomed Inform 40:5–16

Lussier YA, Liu Y (2007) Computational approaches to phenotyping: high-througput phenomics. Proc Am Thorac Soc 4:18–25

M'ikanatha NM, Lynfield R, Van Beneden CA, de Valk H (2007) Infectious disease surveillance. Blackwell, Oxford

MacLean D, Jones JDG, Studholme DJ (2009) Application of 'next-generation' sequencing technologies to microbial genetics. Nat Microbiol Rev 2009 7:287–296

Majoros WH (2007) Methods for computational gene prediction. Cambridge University Press, Cambridge.

Mansmann U (2005) Genomic profiling: interplay between clinical epidemiology, bioinformatics and biostatistics. Methods Inf Med 44:454–460

McKee KT, Shields TM, Jenkins PR et al (2000) Application of a geographic information system to the tracking and control of an outbreak of shigellosis. Clin Infect Dis 31:728–733

McNeil LK et al (2007) The National Microbial Pathogen Database Resource (NMPDR): a genomic platform based on subsystem annotation. Nucleic Acids Res 35:D347–D353

Médigue C, Moszer I (2007) Annotation, comparison and databases for hundreds of bacterial genomes. Res Microbiol 158:724–736

Meyer F et al (2003) GenDB – an open source genome annotation system for prokaryote genomes. Nucleic Acids Res 31:2187–2195

Michael H, Hogan J, Kel A et al (2008) Building a knowledge base for system pathology. Brief Bioinform 9:518–531

Muzzi A, Masignani V, Rappuoli R (2007) The pan-genome: towards a knowledge-based iscovery of novel targets for vaccines and antibacterials. Drug Discov Today 12:429–439

Navrati V, de Chassey B, Mayniel L et al (2009) VirHostNet: a knowledge base for the management and the analysis of proteome-wide virus-host interaction networks. Nucleic Acids Res 37:D661–D668

Numann E, Prusak L (2007) Knowledge networks in the age of the Semantic Web. Brief Bioinform 8:141–149

Pallen MJ, Wren BW (2007) Bacterial pathogenomics. Nature 449:835–842

Parkhill J, Dougan G, James KD et al (2001a) Complete genome sequence of a multiple drug resistant *Salmonella enterica* serovar Typhi CT18. Nature 413:848–852

Parkhill J, Wren BW, Thomson NR et al (2001b) Genome sequence of *Yersinia pestis*, the causative agent of plague. Nature 413:523–527

Persson J, Vance RE (2007) Genetics-squared: combining host and pathogen genetics in the analysis of innate immunity and bacterial virulence. Immunogenet 59:761–778

Pop M, Salzberg SL (2008) Bioinformatics challenges of new sequencing technology. Trends Genet 24:142–149

Rachman H, Kaufmann SHE (2007) Exploring functional genomics for the development of novel intervention strategies against tuberculosis. Intern J Med Microbiol 297:559–567

Raman K, Kalidas Y, Chandra N (2008) TargetTB: a target identification pipeline for *Mycobacterium tuberculosis* through an interactome, reactome and genome-scale structural analysis. BMC Systems Biol 2:109

Raskin DM et al (2006) Bacterial genomics and pathogen evolution. Cell 124:703–714

Reddy TBK, Riley R, Wymore F et al (2009) TB Database: an integrated platform for tuberculosis research. Nucleic Acids Res 37:499–508

Restif O (2009) Evolutionary epidemiology 20 years on: challenges and prospects. Infect Genet Evol 9:108–123

Rzhetsky A, Seringhaus M, Gerstein M (2008) Seeking a new biology through text mining. Cell 134:9–13

Sakata T, Winzeler EA (2007) Genomics, system biology and drug development for infectious diseases. Mol BioSyst 3:841–848

Samore MH, Bateman K, Alder SC et al (2005) Clinical decision support and appropriateness of antimicrobial prescribing. J Am Med Assoc 294:2305–2314

Sanger F, Air GM, Barrell BG et al (1977) Nucleotide sequence of bacteriophage X174 DNA. Nature 265:687–695

Schattner P (2008) Genomes, browsers and databases. Cambridge University Press, Cambridge.

Schreiber MJ, Ong SH, Holland RCG et al (2007) DengueInfo: a web portal to dengue information resources. Infect Genet Evol 7:540–541

Shendure J, Ji H (2008) Next-generation DNA sequencing. Nature Biotech 26:1135–1145

Sintchenko V, Gallego B (2009) Laboratory-guided detection of disease outbreaks: three generations of surveillance systems. Arch Pathol Lab Med 133:916–925

Sintchenko V, Iredell JR, Gilbert GL (2007) Genomic profiling of pathogens for disease management and surveillance. Nat Microbiol Rev 5:464–470

Sintchenko V, Magrabi F, Tipper S (2007) Are we measuring the right thing? Variables that affect the impact of computerized decision support on patient outcomes: a systematic review. Med Inform Internet Med 32:225–240

Sintchenko V, Coiera E, Gilbert GL (2008a) Decision support systems for antibiotic prescribing. Curr Opin Infect Dis 21:573–579

Sintchenko V, Gallego B, Chung G, Coiera E (2008b) Towards bioinformatics assisted infectious disease control. BMC Bioinform 10:S10

Smarr L, Gilna P, Papadopoulos P et al (2009) Building an OptIPlante collaboratory to support microbial metagenomics. Future Gen Comp Systems 25:124–131

Squires B et al (2008) BioHealthBase: informatics support in the elucidation of influenza virus host-pathogen interactions and virulence. Nucleic Acids Res 36:D497–D503

Stavrinides J, McCann HC, Guttman DS (2008) Host-pathogen interplay and the evolution of bacterial effectors. Cell Microbiol 10:285–292

Stead DA et al (2008) Information quality in proteomics. Brief Bioinform 9:174–188

Stothard P, Wishart DS (2006) Automated bacterial genome analysis and annotation. Curr Opin Microbiol 9:505–510

Suzek BE, Ermolaeva MD, Schreiber M, Salzberg SL (2001) A probabilistic method for identifying start codons in bacterial genomes. Bioinform 17:1123–1130

Tettelin H, Masignani V, Cieslewicz MJ et al (2005) Genome analysis of multiple pathogenic isolates of *Streptococcus agalactiae*: implications for the microbial "pan-genome". Proc Nat Acad Sci USA 102:13950–13955

Thorisson GA, Muilu J, Brookes AJ (2009) Genotype-phenotype databases: challenges and solutions for the post-genomic era. Nat Rev Genet 10:9–18

Turnbaugh PJ et al (2007) The Human Microbiome Project. Nature 449:804–810

Urisman A, Fischer KF, Chiu CY, Kistler AL et al (2005) E-Predict: a computational strategy for species identification based on observed DNA microarray hybridization patterns. Genome Biol 6:R78

Ussery DW, Wassenaar TM, Borini S (2009) Computing for comparative microbial genomics: bioinformatics for microbiologists. Springer-Verlag, London

Van Domselaar GH, Stothard P, Shrivastava S et al (2005) BASys: a web server for automated bacterial genome annotation. Nucleic Acids Res 33:W455–W459

Verberkmoes NC, Russell AL, Shah M et al (2009) Shortgun metaproteomics of the human distal gut flora. ISME J 3:179–189

Whitworth DE (2008) Genomes and knowledge – a questionable relationship? Trends Microbiol 16:512–519

Winnenburg R et al (2006) PHI-base: a new database for pathogen host interactions. Nucleic Acids Res 36:D459–D464

Wu H-J, Wang A H-J, Jennings MP (2008) Discovery of virulence factors of pathogenic bacteria. Curr Opin Chem Biol 12:93–101

Xiang Z, Tian Y, He Y (2007) PHIDIAS: a pathogen-host interaction data integration and analysis system. Genome Biol 8:R150

Yang JY, Yang MQ, Arabnia HR, Deng Y (2008a) Genomics, molecular imaging, bioinformatics, and bio-nano-info integration are synergistic components of translational medicine and personalized healthcare research. BMC Genomics 9(Suppl 2):11

Yang X, Yang H, Zhou G, Zhao G-P (2008b) Infectious disease in the genomic era. Ann Rev Genom Hum Genet 9:21–48

Yan Q (2008) Bioinformatics databases and tools in virology research: an overview. In Silico Biol 8:71–85

Yao J, Lin H, Van Deynze A (2008) PrimerSNP: a web tool for whole-genome selection of allele-specific and common primers of phylogenetically-related bacterial genomic sequences. BMC Microbiol 8:185

Young J, Stevenson KB (2008) Real-time surveillance and decision support: Optimizing infection control and antimicrobial choices at the point of care. Am J Infect Control 36:S67–S74

Zaremba S, Ramos-Santacruz M, Hampton T, Shetty P et al (2009) Text-mining of PubMed abstracts by natural language processing to create a public knowledge base on molecular mechanisms of bacterial enteropathogens. BMC Bioinform 10:177

Zeng D, Chen H, Lynch C, Eidson M, Gotham I (2005) Infectious disease informatics and outbreak detection. In: Chen H, Fuller SS, Friedman C, Hersh W (eds) Medical informatics: knowledge management and data mining in biomedicine. Springer, New York

Zhou F, Olman V, Xu Y (2008) Barcodes for genomes and applications. BMC Bioinform 9:546.

Chapter 2
Bioinformatics of Microbial Sequences

Phil Giffard

2.1 Overview of Prokaryotic Microorganisms

Pathogenic Prokaryotes are cells that are defined by a lack of membrane-bound organelles. It is now clear that this is not an acquired characteristic, and the most recent common ancestor of all life on earth was probably a prokaryote. As a result, prokaryotes do not form a natural group in any meaningful sense, and are generally divided into two fundamental taxonomic units or domains: the Bacteria and the Archaea, with the divergence of these domains being the most ancient bifurcation in ribosomal RNA sequence-based phylogenetic trees. However, despite the profound phylogenetic gulf between the Bacteria and the Archaea, it appears that these domains have much in common regarding genome organisation and population structures. Therefore, for the sake of simplicity and clarity, in this discussion the lower case "bacteria" will be used to refer to prokaryotes in general.

2.1.1 Nature of the Bacterial Genome

The last 50 years have seen enormous advances in the understanding of the nature of the microbial world, with important milestones being the determination of the structure of DNA (Watson and Crick 1953), the unraveling of the genetic code (Crick 1962), the physical characterization of the bacterial chromosome (Cairns 1963), the development of dideoxy DNA sequencing (Sanger et al. 1977), the realization that evolutionary distances between microbial taxa could be inferred from biological sequences and the subsequent inference of a universal phylogenetic tree (Woese and Fox 1977; Woese 1987), the development of non culture-dependent DNA-based approaches to environmental sampling and analysis (Relman 1993), the advent of full genome sequencing (Fleischmann et al. 1995), and the development

P. Giffard
Menzies School of Medical Research, Darwin, NT, Australia

V. Sintchenko (ed.), *Infectious Disease Informatics*,

DOI 10.1007/978-1-4419-1327-2_2, © Springer Science+Business Media, LLC 2010

of "next generation" high throughput DNA sequencing methods (Margulies et al. 2005; Mardis 2008). This has led to the sequence determination of more than 700 complete microbial genomes, and the discovery of many uncultured microbial taxa. Consequently, a considerable degree of insight into the genetic and evolutionary structure of the microbial world has been gained.

The genomics of bacteria has recently been reviewed by Koonin and Wolf (Koonin and Wolf 2008). Bacterial genomes range from approximately 0.5 Mb to 13 Mb in size, with the lower level being difficult to define because of the grey area between endosymbionts and organelles. The genome is usually organized into a single circular chromosome, and a small number of autonomously replicating plasmids. The protein coding sequences are almost entirely devoid of introns, and the distances between the genes are, in general, short. As a result, there is a very strong correlation between genome size and gene number in bacteria, with close to 1,000 genes per Mb. This correlation is much weaker in more complex multicellular organisms, because of their large and variable intronic and other non-coding DNA content. Larger genomes in bacteria are associated with metabolic and morphological versatility, and adaptation to nutritionally sparse and variable environments, such as fresh water and soil. Conversely, small genomes are associated with parasitic lifestyles, which provide a nutritionally enriched and stable environment. Unsurprisingly, the smallest genomes are possessed by obligate intracellular parasites that can access many host-derived molecules.

One insight of immense significance that has resulted from the complete genome sequencing of multiple strains within bacterial species is that many of the latter possess what is now termed a "pan-genome" (Tettelin et al. 2008). Contrary to expectations, it has been found that every new genome sequence contains large numbers of previously unknown genes; so, the total number of genes within a species is much greater than the total number of genes within any individual isolate. The sizes of the pan-genomes are currently an enigma, as for some species; the number of new genes discovered per genome shows no sign of reducing as the total number of known genomes increases. This suggests that the pan-genome may be a very deep well of genetic information. Also, the evolutionary and adaptive significance of pan-genomes is not understood. For instance, it is unknown whether the pan-genome represents an adaptation that adds to the fitness of the community and may be regarded as evidence for group selection, or whether it is the contingent result of the activities of large numbers of mobile genetic elements that are essentially parasites. It may very well be a mixture of both.

2.1.2 Bacterial Evolution and the Universal Tree

Until the second half of the twentieth century, microbiology was a hostage of historically complicated taxonomy, which was based upon insufficient information. A hierarchical taxonomy implied knowledge of phylogeny. Unfortunately, the inference of evolutionary relationships between prokaryotic taxa was at that time virtually impossible, and many of the evolutionary relationships implied by the taxonomic structures of the early to mid twentieth century have since been shown to be incorrect. The basis for this difficulty was the morphological and metabolic simplicity of most prokaryotes. For example, rod shapes and heterotrophic metabolisms are now known to have evolved in diverse lineages; so, these characters are highly homoplastic (i.e. the products of convergent or parallel evolution) and so, evolutionarily uninformative. What was needed was information contained within the prokaryotic cell that was sufficiently complex to make homoplasy highly improbable. Woese and co-workers realized that biological sequences are sufficiently information-rich to be used to reliably infer evolutionary relationships (Woese and Fox 1977; Woese 1987). For many years, the molecules of greatest interest were ribosomal RNA (rRNA). This is because they perform the same function in every cell, and so changes in sequences are not biased by differing functional selection in different taxa. Also, rRNA has domains that evolve at very different speeds. This occurs because the structure of rRNA is maintained by base pairing. The base paired regions evolve extremely slowly because a single base change will disrupt base pairing, and this disrupts the function of the molecule. Therefore only very rare simultaneous double mutations that do not disrupt base pairing are likely to be fixed by evolution. Conversely, the regions of rRNA molecules that are not base-paired evolve much more rapidly. Therefore, a comparison of rRNA sequences can reveal evolutionary relationships over greatly different time scales. The particular value of the rRNA is its ability to reveal the nature of very distant evolutionary relationships and so provide insight into the universal evolutionary tree of life on earth. A landmark review published 21 years ago contained a 16S/18S RNA-based universal phylogenetic tree that clearly demonstrated that from an evolutionary point of view, the great bulk of the earth's biodiversity is composed of unicellular prokaryotes (Woese 1987). Animals and land plants occupied two twigs on the tree, thus eliminating the concept that all life on earth falls into either the animal or plant kingdoms. The 16S/18S RNA-derived universal tree supports the concept of three domains of life – the Bacteria, Archaea and Eukarya, with the Bacteria and Archeae being composed entirely of prokaryotes, and the Bacteria diverging from the Archaea and Eukarya, before the Archaea-Eukarya divergence (Fig. 2.1). An inevitable and counter-intuitive consequence of this model is that Bacteria and Archaea are as evolutionarily unrelated as it is possible for cellular life forms on this planet to be, despite extensive similarities in morphology and physiology. In the last two decades, the universal tree has been considerably expanded. This has been primarily because of the discovery of numerous uncultivated taxa in the course of studies that involve the extraction of rRNA encoding genes directly from environmental samples. The great bulk of these taxa are either Bacteria or Archaea so this work has greatly increased the proportion of known biodiversity that is composed of prokaryotic microorganisms (Fig. 2.2).

Although there is a consensus that the assembly of rRNA sequence-based phylogenies represents an enormous step forward, these studies contain an inherent assumption that is certainly not entirely correct. This assumption is that the evolutionary histories of the rRNA genes used as evolutionary markers are identical to

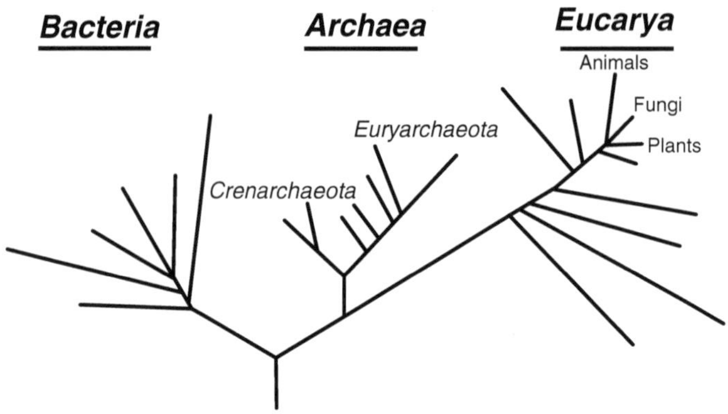

Fig. 2.1 The universal phylogenetic tree as deduced from rRNA sequences. This illustrates that from a phylogenetic viewpoint, the great majority of biodiversity is microbial (Reproduced from Woese (2000) with permission)

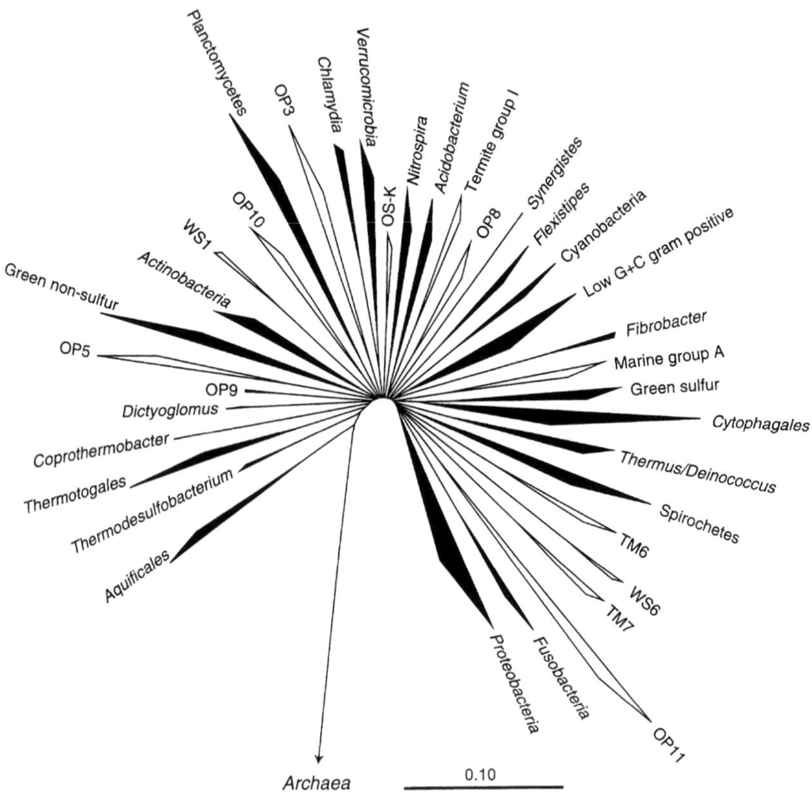

Fig. 2.2 A 16S ribosomal RNA tree illustrating divisions within the domain Bacteria. Divisions are broad taxonomic groupings conceptually similar to phyla in multicellular organisms. A large proportion of these divisions were discovered as a result of direct environmental sampling for rRNA, and have not been cultured (Hugenholtz et al. 1998)

the evolutionary histories of all the other genes on the chromosomes in which they reside. This question has been a subject of considerable debate for more than a decade. The central issue is lateral gene transfer (LGT). LGT (sometimes known as horizontal gene transfer) is the transfer of a gene(s) from one microbial cell to another cell. If LGT occurs between cells that are separated by any significant evolutionary distance, then the event creates a chimeric genome in which not all genes have the same evolutionary history. The worst-case scenario for the validity of an rRNA-based universal tree is that LGT is so common that an rRNA tree simply represents the phylogeny of rRNA genes and says nothing concerning the phylogeny of the cells within which they are found.

2.2 Classification of Prokaryotic Microorganisms

Given that the species remains the fundamental taxonomic unit, it has proven surprisingly difficult to arrive at a consensus as to what defines a bacterial species. In diploid organisms that always reproduce sexually, the accepted definition of a species is a group of organisms in which there is reasonably unimpeded gene flow i.e. the species represents a pool of DNA that is subject to recombination during the production of every new individual or set of identical siblings. However, this definition is difficult to apply consistently to bacteria. This is because there are no known bacteria in which there is genome-wide genetic recombination between individuals at every generation. In multi-cellular diploid organisms that do undergo complete recombination at every generation, the rate of gene flow is purely a function of ecological or geographic factors, with ecological or geographic boundaries leading to speciation. In prokaryotes, the rate and extent of gene flow is a function of the inherent rate of LGT, as well as of ecological or geographic factors. The inherent rate of LGT is highly variable from taxon to taxon, which makes it essentially impossible to delineate species using hard and fast rules regarding gene flow. The current practice is to delineate prokaryotic species on the basis of a combination of phylogeny, as defined by levels of sequence similarity, and phenotype. Phenotype is still given considerable weight, particularly when it involves virulence properties and host range, or in some other way impacts directly upon the interests of humans. A consequence of this is a tendency towards taxonomic splitting where human and livestock pathogens are concerned and taxonomic lumping for everything else. Genetic methods for defining species have long required whole genome hybridization studies, with 70% hybridization being the usual cut-off for delineating a species. This technique is now regarded as outdated, primarily because it is difficult to perform and interpret. There is no current consensus regarding a clear definition of a bacterial species, although there does appear to be some agreement that species should be defined in relation to the population structure as delineated by the sequencing of multiple genes, and the existence of cohesive ecological groups (Hanage et al. 2006; Staley 2006; Fraser et al. 2007; Achtman and Wagner 2008; Fraser et al. 2009).

2.3 Revealing Phylogenies and Population Structures

In groups of organisms that are closely related to each other, LGT may be sufficiently frequent to decouple the evolution of any one gene from the evolution of the genome, and so render a single gene-based classification as an inadequate and misleading depiction of evolutionary relationships within the group. However, an apparent paradox is that in groups of organisms that are extremely closely related, a tree can once again become the appropriate phylogenetic model. This is because in such groups, there may have simply been insufficient time to erase the phylogenetic signal since the existence of the most recent common ancestor for LGT events. Therefore, it is possible to generalize that LGT has the greatest disruptive effect on the phylogenies of groups of organisms that are distantly related enough for many LGT events to have taken place, but not so distantly related that mechanistic and/or ecological barriers to LGT have arisen. It should be recognized that there are some bacterial species in which LGT is so rapid that any position on the genome is more likely to be changed by an LGT event than by point mutation. In such species, the phylogenetic signal is largely erased, no matter what evolutionary time scale is being examined.

The purpose of this section is to describe methods for revealing the population structures and phylogenies of prokaryotes. This discussion incorporates case studies of different species and genera of bacteria, demonstrating that the population structures and extents of LGT differ so much across evolutionary space that there is no standard work flow for understanding the population structure of a species or group of species, and that there may be little choice but to maintain considerable case by case flexibility regarding the general principals of taxonomy.

2.3.1 Methods for Revealing the Extent and Frequency of LGT

At this point, it is important to state that LGT can be divided into two categories. Firstly, it is now clear that that the majority of bacterial species harbor genetic elements that encode functions that mediate or contribute to their own horizontal transfer. Examples of these include temperate bacteriophage, conjugative plasmids, transposable elements, and "islands" that apparently integrate into the genome by site-specific recombination. These elements are by definition evolutionarily homoplastic unless they are completely inactive. However, in the following paragraphs, the focus is upon mechanisms that render all genes in the genome prone to LGT. Mechanisms that can mediate the horizontal transfer of any gene in the genome include the uptake by transformation of DNA released from lysed cells, bacteriophage mediated generalized transduction, and the transfer of DNA by chromosomally integrated conjugative plasmids. A common feature of these mechanisms is that transferred DNA is integrated into the recipient cell by homologous recombination, and this results in the replacement of the recipient alleles with the donor alleles.

The essential principle behind most methods for revealing LGT within groups of bacteria is that LGT results in the phylogenetic signals derived from different genes being inconsistent i.e. it results in homoplasy. One of most useful resources for the study of LGT is multilocus sequence typing (MLST) data. MLST is an approach to microbial typing based upon the sequences of multiple genes, and was first described a decade ago (Maiden et al. 1998). Setting up an MLST scheme for a bacterial species requires the identification of seven widely spaced housekeeping genes. Housekeeping genes are used because they are expected to evolve at a slow and constant rate. Primer sets that may be used for PCR amplification and sequencing of approximately 450 bp fragments of each gene are developed. The method is then published and supported by a web site that stores all the data, and assigns numbers to each variant (allele) of each sequence, and to every unique combination of alleles (sequence type [ST]) that is found within an isolate. MLST web sites are now heavily used for the great majority of main bacterial pathogens (http://www.mlst.net/), with more than 3,800 known STs for *Campylobacter jejuni* and *Neisseria meningitidis*. MLST is explicitly designed to reveal the population structures of bacterial species, and the MLST databases certainly provide great insight into the contribution of LGT to evolution in different bacterial species.

The most straightforward approach to determining the extent of LGT is to take a group of isolates and construct trees on the basis of sequence variation in different genes. Obviously, MLST data is ideal for this purpose. In a non-recombining "clonal" population, the trees will be similar, and in a recombining "non-clonal" population they will be dissimilar (Spratt et al. 2001; Feil et al. 2003).

Although this approach is very useful, it is essentially qualitative in that it does not indicate the relative probabilities that any given nucleotide will undergo an evolutionary change as a result of an LGT event or point mutation. Also, in common with all phylogenetic tree-based methods, variations in the branching order from gene to gene can be due to stochastic effects.

A quantitative variant of this approach is the calculation of the index of association (I_A^S) between MLST loci (Smith et al. 1993; Enright et al. 2001). In a clonal bacterial population, a single gene can serve as the evolutionary marker for the genome, so different genes in the genome are in linkage disequilibrium. It therefore follows that if two isolates from a clonal population differ at one gene, they would likely differ at other genes as well. Conversely, in a completely non-clonal population, different genes are in linkage equilibrium, and whether or not two isolates are different at one gene has no bearing on whether they differ at other genes as well. In consequence, STs from clonal populations will, on average, differ at more loci than STs from non-clonal populations. The LIAN software package (Haubold and Hudson 2000) computes the I_A^S by comparing the variance of actual MLST data with randomized MLST data. The extent to which the variance from the actual data is greater than the variance of randomized data is a measure of linkage, and therefore a measure of the inverse of the extent of LGT.

A remarkably simple and effective approach to assessing the relative frequencies of point mutation and LGT is through the examination of single locus variants of clonal complex (CC) founders in MLST databases (Feil et al. 2000a, b; Feil et al. 2001;

Spratt et al. 2001). It is now apparent that most if not all bacterial species encompass successful clones that become numerically dominant. This numerical dominance makes it possible to detect derivatives of these successful clones that differ at just one MLST locus. These are known as single locus variants (SLVs). The successful clone plus SLVs (and sometimes double locus variants) is known as a CC, and the successful clone is termed a CC founder. In a non-recombining population, the only way that SLVs can arise is by de novo point mutation. In the great majority of instances, this will generate a new single nucleotide polymorphism (SNP), and in so doing will generate a new allele for that locus. Conversely, in a rapidly recombining population, a large proportion of SLVs will be SLVs by virtue of LGT events that have introduced an allele that already exists elsewhere within the species into the CC progenitor. Therefore, an indication of the ratio of LGT and point mutation can be obtained by examining all the known SLVs of a CC progenitor, and determining which variations are due to probable point mutation and which are due to probable LGT events. A possible confounder of such analyses is that MLST alleles may be undiscovered rather than new. However, assuming that the MLST database in question represents a substantial sample of the actual population, this will be a rare event. This is because it can only be concluded that an SLV has arisen by point mutation when the allele that discriminates the SLV from the CC founder is unique in the MLST database, the SNP is not polymorphic anywhere in the species except between the CC founder and the SLV in question, and that the new allele differs from the previous allele at only that SNP. Even if there is a certain percentage of instances in which LGT events are misclassified as point mutation events, and vice versa, this type of analysis can place reliable boundaries upon LGT frequencies. For instance, if the majority of SLVs of a CC progenitor harbors alleles that are common throughout the species, then it can only be concluded that more SLVs arise by LGT than by point mutation.

2.3.2 Methods for Depicting Population Structures and Phylogenies

Phylogenetic Trees. Tree-like diagrams have been used to depict evolutionary relationships for more than a century. A phylogenetic tree displays a single pattern of evolutionary relationships between the taxa concerned. Therefore, an inherent assumption is a lack of homoplasy, although this assumption is rarely completely met. The development and comparison of computational methods for transforming aligned sequences into phylogenetic trees is an enormous field of research and ongoing debate (Nei 1996; Brocchieri 2001), and a comprehensive description of extant methods and their merits is beyond the scope of this discussion.

There are four basic approaches to constructing trees. Distance based methods involve the conversion of the sequence alignment into a matrix of numerical evolutionary distances, and the inference of the tree from that data. There are a number of algorithms for inferring a tree from a distance matrix, of which the most commonly used, is probably the neighbor-joining method (Saitou and Nei 1987). Its popularity is a result of its computational speed, a general acceptance that it produces reasonably good results, the fact that it produces a single tree as the "answer", and its incorporation into popular alignment construction software packages such as the Clustal family (Larkin et al. 2007) and Mega (Kumar et al. 2008).

The parsimony approach to tree inference is the simplest method that directly analyses the actual sequences, rather than numerical values for sequence similarities. Parsimony analyses search for the tree(s) that can account for the observed sequences in the minimum number of evolutionary steps from a hypothetical common ancestor. This method has been used extensively. However, the manner in which it searches evolutionary space is often regarded as oversimplified in that the number of optimal evolutionary pathways that can lead to a given tree is not taken into account. This can lead to a tree that defines a single short evolutionary pathway being preferred over a tree that defines multiple pathways that are only marginally longer. In such a situation, the second tree is probably a better "answer".

Maximum Likelihood methods are regarded as being rigorous and tolerant of wide ranges of sequence divergence, and are also computationally expensive. Likelihood is conditional probability. In this context, it is the probability of observing the actual sequence alignment, given a particular tree topology. The likelihood can be calculated using any plausible model for the evolutionary process. The tree topology that maximizes the likelihood is arrived at by the empirical testing of many different topologies. It is commonplace to incorporate the testing of variations of the parameters for the evolutionary model into machine learning analyses, with the assumption that parameters that maximize the likelihood are preferred.

Bayesian methods for phylogenetic inference have similarities to machine learning methods, but also some subtle but important differences (Yang and Rannala 1997; Holder and Lewis 2003; Ronquist and Huelsenbeck 2003). The basis for these methods is the use of Bayes theorem to convert the likelihood of a particular sequence alignment, given a particular tree topology, into the probability of a given tree topology (the posterior probability), given a particular sequence alignment (the evidence). The calculation of the post-test probability takes into account the probability of that tree topology in the absence of any alignment data (the prior probability), and the probability of the sequence alignment itself, which in general can be regarded as a constant. This allows a search to be made for tree topologies that maximize the post-test probability. The Bayes theorem allows the incorporation of unlimited numbers of parameters of "evidence" and this makes it possible to simultaneously search tree topology space and evolutionary model space in a computationally efficient manner. Essentially all Bayesian methods use the Markov Chain Monte Carlo (MCMC) search algorithm. This entails defining a starting point of a particular tree topology and evolutionary model, and then perturbing this in a random step-wise fashion. New points within the tree and parameter space are accepted if they provide a higher posterior probability than the previous point. There is also a probability that they will be accepted if the posterior probability is less than at the previous point, and this probability is inversely proportional to the reduction in posterior probability from the previous point to the new point. This is an approximation that allows the regions of tree and parameter space with high posterior probabilities to be identified efficiently, while at the same time avoiding becoming stranded in local minima. A very clever and valuable aspect of the MCMC method is that it calculates a probability distribution in the tree and parameter space by determining how often points within that space are visited during the course of the search. The premise is that points visited more frequently represent a higher probability density. This allows trees to be assessed not on the basis of which one provides the highest peak of posterior probability, but on the basis of which one provides the greatest area under the multidimensional curve of the relationship between the posterior probability and parameter space. This therefore takes into account the fact that a tree that can be reached by many reasonable evolutionary pathways from the sequence alignment is preferred over one that can be reached by fewer reasonable evolutionary pathways, even if the peak posterior probability for the second tree is higher. Therefore, Bayesian analysis combines, in a computationally efficient manner, the determinations of the preferred tree, the tree's robustness, and the preferred evolutionary model.

As stated above, an inherent assumption in the inference of a phylogenetic tree is a lack of homoplasy. It is important to bear in mind that a tree can be inferred from essentially any data set, no matter how homoplastic. For example, a tree inferred from MLST data derived from a highly non-clonal species will look like a plausible tree, but will represent the average of the different trees that would have been derived from each locus. It is commonplace to infer trees from non sequence-based genetic data such as the electrophoretic banding patterns derived from pulse field electrophoresis or fluorescent amplified fragment length polymorphism analysis. If this is carried out on isolates from a non-clonal species, it provides some insight regarding the population structure. However, it is difficult to define exactly what this insight is; therefore, these relationships cannot really be termed a phylogeny. Trees from multilocus sequence data or genome-wide electrophoretic analysis of non-clonal species are best regarded as similarity trees rather than phylogenetic trees. There are a number of strategies for calculating and depicting the networks defined by conflicting or homoplastic phyogenetic signals, with split decomposition analysis being the best known (Huson 1998).

eBURST and related algorithms. Feil and co-workers have developed a means of depicting bacterial population structures that is very different to a conventional phylogenetic tree (Feil et al. 2004). It is designed to accommodate two phenomena that are handled poorly by phylogenetic trees: high levels of LGT, and the co-existence of ancestors and descendents. The eBURST software is designed to analyze MLST allele profile data rather than the actual sequence data, and as a result does not take the degree of divergence between different alleles into account. The core of the algorithm is a parsimony-based search for founders of CCs. The premise is that the ancestral ST of a group of closely related STs (i.e. STs that differ at a maximum of two of the seven MLST loci) will be the one that differs at only one locus from maximum number of STs. In this way, CCs can be depicted as founders surrounded by descendents, or SLVs. Elaborations of the algorithm accommodate SLVs themselves being CC founders, with the concomitant linking together of CCs into "super complexes". Also, the reliability of the identification of CC founders can be tested through bootstrapping analysis. An eBURST diagram depicts each ST as a circle, and the size of this can be scaled in proportion to the number of isolates of that ST in the database. CC founders and their SLVs are connected by lines. These lines depict only a subset of the SLV relationships, and it is possible to select an option that turns on the depiction of all SLV relationships, or all double locus variants (DLV).

The obvious disadvantage of eBURST analysis is that it operates over a narrow dynamic range and provides no information regarding the relationships between STs that are not connected to each other, and also says nothing about the similarities between alleles. In addition, whether an isolate is classified as a CC founder or an SLV is subject to strong stochastic effects, with the assignments likely to be very different if, for example, a different seven gene fragments were chosen for MLST analysis. However, there is no doubt that eBURST is a very effective means of gaining an understanding of the general characteristics of the population structures of bacterial species. This is particularly the case when an entire MLST database is depicted in a single diagram, known as the population snapshot. This has considerable power to reveal the extent of LGT, and the extent to which the species has been sampled (Fig. 2.3).

In recent years, there has been considerable interest in the use of sequence repeat containing loci for high-resolution bacterial genotyping and the inference of bacterial population structures. DNA sequence repeats evolve rapidly in essentially all life forms, with the principal mechanisms being slipped strand mis-pairing during DNA replication, and homologous recombination. These methods are known as variable number tandem repeat (VNTR) analysis or multi-locus VNTR analysis (MLVA) (Grissa et al. 2008). A technique with close similarity to eBURST analysis is the inference of minimum spanning trees (MSTs), and this is increasingly being used to analyze MLVA data (Melles et al. 2009). MSTs are well known in graph theory and are the most efficient means of connecting all points on a graph. Efficiency is calculated using weightings of all the possible connections between the points, and these weightings would normally equate to the length of the line. The inference of an MST is in essence an exercise in parsimony. MSTs are similar to the output from eBURST analyses, but with the important difference that all genotypes are connected. Thus MSTs depict a greater range of evolutionary distances than eBURST analyses. In most MLVA methods, the VNTR loci have highly conserved repeating units. In consequence, most of the information from the loci can be extracted by simply sizing the loci by means of capillary electrophoresis of PCR products, rather than by sequence determination, and this is what is usually done. In order to infer an MST from MLVA data of this nature, the similarities between genotypes are determined solely on the basis of how many loci are the same and how many are different. It is not usual to attempt to deduce degrees of similarity at individual loci, because an assumption that similar sized loci are more closely related than more differently sized loci may not be warranted. Elements of the eBURST algorithm may be included in the analysis in order to discriminate between equivalent trees e.g., to decide which are major CC progenitors that should

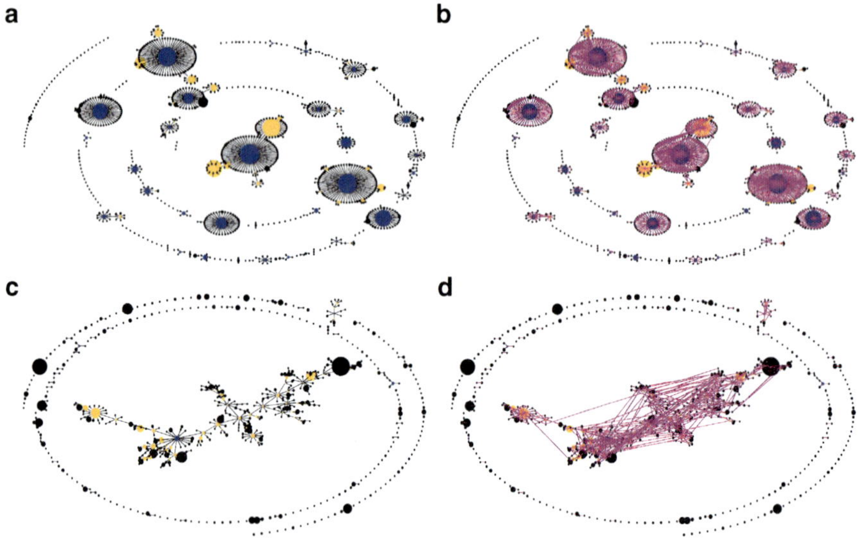

Fig. 2.3 (**a**) An eBURST derived population snapshot of *S. aureus*. Each *filled circle* represents an ST, with the size proportional to the number of isolates of that ST in the MLST database. Large CCs are conspicuous. (**b**) The same diagram as in (**a**), but with all pairs of STs that differ at a single locus connected by *pink lines*. The lack of links between the CCs is evidence for slow gene flow between the CCs. (**c**) An eBURST derived population snapshot of *Burkholderia pseudomallei*, the causative agent of the tropical disease melioidosis The central portion of the diagram consists of a highly linked complex of STs, dissimilar in its structure to anything seen with *S. aureus*. (**d**) The same diagram as in C. but with all pairs of STs that differ at a single locus connected by *pink lines*. It can be seen that the number of linkages within the central complex is very high, suggesting a high rate of LGT. There are numerous instances in which STs that differ at only one locus are located far from each other in the diagram, indicating that the defining of a reliable evolutionary pathway with a rapidly recombining population is very difficult

be directly connected together (Ghosh et al. 2008; Melles et al. 2009) (Fig. 2.4). In the case of the electrophoresis-based methods, the data are encoded in a binary fashion such that each genotype is defined on the basis of which electrophoretic bands it includes, and which it does not include (Melles et al. 2009). The recent popularity of MST analysis is a likely consequence of its inclusion as a function in the Bionumerics software (http://www.applied-maths.com/bionumerics/bionumerics.htm), which is extensively used by microbiologists.

'The Based Upon Repeat Pattern' (BURP) algorithm is also closely related to the eBURST algorithm and is specifically designed to deduce *S. aureus* population structures from variation in the VNTR-containing *spa* gene. In effect, *spa* typing is a single locus VNTR typing method (Mellmann et al. 2007, 2008). The *spa* gene encodes protein A, which is a hypervariable cell surface attached immunoglobulin binding protein. In effect, the *spa* molecular clock has both a minute hand and an hour hand, with the minute hand being the rapid re-arrangement and alteration of the number of repeats, and the hour hand being the generation of novel repeating units by point mutation. The BURP algorithm divides *spa* sequences into CCs on

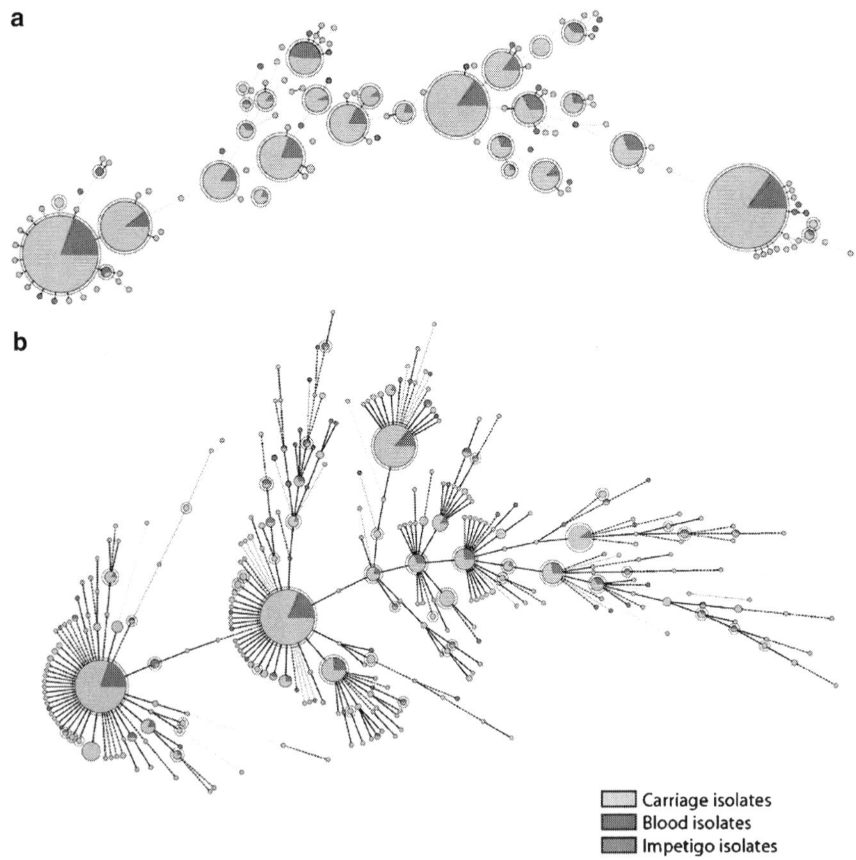

Fig. 2.4 Minimum spanning trees demonstrating that *S. aureus* carriage, blood and impetigo isolates do not fall into different phyogenetic groups. (**a**) The tree is derived from AFLP data. (**b**) The tree is derived from MLVA data (Adapted from Melles et al. 2009)

the basis of the sequences of the repeating units, and uses an eBURST like analysis of the number and arrangement of the repeat units to infer the CC founders. A related and more sophisticated examination of the evolutionary informative power of the *spa* locus has been reported (Agius et al. 2007). This analysis involved testing different alignment algorithms of *spa* sequences so as to determine which generates a distance matrix that is most consistent with MLST data, and it was found that an algorithm that searches for subsequences in common, and corrects for the distance between these subsequences performs best. An interesting aspect of this analysis was the use of multidimensional scaling (MDS) to depict the relationships between multiple *spa* sequences. MDS is in principle very similar to MST construction, but does not involve linking different sequences. Rather, the sequences are placed on a two or three-dimensional chart in such a way that consistency with the distance matrix is maximized.

2.3.3 Comparisons of Entire Genomes

The last five years have seen the emergence of new ultra-rapid methods for genome sequence determination, and these are proving to be a disruptive technology. It is becoming possible to reveal the evolution and population structures within and between bacterial species by comparing complete genomes, rather than genome samples. An example of an application for inferring whole gene phylogenies is OGtree (Jiang et al. 2008). This only requires the GenBank accession numbers of entire genomes, and it will identify a set of orthologous genes that are present in all the genomes, concatenate these, and infer a tree. However, it is currently an open question as to how much more information will be derived from whole genome studies as opposed to genome sampling-based phylogenetic studies. If LGT is not a significant confounding factor, then the analysis of ever more genes rapidly enters a region of diminishing returns, because even a single gene can perform very well as a marker of the whole genome. When LGT is significant, analyzing whole genomes does not result in convergence to the correct tree, because there is no single phylogeny to define a single tree. Rather, analyzing whole genomes will simply add to the complexity of a "network" result. If the data are forced to define a tree, then the tree will be a similarity tree rather than a phylogenetic tree. One way of at least partly circumventing this problem has been developed as a tree inference approach with software named "ClonalFrame" (Didelot and Falush 2007). It appears to be a powerful method for comparing complete genomes, between which there is significant, but not overwhelming LGT. Recently, this research group has introduced a method for mapping LGT using whole genome sequences (Didelot et al. 2009).

2.4 Impact of Advances in Microbial Evolution on the Practice of Microbiology

The task of an analytical microbiologist may be summarized as (1) the determination of what micro-organism(s) are present within an analytical sample, and (2) the derivation of information that that serves the clinical, public health, food safety, environmental or other anthropocentric interest from such microorganisms. The tools that are central to these tasks are diagnostic targets with high sensitivity and specificity for relevant taxonomic groups, marker sets for genetic fingerprinting methods that allow the tracking of significant clones or CCs, and marker sets that allow the inference of clinical or other significant aspects of the phenotype. It is self evident that there are conceptual and practical overlaps between these tools, and that an understanding of the comparative genomics, population biology and gene function are central to their development. The following sections outline current research into the population structures of four species of bacterial pathogens and illustrate the considerable differences between bacterial species, both in terms of the actual population structures, and the clinical and public health imperatives.

2.4.1 *Bacillus anthracis*

Anthrax is a severe disease that primarily affects grazing animals. The causative organism, *B. anthracis*, is a Gram-positive endospore forming bacterium that can remain dormant in the environment for long periods of time. The primary mode of transmission amongst grazing animals is the ingestion of soil that is contaminated with endospores derived from the carcass of an animal that has died of anthrax. Humans can contract anthrax as a result of inhalation, ingestion or cutaneous inoculation of endospores. All three modes of disease are extremely dangerous, although cutaneous anthrax has a lower fatality rate than either gastrointestinal or inhalational anthrax.

B. anthracis has long been seen as a potential bio-weapon. This potential was realized in 2001 with well-known anthrax attacks using endospores in letters mailed to the victims. Because of the enormous imperative to determine the source of the strain used in the attacks, large financial resources were made available to rapidly determine the population structure of *B. anthracis* and to develop appropriate genotyping methodologies. In the course of these studies, the term "microbial forensics" was coined (Keim et al. 2008). The principal underpinning this is that for a genotyping method to provide reliable information in identifying the origin of a point source outbreak, the performance of the genotyping method has to be understood in relation to the population structure of the species. This allows the estimation of the probability of epidemiologically unconnected isolates being of the same genotype.

Early studies of the population structure and diversity of *B. anthracis* quickly revealed that mutational diversity (i.e. the number of SNPs) is extremely low compared with other bacterial species. The lack of diversity is so extreme that there is no MLST scheme for *B. anthracis* – there are simply too few SNPs to make MLST useful. Accordingly, in order to identify more diversity, which could be used to develop genotyping methods, six genomes were completely sequenced and the SNPs defined by these sequences were identified (Read et al. 2002; Keim et al. 2004; Pearson et al. 2004). Analysis of the SNPs revealed that they were not homoplastic and so could be used to infer a single phylogenetic tree. Therefore, LGT does not occur. In order to streamline the SNP-based genotyping, a small subset of the SNPs that defined the tree branch points were identified and named canonical SNPs (canSNPs). Concurrently with this, VNTR loci were identified and methods to interrogate these were developed. A combinatorial typing strategy using both SNPs and VNTRs was developed and named Phylogenetic Hierarchical Resolving Assays using Nucleic Acids (PHRANA) (pronounced "piranha") was developed (Keim et al. 2004). The rationale for this was that the canSNPs divide the species into phylogenetically valid groups, but have insufficient resolving power for most genotyping applications. Conversely, the VNTR loci provide much higher resolution on their own but are also highly homoplastic because of the frequency of events such as gaining one repeating unit by slipped strand mispairing, and then losing it again so as to yield the original sequence. In other words, with

VNTRs, evolution is frequently reversible. A potential serious problem with the use of the SNPs was identified when it was realized that the tree defined by the SNPs is a function of the genome sequences chosen to identify the SNPs in the first place. More specifically, it was reasoned that the strains used would not define all the SNPs in the species, and that if even one un-sequenced strain had diverged from the sequenced strains relatively early in the life of the species, then all of the tree structure in the diverged lineage(s) defined by the un-sequenced strain would be invisible to the SNPs defined by the sequenced strains. This is because the "invisible" structure of the tree is defined by SNPs that are not polymorphic amongst the sequenced strains and so are not identified as SNPs at all. The consequence is that the invisible structure collapses to a point on the tree. This is known as "SNP discovery bias leading to branch collapse". Notwithstanding this issue, the population biology and evolutionary history of *B. anthracis* are now well understood (Van Ert et al. 2007a). The extremely low level of diversity in *B. anthracis* indicates that the species is approximately 20,000 old, which is much younger than other bacterial species. There are three major lineages that are termed A, B and C. Lineages B and C are rare and found in a small number of locations in Africa and Europe. In contrast, lineage A is widely distributed and abundant and appears to have undergone a rapid radiation through Africa/Eurasia between three and six thousand years ago. It is hypothesized that this was connected with the emergence of animal husbandry by humans. Dispersion to the Americas appears to be the result of European colonization, while introduction into Australia may have been the result of the import of a batch of contaminated fertilizer in the mid nineteenth century. Interestingly, this work led to the conclusion that the "Ames" strain of *B. anthracis* was used in the 2001 bioterrorism attacks, and a SNP-based diagnostic for this strain was developed (Van Ert et al. 2007b).

The natural history and evolution of *B. anthracis* provides an excellent illustration of the inherent difficulties in developing a consistent definition of a bacterial species. *B. anthracis* is very closely related to the species *Bacillus cereus* and *Bacillus thuringiensis*. In general terms, *B. anthracis* is defined as the causative agent of anthrax, *B. thuringiensis* is defined as being pathogenic to insects, and *B. cereus* is a food contaminant and agent of food-borne disease. All three are commonly found in soil. The virulence of *B. anthracis* is dependent upon two plasmids. pXO1 encodes the virulence factors: lethal factor, protective antigen and edema factor. pXO2 confers the ability to synthesize a capsule, which also contributes to virulence. As stated in the previous paragraph, *B. anthracis* has very low genetic diversity. MLST studies of the three above species have confirmed that *B. anthracis* is monophyletic (Tourasse et al. 2006). However, *B. cereus* is much more diverse than *B. anthracis*. Therefore, from a phylogenetic point of view, the only "valid" species may be *B. cereus*, with *B. anthracis* being a recently evolved clone or CC within this species, and *B. thuringiensis* being several disparate clones (Kim et al. 2005; Tourasse et al. 2006; Vilas-Boas et al. 2007). The acquisition of a suite of virulence factors has given *B. anthracis* distinctive ecological, epidemiological and virulence properties that seriously impact mankind, and this is the obvious reason for its status as a separate species, despite its recent origin.

2.4.2 *Staphylococcus aureus*

S. aureus is a Gram-positive nonsporulating coccus with a facultative metabolism. It is a major human pathogen that can cause a wide variety of disease states including skin and soft tissue infections, sepsis and pneumonia (Deurenberg and Stobberingh 2008). A significant proportion of *S. aureus* carry the large mobile genetic element SCC*mec* (Katayama et al. 2000). This contains the *mecA* gene which encodes a variant of Penicillin Binding Protein 2 that is refractory to inhibition by β-lactam antibiotics. Strains carrying this element are known as methicillin resistant *S. aureus* (MRSA), and MRSA is regarded as a more serious clinical issue than methicillin sensitive *S. aureus* (MSSA), primarily because of the constrained treatment options. *S. aureus* is a commonly found inhabitant of the human skin and nasal passages, and was for many years regarded primarily as an agent of nosocomial infections. However, the last ten years have seen the emergence of the so-called "community acquired MRSA" (CA-MRSA) (Deurenberg and Stobberingh 2008). CA-MRSA is able to cause infections in individuals who are not associated with health care facilities. CA-MRSA transmission is by skin to skin contact, or practices such as sharing towels, so CA-MRSA can be associated with body contact sports such as rugby and wrestling, the military, prisons, and poor living conditions.

S. aureus was one of the first species to be studied intensively by MLST, and this provided several valuable insights (Enright et al. 2000). Firstly, it was determined that LGT in *S. aureus* is less frequent than in some other bacterial species, but still frequent enough to have a considerable impact on the population structure (Feil et al. 2003). A population snapshot using eBURST yields a highly structured diagram in which the dominant features are large CCs composed of progenitors and significant numbers of SLVs (Feil et al. 2004; Huygens et al. 2006). Approximately 90% of the SLVs appear to exist by virtue of unique alleles, thus indicating that SLVs are formed by mutation in approximately 90% of cases and LGT in approximately 10% of cases. While CCs within a subset of major CCs are linked to each other, most are not. This demonstrates that there is a significant degree of genetic isolation between the CCs. It has been suggested that the major CCs deserve subspecies status (Turner and Feil 2007). MRSA isolates are found in essentially all the major CCs, and also within minor CCs and singletons. This shows that *S. aureus* has either acquired SCC*mec* on many separate occasions, or that SCC*mec* has undergone frequent LGT within *S. aureus*. SCC*mec* displays a very high degree of diversity, especially with respect to gene content (Lina et al. 2006; Kondo et al. 2007), which suggests that separate acquisitions of SCC*mec* are at least partly responsible for the wide distribution of this element within *S. aureus*. MLST analysis of MRSA clinical isolates from hospital acquired infections has shown that in the majority of cases major CCs are internationally distributed (Deurenberg et al. 2007). These in effect represent epidemics or pandemics of strains that are highly effective at colonizing health care facilities and are probably transmitted via the international movement of health care workers. The presence of different SCC*mea* allotypes within the same CC shows that a given CC can contain more than one hospital acquired MRSA clone.

MLST and SCC*mec* allotype analysis has shown that CA-MRSA isolates do not belong to the major hospital acquired clones. Remarkably, multiple clones of CA-MRSA appear to have emerged simultaneously in the last decade (Tristan et al. 2007). The STs of CA-MRSA show no particular relationship to each other, but most CA-MRSA isolates harbor a prophage that encodes the Panton-Valentin leukocidin (PVL toxin), and also a particular truncated SCC*mec* allotype known as SCC*mec* IV. There is evidence that the PVL toxin facilitates the causation of infection in lungs and healthy skin (Gillet et al. 2002; Yamasaki et al. 2005; Labandeira-Rey et al. 2007). However, there has been some debate regarding the reproducibility of animal experiments, and the role of PVL toxin in CA-MRSA infections remains controversial (Diep et al. 2008). While these findings could be regarded as evidence for a current rapid radiation of the PVL toxin encoding prophage into preexisting populations of MRSA, this is probably incorrect. PVL – positive CA-MRSA and CA-MSSA belonging to ST93 have been found to co-exist in the same geographical locations (Huygens et al. 2006), which suggests that in this instance at least, SCC*mec* is in the process of radiating into a preexisting population of PVL-positive MSSA. Interestingly, once established, at least some CA-MRSA clones are capable of explosive intercontinental spread, with the USA300 (ST8) CA-MRSA clone being an excellent example of this (Kennedy et al. 2008). In summary, *S. aureus* may be considered a species in which the dissemination of clinically relevant mobile genes and the dissemination of clones are not congruent. Rather, they constitute a complex counterpoint that underpins the ever-changing nature of the interaction between humans and this species.

Because of the clinical importance of *S. aureus*, and the need to monitor its dissemination at all scales from intercontinental to within health care facilities, this species has been the test bed for a number of new approaches to genotyping. MLST has been extensively used, as have the electrophoresis-based methods and VNTR analyses (Deurenberg and Stobberingh 2008; Melles et al. 2009). Very recently, micro-array based methods targeting mobile genes have shown considerable promise (Monecke et al. 2007). My research group developed a computerized approach to deriving resolution optimized SNP sets from DNA sequence alignments. The measure of resolution is the Simpson's index of Diversity (*D*), which in this case is the probability that any two STs chosen at random will be discriminated by the SNPs. Therefore, *D* optimized SNPs are in effect optimized for discriminating all STs from all STs. It was found that it was possible to derive from the *S. aureus* MLST database a set of eight SNPs that resolved the major *S. aureus* CCs (Huygens et al. 2006; Stephens et al. 2006). Forty-seven genotypes were defined by the SNPs, which is considerably more than nine, the maximum number of genotypes that could be defined by the eight two-state SNPs in a population with no homoplasy. This confirms that *S. aureus* housekeeping genes are indeed subject to LGT. In contrast, it was impossible to identify a SNP set that would efficiently resolve the STs within CCs. It was determined that this was a direct result of the low frequency of LGT. As the majority of SLVs are separated from the CC progenitor by virtue of a unique MLST allele (which by definition will contain a unique SNP allele), each SNP will normally resolve one SLV from the progenitor, but will provide no resolving power elsewhere in the species. This graphically demonstrates that SNPs can be young or old. Very old SNPs have been in existence long enough for both alleles to reach appreciable frequencies, and to be distributed through the species by LGT. They have good resolving power, especially in combination, because with highly homoplastic SNPs the resolving power may potentially increase as the exponential of the number of SNPs. Conversely, young SNPs have a highly skewed allele distribution and have not been subjected to appreciable LGT. They therefore have little resolving power. A search for SNPs that maximize *D* is in effect a search for the very oldest SNPs, and it provides a very small set of SNPs that very efficiently discriminates the major CCs. It may be concluded from this that if the *S. aureus* CCs are regarded as single entities, then they comprise a non-clonal population that is more like a network than a tree.

A recent and significant study examined microevolution within ST5 (Nubel et al. 2008). This was done by identifying SNPs by sequencing large portions of the genome from a number of ST5 isolates, and using this information to define a large SNP set which was then used to genotype many more ST5 isolates. One unexpected finding was a strong association between particular genotypes or lineages, and particular geographical regions. This suggests that international dissemination may not be as rapid as has previously been assumed. It was also found that the SCC*mec* allotype data was highly homoplastic in relation to the SNP data, thus indicating that the number of introductions of SCC*mec* was much higher than previously thought. Finally, the *spa* locus was also shown to be highly homoplastic, thus raising doubts as to its usefulness as the basis for a stand-alone typing method.

2.4.3 *Campylobacter jejuni and Campylobacter coli*

Campylobacter jejuni and *Campylobacter coli* are closely related species that are the most common cause of bacterial gastroenteritis in developed countries. They are small Gram-negative rods that are gastro-intestinal tract commensals found in a wide variety of birds and mammals. It is highly likely that the majority of human disease caused by these species is food-borne and associated with chicken meat or beef. Complete genome analyses have shown that *Campylobacer jejuni/coli* genomes are small (~1.8 megabases), very AT rich, and remarkably deficient in mobile genetic elements of any sort.

There is a combined MLST database for *C. jejuni and C.coli* (http://pubmlst.org/campylobacter/) (Dingle et al. 2001a; Dingle et al. 2005b). This now contains in excess of 3,800 STs. eBURST analysis assigns the majority of these to a small number of large CCs that display complex internal structures (Feil et al. 2004). It is becoming apparent that there are some associations between CC and the primary host species (French et al. 2009). It appears that the frequency of LGT is comparable to or greater than the frequency of point mutation (Dingle et al. 2001a, c; Fearnhead et al. 2005). Since an LGT event typically introduces more than one SNP, this means that any given position in a gene is more likely to evolve by LGT than by mutation.

The relationship between *C. jejuni* and *C. coli* has unusual aspects. Housekeeping gene homologs, including the MLST loci, from *C. jejuni* and *C. coli* are approximately 85% identical. This is not unusual for different species in the same genus, and provides ample justification for the existence of two species. However, the MLST database contains examples of hybrid isolates that contain some alleles typical of *C. jejuni* and some typical of *C. coli*. Therefore, interspecies LGT events take place. More remarkably, there appears to be a particularly frequent LGT from *C. jejuni* to *C. coli* clade one, which is one of three major *C. coli* clades. The presence in *C. coli* of many exact copies of *C. jejuni* alleles, together with the lack, in *C. coli* of variant *C. jejuni* alleles, indicates that this rapid LGT is a recent phenomenon. It has been concluded that *C. jejuni* and *C. coli* are in the process of converging or "de-speciating", and it has been postulated that this is because of changes in the ecology of the species (Sheppard et al. 2008; Wilson et al. 2009). A straightforward explanation for the data is that *C. coli* has recently colonized a niche that is already heavily populated by *C. jejuni*. It is tempting to link this to the development of animal husbandry in recent pre-history.

Calibrating the molecular clocks that define the trees or other diagrams that used to describe bacterial evolution is an inherently difficult task. This is because the only really direct means of calibration involves correlating a speciation event with the fossil record, or directly measuring the rate of evolution in the laboratory. The former is inherently speculative, and the latter may be confounded by differences in evolution speeds between laboratory grown cells and cells in the wild. A rule of thumb that has been extant in this field for the last two decades is that the divergence of *Salmonella* and *Escherichia* coincided with the appearance of mammals, approximately 150 million years ago (Ochman and Wilson 1987). By extrapolation from relative sequence divergences, that would place the divergence of *C. jejuni* and *C. coli* at 10 million years ago. However, a recent and highly provocative analysis by Wilson and co-workers arrives at the conclusion that the *Campylobacter* evolutionary clock runs 1,0002009 times faster than this (Wilson et al.). The basis for this was MLST data from 1,205 isolates obtained over three years at a single health service. The key aspect of the analysis was an "importance sampler" approach to determine if evolution was occurring during those three years. It was concluded that evolution was indeed occurring, and this is what was used to calibrate the evolutionary clock.

2.4.4 *Streptococcus agalactiae*

Streptococcus agalactiae (group B *Streptococcus* (GBS)) is a Gram-positive coccus that is the predominant cause of neonatal sepsis in developed countries, and can also cause a variety of other pathologies. GBS is of particular interest from a population biology and epidemiology point of view because there is strong evidence that it is primarily a cause of mastitis in cattle, and at least some lineages have switched hosts from cattle to humans within the last century. GBS is not a classical zoonosis in which each instance of disease in humans results from a cross species transmission event. Rather, it appears that there have been a small number of cross species transmission events, and these have been followed by dissemination within the human population to the point that GBS carriage by humans is now endemic.

GBS have long been classified into serotypes, on the basis of capsular polysaccharides. More recently, DNA-based methods for revealing the serotype on the basis of the presence or absence of capsule synthesis genes have been developed. This approach has been enhanced through the addition of genetic assays for the presence of mobile genetic elements and genes encoding surface proteins. These methods make use of an efficient "reverse line blot" technique which is in essence a membrane based array, in which the probes are anchored to the membrane, and the labeled analyte DNA is hybridized to the probes (Kong et al. 2005; Sun et al. 2005; Kong and Gilbert 2006).

An MLST scheme for GBS has been constructed (Jones et al. 2003). This has revealed that that the percentage of variable sites at the individual loci is only 1.5–2.5%, which is much less than other bacterial species. This indicates a recent common ancestry for extant lineages. Evidence suggests that LGT of housekeeping genes is a frequent occurance (Bisharat et al. 2004; Honsa et al. 2008). An eBURST population snapshot of the species reveals one large "super complex" of interlinked CCs, of which the largest are CC19, CC1 and CC10, and two other major CCs which are CC17 and CC23. There have been several reports in the literature that ST17 represents a hypervirulent clone that has been very recently acquired from a bovine source. The primary evidence for this was that most bovine isolates belong to CC17 (Bisharat et al. 2004), and a study of the distribution of mobile elements has lent support to this (Hery-Arnaud et al. 2007). However, the current focus on CC17 begs the question as to the origins of non-CC17 strains in humans. Despite statistical evidence for the greater virulence of CC17 (Lin et al. 2006), the incidence of invasive disease caused by non-CC17 strains is far from insignificant (Bohnsack et al. 2008). Therefore, given that GBS disease in humans is a recently emerged phenomenon, it is difficult to rule out the notion that all GBS subpopulations are the result of recent cross species transmission events into humans, and that the identification of CC17 as a "bovine" lineage may be a consequence of its prevalence in the bovine populations that have been subjected to sampling. Recent evidence for this is the finding that GBS CC26 infects both cattle and humans in South and South-East Brazil (Oliveira et al. 2006). This issue remains to be fully resolved. It is not a trivial issue, as population genetics and clinical evidence for enhanced virulence will inevitably underpin large-scale comparative genomics-based methods for identifying enhanced virulence determinants. Therefore, the findings of enhanced virulence need to be correct.

Finally, the study of bacterial pathogens normally has a highly anthropogenic focus, with human disease being of primary concern. However, the abundance of humans on this planet means that instances of diseases, in which there is cross species transmission from humans to animals, are likely to be very frequent but rarely noticed. One interesting example of a likely anthroponosis is an outbreak of GBS caused necrotizing fasciitis in farmed juvenile estuarine crocodiles in Northern Australia.

Genetic analysis revealed that all isolates were CC23, and their serotype and complement of mobile genetic elements were typical for the human isolates of that CC (Bishop et al. 2007). The simplest explanation for these findings is that the crocodiles had acquired a human GBS strain that is not of enhanced virulence in humans, but is highly virulent to crocodiles.

2.5 Concluding Remarks

Recent research has shown that populations of Bacteria and Archaea, like Shakespeare's rose that is a rose by any other name, simply are what they are. The highly variable nature and extent of LGT means that these life forms by and large refuse to conform to generalizations regarding diversity and gene flow, and therefore mightily resist the development of workable and consistent rules and conventions for taxonomy. The fourteen years that have elapsed since the publication of the first complete bacterial genome sequence have seen not just a massive accumulation of data, but a series of surprises, with the biggest surprise probably being the size of the accessory genome in any given cell, and the seemingly infinite extent of the pan-genome that is the aggregate gene content of all genomes in a species. The advent of high throughput sequencing means that we are entering a new era in which there will be many thousands of known complete genome sequences. The meaningful analysis of these data will be a conceptual and technical challenge, but one that is certain to be met. It will allow the determination of the phylogenies of thousands of genes, and the use of that information to place extant cells into a network of stunning complexity.

References

Achtman M, Wagner M (2008) Microbial diversity and the genetic nature of microbial species. Nat Rev Microbiol 6(6):431–440

Agius P, Kreiswirth B et al (2007) Typing *Staphylococcus aureus* using the spaspa gene and novel distance measures. IEEE/ACM Trans Comput Biol Bioinform 4(4):693–704

Bisharat N, Crook DW et al (2004) Hyperinvasive neonatal group B streptococcus has arisen from a bovine ancestor. J Clin Microbiol 42(5):2161–2167

Bishop EJ, Shilton C et al (2007) Necrotizing fasciitis in captive juvenile *Crocodylus porosus* caused by *Streptococcus agalactiae*: an outbreak and review of the animal and human literature. Epidemiol Infect 135(8):1248–1255.

Bohnsack JF, Whiting A et al (2008) Population structure of invasive and colonizing strains of *Streptococcus agalactiae* from neonates of six U.S. Academic Centers from 1995 to 1999. J Clin Microbiol 46(4):1285–1291.

Brocchieri L (2001) Phylogenetic inferences from molecular sequences: review and critique. Theor Popul Biol 59(1):27–40.

Cairns J (1963) The bacterial chromosome and its manner of replication as seen by autoradiography. J Mol Biol 6:208–213.

Crick FH (1962) The genetic code. Sci Am 207:6–74
Deurenberg RH, Stobberingh EE (2008) The evolution of *Staphylococcus aureus*. Infect Genet Evol 8(6):747–763
Deurenberg RH, Vink C et al (2007) The molecular evolution of methicillin-resistant *Staphylococcus aureus*. Clin Microbiol Infect 13(3):222–235
Didelot X, Darling A et al (2009) Inferring genomic flux in bacteria. Genome Res 19(2):306–317
Didelot X, Falush D (2007) Inference of bacterial microevolution using multilocus sequence data. Genetics 175(3) 1251–1266
Diep BA, Palazzolo-Ballance AM et al (2008) Contribution of Panton-Valentine leukocidin in community-associated methicillin-resistant *Staphylococcus aureus* pathogenesis. PLoS One 3(9):e3198
Dingle KE, Colles FM et al (2005a) Sequence typing and comparison of population biology of *Campylobacter coli* and *Campylobacter jejuni*. J Clin Microbiol 43(1):340–347
Dingle KE, Colles FM et al (2001b) Multilocus sequence typing system for *Campylobacter jejuni*. J Clin Microbiol 39(1):14–23
Dingle KE, Van Den Braak N et al (2001) Sequence typing confirms that *Campylobacter jejuni* strains associated with Guillain-Barre and Miller-Fisher syndromes are of diverse genetic lineage, serotype, and flagella type. J Clin Microbiol 39(9):3346–3349
Enright MC, Day NP et al (2000) Multilocus sequence typing for characterization of methicillin-resistant and methicillin-susceptible clones of *Staphylococcus aureus*. J Clin Microbiol 38(3):1008–1015
Enright MC, Spratt BG et al (2001) Multilocus sequence typing of *Streptococcus pyogenes* and the relationships between *emm* type and clone. Infect Immun 69(4):2416–2427
Fearnhead P, Smith NG et al (2005) Analysis of recombination in *Campylobacter jejuni* from MLST population data. J Mol Evol 61(3):333–340
Feil EJ, Cooper JE et al (2003) How clonalclonal is *Staphylococcus aureus*? J Bacteriol 185(11):3307–3316
Feil EJ, Enright MC et al (2000a) Estimating the relative contributions of mutation and recombination to clonalclonal diversification: a comparison between *Neisseria meningitidis* and *Streptococcus pneumoniae*. Res Microbiol 151(6):465–469
Feil EJ, Holmes EC et al (2001) Recombination within natural populations of pathogenic bacteria: short-term empirical estimates and long-term phylogenetic consequences. Proc Natl Acad Sci USA 98(1):182–187
Feil EJ, Li BC et al (2004) eBURSTeBURST: inferring patterns of evolutionary descent among clusters of related bacterial genotypes from multilocus sequence typing data. J Bacteriol 186(5):1518–1530
Feil EJ, Smith JM et al (2000b) Estimating recombinational parameters in *Streptococcus pneumoniae* from multilocus sequence typing data. Genetics 154(4):1439–1450
Fleischmann RD, Adams MD et al (1995) Whole-genome random sequencing and assembly of *Haemophilus influenzae* Rd. Science 269(5223):496–512
Fraser C, Alm EJ et al (2009) The bacterial species challenge: making sense of genetic and ecological diversity. Science 323(5915):741–746
Fraser C, Hanage WP et al (2007) Recombination and the nature of bacterial speciation. Science 315(5811):476–480
French NP, Midwinter A et al (2009) Molecular epidemiology of *Campylobacter jejuni* isolates from wild-bird fecal material in children's playgrounds. Appl Environ Microbiol 75(3):779–783
Ghosh R, Nair GB et al (2008) Epidemiological study of *Vibrio cholerae* using variable number of tandem repeats. FEMS Microbiol Lett 288(2):196–201
Gillet Y, Issartel B et al (2002) Association between *Staphylococcus aureus* strains carrying gene for Panton-Valentine leukocidin and highly lethal necrotising pneumonia in young immunocompetent patients. Lancet 359(9308):753–759
Grissa I, Bouchon P et al (2008) On-line resources for bacterial micro-evolution studies using MLVAMLVA or CRISPR typing. Biochimie 90(4):660–668

Hanage WP, Fraser C et al (2006) Sequences, sequence clusters and bacterial species. Philos Trans R Soc Lond B Biol Sci 361(1475):1917–1927

Haubold B, Hudson RR (2000) LIAN 3.0: detecting linkage disequilibrium in multilocus data. Linkage Anal Bioinform 16(9):84784–84788

Hery-Arnaud G, Bruant G et al (2007) Mobile genetic elements provide evidence for a bovine origin of clonalclonal complex 17 of *Streptococcus agalactiae*. Appl Environ Microbiol 73(14):4668–4672

Holder M, Lewis PO (2003) Phylogeny estimation: traditional and BayesianBayesian approaches. Nat Rev Genet 4(4):275–284

Honsa E, Fricke T et al (2008) Assignment of *Streptococcus agalactiae* isolates to clonalclonal complexes using a small set of single nucleotide polymorphisms. BMC Microbiol 8:140

Hugenholtz P, Goedel BM et al (1998) Impact of culture-independent studies on the emerging phylogenetic view of bacterial diversity. J Bacteriol 180(18):4765–4774

Huson DH (1998) SplitsTree: analyzing and visualizing evolutionary data. Bioinform 14(1):68–73

Huygens F, Inman-Bamber J et al (2006) *Staphylococcus aureus* genotyping using novel real-time PCR formats. J Clin Microbiol 44(10):3712–3719

Jiang LW, Lin KL et al (2008) OGtreeOGtree: a tool for creating genome trees of prokaryotes based on overlapping genes. Nucleic Acids Res 36(Web Server issue):W475–W480

Jones N, Bohnsack JF et al (2003) Multilocus sequence typing system for group B streptococcus. J Clin Microbiol 41(6):2530–2536

Katayama Y, Ito T et al (2000) A new class of genetic element, staphylococcus cassette chromosome mec, encodes methicillin resistance in *Staphylococcus aureus*. Antimicrob Agents Chemother 44(6):1549–1555

Keim P, Pearson T et al (2008) Microbial forensics: DNA fingerprinting of *Bacillus anthracis* (anthrax). Anal Chem 80(13):4791–4799

Keim P, Van Ert MN et al (2004) Anthrax molecular epidemiology and forensics: using the appropriate marker for different evolutionary scales. Infect Genet Evol 4(3):205–213

Kennedy AD, Otto M et al (2008) Epidemic community-associated methicillin-resistant *Staphylococcus aureus*: recent clonalclonal expansion and diversification. Proc Natl Acad Sci USA 105(4):1327–1332

Kim K, Cheon E et al (2005) Determination of the most closely related bacillus isolates to *Bacillus anthracis* by multilocus sequence typing. Yale J Biol Med 78(1):1–14

Kondo Y, Ito T et al (2007) Combination of multiplex PCRs for staphylococcal cassette chromosome *mec* type assignment: rapid identification system for *mec, ccr*, and major differences in junkyard regions. Antimicrob Agents Chemother 51(1):264–724

Kong F, Gilbert GL (2006) Multiplex PCR-based reverse line blot hybridization assay (mPCR/RLB) - a practical epidemiological and diagnostic tool. Nat Protoc 1(6):2668–2680

Kong F, Ma L et al (2005) Simultaneous detection and serotype identification of *Streptococcus agalactiae* using multiplex PCR and reverse line blot hybridization. J Med Microbiol 54(Pt 12):1133–1138

Koonin EV, Wolf YI (2008) Genomics of bacteria and archaea: the emerging dynamic view of the prokaryotic world. Nucleic Acids Res 36(21):6688–6719

Kumar S, Nei M et al (2008) MEGA: a biologist-centric software for evolutionary analysis of DNA and protein sequences. Brief Bioinform 9(4):299–306

Labandeira-Rey M, Couzon F, et al (2007) *Staphylococcus aureus* Panton-Valentine leukocidin causes necrotizing pneumonia. Science 315(5815):1130–1133

Larkin MA, Blackshields G et al (2007) Clustal W and Clustal X version 2.0. Bioinform 23(21):2947–2948

Lin FY, Whiting W et al (2006) Phylogenetic lineages of invasive and colonizing strains of serotype III group B Streptococci from neonates: a multicenter prospective study. J Clin Microbiol 44(4):1257–1261

Lina, G., Durand G et al (2006) Staphylococcal chromosome cassette evolution in *Staphylococcus aureus* inferred from ccr gene complex sequence typing analysis. Clin Microbiol Infect 12(12):1175–1184

Maiden MC, Bygraves JA et al (1998) Multilocus sequence typing: a portable approach to the identification of clones within populations of pathogenic microorganisms. Proc Natl Acad Sci USA 95(6):3140–3145

Mardis ER (2008) Next-generation DNA sequencing methods. Ann Rev Genomics Hum Genet 9:387–402

Margulies M, Egholm M et al (2005) Genome sequencing in microfabricated high-density picolitre reactors. Nature 437(7057):376–380

Melles DC, Schouls L et al (2009) High-throughput typing of *Staphylococcus aureus* by amplified fragment length polymorphism (AFLPAFLP) or multi-locus variable number of tandem repeat analysis (MLVA) reveals consistent strain relatedness. Eur J Clin Microbiol Infect Dis 28(1):39–45

Mellmann A, Weniger T et al (2008) Characterization of clonalclonal relatedness among the natural population of *Staphylococcus aureus* strains by using spaspa sequence typing and the BURPBURP (based upon repeat patterns) algorithm. J Clin Microbiol 46(8):2805–2808

Mellmann A, Weniger T et al (2007) Based Upon Repeat Pattern (BURP): an algorithm to characterize the long-term evolution of *Staphylococcus aureus* populations based on spaspa polymorphisms. BMC Microbiol 7:98

Monecke S, Berger-Bachi B et al (2007) Comparative genomics and DNA array-based genotyping of pandemic *Staphylococcus aureus* strains encoding Panton-Valentine leukocidin. Clin Microbiol Infect 13(3):236–249

Nei M (1996) Phylogenetic analysis in molecular evolutionary genetics. Ann Rev Genet 30:371–403

Nubel U, Roumagnac P et al (2008) Frequent emergence and limited geographic dispersal of methicillin-resistant *Staphylococcus aureus*. Proc Natl Acad Sci USA 105(37):14130–14135

Ochman H, Wilson AC (1987) Evolution in bacteria: evidence for a universal substitution rate in cellular genomes. J Mol Evol 26(1–2):74–86

Oliveira IC, de Mattos MC et al (2006) Genetic relatedness between group B streptococci originating from bovine mastitis and a human group B Streptococcus type V cluster displaying an identical pulsed-field gel electrophoresis pattern. Clin Microbiol Infect 12(9):887–893

Pearson T, Busch JD et al (2004) Phylogenetic discovery bias in *Bacillus anthracis* using single-nucleotide polymorphisms from whole-genome sequencing. Proc Natl Acad Sci USA 101(37):13536–13541

Read TD, Salzberg SL et al (2002) Comparative genome sequencing for discovery of novel polymorphisms in *Bacillus anthracis*. Science 296(5575):2028–2033

Relman DA (1993) The identification of uncultured microbial pathogens. J Infect Dis 168(1):1–8

Ronquist F, Huelsenbeck JP (2003) MrBayes 3: Bayesian phylogenetic inference under mixed models. Bioinform 19(12):1572–1574

Saitou N, Nei M (1987) The neighbor-joining method: a new method for reconstructing phylogenetic trees. Mol Biol Evol 4(4):406–425

Sanger F, Nicklen S et al (1977) DNA sequencing with chain-terminating inhibitors. Proc Natl Acad Sci USA 74(12):5463–5467

Sheppard SK, McCarthy ND et al (2008) Convergence of *Campylobacter* species: implications for bacterial evolution. Science 320(5873):237–239

Smith JM, Smith NH et al (1993) How clonalclonal are bacteria? Proc Natl Acad Sci USA 90(10):4384–4388

Spratt BG, Hanage WP et al (2001) The relative contributions of recombination and point mutation to the diversification of bacterial clones. Curr Opin Microbiol 4(5):602–606.

Staley JT (2006) The bacterial species dilemma and the genomic-phylogenetic species concept. Philos Trans R Soc Lond B Biol Sci 361(1475):1899–1909

Stephens AJ, Huygens F et al (2006) Methicillin-resistant *Staphylococcus aureus* genotyping using a small set of polymorphisms. J Med Microbiol 55(Pt 1):43–51

Sun Y, Kong F et al (2005) Comparison of a 3-set genotyping system with multilocus sequence typing for *Streptococcus agalactiae* (Group B Streptococcus). J Clin Microbiol 43(9):4704–4707

Tettelin H, Riley D et al (2008) Comparative genomics: the bacterial pan-genomepan-genome. Curr Opin Microbiol 11(5):472–477

Tourasse NJ, Helgason E, et al (2006) The *Bacillus cereus* group: novel aspects of population structure and genome dynamics. J Appl Microbiol 101(3):579–93

Tristan A, Bes M et al (2007) Global distribution of Panton-Valentine leukocidin-positive methicillin-resistant *Staphylococcus aureus*, 2006. Emerg Infect Dis 13(4):594–600

Turner KM, Feil EJ (2007) The secret life of the multilocus sequence type. Int J Antimicrob Agents 29(2):129–135

Van Ert MN, Easterday WR et al (2007a) Global genetic population structure of *Bacillus anthracis*. PLoS One 2(5):e461

Van Ert MN, Easterday WR et al (2007b) Strain-specific single-nucleotide polymorphism assays for the *Bacillus anthracis* Ames strain. J Clin Microbiol 45(1): 47–53

Vilas-Boas GT, Peruca AP et al (2007) Biology and taxonomy of *Bacillus cereus, Bacillus anthracis*, and *Bacillus thuringiensis*. Can J Microbiol 53(6):673–687

Watson JD, Crick FH (1953). The structure of DNA. Cold Spring Harb Symp Quant Biol 18:123–131

Wilson DJ, Gabriel E et al (2009) Rapid evolution and the importance of recombination to the gastroenteric pathogen *Campylobacter jejuni*. Mol Biol Evol 26(2):385–397

Woese CR (1987) Bacterial evolution. Microbiol Rev 51(2):221–371

Woese CR (2000) Interpreting the universal phylogenetic tree. Proc Natl Acad Sci USA 97(15):8392–8396

Woese CR, Fox GE (1977) Phylogenetic structure of the prokaryotic domain: the primary kingdoms. Proc Natl Acad Sci USA 74(11):5088–5090

Yamasaki O, Kaneko J et al (2005) The association between *Staphylococcus aureus* strains carrying panton-valentine leukocidin genes and the development of deep-seated follicular infection. Clin Infect Dis 40(3):381–385

Yang Z, Rannala B (1997) BayesianBayesian phylogenetic inference using DNA sequences: a Markov Chain Monte Carlo method. Mol Biol Evol 14(7):717–724

Chapter 3
Mining Databases for Microbial Gene Sequences

Richard Christen

3.1 Introduction

The advent of DNA based technologies, particularly the polymerase chain reaction (PCR), was an important turning point in microbiology and a revolution in the diagnostics of pathogens. In the last two decades, the range of molecular targets that can be used for PCR-based testing has grown tremendously. It is estimated that between ten and twenty percent of clinical isolates are novel microorganisms that defy phenotype-based identification, leading to the misidentification of rare isolates or new strains (Clarridge 2004). Evidence suggests that real time PCR can not only detect the presence or absence of the target pathogen but also quantify the microbial load in a sample.

Highly accurate diagnostic assays are characterized by their effectiveness and ability to detect target microorganisms without interference from non-target species. Sensitivity, exhaustivity and specificity are the proxies of effectiveness. Specificity indicates how the method is affected by non-target species, and a bad specificity may result in false positive reactions. Sensitivity reflects the number of cells required for detection, which may otherwise lead to a false negative response. Exhaustivity is affected by mutant alleles of the gene, which may escape the detection system. To be specific, divergent gene regions are the best choice for designing primers; but, to be exhaustive, one should target more conserved domains, or design degenerated primers, as some alleles may bear mutations in divergent domains.

In the last decade, the 16S rRNA gene sequence has become the "gold standard" platform for microbial identification, as well as the technical basis for modern bacterial taxonomy and for the discovery of novel bacteria in clinical microbiology laboratories. 16S rRNA gene sequencing is particularly important in the case of bacteria with unusual phenotypic profiles, rare bacteria, slow-growing bacteria, uncultivable bacteria and culture-negative infections. As a result, hundreds of new

R. Christen
University of Nice Sophia-Antipolis, and Institute of Developmental Biology and Cancer, Parc Valrose, Centre de Biochimie, Nice, France

V. Sintchenko (ed.), *Infectious Disease Informatics*,
DOI 10.1007/978-1-4419-1327-2_3,

bacterial species and tens of genera have been discovered from human specimens during the first years of the twenty-first century (Clarridge 2004; Hall et al. 2003; Luna et al. 2007; Woo et al. 2008). However, there are some genera in which 16S rRNA gene sequences do not differ much between species. Also, some species such as *Escherichia coli* are not always pathogenic, depending upon the presence of pathogenicity genes. Therefore, targeting well-chosen pathogenicity genes is also becoming a standard procedure. Virulence genes are often an appropriate target, because selectivity is easier to address when related non-pathogenic species do not bear the gene. As a result, it is easier to discriminate pathogenic strains from non-pathogenic ones (Bielaszawska et al. 2007a, b; Orth et al. 2007).

Although commonly regarded as a non pathogenic commensal of the gastrointestinal tract, *E. coli* can be an important bacterial pathogen. Several strains have acquired specific virulence factors that are the causes for a variety of intestinal and extra intestinal diseases. These strains are leading causes of morbidity and mortality, especially in developing countries (Kaper et al. 2004). Currently, these *E. coli* strains can be grouped into major categories (Nataro and Kaper 1998; Gyles 2007) depending upon the pathogenicity genes they bear or express. Enterohaemorrhagic *E. coli* (EHEC), a subgroup of Shiga toxin-producing *E. coli* (Mora et al. 2007), has become increasingly important as a human pathogen in developed countries. EHEC is able to cause serious food-borne intestinal diseases that can be followed by extra intestinal sequelae such as the hemolytic-uremic syndrome (HUS) (Creuzburg and Schmidt 2007), which can cause acute renal failure in children. HUS is mainly caused by the production of verocytotoxins (VT1 or VT2) or Shiga-like toxins (*stx1* or *stx2*), which are different names for what is essentially one toxin (Huang et al. 1987; Wani et al. 2007), by EHEC. The *stx* genes are usually located in the genomes of bacteriophages and are expressed during the phage life cycle (Creuzburg and Schmidt 2007). Additional potential virulence factors include cytolysins (haemolysin, *hly*), serine proteases (EspP), lymphotoxins (Efal) and adhesins (intimin) (Wani et al. 2007), among others. More than 500 different serogroups of *E. coli* have been reported to produce Shiga toxins (Allison 2007), and they can carry a wide variety of combinations of virulence factors (Praget et al. 2005). To simplify and accelerate differential diagnosis, multiplex PCR assays have been developed for the simultaneous detection and differentiation of the major categories of intestinal pathogenic *E. coli* strains (Muller et al. 2007). However, a good detection system is not trivial to achieve because many variant alleles can be present in a given gene. For example, few variants of the *stx1* gene but more than 20 variants of the *stx2* gene have been described (Gourmelon et al. 2006).

Since many detection systems for different target genes have already been published, it is often a good idea to collect and assess published sequences and primers before proceeding to experiments or trying to design new oligomers. However, there are a large number of publications in clinical microbiology literature that can be utilized for the extraction of published primers. For example, many papers include the words "identification" or "detection" in the title or abstract, as well as MESH terms for the *E. coli* species. The number of such publications in PubMed has been increasing steadily each year (Fig. 3.1). There are a total of 18,369 publications (Fig. 3.1), which include three papers dating from 1935 and 1936 (Moldavan

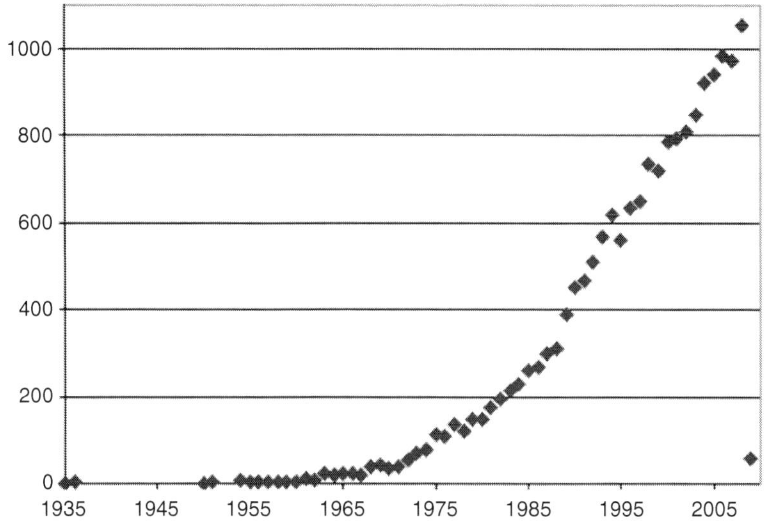

Fig. 3.1 Count of publications added to PubMed each year concerning identifications of *E. coli*

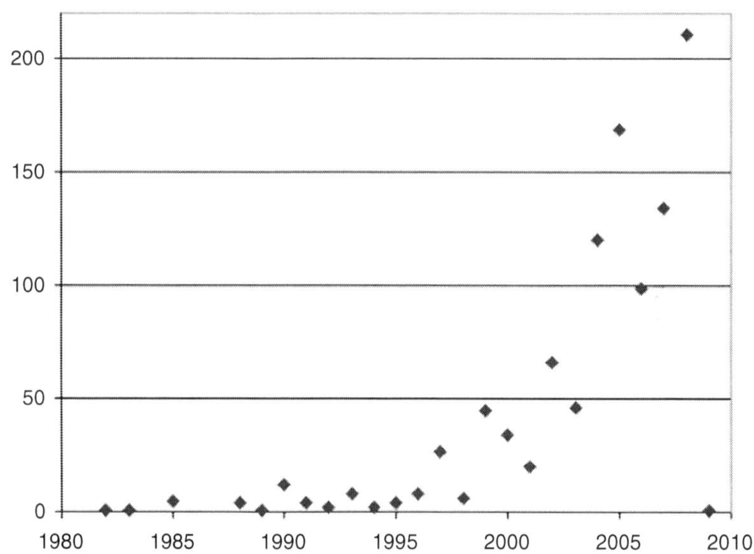

Fig. 3.2 Counts of entries for genes of toxins in *E. coli* submitted each year to the public databases

1935; Black and Klinger 1936; Griffiths and Fuller 1936). These were scanned because experts had identified them as key papers.

Similarly, there has been a tremendous increase in the number of available sequences in the public databases. Figure 3.2 shows the number of entries for *E. coli* concerning toxin-coding genes. There was only a single deposit for 2009 at the time of submission of this chapter. Unfortunately, retrievals of the datasets can be difficult and analyses of the collected data almost impossible, without the use of dedicated computational tools.

Despite the fact that there are multiple web servers and programs freely available for download, *in silico* analyses remain problematic largely because of the variety of data formats that are used to store sequences in public databases. These analyses can also be challenging because sequence data is often stored in pdf files as part of a publication. This chapter aims to:

- Critically review the problems involved at each step of *in silico* analyses, and to
- Propose programmatic solutions for the problems encountered.

3.2 Retrieval of Target Sequences

3.2.1 Retrieval by Similarity

In order to retrieve similar sequences, a common practice is the use of a similarity search (BLAST, for example) using a known sequence as a query sequence. This is often not the most appropriate approach for three reasons. First, it is very difficult to place parameters upon the search in order to retrieve only the homologous sequences. Similar but not homologous sequences (i.e. a different gene) can also be retrieved and these then have to be eliminated by a tedious manual selection. Second, some very divergent alleles can be extremely difficult to retrieve. Finally, public BLAST interfaces allow the retrieval of every matched sequence, but this retrieves the complete entry (i.e. the complete sequence submitted) and not the particular gene sequence contained within a (much) larger piece of DNA. For example, using the sequence for the *hlyA* gene (2,997 nt) contained in record AB011549 easily retrieves 50 sequences (NCBI blast, default query) that can be saved under FASTA format. Doing this readily retrieves the AB011549 accession number, but the associated sequence is 92,721-nt long. A sequence of 165,548 nt also corresponds to a complete plasmid (AY258503). These very long sequences are extremely difficult to align using multiple sequence alignment tools and are very difficult to deal with in further analyses.

3.2.2 Retrieval by Keywords

The second approach is to try to retrieve these gene sequences using annotations contained in their features. For example the gene *hlyA* is identified in entry AB011549 by these annotations:

```
FT          CDS      16612..19608
FT                   /transl_table=11
FT                   /gene="hlyA"
FT                   /product="Hemolysin A"
FT                   /protein_id="BAA31774.1"
```

These annotations describe this particular CDS (Coding Sequence) to encode the gene *hlyA* for a protein "Hemolysin A." The annotations also describe that this sequence spans positions 16612 to 19608 of the complete entry (92,721-nt long). Three public tools are designed to retrieve sequences according to keywords.

Entrez: Entrez at NCBI is without doubt the most popular tool among biomedical scientists. Using the query "Escherichia [Organism] AND hlya" retrieves 51 sequences. However, display in FASTA format demonstrates that only the complete entries (sequences) can be downloaded.

SRS: A second query system is that of SRS at EBI (and elsewhere). Using the "Extended Query Form" and the EMBL database, one can query for "Escherichia" as the organism name and "hlya" as a gene. This retrieves 36 entries (40 if "hlya" is used to search the "FTDescription" field). Corresponding sequences can be retrieved using the format "FastaFtSeqs." Corresponding complete entries are also easily retrieved using the "Link to EMBL" and then the display option "Complete entries" (see below).

```
ID     AB011549_18;    parent: AB011549
AC     AB011549;
FT     CDS             16612..19608
FT                     /transl_table=11
FT                     /gene="hlyA"
FT                     /product="Hemolysin A"
FT                     /db_xref="GOA:Q46716"
FT                     /db_
                         xref="InterPro:IPR001343"
FT                     /db_
                         xref="InterPro:IPR003995"
FT                     /db_
                         xref="InterPro:IPR013550"
FT                     /db_xref="UniProtKB/
                         TrEMBL:Q46716"
FT                     /protein_id="BAA31774.1"
```

An important problem is the fact that SRS does not accept complex keywords, which are combinations of words. Not all sequences are annotated with the gene symbol (see below), and have to be queried using their "product name," such as "Hemolysin A," which contains a space (Crose et al. 2006). For example the query below returns no entries:

```
Query:"([embl-Organism:Escherichia]>([embl-
  FtDescription:Hemolysin A]))"
```

It has to be reformulated as:

```
Query:"([embl-Organism:Escherichia]>(([embl-
  FtDescription:Hemolysin]&[embl-FtDescription:A])))"
```

This query returns 47 entries, but is not as precise as desired because it also returns sequences annotated as "A Hemolysin….." As shown below, this situation would be even more problematic for the *stx* gene products which are annotated using several words.

Finally, SRS is not able to download very large datasets.

ACNUC: Another tool that can be used on-line or through a dedicated client is ACNUC (http://pbil.univ-lyon1.fr/databases/acnuc/acnuc.html). It allows queries on keywords with spaces and has no problem with large datasets. Even though ACNUC easily extracts sequences that are annotated by keywords, not every public sequence is easily retrieved. There are three main reasons: (1) not every sequence is duly annotated for a gene name, (2) gene names and, more particularly, product names (e.g., protein descriptions) are often found with quite different wordings (Table 3.1) if not with misspelling, and finally (3) quite a large number of sequences are misdescribed. However, the most important problem is the large variation in the way gene products are annotated. To demonstrate this problem, a program was written to identify the main variants for the products of the gene *stx2a* within a response to a query such as "sp = Escherichia AND k = @toxin@" using ACNUC.

Variations in names and spellings, a variable use of upper or lower case, and the use of either 1 (2) or I (II) for describing a subunit are very common. For *stx* gene products, a total of 303 alternate descriptions were found (see examples in Table 3.1). Obviously, there is a large number of different descriptions used to identify the same protein. Gene names are less variable (Table 3.2), but the problem is that a large number of entries do not have an annotation for gene names.

Table 3.1 Examples of variations in the annotations of the gene product for the *stx2A* gene (the most frequent observations from a longer list)

Shiga toxin 2 A subunit	38
shiga toxin 2A subunit	22
Shiga toxin 2 subunit A	21
verocytotoxin 2 variant A subunit	18
Shiga toxin 2 A-subunit	17
variant shiga-like toxin II VT subunit A	13
Shiga toxin II subunit A	11
shiga toxin 2e A	8
variant Shiga toxin type 2 A subunit	7
shiga toxin 2d activatable subunit A	7
Shiga toxin 2A subunit	7
verocytotoxin 2 subunit A	6
Shiga toxin 2c A unit protein	5
SLT-IIeA	4
Shiga toxin 2 variant d A subunit	2

Table 3.2 Examples of alternate symbols used to annotate the gene *stx2A* (excerpts)

Stx2A	49
stxA2	19
Stx2 A-subunit	16
Stx2e A	9
Vtx2A	7
Stx2dA	7
Stx2cA	5
stxA2d	4

A tedious revision of the list provided 68 gene product descriptions (see except in Table 3.1), which allowed the retrieval of the corresponding *stx2A* gene sequences. However, if we compare keyword and BLAST retrievals, we discover that 26 of these sequences were not found by BLAST, while some sequences retrieved using BLAST were not in the FASTA file. BLAST did not retrieve short sequences or very divergent sequences and keyword searching did not retrieve mis-annotated sequences or sequences annotated with as yet unidentified keywords.

3.2.3 The Brute Force Approach: By Keywords

Using ACNUC, the query "sp = Escherichia and k = @toxin@" returned 515 records corresponding to 1,013 different sequences. It is then possible to use a program such as Cd-hit-est (Li and Godzik 2006) to cluster these sequences quickly without alignments. Such clustering at 20 different levels of similarity took only a few minutes. After briefly looking at the results, the analysis done at 80% similarity using words of length 4 was selected. A specific program allowed the quick identification of 87 different keywords used to annotate genes or gene product sequences of the *stx2A* gene. Examples of the most common annotations for the gene product of the *stx2A* gene are listed below (all descriptions turned to lower case).

stx2a	50
shiga toxin 2 a subunit	39
shiga toxin 2a subunit	29
shiga toxin 2 subunit a	22
stxa2	19
verocytotoxin 2 variant a subunit	18
shiga toxin 2 a-subunit	17
stx2 a-subunit	16
shiga toxin ii subunit a	11
stx2e a	9
shiga toxin 2e a	8
variant shiga toxin type 2 a subunit	7
shiga toxin 2d activatable subunit a	7
verocytotoxin 2 subunit a	6
shiga toxin 2c a unit protein	5
shiga toxin 2e subunit a	4

A careful examination of this list indicated that some annotations were not good enough to retrieve *stx2A* only. There were clear mis-annotations of some sequences (see below for *stx2B* annotated as *stx2A*).

Clusters duly identified were merged, providing 237 sequences contained in 234 entries only. One might suspect that some sequences are not truly *stx2A*, since this is a single copy gene. The pool of these sequences was reduced to 122 unique long sequences that were then aligned using Muscle (Edgar 2004) (some very short sequences were removed). Among the sequences that could not be properly aligned were sequences with accession numbers U41251, U41259, U41249, U41248, U41253 and U41257. A BLAST query quickly identified that they in fact corresponded to the B subunit of the toxin, a typical example of the errors that occur when many sequences are submitted. They were removed from the analysis. Accession numbers U41253, and AJ271139 were similarly found to be wrongly annotated. Finally, 106 sequences were kept for building a phylogenetic tree using the BioNJ methods with the Kimura 2-parameters corrections and using only positions of the alignment containing no indels.

3.2.4 The Brute Force Approach: By Similarity

Using ACNUC it was easy to extract the 148,465 CDS sequences available for *E. coli*. These sequences were formatted as a BLAST database and a complete *stx2A* sequence (obtained using the keyword search above) was used as a query sequence to produce an output file in XML format. The output file was analyzed to extract sequences (longer than 250-nt long) that had at least one local alignment (hsp) of more than 40 nucleotides with the query sequence (this was a very crude but easily implemented filter). The resulting 309 sequences were either aligned using ClustalW or clustered by Cd-hit-est (at 80% similarity, parameters: –n 3 –sc 0.80).

The CD-HIT analysis resulted in 4 clusters only, with a cluster of 72 sequences identified as *stx1A* sequences, for example AB015056 (*stx1* is known to share a high percentage of similarity with *stx2*). The complete analysis, done in less than 20 min, yielded 237 true *stx2A* sequences, 115 of which were unique and the longest of which contained some shorter ones. This method appears to be the fastest and easiest way to retrieve every gene sequence. It can be summarized in the following steps:

- ACNUC:
  `sp=Escherichia and t=CDS and not t=id ==>` 148,465 sequences in 20 s. Download time in FASTA format is less than 1 min.
- BLAST, download OS specific standalone version from: http://www.ncbi.nlm.nih.gov/BLAST/download.shtml.
- Format the FASTA file:
  `formatdb.exe -i demo.fasta -p F` (less than 15 s).

- BLAST a single sequence on this database then ask for xml output:
  ```
  blastall.exe -p blastn -i AB030484.fasta -d demo.fasta
    -a 2 -v 1500 -b 1500 -F F -W 7 -e 100 -o out.xml -m 7
  ```
  (results in about 5 s).

3.3 Retrieval of Published Primers

In order to build a bibliographic database while also looking for already published and validated primers, a search of articles in PubMed using Entrez at NCBI is probably the best and the most commonly used strategy. Such queries are often done by scientists using one or two keywords only. However, it is much more efficient in terms of exhaustivity and specificity to combine keywords using Boolean OR and AND, and to eventually use the "Limits" tab.

3.3.1 A Note of Caution About PubMed Queries

Almost every publication in the medical field is referenced in PubMed, but searches are not always simple; searches are done for words contained in the title (and abstract) as well as the MESH terms, and Entrez often reformulates the queries. Let's, for example, examine a search for every paper dealing with *stx* genes and "identification or detection." If one pastes the following line into the Entrez query box for PubMed:

```
(stx OR stx1 OR stx1a OR stx2a OR stx1b OR stx2b) AND
  (identification OR detection)
```

This query retrieved 314 PMIDs (314 different publications). However, using the tab "details" to see how Entrez really formulated the search showed:

```
(stx[All Fields] OR stx1[All Fields] OR stx1a[All
Fields] OR stx2a[All Fields] OR stx1b[All Fields] OR
stx2b[All Fields]) AND (("identification (psychology)"
[MeSH Terms] OR ("identification"[All Fields] AND
"(psychology)"[All Fields]) OR "identification (psycho-
logy)"[All Fields] OR "identification"[All Fields])
OR detection[All Fields])
```

Entrez added keywords searches in the field of psychology. Note also that a very simple search such as:

```
(detection OR identification)AND O157:H7
```

does not work, the query is reformulated by Entrez as a query which finds no match:

`AND (O157[All Fields] AND H7[All Fields]`

It has to be rewritten

`(detection OR identification)AND "O157:H7",`

which retrieves 779 items.

A number of queries to PubMed below were tested (in January 2009) for their sensitivity and specificity to retrieve appropriate publications, targeting three of the genes mentioned earlier: *eae*, *stx* and *hlya*. For the *stx* gene and because of the keyword search done earlier, we know that we need to use alternate words.

(**intimin OR eae**) AND (escherichia OR "E. coli"): 903.
(**intimin OR eae**) AND (escherichia OR "E. coli") AND (detection OR identification): 204.
(shiga OR stx OR verotoxin or vtx) AND (escherichia OR "E. coli"): 3376.
(stx OR vtx OR stx1 OR stx2 OR stx1a OR stx2a OR stx1b OR stx2b OR verotoxin OR shiga) AND (escherichia OR "E. coli"): 3448.
(shiga OR stx OR verotoxin) AND (escherichia OR "E. coli") AND (detection OR identification): 685.
(hlya OR hemolysin) AND (escherichia OR "E. coli"): 1990.
(hlya OR hemolysin) AND (escherichia OR "E. coli") AND (detection OR identification): 193.
(escherichia OR "E. coli") AND (detection OR identification): 18,369.
(escherichia OR "E. coli") AND (detection OR identification) AND (pcr) AND (pathogen OR pathogens OR pathogenic): 587

According to these results, the *stx* genes seem to be the most studied among the three genes investigated. When faced with the large number of results to analyse, it would be tedious and probably much too time consuming to download and read each article to look for appropriate primers. The results of queries were thus saved as files under XML format, and a program was written to extract each PMID and the links to the full text articles on the publisher's sites. PMIDs are easily located in lines of XML files such as:<PMID>19086378</PMID>. Journals in which these articles have been published are also easily retrieved in lines such as<ISOAbbreviation>Can. J. Vet. Res.</ISOAbbreviation>. Finally, the authors, year, abstract, and page numbers are also very easily retrieved. However, the links to the full text are not included in the XML files. They can be retrieved from NCBI using the facility "EUtils." The EUtils tool "ELink," for example, checks for the existence of a hyperlink to the appropriate journal (using the option "cmd = prlinks"). It was found that this procedure is not always effective. The full information about every external link attached to a PMID was retrieved using ELink and the option "cmd = llinks," which allowed the user to obtain the proper link to the publisher's site. Unfortunately, this web page is not usually a direct link to the full text or the pdf file but rather a link to an abstract that also displays a "link to pdf" button somewhere in the page. In some cases, the link to the pdf is easily derived from the abstract URL (page address), but this is not always the case. In some cases, the web page must be analyzed to find out the proper link. Finally, a "fake" web browser was built to automatically retrieve each pdf file. Appropriate rules had to be set to analyse the site of each publisher. Some journals did not have a web site (in a query, 568 out of 2,058 references had no web link according to NCBI), some sites required cookies (which were then faked), some were not freely accessible and some journals required a login to access recent articles. As a result, only freely available documents where used.

Using this procedure and the queries described above, more than 1,000 pdf files were retrieved. One should acknowledge that "fake" browsers may slow down successive queries to the same journal by 60 s or more, so as not to be identified as a (malicious) robot. Successive queries at EUtils should also be delayed and a proper email address should be included to avoid problems. The wget command (LINUX or Cygwin for MS Windows) can be an alternative, but since it retrieves the entire site, free space on the hard disk may become a problem, as may denial from the target site.

3.3.2 Primer Extraction

Files in pdf format use a specific language to describe the document's contents for printing. They are also binary encoded and quite difficult to read programmatically. It is possible to use Adobe Acrobat Reader to extract a pdf file as a text (ASCII) file. However, this is not very realistic for extracting hundreds of documents, and this Reader is not scriptable (it cannot be used to automatically read a series of pdf files and save them as text). There are a number of freely downloadable programs that can do this. Xpdf, which is licensed under the GNU General Public License, can be used across platforms (MS Windows and UNIX) and can be found at http://www.foolabs.com/xpdf/home.html. In our experience, it is effective but is not foolproof, as is illustrated below. Python PDF toolkits such as pyPdf or PDFlib can also be employed to extract text.

Using this program and the *stx* genes, 823 out of 882 pdf files could be extracted. The remaining files either required a login to extract the text or were somehow corrupted during the automated download. A list of 85 primers that were listed in publications and targeted at least one of the *stx2A* gene sequences, could be extracted (for examples see Figs. 3.3 and 3.4).

Table 3: PCR primers used for the identification and characterization of attaching and effacing

Target	Primer	Ollgonucleotide sequence (5'-3')
astA	Eastl la	CCA TCA ACA CAG TAT ATC CGA
	Eastl lb	GGT CGC GAG TGA CGG CTT TGT
bfpA	EPI	AAT GGT GCT TGC GCT TGC TGC
	EP2	GCC GCT TTA TCC AAC CTG GTA
eae	SKI	CCC GAATTC GGC ACA AGC ATA AGC
	SK2	CCC GGATCC GTC TCG CCA GTA TTC G
eae-α	SKI-LP2	CCC GAATTC TTA TTT TAC ACA AGT GGC
eae-γ	SKI-LP3	CCC GAATTC TTC TTT TAC ACA AAC CGC
eae-β	SKI-LP4	CCC GTGATA CCA GTA CCA ATT ACG GTC
eae-ε	SKI-LP5	AGC TCA CTC GTA GAT GAG GGC AAG CG
eae-ζ	SKI-LP6B	TAG TTG TAC TCC CCT TAT CCC
eae-ι	SKI-LP7	TTT ATC CTG CTC CGT TTG CT
eae-η	SKI-LP8	TAG ATG ACG GTA GAC
eae-κ	SKI-LP10	GGC ATT GTT ATC TGT TGT CT
eae-θ	SKI-LP11B	GTT GAT AAC TCC TGA TAT TTT A
EAF	EAFI	CAG GGT AAA AGA AAG ATG ATA A
	EAF25	TAT GGG GAC CAT GTA TTA TCA
fliC	FliC up	CAA GTC ATT ATT AC(AC) AAC AGC C
	FliC down	GAC AT(AG) TT(AG) GA(AGC) ACT TC(GC) GT
stx	VT I	ATT GAGCAA AAT AAT TTA TAT GTG
	VT 2	TGA TGA TGG CAA TTC AGT AT
tir	*tir*-R	TAA AAGTTC AGA TCT TGA CAT
tir Y-P	tirYA74-F	CAT ATT TAT GAT GAG GTG GCT C
tir S	tisS478-F	TCT GTT CAG AAT ATG GGG AAT A

Fig. 3.3 The original description of primers in the pdf file (as reported in Frohlicher et al. 2008)

Table 3: PCR primers used for the identification and characterization of attaching and effacing Escherichia coli strains

Target astA bfpA eae eae– eae– eae– eae– eae– eae– eae– eae– eae– EAF flic stx tir tir Y–P tir S

Primer East11a East11b EP1 EP2 SK1 SK2 SK1-LP2 SK1-LP3 SK1-LP4 SK1-LP5 SK1-LP6B SK1-LP7 SK1-LP8 SK1-LP10 SK1-LP11B EAF1 EAF25 Flic up Flic down VT1 VT2 tir–R tirY474–F tisS478–F

oligonucleotide sequence (5'–3') CCA TCA ACA CAG TAT ATC CGA GGT CGC GAG TGA CGG CTT TGT AAT GGT GCT TGC GCT TGC TGC GCC GCT TTA TCC AAC CTG GTA CCC GAA TTC GGC ACA AGC ATA AGC CCC GGA TCC GTC TCG CCA GTA TTC G CCC GAA TTC TTA TTT TAC ACA AGT GGG CCC GAA TTC TTC TTT TAC ACA AAC CGC CCC GTG ATA CCA GTA CCA ATT ACG GTC AGC TCA CTC GTA GAT GAC GGC AAG CG TAG TTG TAC TCC CCT TAT CCC TTT ATC CTG CTC CGT TTG CT TAG ATG ACG GTA GAC GGG ATT GTT ATC TGT TGT CT GTT GAT AAC TCC TGA TAT TTT A CAG GGT AAA AGA AAG ATG ATA A TAT GGG GAC CAT GTA TTA TCA CAA GTC ATT ATT AC(AC) AAC AGC C GAC AT(AG) TT(AG)GA(AGC) ACT TC(GC) GT ATT GAG CAA AAT AAT TTA TAT GTG TGA TGA TGG CAA TTC AGT AT TAA AAG TTC AGA TCT TGA CAT CAT ATT TAT GAT GAG GTC GCT C TCT GTT GAG AAT ATG GGG AAT A

Reference 27 5 22 25 25 22 22 25 25 25 25 25 27 5 27 26

Fig. 3.4 Text extracted from the pdf file shown in Figure 3.3. In this case, it was not possible to readily obtain the proper primer's sequences from the text

3.4 Assessing Primers

We can now use the extracted primers and the aligned gene sequences to assess the different primers using, for example, the OHM server located at http://bioinfo.unice.fr/ohm. Because the exact extraction of primers is difficult and can be misleading (for reasons described above), only primers that had been found in at least two publications were kept to avoid errors or partial extractions. This analysis demonstrated that most of the retrieved primers were reverse primers, because they were often used in combination with a forward primer located in the *stx2B* gene. These results also suggested the poor design of many primers, as they failed to hybridize to recently described variant alleles. As a case study and because we could not present the entire results here, we have focused on a few primers that have been used in recent studies (Beutin et al. 2007; Dhanashree and Mallya 2008; Islam et al. 2008; Kobayashi et al. 2009; Mansouri-Najand and Khalili 2007), to which we also added primers validated by a widely used diagnostic reference center (Manual of Diagnostic Tests and Vaccines for Terrestrial Animals. CHAPTER 2.9.11. http://www.oie.int/eng/normes/mmanual/A_00001.htm 1). The list of primer local identifiers, sequences and the number of respective publications that were found to contain them, as well as the number of hits retrieved using Google searches, are shown below:

1	Tbody>GCTCTGGATGCATCTCTGGT	(26)	30
2	CTGGTGGTGTATGATTAATA	(26)	12
3	AGATTGGGCGTCATTCACTGGTTG	(4)	9
4	TACTTTAATGGCCGCCCTGTCTCC	(4)	9
5	CCACATCGGTGTCTGTTATTAACCACACC	(24)	2
6	GCAGAACTGCTCTGGATGCATCTCTGGTC	(24)	2
7	TCCATGACAACGGACAGCAG	(1)	1
8	GCTTCTGCTGTGACAGTGAC	(1)	1
9	GGCACTGTCTGAAACTGCTCC	(30)	51
10	TCGCCAGTTATCTGACATTCTG	(30)	44
11	CCATGACAACGGACAGCAGTT	(13)	32
12	CCTGTCAACTGAGCAGCACTTTG	(13)	16

One of the results returned by the OHM server is the position and strand of each primer (Fig. 3.5), a useful visual representation to select primers providing an amplicon of the requested size.

Among other data provided by this server is a series of files to be used in combination with TreeDyn (Chevenet et al. 2006), allowing the production of informative figures such as in Fig. 3.6. In this example, predicted melting temperatures for each sequence of the phylogenetic tree are displayed for each primer as a column of the heat map (colored squares); melting temperatures are translated into colors from yellow (58°C) to light blue (41°C). A grey square is used when the Tm is

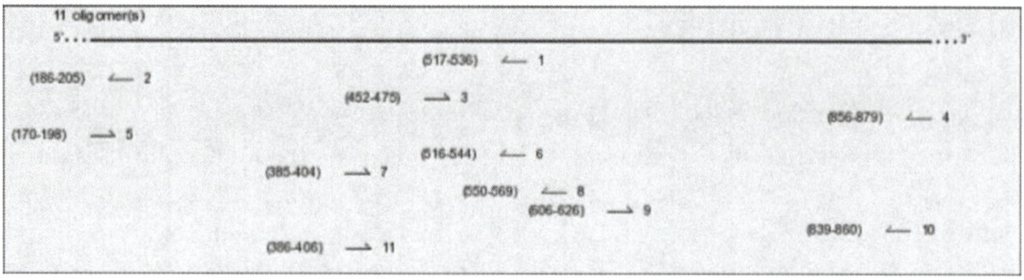

Fig. 3.5 Positions and strands of each primer in the *stx2A* gene (aligned) sequences

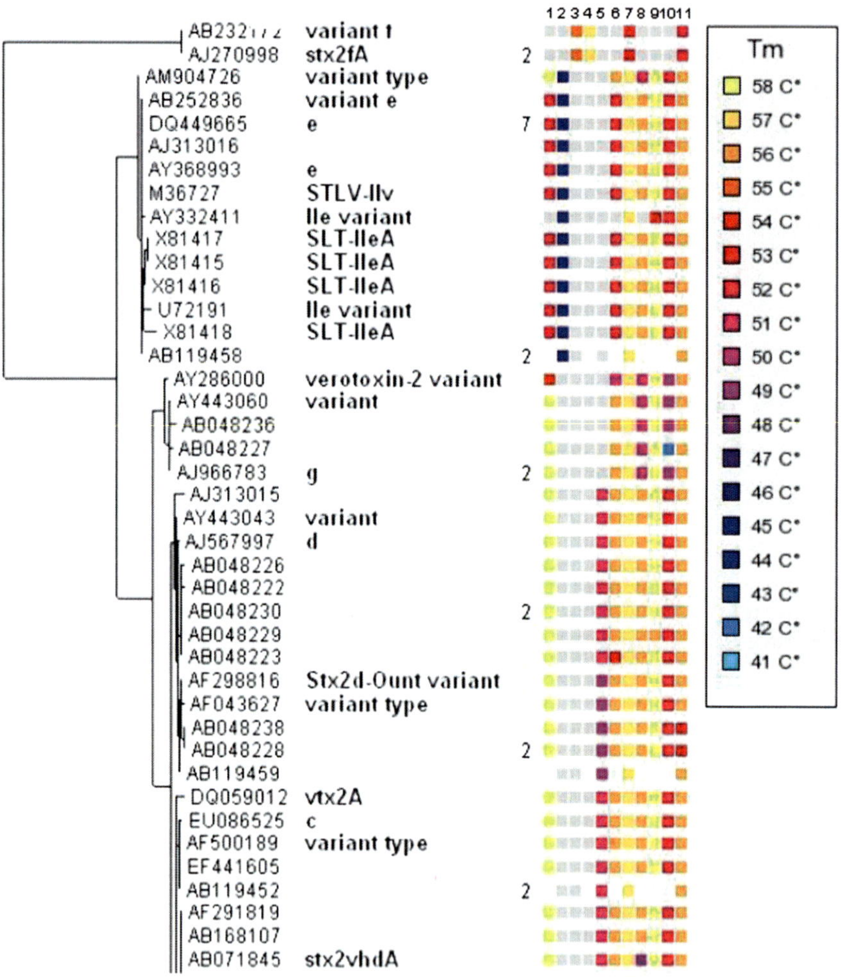

Fig. 3.6 Phylogenetic tree and heat map of the primers (excerpt of a larger figure). *From left to right*: tree derived using the bionj algorithm, annotations of gene products (as found in complete entries), numbers of exactly similar sequences contained in each sequence analyzed, heat map and color coding

below 40°C, and a white square is used when the primer is located outside of a sequence (a gene not completely sequenced).

Figure 3.6 highlights that the *stx2A* gene sequences can be grouped in several distinct clusters, from the top of the tree: variant f, variant e, variant g, and finally the rest of the sequences (the tree shown is the result of a very fast and not definitive phylogenetic analysis, in part because sequences of quite different lengths were included).

Primer 12 could not be found in any sequence; a BLAST query showed no hit (the closest hit was with *Streptococcus agalactiae* NEM316 complete genome, with only 19/23 identities). This primer had first been published in 1999 (Fagan et al. 1999). Primers 3 and 4 had been designed to amplify variant f sequences only. These variants could be also amplified by primers 7 and 11. The best amplifications of the rest of the gene sequences could be seemingly obtained using primers 7, 9 and 11 only. All of the other primers failed on a variable number of sequences.

The OHM server provides a summary file, which is an easy way to estimate the exhaustivity of primers, as well as the possible modifications required to improve a given primer. The identification of sequences that have mismatches with a given primer is presented below. The predicted T_m (the first column), the sequence of the possible variant genes (second column), and the number of such sequences ("f variant" indicates the two sequences of the f cluster) are given to each primer.

7	tccatgacaacggacagcag	F	
57.4	....	112	
52.75	................g.......	3	f variant
9	ggcactgtctgaaactgctcc		F
57.58	....		94
55.83	...................c....		1
53.86	....................m...		1
52.68	.................a......		1
52.19	..............a.........		1
50.4	....t...................		1
50.27	..................a.....		1
35.01	......g......gg at......	3	f variant
11	ccatgacaacggacagcagtt		F
55.97	....		107
54.17	.......................a		1
53.35	......................c.		4
51.59	..........g.............	3	f variant

This illustrates in detail how each of these 3 primers matches the 115 sequences used as references. Except for primer 7, the other primers might present some difficulties with a number of variant alleles. It is of note that these three primers are all reverse primers, and that, in combination with a forward primer located in the B subunit, they will amplify only part of the *stx2A* gene (see Fig. 3.5). Interestingly, primer 7 (Manual of Diagnostic Tests and Vaccines for Terrestrial Animals. CHAPTER 2.9.11. http://www.oie.int/eng/normes/mmanual/A_00001.htm 1) is almost identical to primer 11 (Dhanashree and Mallya 2008).

3.5 Concluding Remarks

This study has highlighted three different aspects of a bioinformatics analysis of primers published in articles and designed to amplify a pathogenicity gene. The *stx2A* gene was chosen as an illustration because it has been the subject of many publications and because it has been described as having many alleles in the literature. Our analysis emphasized several problematic aspects, namely the retrieval of the gene sequences, the retrieval of the primers, and the reliability of published results. There are, however, three challenges: (1) the EMBL/GenBank database uses a format that, in most part, dates from the 1980s and is now difficult to use considering the growing number of sequences, their size and their complexity, (2) the suboptimal annotation of many microbial sequences, and (3) the fact that there is a community effort to standardize genes and gene products nomenclature, and to create dedicated ontologies (see also Chap. 19).

Gene sequence retrieval. There are presently two main problems that impede reliable sequence retrieval. First, despite the fact that a search by sequence annotation is easily done using a tool such as ACNUC, the present format of the EMBL/GenBank database makes such retrieval difficult. Indeed, gene symbols and gene product annotations are not standard - using an *ad hoc* program, it is feasible and reasonably fast to collect a series of alternate keywords. However, collecting every alternate keyword proved to be laborious. As there is no standard for naming genes and gene products, some keywords cannot be easily retrieved, as for example:

```
EU999150 (Escherichia coli strain R1388 stx2cA upstream
  region), gene symbol: "stx2cA", product annotation:
  "Shiga toxin 2c A unit protein".
```

A more elaborate strategy, for example using regular expressions, could be envisioned, but without any rule for annotations, this is a difficult and time consuming task, with no proof that every available sequence has been collected in the end.

Second, it is almost impossible to use a sequence similarity search as provided by the main public servers, because most gene sequences are "buried" within larger sequences (complete genomes or plasmids) which the present public BLAST tools cannot extract. One solution presented here was to build a local BLAST database of every CDS sequence for a given taxon and then use a stand-alone similarity search against such a database. This is done quickly, the calculation is very fast and the analysis of the (XML) output can be done in a reasonable amount of time (i.e.

determination of the threshold levels to retain a sequence as a true hit). This is by far the best solution both in terms of time and ease.

Primers sequence retrieval. Any query on PubMed quickly demonstrates that a large number of references are retrieved when one tries to gather publications describing primers used in PCR experiments (and this would also be true for many other queries). Retrieval of full texts in pdf format can be done for a large number of papers in a reasonable time using a dedicated robot. However, the pdf format is not well suited to the automated extraction of information. As a result, much information is either lost or retrieved in a corrupted form. It would be preferable to retrieve a full text in html format or in XML format yet this is very seldom the case, and is further complicated by the inclusion of specific tags to identify the most important data types such as gene or protein names, oligomers, etc. Despite these limitations, it was possible to quickly retrieve a large number of primers, but the automated extraction of other information (such as gene and protein names, for example) has presented an even greater challenge.

Using primers extracted from recently published papers, it was demonstrated that:

- None of these studies employed even a crude bioinformatics analysis to evaluate the exhaustivity and specificity of every primer used. As a result, many of these primers lack exhaustivity because they do not take into account variant alleles described in other publications.
- Some recent studies used primers designed and published up to a decade ago, when the number of available sequences was very small.
- Recent publications have used a combination of primers for the detection of the different variant alleles (Beutin et al. 2007; Wani et al. 2007; Zheng et al. 2008), but none have reported an analysis of the specificity or sensitivity of these primers.

Similar conclusions can be drawn from the retrieval of sequences and the analysis of primers for every other gene tried. Hence, we suggest that *in silico* validation (and the design of new primers when required) should be a required step preceding bench experiments. Reliable design software tools and, perhaps more importantly, the knowledge of its use and the retrieval of every known sequence, are critical for the successful design of new oligomers. Most of the work reported here was done using publicly available tools such as ACNUC (Gouy and Delmotte 2008), SeaView (Galtier et al. 1993), Muscle (Edgar 2004) or Clustal (Larkin et al. 2007), BLAST (Altschul et al. 1990) and phylogeny programs. The subsequent bioinformatics analysis did not require professional skills in computing, and was done using a language that is easy to learn (Python) and using simple tests and iterations. Starting from scratch, it would take less than a week for somebody who already knows programming to retrieve the relevant sequences and primers (at least those used in the past few years), evaluate them, and suggest improvements. Given how expensive wet bench experiments are, it is suggested that laboratories should invest in bioinformaticians.

References

Allison HE (2007) Stx-phages: drivers and mediators of the evolution of STEC and STEC-like pathogens. Future Microbiol 2:165–174

Altschul SF, Gish W, Miller W, Myers EW, Lipman DJ (1990) Basic local alignment search tool. J Mol Biol 215:403–410

Beutin L, Miko A, Krause G, Pries K et al (2007) Identification of human-pathogenic strains of Shiga toxin-producing *Escherichia coli* from food by a combination of serotyping and molecular typing of Shiga toxin genes. Appl Environ Microbiol 73:4769–4775

Bielaszewska M, Kock R, Friedrich AW, von Eiff C et al (2007a) Shiga toxin-mediated hemolytic uremic syndrome: time to change the diagnostic paradigm? PLoS One 2:e1024

Bielaszewska M, Zhang W, Mellmann A, Karch H (2007b). Enterohaemorrhagic *Escherichia coli* O26:H11/H-: a human pathogen in emergence. Berl Munch Tierarztl Wochenschr 120:279–287

Black LA, Klinger ME (1936) A comparison of media for the detection of *Escherichia-Aerobacter*. J Bacteriol 31:171–179

Chevenet F, Brun C, Bañuls A-L, Jacq B, Christen R (2006) TreeDyn: towards dynamic graphics and annotations for analyses of trees. BMC Bioinform 7:439–448

Clarridge JE, III (2004) Impact of 16S rRNA gene sequence analysis for identification of bacteria on clinical microbiology and infectious diseases. Clin Microbiol Rev 17:840–862

Creuzburg K, Schmidt H (2007) Shiga toxin-producing *Escherichia coli* and their bacteriophages as a model for the analysis of virulence and stress response of a food-borne pathogen. Berl Munch Tierarztl Wochenschr 120:288–295

Croce O, Lamarre M, Christen R (2006) Querying the public databases for sequences using complex keywords contained in the feature lines. BMC Bioinform 7:45

Dhanashree B, Mallya PS (2008) Detection of shiga-toxigenic *Escherichia coli* (STEC) in diarrhoeagenic stool & meat samples in Mangalore, India. Indian J Med Res 128:271–277

Edgar RC (2004) MUSCLE: multiple sequence alignment with high accuracy and high throughput. Nucleic Acids Res 32:1792–1797

Fagan PK, Hornitzky MA, Bettelheim KA, Djordjevic SP (1999) Detection of shiga-like toxin (stx1 and stx2), intimin (eaeA), and enterohemorrhagic *Escherichia coli* (EHEC) hemolysin (EHEC hlyA) genes in animal feces by multiplex PCR. Appl Environ Microbiol 65:868–872

Frohlicher E, Krause G, Zweifel C, Beutin L, Stephan R (2008) Characterization of attaching and effacing *Escherichia coli* (AEEC) isolated from pigs and sheep. BMC Microbiol 8:144

Galtier N, Gouy M, Gautier C (1993) SeaView and Phylo_win, two graphic tools for sequence alignment and molecular phylogeny. Comput Appl Biosci 12:543–548

Gourmelon M, Montet MP, Lozach S, Le Mennec C et al (2006). First isolation of Shiga toxin 1d producing *Escherichia coli* variant strains in shellfish from coastal areas in France. J Appl Microbiol 100:85–97

Gouy M, Delmotte S (2008) Remote access to ACNUC nucleotide and protein sequence databases at PBIL. Biochimie 90:555–562

Griffiths FP, Fuller JE (1936). Detection and significance of *Escherichia coli* in commercial fish and fillets. Am J Public Health Nations Health 26:259–264

Gyles CL (2007) Shiga toxin-producing Escherichia coli: an overview. J Anim Sci 85:E45–62

Hall L, Doerr KA, Wohlfiel SL, Roberts GD (2003) Evaluation of the MicroSeq system for identification of mycobacteria by 16S ribosomal DNA sequencing and its integration into a routine clinical mycobacteriology laboratory. J Clin Microbiol 41:1447–1453

Huang A, Friesen J, Brunton JL (1987) Characterization of a bacteriophage that carries the genes for production of Shiga-like toxin 1 in *Escherichia coli*. J Bacteriol 169:4308–4312

Islam MA, Mondol AS, de Boer E, Beumer RR et al (2008) Prevalence and genetic characterization of shiga toxin-producing *Escherichia coli* isolates from slaughtered animals in Bangladesh. Appl Environ Microbiol 74:5414–5421

Kaper JB, Nataro JP, Mobley HL (2004) Pathogenic *Escherichia coli*. Nat Rev Microbiol 2:123–140

Kobayashi H, Kanazaki M, Hata E, Kubo M (2009) Prevalence and characteristics of eae- and stx-positive strains of *Escherichia coli* from wild birds in the immediate environment of Tokyo Bay. Appl Environ Microbiol 75:292–295

Larkin MA, Blackshields G, Brown NP, Chenna R et al (2007) Clustal W and Clustal X version 2.0. Bioinform 23:2947–2948

Li W, Godzik A (2006) Cd-hit: a fast program for clustering and comparing large sets of protein or nucleotide sequences. Bioinform 22:1658–1659

Luna RA, Fasciano LR, Jones SC, Boyanton BL Jr et al (2007) DNA pyrosequencing-based bacterial pathogen identification in a pediatric hospital setting. J Clin Microbiol 45:2985–2992

Mansouri-Najand L, Khalili M (2007) Detection of shiga-like toxigenic *Escherichia coli* from raw milk cheeses produced in Kerman-Iran. Vet Arch 77:515–522

Moldavan A (1935) A modified technic for the detection of the *Escherichia-Aerobacter* Group in milk. Am J Public Health Nations Health 25:1032–1033

Mora A, Blanco M, Blanco JE, Dahbi G, et al (2007) Serotypes, virulence genes and intimin types of Shiga toxin (verocytotoxin)-producing Escherichia coli isolates from minced beef in Lugo (Spain) from 1995 through 2003. BMC Microbiol 7:13

Muller D, Greune L, Heusipp G, Karch H et al (2007) Identification of unconventional intestinal pathogenic *Escherichia coli* isolates expressing intermediate virulence factor profiles by using a novel single-step multiplex PCR. Appl Environ Microbiol 73:3380–3390

Nataro JP, Kaper JB (1998) Diarrheagenic *Escherichia coli*. Clin Microbiol Rev 11:142–201

Orth D, Grif K, Khan AB, Naim A et al (2007) The Shiga toxin genotype rather than the amount of Shiga toxin or the cytotoxicity of Shiga toxin in vitro correlates with the appearance of the hemolytic uremic syndrome. Diagn Microbiol Infect Dis 59:235–242

Prager R, Annemuller S, Tschape H (2005) Diversity of virulence patterns among shiga toxin-producing *Escherichia coli* from human clinical cases-need for more detailed diagnostics. Int J Med Microbiol 295:29–38

Wani SA, Hussain I, Nabi A, Fayaz I, Nishikawa Y (2007) Variants of eae and stx genes of atypical enteropathogenic *Escherichia coli* and non-O157 Shiga toxin-producing *Escherichia coli* from calves. Lett Appl Microbiol 45:610–615

Woo PC, Lau SK, Teng JL, Tse H, Yuen KY (2008) Then and now: use of 16S rDNA gene sequencing for bacterial identification and discovery of novel bacteria in clinical microbiology laboratories. Clin Microbiol Infect 14:908–934

Zheng J, Cui S, Teel LD, Zhao S et al (2008) Identification and characterization of Shiga toxin type 2 variants in *Escherichia coli* isolates from animals, food, and humans. Appl Environ Microbiol 74:5645–5652.

Chapter 4
Comparative Genomics of Pathogens

Elena P. Ivanova, Arkadiy Kurilenko, Feng Wang, and Russell J. Crawford

4.1 Introduction

Due to recent advances in molecular methods for the rapid detection of pathogenic bacteria, DNA-sequencing technologies and computational biology, comparative genomics has become a valuable tool, not only for the identification of a wide range of infectious agents but also for pathogen genotyping, the prediction of virulence, and resistance to antibiotics (Barken et al. 2007; Fournier et al. 2007). The rapid detection of slow growing or fastidious microorganisms has become possible as a result of the development of molecular assays, including different array-based technologies and novel methodologies (Barken et al. 2007; Neonakis et al. 2008). The importance of these modern sequence-based tools in the monitoring of known pathogens, the development of a new generation of vaccines, and the tracing of the origin of new infectious diseases cannot be overstated (Rappuoli 2004; Kaushik and Sehgal 2008).

Comparative genomics provides a unique opportunity for the microbial culture and independent detection of fastidious microorganisms or even groups of microorganisms. It also enables the accurate identification of pathogenic varieties of bacteria as well as the estimatation of the biodiversity of a bacterial population and its functional metabolic features (Rappuoli 2004; Barken et al. 2007; Fournier et al. 2007; Neonakis et al. 2008; Kaushik and Sehgal 2008). These new nucleic acid-based procedures are more rapid and precise when compared to the often time-consuming conventional diagnostics. Novel technologies enabling the sequencing of the whole bacterial genome of a single cell, or the rapid sequencing of the whole bacterial genome in just a few days, allows for routine methods to be replaced with faster and more accurate analyses, which may be less prone to human error.

E.P. Ivanova (✉)
Swinburne University of Technology, Melbourne, Victoria, Australia

V. Sintchenko (ed.), *Infectious Disease Informatics*,
DOI 10.1007/978-1-4419-1327-2_4,

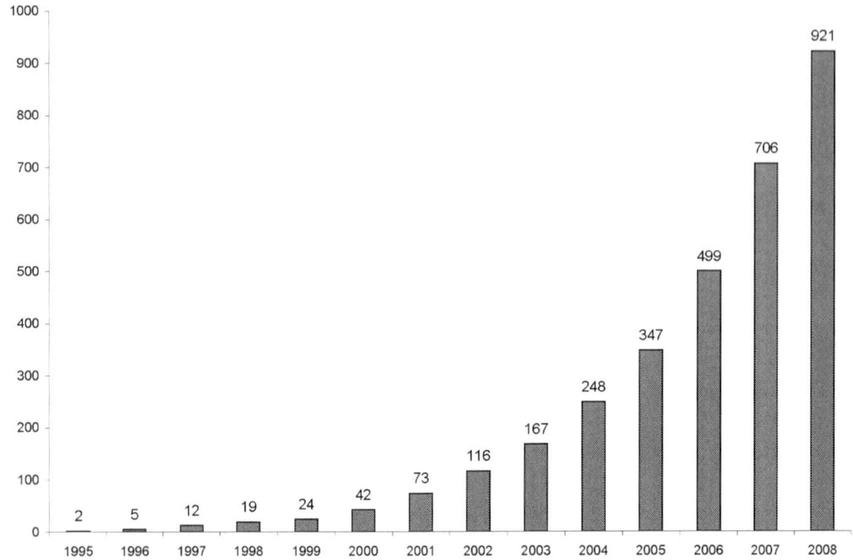

Fig. 4.1 Dynamics of complete microbial genome sequencing projects (as on January 2009)

The data accumulated in the area of molecular identification is growing at a great pace, with the number of sequenced bacterial genomes each year having increased more then a hundredfold in the last decade (Fournier et al. 2007; Liolios et al. 2008).

According to the Genomes Online Database (http://www.genomesonline.org/), as on10 January 2009, the sequencing of 921 genomes had been completed and published (Fig. 4.1), and 3376 genome projects were ongoing, including 2,261 bacterial, 1,014 eukaryotic, and 100 archaeal projects.

4.2 Tools for Microbial Classification and Identification of Pathogens

Comparative genomics can be defined as a scientific discipline focusing on the study of relationships between the genomes of different species. The first completely sequenced genome was obtained in 1977 by Fred Sanger with co-workers. This genome was that of the bacteriophage Φ-X174, which was just 5,368 bp in size (Sanger et al. 1977). This research opened a new era in genomics, which has exploded now: almost 18 years later. In 1995, advances in sequencing technology, such as the automation of the process, together with the appreciable cost reductions, made the sequencing of whole microbial genomes possible (Hall 2007). The first complete microbial genome of *Haemophilus influenzae* was sequenced and completely decoded at the Institute for Genomic Research (Rockville, MD, USA).

The data, which included 1,830,137 bp of DNA and 1,743 predicted genes, laid out the full genetic complement of a bacterial organism for the first time (Fleischmann et al. 1995). Within five years of that publication, numerous other bacteria were sequenced, including *Mycobacterium tuberculosis,* one of the most important human bacterial pathogens (Cole et al. 1998), *Escherichia coli* (Blattner et al. 1997), and the first archaeon, *Archaeoglobus fulgidus* (Klenk et al. 1997). This marked the beginning of a boom in genome sequencing projects across the globe. Notably, most of these bacterial genome projects were funded with the intent to utilize the data in biomedical applications (Fig. 4.2). After genome annotation and gene function identification, what is important is that specific phenotypic traits may be deduced from the genotype (Fournier et al. 2007). The information obtained is useful for serological applications, for the development of specific culture media, for the identification of antibiotic resistance mechanisms, virulence factors, and for the exploration of host–pathogen interactions (Fournier et al. 2007).

Due to the continuing development and improvement of high-throughput sequencing technologies and the computational power available for the assembly of

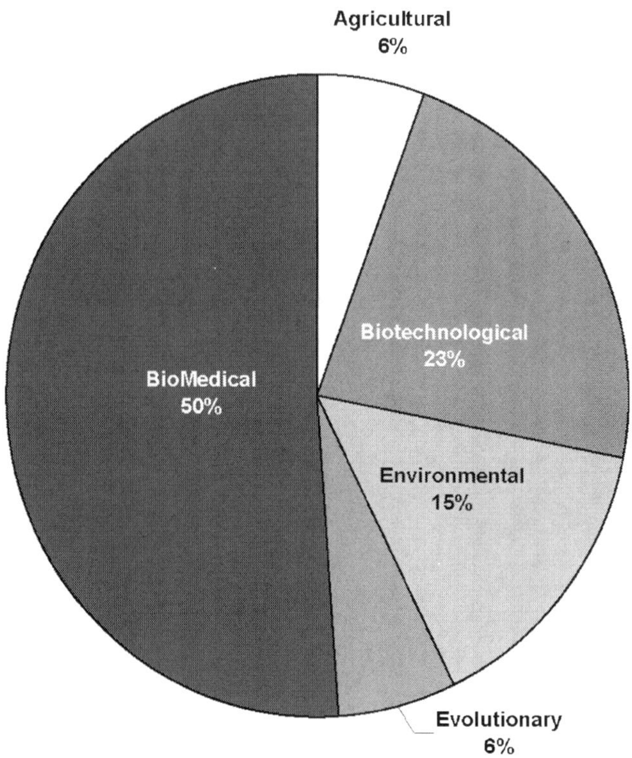

Fig. 4.2 Funding relevance of bacterial genome projects (Modified from Woodford and Johnson 2004)

sequences what was mainly a fundamental discipline of genomics to begin with, has now proven itself to be an irreplaceable research tool for various aspects of clinical microbiology (Hall 2007).

4.2.1 Sequencing of Selected Genes and Genomes

DNA sequencing is a primary technique in genomics and is the only source of genetic information about any kind of life form. The value of these data cannot be overestimated because of their potential application in the development of molecular methods for the classification of microorganisms, the identification of pathogenic species in a wide range of taxa (strain, species, genus, phylum), the detection of virulent genes and mutations responsible for antibiotic resistance, and the development of new vaccines and drugs (Kaushik and Sehgal 2008).

Several sequencing technologies have been developed, including first generation classic Sanger sequencing and the modern second-generation high-throughput sequencing platforms that have become available as a result of the following technologies: the 2007 Genome Sequencer 20/FLX (commercialized by 454/Roche); the "Solexa 1G" (later named "Genome Analyzer" and commercialized by Illumina/ Solexa); and the SOLiD system (commercialized by Applied Biosystems) (Hall 2007). So-called 'third-generation' sequencers (single molecule-SBS) such as Helicos tSMS, and PacBio SMRT, together with modifications of the Nanopore based sequencing system and the ZS Genetics TEM, are expected to be available in the next few years (Gupta 2008).

Sequencing data obtained from certain separate genes, complete bacterial genomes, and whole shotgun sequences have been accumulated in public databases over the last two decades. Three major public databases include the US National Center for Biotechnology Information (http://www.ncbi.nlm.nih.gov/), the DNA database of Japan (http://www.ddbj.nig.ac.jp/), and the European Bioinformatics Institute (http://www.ebi.ac.uk/). Free public access to the data has allowed the utilization of genomic information in a wide range of comparative genomic and metagenomic sequence analysis techniques. A great advantage of sequence-based methods is that their results can be easily compiled into databases and subsequently compared between laboratories.

Many of these sequence-based approaches involve the comparative analysis of 16S rRNA gene sequences for the identification of novel isolates. The 16S rRNA gene is highly conserved among all microorganisms, is of suitable length (about 1,500 bp) for bioinformatics analysis and is an excellent molecule for discerning evolutionary relationships among prokaryotic organisms (Barken et al. 2007). The genotypic and phylogenetic identification of newly isolated microorganisms has incorporated data from genome sequences or has been designed on the basis of genome sequences. Of the impressive number (up to 100,000) of sequences available in public databases, 16S rRNA gene sequences can be used for the comparative genomic analysis of these genomes, making this analysis a very powerful tool for the identification of microorganisms (Amor et al. 2007).

It would be reasonable to suggest that 16S rRNA gene sequencing has become a 'gold standard' tool in the modern classification of microorganisms (Barken et al. 2007; Sidarenka et al. 2008). This gene is present in all prokaryotes and encodes the same product. Mutations occur randomly and are, by and large, not subject to selective forces (Woese 1987). The 16S rRNA gene contains many domains, some of which are conserved and others of which are variable (Woese 1987). Often, even the partial sequencing of the 16S rRNA gene is sufficient to discriminate between species of bacteria. For example, it has been shown that the sequencing of the 5′ end of 16S rRNA is sufficient to allow the species level identification of most clinically relevant *Mycobacterium* isolates (Tortoli 2003).

Moreover, 16S rRNA gene sequencing is often a more accurate bacterial identification method than phenotypical methods based on biochemical analysis (Barken et al. 2007). It was recently demonstrated that the sequencing of some other genes that are less conservative than the 16S rRNA gene may provide sufficient data for species identification (O'Sullivan 2000). It is, however, important to note one of shortcomings associated with bacterial identification techniques based on the comparison of 16S rRNA sequences. This approach is effective for well-resolved species, but may not always be sufficient to establish the species identity on newly diverged species (Barken et al. 2007). In the case of *Mycobacteria,* for example, some strains like *Mycobacterium chelonae* and *M. abscessus,* while showing almost identical 16S rRNA gene sequence, have only about 35% sequence identity when their chromosomes are compared (Tortoli 2003). Other disadvantages associated with bacterial identification based on 16S rRNA gene sequences include a high risk of contamination, difficulties associated with polymicrobial specimens, and insufficient discriminatory power for closely related species (Barken et al. 2007).

More recent advances in this field include the development of genotyping methods based on the comparative analysis of DNA sequences of several 'house-keeping' genes, such as the genes associated with certain surface and heat-shock proteins, in addition to the 16S rRNA gene and the non-coding conservative parts of DNA. These, allow phylotyping to be more precise and allow the identification of microorganisms up to a subspecies or even isolates levels (Fournier et al. 2007). The comparative resolution of sequencing-based methods is shown in Fig. 4.3.

Sequence data for specific loci (e.g., genes for virulence, pathogenicity, drug resistance, etc) from different strains of the same species have revealed variability in a specific gene, such as single-nucleotide polymorphisms (Singh et al. 2006). The resolution of this method differs with regard to the gene being targeted. The single-locus sequence typing (SLST) approach involves the analysis of a particular region of the targeted gene, which is polymorphic. A demonstration using a *S. aureus* staphylococcal protein A gene showed that SLST appears to be a very robust technique, with benefits including its throughput rate, ease of use and interpretation. At present, however, no SLST protocol has emerged as a clear stand-alone method for epidemiologic typing (Koreen et al. 2004; Shopsin et al. 1999; Stampone et al. 2005).

Multilocus sequence typing (MLST) is considered as a generic typing method that provides reproducible results, is reliable, relatively inexpensive, and allows a high rate of throughput (Brehony et al. 2007). This method utilizes a larger,

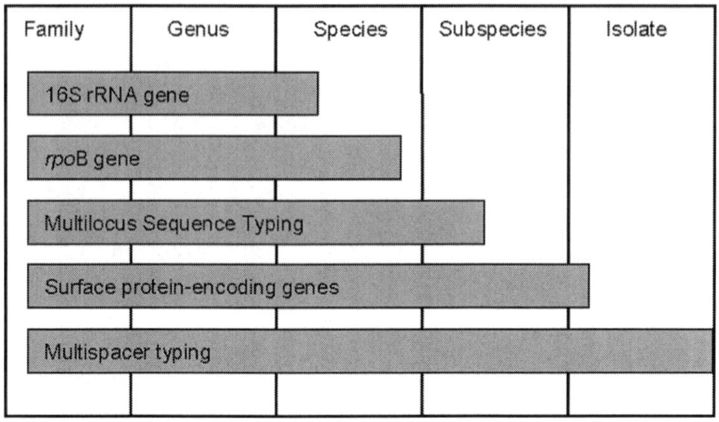

Fig. 4.3 Comparative resolution of sequence based methods of microbial identification

and potentially more representative, portion of the genome than SLST. MLST was based upon the principles of such molecular methods as Multilocus Enzyme Electrophoresis (MLEE), but has also exploited high-throughput nucleotide sequencing and data dissemination via the Internet (Urwin and Maiden 2003). As with SLST, MLST demonstrates the potential of sequence-based typing to generate consistent, reproducible isolate profiles that are highly amenable to standardization and database cataloging. However, MLST is less suited to clinical settings due to the expense, labor, and time involved in surveying multiple (usually seven or eight) genes and the corresponding approximately 2,500 bp or so of sequence to allow real time differentiation between multiple isolates (Devulder et al. 2005; Singh et al. 2006; Brehony et al. 2007; Mignard and Flandrois 2008).

However, sequencing is not without limitations. Unfortunately, the direct sequencing of DNA still remains a long-term and expensive procedure, which limits its application in clinical practice analysis. The identification of multiple pathogens and the determination of the abundance of different organisms is not possible using this method and require the implementation of additional techniques such as in-situ hybridization or the use of probe-arrays (Barken et al. 2007).

4.2.2 DNA Hybridization-Based Approaches

Originally, Fluorescent In-Situ Hybridization or FISH was designed as a tool for the identification, visualization and localization of microorganisms in environmental samples (Amann et al. 2001). Currently, this technique has been applied in many fields of microbiology as a rapid and direct method allowing the detection of both culturable and non-culturable species (Amann et al. 1990; Daims et al. 2001; Kempf et al. 2000; Peters et al. 2006; Poppert et al. 2002; Thurnheer et al. 2004).

The striking feature of FISH is its employment of specific probes. The probe/s are short oligonucleotides with a fluorochrome molecule/s attached to the 3′ or the 5′-end. They can be highly specific by hybridizing with a complementary sequence on the rRNA on the species/subspecies or genus level. A visualization and localization of target microorganisms is achieved by using fluorescent microscopy or confocal laser scanning microscopy (CLSM) (Barken et al. 2007).

Despite its simplicity, high sensitivity and specificity, FISH has some shortcomings that are mostly associated with possible unspecific bonding. This results from insufficient washing after hybridization, insufficient fixation of the bacteria prior to hybridization or ineffective penetration of the probe during hybridization. Other factors, such as low rRNA content in cells or auto fluorescence, may also affect the accuracy of the fluorochrome signal reading. In addition, the photo bleaching of the samples may decrease signal intensity (Moter and Gobel 2000; Wagner et al 2003).

In a number of studies, FISH has been used to investigate the distribution and spatial organization of *Pseudomonas aeruginosa* in sputum samples and lung expectorates from patients (Hogardt et al. 2000). *P. aeruginosa* and *Achromobacter xylosoxidans* cells were accurately identified in 2.5 hours (Wellinghausen et al 2006). Another area in which FISH has been successfully employed is the monitoring of the microbial colonization of biofilms (Poulsen et al. 1993). A similar approach was applied to biofilms developing in the human body in connection with implants (Sunde et al 2003). The visualization of periodontitis causative bacteria in a biofilm was achieved using the combination of FISH and CLSM (Wecke et al. 2000). Bacterial diversity in the oral cavity has also been examined using FISH (Fredricks et al. 2005). Some uncultivable causative species have been identified and their localization in periapical lesions has been confirmed using a combination of FISH and CLSM (Sunde et al. 2003).

FISH has also been applied for the detection of non-cultivable pathogens in the blood stream (Kempf et al. 2000). Søgaard et al. have designed highly specific peptide nucleic acid (PNA) probes allowing the rapid detection of infectious bacteria (Sogaard et al. 2005). Poppert et al. were able to identify *Neisseria meningitides* directly in cerebrospinal fluid using a combination of real time PCR and FISH (Poppert et al. 2005)

DNA microarrays have emerged as a high throughput assay for bacteria genotyping (Cassone et al. 2007). Bacterial DNA microarrays for clinical microbiology are built upon the increasing amount of sequence information available in public databases. The advantages of this technique include its ability to simultaneously compare strains at the whole-genome level, its sensitivity to detect subtle differences, and the automation of microarray handling. As a result of decreased costs, DNA micro-arrays are now becoming useful tools in routine clinical laboratories. For example, a new low-cost micro-array was developed by Clondiag Chip Technologies, Germany (Barken et al. 2007). Specific to certain microorganisms, oligonucleotide probes can be immobilized at the bottom of a plastic tube (AT DNA microarray chip). An AT biochip based on 13 SNP's that have been shown to be present in different genomic regions in *Pseudomonas aeruginosa* has already been designed (Jelsbak et al. 2007; Morales et al. 2004). Another AT chip was designed for the rapid detection and identification of *Chlamydia* and *Chlamydiaceae* in clinical samples (Sachse et al. 2005).

The essence of DNA microarrays is the use of dot blot hybridization. Since this is performed in a small and highly parallel format, multiple targets in the same assay can be identified (Barken et al. 2007). It is important to note that several techniques are used for the deposition of probes for DNA microarrays. Direct photolithographic immobilization is most commonly used (Bryant et al. 2004). Oligonucleotide probes usually consist of up to 20,000 of 300–800 bp shotgun fragments or PCR products of one or several whole genome(s), specific sets of genes, or as many as 600,000 of 50–70 bp oligonucleotides immobilized onto glass slides (Fournier et al. 2007). The signal intensity is quantified to the amount of the hybridized sample. PCR fragments can also be used in DNA-microarrays: they generate strong signals and small sequence differences are hardly detectable. Short oligonucleotide probes are able to detect single base mutations but require the careful optimization of hybridization conditions (Cassone et al. 2007).

There are several applications where the DNA microarray can be used directly for diagnostic purposes: pathogen detection (e.g., *Neisseria meningitidis*), the identification of vaccine candidates (e.g., in *Vibrio cholerae*), the identification of virulence genes and genes encoding antibiotic resistance (e.g., *Streptococcus pneumoniae*, MRSA), and the genotyping of bacterial strains, such as in *Campylobacter jejuni*, *Escherichia coli*, *Francisella tularensis*, *Helicobacter pylori*, *Listeria monocytogenes*, *Pseudomonas aeruginosa*, *Staphylococcus aureus*, *Streptococcus pneumoniae*, *Vibrio cholerae*, and *Yersinia pestis* in molecular epidemiological studies (Fournier et al. 2007; Neonakis et al. 2008). Chang et al. (2008) used genome-probing microarrays together with the digital multiple displacement amplification (MDA) of DNA from single uncultivated bacterial cells for microbial detection and microbial diversity assessment.

Specific sequences of regions from the 16S rRNA and *rpoB* loci for mycobacteria have been identified and synthesized. Out of seventy mycobacterial isolates belonging to 27 species and the 15 Rifampicin-resistant *M. tuberculosis* strains used in this study, all of the *rpoB* mutant alleles and 26 species were correctly identified (Neonakis et al. 2008). Stavrum with colleagues successfully estimated the genomic diversity of *M. tuberculosis* isolates using whole-genome arrays (Stavrum et al. 2008).

Single nucleotide polymorphism (SNP) microarrays can be applied for the detection of a limited set of genetic polymorphisms found in some *Staphylococcus aureus* strains, and in methicillin-resistant *S. aureus* (MRSA) strains, in particular. Korczak et al. have surveyed virulence factors of different *E. coli* strains isolated from patients with neonatal meningitis, urinary tract infections and enterohemorrhagic syndrome. The authors identified 32 probes associated with the different pathotypes (Korczak et al. 2005). Stabler et al. constructed a *Neisseria* microarray incorporating all of the genes of four different *Neisseria* species and the localized genes conserved for *N. miningitidis* serogroup B strains. The identified combination of virulence-associated genes is useful for the detection of pathogenic *N. miningitidis* strains (Stabler et al. 2005).

4.2.3 Polymerase Chain Reaction (PCR)-Based Approaches

Numerous techniques based on PCR for the amplification of certain parts of the bacterial genome with complementary oligonucleotide primers are precise, sensitive and rapid (Barken et al. 2007). These assays generally allow the detection of the presence of pathogenic microorganisms, including fastidious and slow growing ones, directly in clinical specimens. They also allow the samples to be tested for antimicrobial resistance genes, and have many other applications with wide range specificity (Yang and Rothman 2004). These methods improve the accuracy and timeliness/aptness of tuberculosis diagnosis, and can also be used to detect new or emerging infections (Barken et al. 2007). Several studies, including one focusing on the gastric pathogen *Helicobacter pylori*, have successfully applied this methodology. Different multiplex PCR assays as well as the random amplified polymorphic DNA-PCR (RAPD-PCR) (Akopyanz et al. 1992a; Krogfelt et al. 2005) and restriction fragment length polymorphism-PCR (RFLP-PCR) (Akopyanz et al. 1992b) have been used for the identification of *vac*A and *cag*A genes of *H. pylori* (Monstein and Ellnebo-Svedlund 2002). Several commercial applications of multiplex PCR followed by hybridization to a DNA strip are available for the detection of multidrug resistant *M. tuberculosis* (Hillemann et al. 2005; Palomino 2005).

Real time PCR (RT-PCR) technology is an appealing alternative to the conventional culture-based or immunoassay-based testing methods used for diagnosing many infectious diseases (Espy et al. 2006). RT-PCR combines PCR chemistry with the fluorescent probe detection of an amplified product in the same reaction vessel (Heid et al. 1996) and is faster than conventional PCR. Importantly, RT-PCR reduces the risk of contamination with amplified nucleic acids because the analysis is performed in a closed vessel (Valasek and Repa 2005).

Two different detection methods based either on fluorescent stain (e.g., SYBR Green) or fluorescent resonance energy transfer (FRET) probes are utilized in this technology (Heid et al. 1996). The FRET probes are more sensitive and specific that SYBR Green ones. FRET hybridization probes consist of the upstream probe with the fluorescent dye at their 3′ end and the downstream probe with an acceptor dye at their 5′ end. The downstream probe is phosphorylated at its 3′ end to prevent it from being used by the Taq polymerase during PCR amplification (Espy et al. 2006; Ota et al. 1998). Currently, three different FRET probes are used: 5′ nuclease probes (TaqMan), molecular beacons, and FRET hybridization probes. TaqMan probes secure an increasing abundance of fluorescence after each PCR cycle due to the 5′ nuclease activity of the Taq polymerase (Espy et al. 2006; Heid et al. 1996). Molecular beacons carry both a fluorophore and a quencher; a probe sequence is embedded within two complementary 5-nucleotide-long arm sequences. A fluorescent signal is generated after hybridization as the fluorescent dye and the quencher are separated (Tyagi and Kramer 1996).

The RT PCR technique was adopted in several studies. In 1985 Dutka-Malen et al. were one of the first groups who reported data where an RT-PCR assay was applied for the simultaneous detection of *vanA* and *vanB* genes responsible for the resistance to vancomycin (Dutka-Malen et al. 1995; Palladino et al. 2003). Uhl et al employed LightCycler RT-PCR (Roche Diagnostics) for the detection of group A streptococci (GAS), arguing that this method was much better than the antigen detection methods since it allowed the results to be obtained on the same day as the analysis so that an appropriate antimicrobial treatment could be promptly applied (Uhl et al. 2003). Blome et al. successfully used quantitative real time PCR to evaluate the presence and numbers of specific species as well as the total bacterial load in teeth with endodontic infections (Blome et al. 2008).

The multiplex RT PCR is a useful alternative to RT PCR. The multiplex RT PCR was used for quantitative detections of *Mycobacterium* species, *M. tuberculosis*, *M. avium*, *M. bovis*, *M. abscessus*, *M. chelonae* and *M. ulcerans*, as well as for the detection of drug-resistant isolates from clinical specimens or laboratory cultures (Deepak et al. 2007). The sensitivity of a multiplex RT PCR with the detection of six specific virulence genes was demonstrated by Yang et al. (2007), who showed that RT PCR considerably reduced the high false-positive rate.

Often, RT PCR is found to be a superior technique with regard to sensitivity and, more importantly, with regard to turn-around and hands-on time. This technique requires only a few hours, whilst routine detection using selective media takes days. Warren et al. described a sensitive and specific test for the rapid detection of MRSA directly from nasal swab specimens. This test is based upon real time PCR using a molecular beacon probe. The time from sampling to having the result was less than two hours (Warren et al. 2004).

The limitations of RT-PCR are common to most PCR technologies and they include the possibility of the inhibition of the polymerase by the presence of certain compounds, and the risk of detecting contaminating DNA due to the high sensitivity of the method (Barken et al. 2007).

4.2.4 Pyrosequencing-Based Approaches

Pyrosequencing is a relatively recent DNA sequencing technology that is based on the sequencing-by-synthesis principle. This method is suitable for determining relatively short sequences of 20–60 bp per read in a rapid, high-throughput and semi-automated format. One of the important advantages of this method is the low cost of analyses. The analysis of between 10,000 and 50,000 samples per day may cost approximately 20–30 cents per sample (Ronaghi and Elahi 2002).

The four enzymes implemented in the pyrosequencing system are the Klenow fragment of DNA Polymerase I, ATP sulfurylase, Luciferase and Apyrase. The reaction mixture also contains the enzyme substrates adenosine phosphosulfate and d-luciferin, as well as the single stranded sequencing DNA template with an annealed specific primer to be used as the starting material for the DNA polymerase. The four nucleotides are added one at a time, iteratively, in a cyclic manner, resulting in a cascade of enzymatic reactions that generates visible light. The photons are then captured by a CCD camera (Ahmadian et al. 2006).

A characteristic feature of pyrosequencing is the sequencing of at least 20 bases, which has led to an array of applications associated with the detection of unknown polymorphic positions, the detection of molecular markers of resistance, and microbial typing (Ahmadian et al. 2006; Deyde et al. 2009). Some limitations of pyrosequencing, however, flow from this characteristic feature. Longer read-lengths are necessary to distinguish closely related species. The use of multiple group-specific sequencing primers suggested by Gharizadeh et al. maybe useful to overcome the read-length limitation in 16S rRNA studies (Gharizadeh et al. 2003).

Multiple Displacement Amplification (MDA) is an emerging technique that allows the amplification of the whole bacterial genome of a single cell (Lasken 2007; Chang et al. 2008). As is the case with all traditional nucleic-acid based methods, MDA provides a unique opportunity to access the genetic information of non-culturable bacteria on the level of a single cell as it operates with femtogram masses of DNA from the sample. The application of the 454 Life Sciences pyrosequencing and MDA reactions to a single *E. coli* cell genome has enabled the successful mapping of more than 95% of sequence reads and more than 99% of contigs (Chang et al. 2008).

Pyrosequencing has been used to scan for undefined mutations. For example, it was demonstrated that a set of sequencing primers covering the 4 exons of the p53 gene can be used to detect unknown mutations (Garcia et al. 2000).

4.3 Metagenomics: Principles and Perspectives

In 1985, Pace and colleagues proposed the direct analysis of 16S rRNA gene sequences to describe the microbial diversity in an environmental sample without the traditional culture (Lane et al. 1985). The recent development of advanced high-throughput DNA sequencing and computational technologies gave birth to a new discipline, metagenomics, which focuses on the culture-independent genomic analysis of microbial communities in a particular environmental niche (Handelsman et al. 1998). Metagenomics is concerned with the direct isolation of DNA from a defined habitat, followed by the cloning of the complete genomes of the entire microbial population from the particular environment in a surrogate host, such as *E. coli*, and the DNA sequencing of the resulting enriched fragments (Langer et al. 2006). The results obtained would then be subjected to an evaluation of the phylogenetic affiliation and functional diversity within a microbial community (Riesenfeld et al. 2004; Sharma et al. 2008; Riggio et al. 2008; Wommack et al. 2008). Metagenomic research is often based on the computational analysis of 16S rDNA sequence libraries (Huson et al. 2007). For example, the Random Sequences Read (RSR) approach is based on a genome comparison analysis that does not use particular genes, but rather metagenomic libraries created for the whole bacterial population. It has been found that direct comparison with complete genomes or with Whole Genome Shotgun sequences (WGS) from a database delivers more promising results. This method enables the accurate detection of all submitted known sequences in the databases by BLASTing random genomic sequences with an average size of 615 bp, allowing an indication of the presence of new strains. Metagenomic approaches like RSR show great potential due to the constantly growing number of genomes that have been sequenced and subsequently submitted to the public databases. This methodology offers a faster and cheaper alternative to 16S rDNA sequence library analysis (Manichanh et al. 2008).

Current applications of metagenomics are focused on the characterization of microbial communities (Riesenfeld et al. 2004; Harris et al. 2007; Riggio et al. 2008; Sharma et al. 2008). One example is the detection of lactic acid bacteria in prostate tissue core samples (Amor et al. 2007). Sfanos and colleagues (2008) have extracted DNA from tissue samples from 200 patients. Using organism-specific PCR the presence of *Chlamydia trachomatis*, *Propionibacterium acnes*, *Trichomonas vaginalis*, BK virus, Epstein-Barr virus, human cytomegalovirus, human papillomavirus, and xenotropic murine leukemia-related virus was tested. Eckburg and colleagues assessed the diversity of human intestinal microbial flora by analyzing metagenomic libraries of 13,355 prokaryotic 16S rDNA sequences from multiple colonic mucosal sites and stool samples (Eckburg et al. 2005). Gill et al. (2006) analyzed ~78 million base pairs of unique DNA sequence and 2,062 polymerase chain reaction-amplified 16S ribosomal DNA sequences from the human distal gut microbiome.

4.4 Emerging DNA Sequencing Technologies

Innovative, high-throughput sequencing technologies employ a computer-based analysis (with a subsequent assembly) of relatively short-read sequences in a parallel manner. A comparative evaluation of the currently available DNA sequencing techniques is given in Table 4.1. Each technique is briefly discussed below.

The Sanger sequencing method, invented almost 30·years ago, underwent significant improvements over the years as a result of the development of highly automated template preparation pipelines (Hudson 2007; Sharma et al. 2008). This improvement increased the average length of a sequence that could be read from approximately 450·bases to 850·bases, thus reducing the cost of analysis. Nonetheless, this technology has limitations: the cost of sequencing remains relatively high (US$25 per human-size genome) and only large sequencing centers can maintain and occupy parts of sequencing machines conducting more than ten runs per day (Hudson 2007). Another step forward for Sanger sequencing was polyacrylamide gel capillary electrophoresis, which produced high quality DNA sequences with long read lengths of up to 1,000 bp in a few hours. As a result, the 384 capillary sequencers (MegaBACE™ 4000 DNA Analysis System) can generate over 2.8 million bases of sequence data in 24 h. The current top model is the 1,024-capillary "Monster CAE" sequencer, developed at Stanford, USA in collaboration with the University of California Berkeley, USA (Sharma et al. 2008).

Table 4.1 Modern DNA sequencing technologies

Technology	Read length	Throughput	Cost per human-size genome (US$)
Automated Sanger sequencing	Up to 900 bp	96 kb per 3 hours run	25,000,000
454 pyrosequencing	240–400 bp	80–120 Mb per 4 h run	1,000,000
SOLiD	~35 bp	1–3 Gb per 8 days run	60,000
Solexa	~35 bp	1 Gb per 2–3 days run	60,000
tSMS	~30 bp	60 Mb per < 1 h run	70,000
SMRT	Up to 100,000 bp	Presumably 1 Gb per < 1 h run	Presumably < 1,000
Nanopore sequencing	Presumably hundreds of thousands bp	Presumably 1Gb per ~20 h run	Presumably < 1,000
TEM sequencing	Presumably hundreds of thousands bp	Presumably 1 Gb per ~14 h run	Presumably < 1,000

Three highly computerized technologies for DNA sequencing have become commercially available: the 454 Pyrosequencing, Sequencing by Oligonucleotide Ligation and Detection (SOLiD) and the Illumina/Solexa approach. All of these utilize high-throughput; massively parallel sequencing, making outputs relatively inexpensive. These technologies use similar principles for sample preparation based on *in vitro* amplification of DNA. These techniques, however, significantly differ from each other in the sequencing and detection stages.

The 454 Pyrosequencing method developed in 2005 (Margulies et al. 2005) was commercialized and distributed by 454 Life Sciences (Branford, CT, USA). After the amplification stage, micro-beads are placed individually into pico litre flow-cells on a special plate containing up to 400,000 cells. This step is followed by simultaneous pyrosequencing reactions. The average read length is over 250 base units. The raw base error rate is below 0.5%, making the 454 Pyrosequencing suitable for genome re-sequencing and for the *de novo* sequencing of bacterial or even eukaryotic genomes (Hudson 2007).

The SOLiD technology has been widely deployed by Applied Biosystems (www.appliedbiosystems.com) and is based on an enhancement of the 'colony sequencing' ligation chemistry (Shendure et al. 2005). It employs a fluorescence microscope as a detector. The amplification products are transferred onto a glass surface where sequencing occurs by a sequential cycle of hybridization and ligation, with sixteen dinucleotide combinations labelled by four different fluorescent dyes. Using the specific four dye encoding scheme, each position is effectively probed twice, and the identity of the nucleotide is determined by analyzing the color that results from two successive ligation reactions. This encoding scheme enables the distinction between a sequencing error and a sequence polymorphism. This is a promising technology due to its high throughput rate and low cost, in spite of its shorter read length (about 35 base pairs) compared with the 454 Pyrosequencing method (Hall 2007). The newly released SOLiD instrument is capable of producing 1–3 Gb of sequence data in 35-bp reads per an eight-day run (Morozova and Marra 2008).

Another method for massively parallel sequencing by synthesis from amplified fragments is the Illumina/Solexa approach, which has been developed by a company with the same name (http://www.illumina.com/). Solexa sequencing differs from SOLiD or 454 sequencing as it amplifies the DNA on a solid surface, and then starts the synthesis by incorporating modified nucleotides linked to four different colored dyes. It is a cheap, high-throughput technology: The Illumina "1G genome analyzer" is capable of producing at least 1 Gb of sequence in 2–3 days, by generating at least 35-base reads. Unfortunately, because of the use of modified DNA polymerases and reversible terminators, substitution errors have been noted in Illumina sequencing data (Hudson 2007; Morozova and Marra 2008).

True single-molecule sequencing (tSMS) is a technology where the target DNA is used for the construction of a library of poly(dA)-tailed templates, which pair with millions of poly(dT)-oligonucleotides that are anchored to a glass cover-slip (Helicos Biosciences, Cambridge, MA, USA; http://www.helicosbio.com/). The position of each of the individual poly(dT) oligos is fixed on the cover slip. The sequence of each poly(dA)-tailed fragment is determined by adding nucleotides labeled with the fluorescent cyanine dye Cy5 – in a cyclic manner, one nucleotide at a time. This method is highly parallel, and on a 25-mm square it is possible to sequence 12 million templates simultaneously, so it is expected that 60·million bases of information per 'run' will be generated (Hall 2007; Gupta 2008).

Single-molecule real-time sequencing (SMRT) is another proprietary approach developed by Pacific Biosciences (PacBio, Menlo Park, CA, USA; http://www.pacificbiosciences.com). It involves the use of so-called SMRT chips, each made up of a 100-nm thick metal film and containing thousands of 10–50 nm cavities, each with a DNA polymerase molecule attached at the bottom. The reaction of DNA is synthesized from a single-stranded DNA molecule template visualized using four different fluorophore-labeled nucleotides, where the label is attached to the phosphate group. When a nucleotide is incorporated during DNA synthesis, the attached fluorophore lights up due to the laser-beam-mediated illumination of a minimal detection volume (20 zeptoliters). This allows the identification of each incorporated nucleotide. This set-up enables nucleotides to be incorporated at a speed of ten bases per second, giving rise to a chain of thousands of nucleotides in length within minutes. This simultaneous and continuous detection occurs in real-time, which facilitates the determination of thousands of sequences, each sequence thousands of bases long. PacBio claims that, by 2013, the technology will be able to give a 'raw' human genome sequence in less than 3 min, and a complete high-quality sequence in just 15 min (Gupta 2008).

The idea of nanopore sequencing involves the use of a very thin membrane that contains nanopores (channels of ~1.5–2 nm in diameter). The negatively charged target single-stranded DNA travels through the nanopore towards the positive charge, generating a change in the electrical conductivity of the membrane, which produces a measurable current in the range of picoamperes. However, the nanopore sequencing approach is still in the early stages of development, and is undergoing changes to overcome problems with the resolution of individual base reading. Nanopore sequencing is expected to be high throughput, give an almost unlimited read length, and cost only US$100 per a human-size genome or less than US$20 per an average bacterial genome (Rhee and Burns 2006; Ryan et al. 2007; George Weinstock, personal communication, May 2009).

The SMS platform is developed by ZS Genetics (ZSG; North Reading, MA, USA; http://zsgenetics.com/). This is a direct reading of DNA sequences using a specialized transmission electron microscope (TEM). Natural DNA is transparent when viewed with TEM because of the low atomic number of its compounds. ZS Genetics technology involves the linearization of the target DNA molecule, followed by the synthesis of a complementary strand, whereby three of the four bases are labeled with heavy atoms (e.g., iodine or bromine) that make the DNA heavier and visible under TEM. Thus, when the resulting complementary strand is observed under TEM, the four bases can be discriminated by the size and intensity of the relevant dots. ZSG declares that it can achieve read lengths of around 5,000–7,000 bases, and claims that it will be able to produce an exponential increase in the sequencing potential with future improved versions of this technology (Gupta 2008).

4.5 Conclusions

By identifying a spectrum of bacteria that are associated with infectious processes, corrective measures that would alter microbial communities via diet, drugs, or probiotics (live bacterial cultures) may be designed and implemented. Further advances in the comparative genomics of bacterial genomes will provide the basis for mechanistic hypotheses about the roles that microbes play in many diseases that have been associated with microorganisms (e.g., colonic cancer, periodontal disease, autism, and obesity) (Worthen et al. 2006; Guan et al. 2007; Kikuchi and Graf 2007). Hypothesis testing will demand new conceptual frameworks and tools for investigating community genetics, as well as new models of host-pathogen interactions (Fernandez et al. 2000; Relman and Falkow 2001; Yu and Chu 2005; Handelsman 2008). Further research and investment in metagenomics as one of the leading disciplines for microbial community analysis will undoubtedly benefit human health and biomedical science.

References

Ahmadian A, Ehn M, Hober S (2006) Pyrosequencing: history, biochemistry and future. Clinica Chimica Acta 363:83–94

Akopyanz N, Bukanov NO, Westblom TU et al (1992a) DNA diversity among clinical isolates of *Helicobacter pylori* detected by PCRbased RAPD fingerprinting. Nucleic Acids Res 20:5137–5142

Akopyanz N, Bukanov NO, Westblom TU et al (1992b) PCR-based RFLP analysis of DNA sequence diversity in the gastric pathogen *Helicobacter pylori*. Nucleic Acids Res 20:6221–6225

Amann R, Fuchs BM, Behrens S (2001) The identification of microorganisms by fluorescence in situ hybridisation. Curr Opin Biotechnol 12:231–236

Amann RI, Krumholz L, Stahl DA (1990) Fluorescent-oligonucleotide probing of whole cells for determinative, phylogenetic, and environmental studies in microbiology. J Bacteriol 172:762–770

Amor KB, Vaughan EE, De Vos WM, Amor (2007) Advanced molecular tools for the identification of lactic acid bacteria. J Nutr 137:741S–747S

Barken KB, Haagensen JAJ, Tolker-Nielsen T (2007) Advances in nucleic acid-based diagnostics of bacterial infections. Clin Chim Acta 384:1–11

Blattner FR, Plunkett III G, Bloch CA et al (1997) The complete genome sequence of *Escherichia coli* K-12. Science 277:1453–1474

Blome B, Braun A, Sobarzo V et al (2008) Molecular identification and quantification of bacteria from endodontic infections using realtime polymerase chain reaction. Oral Microbiol Immunol 23:384–390

Brehony C, Jolley KA, Maiden MCJ (2007) Multilocus sequence typing for global surveillance of meningococcal disease. FEMS Microbiol Rev 31:15–26

Bryant PA, Venter D, Robins-Browne R et al (2004) Chips with everything: DNA microarrays in infectious diseases. Lancet Infect Dis 4:100–111

Cassone M, Giordano A, Pozzi G (2007) Bacterial DNA microarrays for clinical microbiology: the early logarithmic phase. Front Biosci 12:2658–2669

Chang H-W, Sung Y, Kim K-H et al (2008) Development of microbial genome-probing microarrays using digital multiple displacement amplification of uncultivated microbial single cells. Environ Sci Technol Environ 42:6058–6064

Cole ST, Brosch R, Parkhill J et al (1998) Deciphering the biology of *Mycobacterium tuberculosis* from the complete genome sequence. Nature 393:537–544

Daims H, Ramsing NB, Schleifer KH et al (2001) Cultivationindependent, semiautomatic determination of absolute bacterial cell numbers in environmental samples by fluorescence in situ hybridization. Appl Environ Microbiol 67:5810–5818

Deepak SA, Kottapalli KR, Rakwal R et al (2007) Real-Time PCR: Revolutionizing detection and expression analysis of genes. Current Genom 8:234–251

Devulder G, de Montclos MP, Flandrois JP (2005) A multigene approach to phylogenetic analysis using the genus *Mycobacterium* as a model. Intern J Syst Evol Microbiol 55:293–302

Deyde VM, Okomo-Adhiambo M, Sheu TG et al (2009) Pyrosequencing as a tool to detect molecular markers of resistance to neuraminidase inhibitors in seasonal influenza A viruses. Antiviral Res 81:16–24

Dutka-Malen S, Evers S, Courvalin P (1995) Detection of glycopeptide resistance genotypes and identification to the species level of clinically relevant enterococci by PCR. J Clin Microbiol 33(1):24–27

Eckburg PB, Bik EM, Bernstein C et al (2005) Microbiology: diversity of the human intestinal microbial flora. Science 308(5728):1635–1638

Espy MJ, Uhl JR, Sloan LM et al (2006) Real-Time PCR in clinical microbiology: applications for routine laboratory testing. Clin Microbiol Rev 19(1):165–256

Fleischmann RD, Adams MD, White O et al (1995) Whole-genome random sequencing and assembly of *Haemophilus influenzae* Rd. Science 269:496–512

Fournier P-E, Drancourt M, Raoult D (2007) Bacterial genome sequencing and its use in infectious diseases. Lancet Infect Dis 7:711–723
Fredricks DN, Schubert MM, Myerson D (2005) Molecular identification of an invasive gingival bacterial community. Clin Infect Dis 41:e1–e4
Garcia CA, Ahmadian A, Gharizadeh B et al (2000) Mutation detection by pyrosequencing: sequencing of exons 5 - 8 of the p53 tumor suppressor gene. Gene 253:249–257
Gharizadeh B, Ohlin A, Molling P et al (2003) Multiple group-specific sequencing primers for reliable and rapid DNA sequencing. Mol Cell Probes 17:203–210
Gupta PK. (2008) Single-molecule DNA sequencing technologies for future genomics research. Trends Biotech 26(11):602–611
Hall N (2007) Advanced sequencing technologies and their wider impact in microbiology. J Exp Biol 210:1518–1525
Handelsman J, Rondon MR, Brady SF et al (1998) Molecular biological access to the chemistry of unknown soil microbes: a new frontier for natural products. Chem Biol 5:R245–R249
Harris JK, De Groote MA, Sagel SD et al (2007) Molecular identification of bacteria in bronchoalveolar lavage fluid from children with cystic fibrosis. Microbiol 104(51):20529–20533
Heid CA, Stevens J, Livak KJ et al (1996) Real time quantitative PCR. Genome Res 6:986–994
Hillemann D, Weizenegger M, Kubica T et al (2005) Use of the genotype MTBDR assay for rapid detection of rifampin and isoniazid resistance in *Mycobacterium tuberculosis* complex isolates. J Clin Microbiol 43:3699–3703
Hogardt M, Trebesius K, Geiger AM et al (2000) Specific and rapid detection by fluorescent in situ hybridization of bacteria in clinical samples obtained from cystic fibrosis patients. J Clin Microbiol 38:818–825
Hudson ME (2007) Sequencing breakthroughs for genomic ecology and evolutionary biology. Mol Ecol Res 8(1):3–17
Huson DH, Auch AF, Qi J et al (2007) MEGAN analysis of metagenomic data. Genome Res 17:377–386
Jelsbak L, Johansen HK, Frost AL et al (2007) Molecular epidemiology and dynamics of *Pseudomonas aeruginosa* population in cystic fibrosis lungs. Infect Immun 75:2214–2224
Kaushik DK, Sehgal D (2008) Developing antibacterial vaccines in genomics and proteomics era. Scand J Immunol 67(3):245–252
Kempf VA, Trebesius K, Autenrieth IB (2000) Fluorescent in situ hybridization allows rapid identification of microorganisms in blood cultures. J Clin Microbiol. 38:830–838
Klenk HP, Clayton RA, Tomb JF et al (1997) The complete genome sequence of the hyperthermophilic, sulphate-reducing archaeon *Archaeoglobus fulgidus*. Nature 390:364–370
Korczak B, Frey J, Schrenzel J et al (2005) Use of diagnostic microarrays for determination of virulence gene patterns of *Escherichia coli* K1, a major cause of neonatal meningitis. J Clin Microbiol 43:1024–1031
Koreen L, Ramaswamy SV, Graviss EA et al (2004) *Spa* typing method for discriminating among *Staphylococcus aureus* isolates: implications for use of a single marker to detect genetic micro- and macrovariation. J Clin Microbiol 42:792–799
Krogfelt KA, Lehours P, Megraud F (2005) Diagnosis of *Helicobacter pylori* infection. Helicobacter 10(Suppl1):5–13
Lane DJ, Pace B, Olsen GJ et al (1985) Rapid determination of 16S ribosomal RNA sequences for phylogenetic analyses. Proc Natl Acad Sci USA 82:6955–6959
Lasken RS (2007) Single-cell genomic sequencing using multiple displacement amplification. Curr Opin Microbiol 10(5):510–516
Manichanh C, Chapple CE, Frangeul L et al (2008) A comparison of random sequence reads versus 16S rDNA sequences for estimating the biodiversity of a metagenomic library. Nucleic Acids Res 36(16):5180–5188
Margulies M, Egholm M, Altman WE et al (2005) Genome sequencing in microfabricated high-density picolitre reactors. Nature 437:376–380

Mignard S, Flandrois J-P (2008) A seven-gene, multilocus, genus-wide approach to the phylogeny of mycobacteria using supertrees. Intern J System Evol Microbiol 58:1432–1441

Monstein HJ, Ellnebo-Svedlund K (2002) Molecular typing of *Helicobacter pylori* by virulence-gene based multiplex PCR and RT-PCR analysis. Helicobacter 7:287–296

Morales G, Wiehlmann L, Gudowius P et al (2004) Structure of *Pseudomonas aeruginosa* populations analyzed by single nucleotide polymorphism and pulsed-field gel electrophoresis genotyping. J Bacteriol 186:4228–4237

Moter A, Gobel UB (2000) Fluorescence in situ hybridization (FISH) for direct visualization of microorganisms. J Microbiol Methods 41:85–112

Neonakis IK, Gitti Z, Krambovitis E et al (2008) Molecular diagnostic tools in mycobacteriology. J Microbiol Methods 75(1):1–11

O'Sullivan DJ (2000) Methods for analysis of the intestinal microflora. Curr Iss Intest Microbiol 1:39–50

Ota N, Hirano K, Warashina M et al (1998) Determination of interactions between structured nucleic acids by fluorescence resonance energy transfer (FRET): selection of target sites for functional nucleic acids. Nucleic Acids Res 26:735–743

Palladino S, Kay ID, Costa AM et al (2003) Real-time PCR for the rapid detection of *van*A and *van*B genes. Diagn Microbiol Infect Dis 45:81–84

Palomino JC (2005) Nonconventional and new methods in the diagnosis of tuberculosis: feasibility and applicability in the field. Eur Respir J 26:339–350

Peters RP, Savelkoul PH, Simoons-Smit AM et al (2006) Faster identification of pathogens in positive blood cultures by fluorescence in situ hybridization in routine practice. J Clin Microbiol 44:119–123

Poppert S, Essig A, Marre R et al (2002) Detection and differentiation of *chlamydiae* by fluorescence in situ hybridization. Appl Environ Microbiol 68:4081–4089

Poppert S, Essig A, Stoehr B et al (2005) Rapid diagnosis of bacterial meningitis by real-time PCR and fluorescence in situ hybridization. J Clin Microbiol 43:3390–3397

Poulsen LK, Ballard G, Stahl DA (1993) Use of rRNA fluorescence in situ hybridization for measuring the activity of single cells in young and established biofilms. Appl Environ Microbiol 59:1354–1360

Rappuoli R (2004) From Pasteur to genomics: progress and challenges in infectious diseases. Nat Med 10(11):1177–1185

Rhee M, Burns MA (2006) Nanopore sequencing technology: research trends and applications. Trends Biotechnol 24:580–586

Riesenfeld CS, Schloss PD, Handelsman J (2004) METAGENOMICS: genomic analysis of microbial communities. Annu Rev Genet 38:525–552

Riggio MP, Lennon A, Rolph HJ et al (2008) Molecular identification of bacteria on the tongue dorsum of subjects with and without halitosis. Oral Dis 14(3):251–258

Ronaghi M, Elahi E (2002) Pyrosequencing for microbial typing. J Chromatog B 782(1–2):67–72

Ryan D, Rahimi M, Lund J et al (2007) Towards nanoscale genome sequencing. Trends Biotechnol 25:385–389

Sachse K, Hotzel H, Slickers P et al (2005) DNA microarraybased detection and identification of *Chlamydia* and *Chlamyophila* spp. Mol Cell Probes 19:41–50

Sanger F, Air GM, Barrell BG et al (1977) Nucleotide sequence of bacteriophage phi X174 DNA. Nature 265:687–695

Sharma P, Kumari H, Kumar M et al (2008) From bacterial genomics to metagenomics: concept, tools and recent advances. Indian J Microbiol 48(2):173–194

Shendure J, Porreca GJ, Reppas NB et al (2005) Accurate multiplex polony sequencing of an evolved bacterial genome. Science 309:1728–1732

Shopsin B, Gomez M, Montgomery SO et al (1999) Evaluation of protein A gene polymorphic region DNA sequencing for typing of *Staphylococcus aureus* strains. J Clin Microbiol 37:3556–3563

Sidarenka AV, Novik GI, Akimov VN (2008) Application of molecular methods to classification and identification of bacteria of the genus *Bifidobacterium*. Microbiol 77(3):251–260

Singh A, Goering RV, Simjee S et al (2006) Application of molecular techniques to the study of hospital infection. Clin Microbiol Rev 19(3):512–530

Sogaard M, Stender H, Schonheyder HC (2005) Direct identification of major blood culture pathogens, including *Pseudomonas aeruginosa* and *Escherichia coli*, by a panel of fluorescence in situ hybridization assays using peptide nucleic acid probes. J Clin Microbiol 43:1947–1949

Stabler RA, Marsden GL, Witney AA et al (2005) Identification of pathogenspecific genes through microarray analysis of pathogenic and commensal *Neisseria* species. Microbiol 151:2907–2922

Stampone L, Del Grosso M, Boccia D et al (2005) Clonal spread of a vancomycin-resistant *Enterococcus faecium* strain among bloodstream-infecting isolates in Italy. J Clin Microbiol 43:1575–1580

Stavrum R, Valvatne H, Bø TH et al (2008) Genomic diversity among Beijing and non-Beijing *Mycobacterium tuberculosis* isolates from Myanmar. PLoS One 3(4):e1973

Sunde PT, Olsen I, Gobel UB et al (2003) Fluorescence in situ hybridization (FISH) for direct visualization of bacteria in periapical lesions of asymptomatic root-filled teeth. Microbiol 149:1095–1102

Thurnheer T, Gmur R, Guggenheim B. (2004) Multiplex FISH analysis of a sixspecies bacterial biofilm. J Microbiol Methods 56:37–47

Tortoli E (2003) Impact of genotypic studies on mycobacterial taxonomy: the new mycobacteria of the 1990s. Clin Microbiol Rev 16(2):319–354

Tyagi S, Kramer FR (1996) Molecular beacons: probes that fluoresce upon hybridization. Nat Biotechnol 14:303–308

Uhl JR, Adamson SC, Vetter EA et al (2003) Comparison of LightCycler PCR, rapid antigen immunoassay, and culture for detection of group A streptococci from throat swabs. J Clin Microbiol 41:242–249

Urwin R, Maiden MC (2003) Multi-locus sequence typing: a tool for global epidemiology. Trends Microbiol 11:479–487

Valasek MA, Repa JJ. (2005) The power of real-time PCR. Adv Physiol Educ 29:151–159

Wagner M, Horn M, Daims H (2003) Fluorescence in situ hybridisation for the identification and characterisation of prokaryotes. Curr Opin Microbiol 6:302–309

Warren DK, Liao RS, Merz LR et al (2004) Detection of methicillin-resistant *Staphylococcus aureus* directly from nasal swab specimens by a real-time PCR assay. J Clin Microbiol 42:5578–5581

Wecke J, Kersten T, Madela K et al (2000) A novel technique for monitoring the development of bacterial biofilms in human periodontal pockets. FEMS Microbiol Lett 191:95–101

Wellinghausen N, Wirths B, Poppert S (2006) Fluorescence in situ hybridization for rapid identification of *Achromobacter xylosoxidans* and *Alcaligenes faecalis* recovered from cystic fibrosis patients. J Clin Microbiol 44:3415–3417

Woese CR (1987) Bacterial evolution. Microbiol Rev 51:221–271

Wommack KE, Bhavsar J, Ravel J (2008) Metagenomics: read length matters. Appl Environ Microbiol 74(5):1453–1463

Woodford N, Johnson AP (2004) Genomics, proteomics, and clinical bacteriology. Humana Press, Totowa, NJ

Yang J-R, Wu F-T, Tsai J-L et al (2007) Comparison between O serotyping method and multiplex real-time PCR to identify diarrheagenic *Escherichia coli* in Taiwan. J Clin Microbiol 45(11):3620–3625

Yang S, Rothman RE (2004) PCR-based diagnostics for infectious diseases: uses, limitations, and future applications in acute-care settings. Lancet Infect Dis 4:337–348

Chapter 5
Systems Microbiology: Gaining Insights in Transcriptional Networks

Riet De Smet[1], Karen Lemmens[1], Ana Carolina Fierro, and Kathleen Marchal

5.1 Systems Microbiology: Introduction

During the past decades, a tremendous evolution in molecular techniques enabled the measurement of the different biological components and their interactions on a genome-wide scale, giving rise to genome-wide data sets. With the advent of these "omics" data, molecular biology has evolved from a rather data-poor to an extremely data-rich research area. Whereas traditional molecular biology focused on the study of individual genes or small sets of genes, systems biology studies the organism at a more global level. A systems biology approach aims at understanding the mechanisms of signal transduction and molecular interactions that give rise to the observed behavior, i.e., understanding the underlying regulatory network (Kitano 2002). Two different approaches for inferring regulatory networks can be distinguished. *Top-down* network inference methodologies reconstruct the regulatory networks by mining and integrating complementary omics data. They usually start from scratch and do not require expert knowledge about the relationships between the molecular components to infer a network. *Bottom-up* network inference on the other hand starts from an expert model of known interactions between molecular entities as described in literature and curated databases. These models are subsequently used to simulate cellular behavior or to predict the outcome of a perturbation experiment. Inconsistencies between observed data and simulations point at deficiencies in the current network structure and outline hypotheses of novel interactions that can better explain the observations (Bruggeman and Westerhoff 2007; De Keersmaecker et al. 2006).

K. Marchal (✉)
Department of Microbial and Molecular Systems, K.U. Leuven, Belgium

[1] These authors contributed equaly to this work.

V. Sintchenko (ed.), *Infectious Disease Informatics*,
DOI 10.1007/978-1-4419-1327-2_5,

This chapter zooms in on the top-down reconstruction of a particular part of the regulatory network, i.e., the transcriptional regulatory network (TRN). This TRN can be represented as a graph in which the nodes are the genes, and the directed edges (edges with a defined direction) indicate that the first gene codes for a transcription factor that regulates the second gene (Fig. 5.1). This TRN is highly condition-dependent: some regulator-gene interactions might be present in some experimental conditions, but absent in others (Luscombe et al. 2004; Van den Bulcke et al. 2006a). Although the TRN forms only a fraction of the total regulatory system, it represents a major level of regulation in prokaryotes: It allows bacteria to alter their gene expression and to adapt to novel environmental conditions. For prokaryotes, efforts to collect all available information on experimentally verified regulator–target interactions resulted in the development of databases such as RegulonDB (Gama-Castro et al. 2008), Ecocyc (Keseler et al. 2009), or EcoliHub (http://www.ecolicommunity.org/) for *Escherichia coli* or DBTBS (Sierro et al. 2008) for *Bacillus subtilis*.

Bacteria have been studied as model organisms in molecular biology for decades, and they are also currently emerging as model organisms for systems biology. In this chapter, we therefore focus on the inference of the bacterial TRN. We will start by describing genome-wide data sources related to the TRN and subsequently discuss methods for the analysis of these "omics" data in prokaryotes. Finally, we will describe how systems biology can contribute to revealing novel drug targets in bacteria.

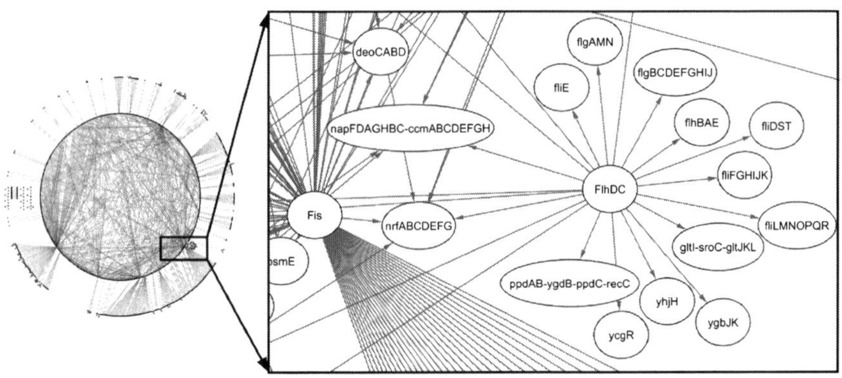

Fig. 5.1 The transcriptional network of *E. coli*. The transcriptional network of *E. coli* is derived from all experimentally verified interactions available in RegulonDB (version 6.2). There are 824 nodes and 1,334 edges present in this network. The nodes consist of the operons and the regulators regulating these operons. An edge between a regulator and operon indicates that the operon is regulated by the regulator. Regulator FlhDC, for instance, regulates several operons involved in flagella biogenesis, such as *fliFGHIJK* but also non-flagella operons, such as the glutamate ABC transporter *gltI-sroC-gltJKL*. This type of network representation ignores the condition dependency of the interactions, i.e., all interactions are shown although it may happen that two interactions never "appear" in the same experimental condition

5.2 High–Throughput Data Sources

In this section we describe "omics" data that give direct information on the TRN and that are used by current methods for network reconstruction.

5.2.1 Expression Data

Microarrays have become the main technology for large scale gene expression profiling: By quantitatively measuring mRNA molecules, they give a snapshot of the level and condition dependency of the transcriptional activity.

Different microarray platforms exist, including Affymetrix, Agilent, Codelink, or in-house microarrays [see Sasik et al. (2004) for a review]. Each different platform requires its own optimized sample preparation, labeling, hybridization, and scanning protocol, and concomitantly also a specific normalization procedure. Normalization of the raw, extracted intensities aims to remove consistent and systematic sources of variation to ensure comparability of the measurements, both within and across arrays.

Microarray experiments are made publicly available in specialized databases such as the Gene Expression Omnibus (Barrett et al. 2007), Stanford microarray database (Demeter et al. 2007), or ArrayExpress (Parkinson et al. 2007). To ensure exchangeability of these data, data submitted to these databases should be compliant with he "Minimum Information About a Microarray Experiment (MIAME)" standard (Brazma et al. 2001). The MIAME standard enforces a careful description of the conditions under which the microarray experiment was performed, such as the genetic background of the used strains, the used media, growth conditions, triggering factors, etc. However, it does not specify the format in which this meta-information should be presented. As a result, extracting data and information from these public microarray databases remains tedious and relies largely on manual curation: information is not only stored in different formats and data models, but is also redundant, incomplete, and/or inconsistent. To fully exploit the large resource of information offered by these public databases, species-specific compendia that combine all of the experiments on one particular organism in a semi-automated process are constructed. Single-platform compendia combine all data on a particular organism that were obtained from one specific platform. Most single-platform compendia focus on Affymetrix data as this is considered one of the more robust and reproducible platforms (Irizarry et al. 2005; Bammler et al. 2005). The Many Microbe Microarrays Database (M3D) (Faith et al. 2007), for instance, offers Affy-based compendia for three microbial organisms. Cross-platform compendia, on the other hand, include data from different platforms and require more specialized normalization procedures to combine data from both one and two channel microarrays (Lemmens et al. 2009).

5.2.2 Regulator-Target Interaction Data

In addition to expression data, data on the interaction between a regulator and its target, like regulatory motif data or ChIP-chip data, also provide important information on the transcriptional regulation of genes. Regulatory motifs are short, conserved DNA-sequences present in the promoter region of a gene. They are the tags that are recognized by the regulators and hence, play a very important role in the TRN. Although the binding of a regulator to a regulatory motif is condition dependent, the motif itself is inherently present in the DNA sequence. Therefore, motifs of an organism can be identified independent of the experimental conditions. Specialized databases such as TRANSFAC (Matys et al. 2006), RegulonDB (Gama-Castro et al. 2008), DBTBS (Sierro et al. 2008), ProDoric (Grote et al. 2009), or TractorDB (Perez et al. 2007) contain information on the regulatory motifs of diverse microorganisms.

Because the empirical validation of binding sites is laborious, computational methods have been developed to identify regulatory motifs. The de novo motif identification methods aim at identifying regulatory motifs from scratch, without any prior knowledge about the motif structure. These methods search for sequence tags that are statistically overrepresented in a set of co-regulated genes as compared to a set of unrelated genes. Their search methods are based on word counting or make use of advanced statistical procedures (Tompa et al. 2005). Alternatively, motif models can be compiled on the basis of lists of experimentally verified binding sites. Motif models, resulting from de novo or supervised motif detection can subsequently be used to perform a genome-wide screen of all intergenic regions of the organism. As such, additional genes that contain the motif in their promoter region can be identified, and thus additional target genes of the corresponding regulator can be indirectly discovered (Hertzberg et al. 2005; Marchal et al. 2004).

In addition to this indirect information about a regulator target interaction, direct physical interactions can be identified by chromatin immunoprecipitation (or ChIP). In the ChIP method, a regulator that is bound to the DNA will be fixated on the DNA. The DNA parts to which the regulator was bound can then be identified, for instance with qPCR. In genome-wide protocols, the qPCR step is replaced by a hybridization step to a microarray (ChIP on chip or ChIP-chip method) or more recently by massive parallel sequencing (Solexa) (Ren et al. 2000; Orlando 2000; Laub et al. 2002). The binding of a particular regulator in a specific environmental condition to the promoter region of a gene can as such be investigated on a genome-wide scale in one single experiment. In *E. coli*, for instance, the regulatory interactions of CRP, FNR, MelR, Lrp, and the nucleoid associated transcription factors IHF, Fis, and H-NS were investigated (Grainger et al. 2005; Grainger et al. 2006, 2007; Cho et al. 2008a, b). ChIP-chip data are also available for *Caulobacter cresentus* (Laub et al. 2002), *B. subtilis* (Molle et al. 2003a, b; Ben Yehuda et al. 2005) and *Salmonella typhimurium* (Thijs et al. 2007; Lucchini et al. 2006; Navarre et al. 2006).

ChIP-chip data are condition dependent: only under the appropriate conditions a regulator will bind to the promoter region of its target gene, for instance when the regulator is available and active and the promoter region of the target gene is accessible. The lack of a physical interaction between a regulator and a target gene under a particular condition therefore does not exclude the possibility that this interaction exists under a different set of conditions. On the other hand, the physical interaction of a regulator with its target gene does not necessarily imply that this regulator will also regulate the expression of the gene in this condition. In the case of combinatorial control, for instance, an additional regulator might be required. Alternatively, the presence of a particular ligand or metabolite may be needed for the activation of the regulator. For these biological reasons and the fact that ChIP-chip is a high-throughput technology, ChIP-chip data might contain both false negatives and false positives.

5.3 Reconstruction of Transcriptional Networks

5.3.1 Reconstructing from "Omics" Data

A major challenge of top-down systems biology is to reconstruct the underlying TRN that explain these heterogeneous data. Here we focus on computational strategies that have been developed to this end. Because of specific data properties and the intricacies of the network structure, network reconstruction is not a trivial task. Ideally, inference methods should not only reproduce the current knowledge on the TRN, but should also provide new high confidence testable hypotheses about the system under study.

The first challenge of inferring networks is the fact that current high-throughput technologies have mainly highlighted one aspect of transcription regulation: data on interactions between transcription factors and their targets is primarily available in the shape of gene expression data and ChIP-chip experiments. However, whereas transcription factors constitute an important means of transcription regulation, several other factors such as chromatine structure (Blot et al. 2006), small non-coding RNAs (Waters and Storz 2009), and metabolites (Shi and Shi 2004) are known to influence gene expression. As the effects of most of these factors on transcription regulation have not been measured, they cannot be taken into account. This inevitably results in an oversimplification of the biological reality. Consequently, most existing computational approaches focus on the interaction between transcription factors and respective genes and the influence of the other regulatory influences mentioned above can only be conjectured from the data.

Even when only focusing on interactions between transcription factors and their target genes, the reverse-engineering problem remains notoriously complex. In particular, in a certain biological system we would like to infer the regulators that govern the expression changes within individual genes. For *E. coli*, for instance, 4,500 genes have to be linked to about 300 known and predicted regulators (Babu and Teichmann 2003). These regulators might act independently from one another or together to elicit a certain transcriptional response. Consequently, for each of the genes the theoretical number of combinations that need to be evaluated in terms of sets of transcription factors that might explain the gene's expression behavior is prohibitively large. In addition, the number of interactions that can be inferred exceeds the number of independent measurements for the activity of the transcription factors (samples). Therefore, the problem is underdetermined, i.e., different possible solutions exist that all explain the data equally well. Extracting the biologically true predictions from this large list of possible solutions is not trivial.

Furthermore, high-throughput data are characterized by a low signal-to-noise ratio, which further complicates the problem of network inference from these data. Yue et al. (2001), for example, reported a variation in expression of over 10% between non-differentially expressed genes on different microarrays. In addition to this measurement noise, variability in gene expression can occur due to the stochastic nature of gene regulation (Kaern et al. 2005). Hence, this natural source of variability further aggravates discovery of true underlying networks from popular high-throughput techniques, such as microarray and ChIP-chip experiments, which measure gene regulation across a whole populations. Each of the published methods deals with this problem of under-determination differently, often using knowledge on the likely layout of the TRN such as modularity, sparseness, and so on to shift the solution space to the most biologically relevant solutions.

5.3.2 Benchmarking Algorithms

A sense of reliability in the reconstructed networks can only arise from an understanding of the limitations of the algorithms. Getting a good insight into the behavior of an algorithm can only be obtained by benchmarking against a known network. For network inference, a positive golden standard is used, i.e., the collection of all known interactions in a particular organism (Stolovitzky et al. 2007), such as the well-curated *E. coli* network present in RegulonDB (Gama-Castro et al. 2008). By means of this network, algorithmic performance can be assessed using measures such as precision (i.e., the proportion of inferred interactions which is correct according to the positive golden standard) and recall (i.e., the proportion of interactions in the positive golden standard that could be inferred using a certain method). Moreover, measures such as precision and recall do not take into account the number of predictions made by a certain method that do involve interactions not present in RegulonDB and which therefore cannot be validated. Indeed, although positive "golden" standard provides us with information on the proportion of inferred true positive interactions between a regulator and its targets, there exists no negative "golden" standard (a set of curated interactions that can never occur) for TRNs. As a result, we can never assess to what extent novel predictions consist of as yet uncharacterized true interactions or false positive predictions.

The only proper validation strategy is to perform wet-lab experiments on a sufficiently large set of predicted interactions and predicted absent interactions to confirm or deny the presumed interaction. It is clear that such an approach is impractical, time-consuming, and sometimes unfeasible (Van den Bulcke et al. 2006a).

Because of these limitations of validation on standard networks and experimental data, in silico networks or simulated data are often used as a first step to verify algorithmic performance. Several efforts have been made to mimic experimental data from an in silico network which is as close to the biological truth as possible (Van den Bulcke et al. 2006b). Hence, in silico networks have the advantage that the underlying network is exactly known. However, even the most biologically inspired in silico networks cannot cover all of the intricacies of real experimental data. They may for instance not account for noise in experimental measurement, or for the multilayered aspect of gene regulation (Stolovitzky et al. 2007). Despite these limitations of simulated data they are still useful in unveiling some qualitative properties of the algorithm under test (e.g., noise robustness, sensitivity, optimality of the proposed solution) and in tuning the parameter settings to some extent (Van den Bulcke et al. 2006b).

While the different means of measuring algorithmic performance described here are useful in guiding computational biologists in improving certain algorithms, they are often of limited use to biologists searching for the most appropriate algorithm for their own research. Benchmarking results are often misleading as they are commonly biased towards a certain aspect in which one algorithm is better than the other and rarely give a global assessment. Most current performance measures give limited or no information on the conceptual differences between the compared methods. Therefore, we will focus on how conceptual differences between methods result in retrieving different but complementary aspects of the inferred networks and how they can influence our choice for a particular method.

5.3.3 Which Method to Choose for Network Reconstruction?

In Fig. 5.2 and Box 5.1 a selection of prominent network inference methods that have been applied to study prokaryotic TRNs is provided. Most of these methods have been benchmarked on *E. coli*. For a more detailed description of these methods and their applications in prokaryotes we refer to Box 5.1. In the following we outline different global concepts of transcriptional network inference methods. Each of the methods contrasted in Box 5.1 combines one or more of these concepts and can be subdivided accordingly.

An inference method can either use only microarray data to learn the TRN or can integrate several data sources (then it is called an integrative method). Stochastic LeMoNe (Joshi et al. 2009), CLR (Faith et al. 2007), Inferelator (Bonneau et al. 2006), and SIRENE (Mordelet and Vert 2008) are all methods that learn the network from microarray data, whereas DISTILLER (Lemmens et al. 2009), cMonkey (Reiss et al. 2006), de Hoon et al. (de Hoon et al. 2004), and SEREND (Ernst et al. 2008) are examples of integrative approaches.

Methods can be supervised or unsupervised. Supervised methods take advantage of known information to infer novel predictions while unsupervised methods do not and rely on the data only. Examples of supervised methods are SEREND, de Hoon et al. (2004), and SIRENE.

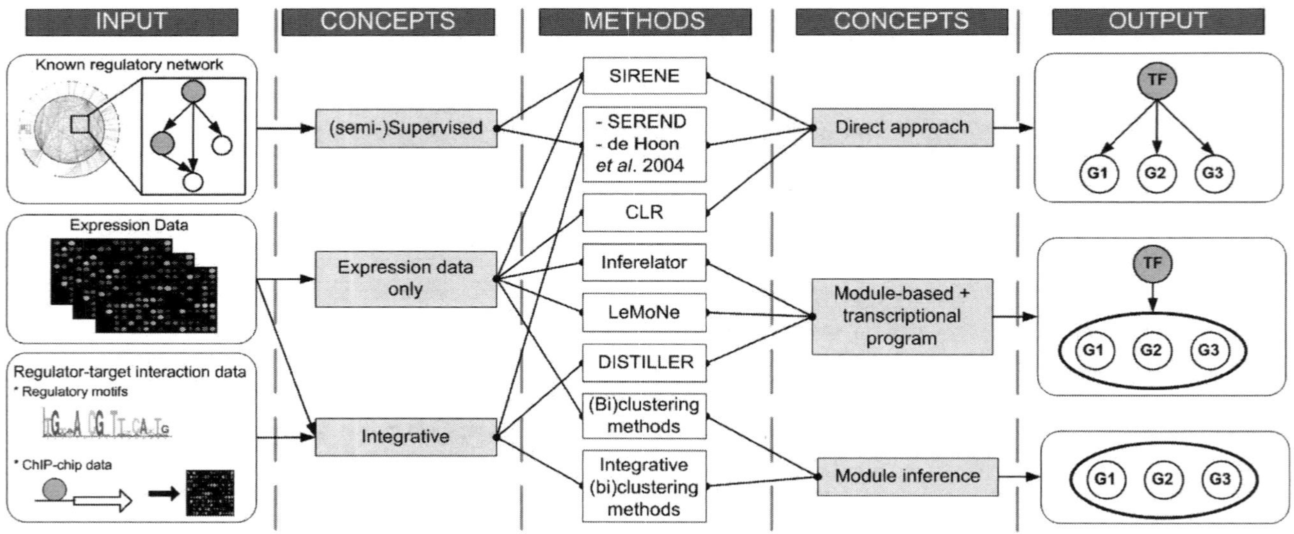

Fig. 5.2 Overview of the different methods for network reconstruction. Methods can be categorized according to the input data. Some methods make use of only microarray data (microarray data only), while others use additional information such as the known network ((semi)-supervised) or data on the interaction between a regulator and its target (integrative methods). Methods can also be divided according to the output they generate. Module inference methods will only detect sets of co-expressed genes or modules, whereas other network inference methods focus on the interactions between regulators and targets. Direct methods infer the regulatory program for each gene individually while module-based inference methods identify modules prior to identifying the regulatory program

Box 5.1 Overview of network inference methods

Name methods; data set analyzed	Description methods; results
CLR (Context Likelihood of Relatedness) (Faith et al. 2007); *E. coli* (445 arrays)	The CLR algorithm starts from a list of putative and known regulators and scores for each regulator. The interaction with all genes in the data set is based on similarity between expression profiles of the regulator and the potential target genes, assessed by mutual information scores. Statistical significance of the mutual information score for a certain regulator–target interaction is evaluated against a background distribution consisting of all mutual information scores for interactions involving that regulator or target gene. CLR was applied to an *E. coli* gene expression compendium and a list of 328 known and putative regulators. At a 60% precision (with respect to the known transcriptional regulatory network in RegulonDB) CLR identified 1,079 interactions of which 338 were known and 741 novel. Of the 741 predicted interactions, 21 novel interactions for three different regulators could be confirmed by ChIP. Furthermore, using qPCR a regulatory link between central metabolism and iron transport could be confirmed.
SIRENE (Supervised Inference of Regulatory Networks) (Mordelet and Vert **2008**); *E. coli* (445 arrays)	SIRENE takes advantage of the known network in RegulonDB to tackle the network inference problem in a supervised setting. The method uses co-expression between targets of the same regulator as induction principle. Hence, gene expression data are used to construct for each regulator a binary classifier, which discriminates between genes known to be regulated by the transcription factor and genes known not to be regulated by the transcription factor. On the basis of expression data, this regulator-specific classifier can then be used to predict regulation of the remaining genes, i.e. genes not used for classifier construction. Specifically, the output consists of a list of ranked genes with higher scores reflecting a higher probability of regulation by the transcription factor according to the constructed classifier. SIRENE was applied on the same data set as CLR and was shown to predict six times more known regulations at a 60% precision level than CLR.

(continued)

Box 5.1 (continued)

Name methods; data set analyzed	Description methods; results
Stochastic LeMoNe (Joshi et al. 2009a); Yeast (Joshi et al. 2009); *E. coli* (445 arrays) (Michoel et al. 2009)	Stochastic LeMoNe combines a clustering approach with the inference of a regulatory program in order to infer the regulatory network from gene expression data. A probabilistic clustering method is used to partition the genes in modules of co-expressed genes. This clustering approach does not only assign genes to clusters but also partitions the conditions within one cluster in such a way that the cluster genes within one condition partition are either all up-or down-regulated. The Stochastic LeMoNe algorithm uses the condition partitions obtained in the clustering approach to predict the regulatory programs for each module. A list of known and putative regulators is taken as input and regulators for which the expression profile best explains the obtained condition partitions are assigned to the modules. Regulators can be assigned to just some of the condition partitions, explaining the module genes' expression behavior under only those conditions, or to all conditions in the data set. Application of Stochastic LeMoNe to an *E. coli* gene expression compendium (Michoel et al. 2009), using a list of 316 known and putative regulators as input, resulted in 53 regulators assigned to 62 modules, comprising 1,079 predicted interactions. Five hundred and ninety four of these interactions are present in RegulonDB (29% precision).
Inferelator (+cMonkey) (Bonneau et al. 2006; Reiss et al. 2006); *Halobacterium salinarum* (268 arrays)	Inferelator starts from a set of biclusters. These biclusters can be obtained from cMonkey, which employs an integrative biclustering approach; sets of condition dependent co-regulated genes are identified by using gene expression data, co-occurrence of *cis*-acting motifs and the presence of highly connected subgraphs in metabolic and functional association networks. Subsequently, Inferelator uses regression-like techniques to assign regulators to the obtained modules; the regulator of which the expression profile best explains the average gene expression profile of the module genes gets assigned to the module. Inferelator explicitly models time experiments, taking advantage of its inherent ability to learn causal relations. The method does not only search for regulators that explain the expression behavior of the

	genes, but also for environmental factors that might underlie changes in gene expression. The model accounts for simple interactions between a maximum of two regulators or environmental factors in the shape of AND, OR, or XOR relationships. The inferelator method was applied to a gene expression compendium for the archeon *Halobacterium salinarun*. Using this method they predicted 80 transcription factors for 500 genes and also predicted some of the metabolites controlling several pathways through the usage of environmental factors. ChIP-chip experiments and knock-out experiments were performed to illustrate that the obtained network can serve as a reliable blueprint for the *Halobacterium salinarun* transcriptional regulatory network.
de Hoon et al. (de Hoon et al. 2004); *B. subtilis* (174 arrays)	This is a supervised approach which assigns targets to regulators based on similarity in expression and motifs with known targets of the regulator. For each transcription factor separate classifiers for the gene expression data and motif data are built taking advantage of knowledge on known targets. The two classifiers are properly balanced and combined into one discriminatory classifier that can be used to classify genes with unknown regulation. The method was applied to a compendium of *B. subtilis* gene expression data and known motifs for several *B. subtilis* sigma factors in order to predict new targets for those sigma factors.
SEREND (SEmi-supervised Regulatory Network Discoverer) (Ernst et al. 2008); *E. coli* (445 arrays)	SEREND is a (semi-)supervised approach which incorporates gene expression data with motif data. It essentially follows the same approach taken by de Hoon et al. (2004) in that it constructs separate classifiers for each data source and subsequently combines these classifiers into one meta-classifier. SEREND not only predicts regulation by a certain transcription factor, but also predicts the sign of the interaction (activator, repressor, or dual regulator). SEREND was applied to an *E. coli* gene expression compendium taking all interactions in EcoCyc, a position weight matrix and the gene expression data as input. Validation of new predictions was performed by comparing predicted targets for several global regulators against the targets predicted by ChIP-chip experiments.

(continued)

Box 5.1 (continued)

Name methods; data set analyzed	Description methods; results
DISTILLER (Data Integration System To Identify Links in Expression Regulation) (Lemmens et al. 2009); *E. coli* (870 arrays)	DISTILLER is a deterministic integrative approach that searches for modules. DISTILLER uses an efficient search strategy derived from itemset mining approaches to exhaustively search for all possible solutions that correspond to predefined criteria. This approach is used to search for modules of condition dependent-co-expressed genes that share the same motif instances. Since the method outputs all results corresponding to the predefined search-criteria, the resulting set of modules is often very large and partially redundant. Therefore DISTILLER combines the efficient search strategy with a statistical method to score the significance of the obtained modules, filtering the most relevant and non-redundant solutions.
	DISTILLER was applied to an *E. coli* gene expression compendium and a weight matrix for 67 known regulators in *E. coli* from RegulonDB in order to study the condition dependent combinatorial regulation in *E. coli* (Lemmens et al. 2009). Seven hundred and thirty two interactions were identified of which 454 could be confirmed by RegulonDB. Additionally, 11 novel interactions for the regulator FNR were tested and could be experimentally verified by ChIP-qPCR.

Learning about the TRN can either focus on the detection of modules (module inference) or on the detection of a regulatory program (regulatory program inference). Module inference methods search for sets of co-expressed genes that are assumed to be under the influence of the same regulatory mechanism. On the other hand, methods for regulatory program inference aim to assign regulators or sets of regulators to their corresponding target genes, thus focusing on the interactions in the network.

Methods that infer regulatory programs can be subdivided into those that infer the program for each gene individually ("direct" network inference methods) and those that perform a module inference step prior to or together with the assignment of the regulatory program. In the latter case, one program is assigned to a complete module at once ("module"-based network inference). CLR and the supervised methods SIRENE, SEREND, and de Hoon et al. (2004) are examples of direct network inference methods. This in contrast to Stochastic LeMoNe, DISTILLER, and Inferelator, which are module-based.

In what follows, we will discuss how these conceptual differences influence the results generated by the different approaches for network inference.

5.3.4 Module Inference: Learning About Co-Expressed Targets

An important property of the TRN is its modularity: the network consists of overlapping modules of functionally related genes that all act in concert under certain environmental cues (Hartwell et al. 1999; Qi and Ge 2006). Module inference methods are useful on their own as they tell us which genes are co-expressed. They can also be an integral part of the module-based network inference methods (see below).

5.3.4.1 From Clustering to Biclustering

To learn about these modules, clustering or biclustering can be used. Clustering methods were among the pioneering methods to mine microarray data (Quackenbush 2001). As clustering approaches find sets of genes that are co-expressed under all conditions of the microarray data set, they are ideal for finding patterns of coexpression in small microarray data sets consisting of similar conditions. However, clustering methods generally ignore the condition dependency of regulatory programs and thus the fact that target genes will only be coexpressed under the conditions in which the regulatory program is active. This is particularly problematic when clustering large heterogeneous gene expression data sets: the presence of conditions in the data set under which the regulatory program is not active will reduce the signal-to-noise level of the data and complicate identifying sets of coexpressed genes. Biclustering methods (Cheng and Church 2000; Lazzeroni and Owen 2002; Murali and Kasif 2003; Getz et al. 2000; Sheng et al. 2003; Bergmann et al. 2003; Dhollander et al. 2007; Madeira and Oliveira 2004; Tanay et al. 2002) deal with this shortcoming of clustering methods by combining a search for coexpressed genes with a condition selection step to identify the conditions under which the genes are coexpressed, i.e., the conditions in which the joint regulatory program of the bicluster genes is active.

Because of the condition dependent nature of the transcriptional regulation, a gene can, depending on the conditions, belong to different pathways and thus modules. Most biclustering algorithms are therefore able to identify overlapping modules, for instance by using different initializations of the algorithm (Ihmels et al. 2004) or an efficient search strategy (Lemmens et al. 2009).

Identifying biclusters in a gene expression data set is a significant challenge, as it is computationally prohibitive to evaluate all possible gene and condition combinations that can co-occur in a bicluster. Different strategies have been developed to limit the search space or the number of combinations, and as a result different methods will produce different outcomes for the same data set (Madeira and Oliveira 2004). For instance, different methods produce constraints on the size of the biclusters that can be obtained, the tightness of co-expression of the bicluster genes, and the optimization strategy they follow. Below we give two different approaches of reducing the search space. In the query-based approach one restricts the biological question to be solved, while in the integrative approach data sources other than microarrays are used to restrict the number of possible combinations.

5.3.4.2 Global vs. Query-Driven Biclustering

Two different ways of approaching the biclustering problem to solve different biological questions, global and query-driven biclustering, can be highlighted. Global biclustering methods (Cheng and Church 2000; Lazzeroni and Owen 2002; Murali and Kasif 2003; Sheng et al. 2003; Tanay et al. 2002) identify the more dominant patterns in the data set: the optimization problem is formulated to find those patterns that explain most of the data. Such an optimization criterion usually results in large modules of co-expressed genes. These modules seem to biologically correspond to large pathways, involving many genes and responsible for general responses (e.g., modules involved in flagella synthesis, amino acids biosynthesis or DNA damage in *E. coli*). They provide the scientist with a view on the general response of the active TRN. However, these methods often do not identify the more subtle patterns of co-expression, involving only a few genes or conditions that might be of a particular interest to a certain scientist. Query-driven biclustering algorithms (Dhollander et al. 2007; Ihmels et al. 2004; Hibbs et al. 2007) are more appropriate for finding those subtle patterns. They search for genes that are co-expressed in a condition dependent way with a set of genes that are of interest to a certain researcher (also called query-genes). Such a query-driven approach can hence be used to identify additional genes belonging to the same pathway as a set of query-genes. A possible practical application of the query-driven approach is the validation of in vitro experiments, such as ChIP-chip experiments (Fig. 5.3). Finally, as biclustering includes a condition selection, it can hint towards the conditions under which the query-genes are co-expressed and thus the regulator is being transcriptionally active.

Integrative Biclustering: From Co-Expression Towards Co-Regulation

A second way of reducing the number of possible solutions is by complementing the gene expression data with other data, thereby restricting only the solutions to biclusters that correspond to the different data sources. An advantage of this method is that spurious bicluster assignments due to noise can be effectively filtered out, because co-expression between genes will need to be at least partially confirmed by other data sources. Indeed, while genes within the same bicluster are often assumed to be subject to the same set of regulators this is not always the case: a random gene might show a correlated expression with other bicluster genes simply by chance (e.g., due to measurement noise) or genes might only be co-expressed because of indirect effects (Reiss et al. 2006). In this respect, combining the microarray data with a complementary data source on the TRN, such as ChIP-chip

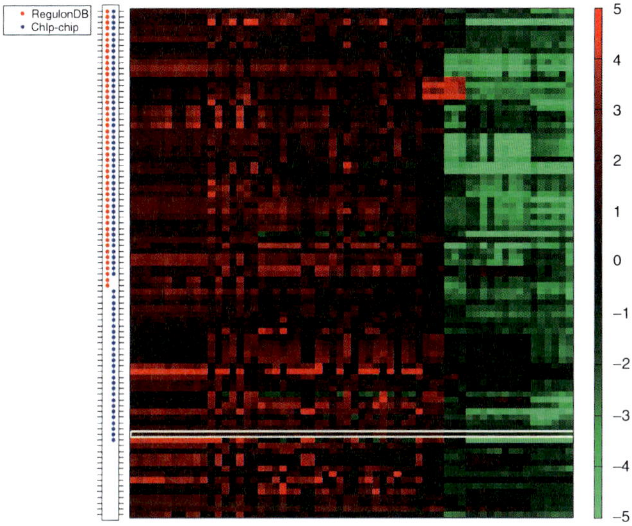

Fig. 5.3 Application of query-driven biclustering to verify predictions made by ChIP-chip experiments. As already mentioned above, ChIP-chip experiments can identify physically interacting, but transcriptional inactive regulator target pairs (Gao et al. 2004). To filter those pairs from the transcriptional active ones, query-driven biclustering can be used to verify which of the identified targets are co-expressed with other known targets of the tested regulator. Those targets are transcriptional active and consequently true positive targets. Moreover, the algorithm can also suggest potential targets that were missed by the ChIP-chip experiment due to the condition-specific nature of this methodology (false negatives). In this figure a heatmap is shown obtained by query-driven biclustering of *pykA* (indicated in the *white rectangle*), a gene shown to be bound by FNR in the ChIP-chip experiment conducted by Grainger et al. (2007). Binding of *pykA* by FNR has not yet been shown to be functional (Gama-Castro et al. 2008). The bicluster obtained using this gene as a query is significantly enriched in known targets of FNR (indicated by the *red dots*) and other genes shown to be bound by FNR in the ChIP-chip experiment (indicated by the *blue dots*) (Grainger et al. 2007). This suggests that regulation by FNR is indeed functional

or motif data, might alleviate the problem. In the latter case the genes within one bicluster are not only co-expressed but also co-regulated (Lemmens et al. 2009; Reiss et al. 2006). We refer to these methods as integrative biclustering methods.

5.3.5 Inference of the Regulatory Program

Although clustering/biclustering methods improve our insight into the biological processes in which genes are involved, they do not give any indication of which transcription factors govern these processes. To this end methods that reconstruct TRNs from "omics" data are more appropriate. In the following paragraphs we will focus on these network inference methods.

5.3.5.1 Regulatory Program Inference from Microarray Data Only vs. Data-Integration

Inference of the regulatory program from microarray data only. Network inference methods (both direct and module based ones) generally extract the regulatory program from a list of potential regulators and/or motifs. For methods that infer the TRN from microarray data only, this assignment is based on similarity in expression between a regulator and its potential targets (Faith et al. 2007; Joshi et al. 2009; Bonneau et al. 2006). Consequently these methods assume that the regulator's expression profile is a proxy for its activity as a transcriptional factor: A regulator is assigned to its potential targets if its expression profile corresponds to that of the assigned target genes(s). While this assumption enables the reconstruction of TRNs from microarray data, it restricts the type of interactions that can be reliably inferred. First, one cannot distinguish between regulators that are assigned to genes because they regulate those genes from regulators that are simply co-expressed with their predicted targets. Moreover, the assumption of targets being co-expressed with their regulators particularly disregards the important role of regulation mechanisms other than the transcriptional one. Herrgard et al (Herrgard et al. 2003) illustrated, for instance, that in *E. coli* only few regulators are co-expressed with their targets. Michoel et al. (2009) observed that methods, such as CLR and Stochastic LeMoNe, mainly infer correct targets for autoregulators and then specifically, autoregulators that only act on a limited number of target genes (specific regulators). Because of the high level of autoregulation in prokaryotes (Thieffry et al. 1998), these methods have a reasonable performance in reconstructing the TRN when applied to prokaryotic systems.

Integrative approaches: inferring the regulatory program by data-integration. Integrative approaches complement gene expression data with other information on the transcriptional interaction between a transcription factor and its target. Consequently, the scope of the predictions can be extended beyond interactions for regulators that are co-expressed with their targets. In DISTILLER for instance an interaction between a regulator and a target is inferred if the target contains a motif for the corresponding regulator in its promoter region. As motifs only give direct information on the link between regulators and target genes, motif information can only help predict the relationship between regulator and target genes for those regulators for which the motif is known. Bonneau et al (Bonneau et al. 2006) tried to overcome this problem by combining both the expression and motif-based assignment of regulators (Inferelator): for regulators with a known motif the regulator can be assigned to a module based on the motif information. For *de novo* motifs or regulators without known motifs, the assignment of regulators is on the basis of gene expression information, as was the case for CLR and Stochastic LeMoNe.

In conclusion, integrative methods (e.g., DISTILLER) or (semi-)supervised approaches (e.g., SIRENE, SEREND and de Hoon et al. (2004)) that do not rely on the co-expression assumption of a regulator are more suitable for the inference of interactions for non-autoregulators or global regulators than the ones that use microarray data only.

5.3.5.2 Module-Based vs. Direct Network Inference

Depending on the output and the working principle, we can distinguish between "direct" and "module-based" network inference methods (Fig. 5.2). Direct methods treat the TRN as a directed graph, in which regulators are assigned to individual target genes. On the other hand, the so-called "module-based" methods exploit the modularity of the TRN (Hartwell et al. 1999; Qi and Ge 2006): rather than assigning the regulatory program to individual genes, they group genes into modules. This first step of data reduction by module inference (see also above) reduces the search space considerably. The method that is used for this first module inference step depends on the network reconstruction method and ranges from clustering (e.g., Stochastic LeMoNe) to integrative biclustering approaches (e.g., DISTILLER and Inferelator). Both "direct" and "module-based" approaches have their advantages and limitations.

In contrast to existing direct approaches, most module-based approaches do not only make predictions on the regulatory interactions but also on the conditions in which a certain regulator is predicted to regulate its target genes. This allows for the generation of hypotheses of the form "Gene X regulates the expression of genes Y under conditions W," which can then be used to guide experimental validation of the predictions, as illustrated in Bonneau et al. (2006) and Lemmens et al. (2009). Whereas some methods include the condition selection step in the module inference step by using a certain biclustering approach (e.g., DISTILLER and Inferelator), others account only for the condition dependency of transcriptional regulation by assigning regulatory programs to clusters of co-expressed genes in a condition dependent way (e.g., Stochastic LeMoNe).

In principle, module-based approaches provide more robustness against noise in the data than direct inference approaches, i.e., a certain gene is only assigned to a regulator if its assignment is confirmed by its co-expression with other targets. This is not the case for direct approaches, as each target is treated individually. In case of regulatory program inference from microarray data only (e.g., Stochastic LeMoNe, Inferelator) this gain in robustness allows for the relaxation of the stringency on the required co-expression between a regulator and its targets but often comes at the expense of loosing flexibility. Indeed, the module inference step in the module-based approach puts constraints on the (bi)clusters in terms of tightness of co-expression, (bi)cluster size and overlap in genes and conditions, restricting the possible solutions to some types of modules only. Consequently, the performance of the module-based algorithms relies heavily on the quality and characteristics of the (bi)clustering method. This trade-off between flexibility and robustness is nicely illustrated in a comparison of the Stochastic LeMoNe algorithm with CLR on the same *E. coli* gene expression compendium (Michoel et al. 2009). Modules produced by Stochastic LeMoNe are generally large and often contain genes that are not so tightly co-expressed. This prevents the identification of correct interactions for certain regulators involved in triggering the expression of just one or few operons. The coarse grained properties of the modules identified by Stochastic LeMoNe, on the other hand, promoted the identification of correct interactions for regulators of large regulons. For CLR, a direct approach, the situation is reversed: the identification of targets for some operonic regulators is no longer impeded by the characteristics of the modules. On the other hand, tighter co-expression between targets and regulators is required because of the lack of a robust (bi)clustering approach, explaining why some of the interactions of the larger regulons identified by Stochastic LeMoNe were not picked up by CLR (Michoel et al. 2009).

5.3.5.3 Supervised vs. Unsupervised Inference of the Regulatory Program

Within the direct approaches for regulatory program assignment we can further distinguish between supervised and unsupervised methods. Supervised approaches make the learning task considerably simpler, as they take advantage of the known TRN to constrain the solutions to targets showing similarities in expression (e.g., SIRENE) and motifs (e.g., de Hoon et al. 2004, SEREND) with known targets of certain regulators. Unsupervised approaches, in contrast, are purely data-driven and not based on resemblance with known information. Because supervised approaches rely on known information, it is not unexpected that several studies showed that supervised approaches outperform their unsupervised counterparts in predicting correct interactions for well-characterized regulators (Mordelet and Vert 2008; de Hoon et al. 2004). However, supervised network inference comes with a few drawbacks.

First, a supervised approach treats network inference as a classification problem and hence needs a curated set of positive and negative interactions in order to discriminate between targets and non-targets of a certain regulator. While positive examples can be found for model organisms like *E. coli* and *B. subtilis* in databases such as RegulonDB and DBTBS, no such set of negative examples exist. Often, genes not known to interact with the specific regulator, which we refer to as the "unknowns," are treated as negatives. However, as our knowledge about the TRNs is still limited, this set of so called negatives might still contain true positives. Consequently, using this set of "unknowns" as negatives to train a classifier, as is for instance the case in SIRENE, might result in a suboptimal performance in discriminating between true and false positive targets. Therefore, in the SEREND framework predictions made on the "unknowns" are incorporated in the training of the classifier itself. To this end an iterative self-training approach is used: to construct a new classifier for the next iteration, "unknowns" that were in a previous iteration predicted to be regulated by a certain transcription factor are added to the set of positives and removed from the negatives. This process is repeated until ultimately no better separation between positives and negatives can be obtained, resulting in a final classifier used for prediction. As SEREND does not only take advantage of known targets but also of predicted targets of the regulator in order to build the final classifier, the method is referred to as semi-supervised. Ernst et al. (2008) showed that such a semi-supervised approach increased the performance of the algorithm significantly over a supervised approach without the self-training step.

Another problem with (semi-)supervised approaches is that they constrain the network inference problem to what is known. As they rely on the presence of a well-curated standard network their usability is mainly restricted to model organisms such as *E. coli* (Mordelet and Vert 2008; Ernst et al. 2008) and *B. subtilis* (de Hoon et al. 2004). Moreover, as each of these methods constructs separate classifiers per regulator on the basis of the known targets for that regulator, reliable new predictions can only be obtained for regulators with many known targets such as global regulators and sigma factors (Mordelet and Vert 2008). To characterize less-studied systems, unsupervised approaches are more appropriate (e.g., Stochastic LeMoNe, CLR, DISTILLER, Inferelator). Bonneau et al. (2007), for instance, used Inferelator on a gene expression compendium in combination with de novo motif detection in order to obtain a blueprint of the TRN of the largely uncharacterized archeon *H. salinarum*.

5.3.6 Data Integration

Whereas integrative methods hold the promise of giving a more comprehensive and reliable view on the TRN, as is illustrated in the paragraphs above, the task of combining heterogeneous data is a tricky one. The most straightforward means of integrating data is by taking either the intersection or the union between the different data sources. The former approach ensures that most spurious interactions due to noise in one of the data sets will be filtered out. This comes at the expense of coverage in real interactions. Specifically, the data source with the lowest sensitivity (i.e., retrieving known interactions) will restrict the sensitivity of the combination of the data sets. Taking the union, on the other hand, guarantees a high coverage of real interactions albeit with a low specificity (i.e., discriminating true positives from false positives). DISTILLER is an example of such an integrative approach in which the intersection of two data sources is taken. Particularly, sets of genes are only retained by the method if they meet both the requirements of condition dependent co-expression and motif sharing.

cMonkey (Reiss et al. 2006), on the other hand, takes a more intermediate approach: user-defined weights reflect the relevance attached to a certain data set in constructing biclusters from the different input data. This has as advantage over the intersection- and union-based approaches in that it enables the detection of biclusters that stress certain data types over others. Motifs, for instance, are often degenerate and ill-defined for many regulators. In a data-integration framework, such as the cMonkey one, it is possible to account for these flaws of motifs by down-weighing the motif data with respect to the gene expression data and hence enabling the identification of biclusters in which the genes are co-expressed but do not all need to contain the same motif within their promoter region. The problem with such an approach is that, at least in an unsupervised setting, there exists to our knowledge no objective way to set the different weights. In a (semi-)supervised setting (e.g., SEREND and de Hoon et al. 2004), however, it is possible to define the corresponding weights of the different data sources objectively on the basis of how well they support known interactions (i.e., the training set) (de Hoon et al. 2004; Ernst et al. 2008).

Depending on the problem that one wants to see solved and the knowledge on the studied network, one method can thus be more appropriate than the other. Intersection-based methods, for instance, will result in few false positives and are hence preferential if further experimental validation is required.

5.3.7 *Prioritization of Predictions*

On the cellular level gene expression is stochastic in nature: not all cells within a population will exhibit the same level of gene expression at the same time (Kaern et al. 2005). This is mainly the case for heterogeneous populations such as biofilms (Stickler 1999). Microarrays typically average out these stochastic effects as they measure the total amount of mRNA in a whole population of cells. Together with noise due to experimental procedures this adds to the uncertainty that comes with gene expression data. The presence of noise in gene expression data can result in the prediction of spurious interactions. In addition, network inference methods generally yield a large number of predictions, often running into the hundreds, which are not always equally meaningful. Therefore, a reliable ranking of the predictions is invaluable with respect to in vitro validation of the in silico predictions.

In (semi-)supervized approaches the significance of predicted interactions is assessed against the known network. Consequently, results are ranked according to how similar predicted targets are in expression (e.g., SIRENE) and/or motifs (e.g., SEREND, de Hoon et al. 2004) related to known targets as contrasted to dissimilarity with assumed non-targets. Therefore, the ranking and subsequently the results of (semi-)supervized approaches are largely biased in terms of what is known. In contrast, unsupervised approaches do not explicitly exploit the information of the known network and thus cannot base their predictions on similarities with known information. To assess the reliability of the predictions they employ statistically motivated and hence more objective measures. Probabilistic methods such as Stochastic LeMoNe, for instance, have a natural way of dealing with this noise by attaching a certain probability to each prediction. The Stochastic LeMoNe algorithm, for instance, employs a stochastic instead of a deterministic optimization (Segal et al. 2003): this means that each time the algorithm is run a different but equally likely solution for the network inference problem will be found. Solutions that are found repeatedly in a certain number of runs of the algorithm are deemed more significant than solutions that occur only once and can be ranked accordingly. In this way spurious predictions due to noise are effectively filtered out. In contrast to probabilistic methods, the outcomes of deterministic methods, such as itemset mining approaches (Zaki and Hsiao 2002) or the relevance networks procedures (Margolin et al. 2006), which serve as a basis for respectively DISTILLER and CLR, do not assign a significance score to the predictions. Therefore, in CLR the relevance networks procedure is extended with an adaptive background correction step to filter out spurious interactions, causing the interactions to be ranked according to a significance score. In DISTILLER, the significance of the resulting modules of the item set mining search strategy is scored by estimating the probability that the same modules and regulatory programs would be selected by chance. In diverse studies both the probabilistic (Michoel et al. 2009) and the deterministic approaches (Faith et al. 2007; Lemmens et al. 2009) discussed here were shown to successfully prioritize known interactions, illustrating the practical usefulness of the different scoring approaches. With respect to the experimental validation of predicted interactions, a method that produces a reliable ranking of the predictions is desirable, as it guides the researcher to the most probable predictions among the abundance of predictions that the different methods produce.

5.4 High-Throughput Data Can Assist in the Search for Novel Drug and Vaccine Targets

As pathogenic bacteria develop more and more resistance against currently used antibiotics, there is a growing need for discovering novel drug and vaccine targets. Systems biology can aid in discovering the mechanisms of action (MoA) of known drugs or in identifying alternative targets for anti-bacterial treatments by completing our knowledge on bacterial physiology and by providing insight in signal transduction networks.

5.4.1 5.4.1 Revealing the Mechanisms of Action

Information on which genes change their expression profile during antibiotics treatment provides insight in the MoA, by revealing which pathways are affected by the treatment (Freiberg et al. 2004; Hutter et al. 2004; Freiberg and Brotz-Oesterhelt 2005). Several studies have therefore measured the effect of antibiotics on gene expression by microarrays (Table 5.1). For instance, Hutter et al. (2004) developed a supervised method that uses microarray data to classify antibacterial test compounds according to their MoA. They constructed a reference expression compendium assessing the response of *B. subtilis* to 37 different antibacterial compounds with known MoA. On the basis of this expression compendium, Hutter et al. revealed correctly the MoA of test compounds that were not used during the classification. In addition to revealing the MoA of an antibacterial compound, such expression studies also help obtaining insight in the defense mechanisms of the affected bacteria.

5.4.2 The Search for Novel Targets

Although based on different mechanisms of action, most currently available antibiotics focus on a bacteriostatic or bacteriocidal activity. As a result, most current

Table 5.1 Overview of expression studies that focus on microarrays measuring the effect of antibiotics

Species	Targets & antibacterial compounds	Study
Bacillus subtilis	Cell wall synthesis (Amoxicillin, Bacitracin, Cefalexin, Cefotaxime, Cefoxitin, Cycloserine, Methicillin, Oxacillin, Penicillin, Phosphomycin, Ristocetin, Vancomycin); DNA topology & synthesis (Ciprofloxacin, Coumermycin, Dapsone, Moxifloxacin, Nalidixic acid, Norfloxacin, Novobiocin, Sulfacetamide, Sulfamethizole, Trimethoprim); Membrane-active compounds (Gramicidin, Monensin, Nigericin, Nitrofurantoin, Polymyxin, Triton); Translation & Protein biosynthesis (Actinonin, Azithromycin, Chloramphenicol, Clarithromycin, Clindamycin, Erythromycin, Fusidic acid, Neomycin, Norvaline, Puromycin, Spectinomycin, Tetracyclin); Fatty acid synthesis (Triclosan, Cerulenin)	12207695 11948165 14707172 15273089 14651641 15273097
Escherichia coli	Cell wall synthesis (Ampicillin); DNA replication, recombination and damage (Norfloxacin, Ofloxacin); Translation & Protein synthesis (Acivicin, 4-Azaleucine, Mupirocin, Kanamycin, Kasugamycin, Puromycin); Transcription (Rifampin); Unknown (Cecropin A)	12499161 14982780 14526028 11344143 12736533
Haemophilus influenzae	DNA topology & synthesis (Ciprofloxacin, Novobiocin)	11156613
Mycobacterium tuberculosis	Fatty acid synthesis (Isoniazid, Thiolactomycin, Triclosan)	10536008 12936993
Pseudomonas aeruginosa	Translation & Protein biosynthesis (Tobramycin)	11677611
Streptococcus pneumoniae	Translation & Protein biosynthesis (Chloramphenicol, Erythromycin, Puromycin, Tetracycline)	12486074

The species on which the microarray experiment was performed (Species), the antibacterial compounds that were tested and their targets (Targets & Antibacterial Compounds), and the PubMed ID (Study) of the experiment are indicated.

bacterial targets have an essential role for bacterial growth or survival. For instance, current antibacterial drugs would attack only around 40 target sites, usually involved in peptidoglycan biosynthesis or gene expression/translation (Bumann 2008). Many more components are expected to be essential for bacterial survival than the currently identified targets. However, Bumann et al. (Bumann 2008) could not find any additional occurring in many related species or broad-spectrum drug targets in a genome wide screening. A possible reason for this is the high level of redundancy in biochemical pathways. A combinatorial strategy in which a drug or combination of drugs can be used to attack multiple redundant targets simultaneously in order to cause lethality may be a solution to circumvent this redundancy. Systems biology can aid in acquiring information on lethality caused by multiple targets. Recently a novel high-throughput technique was developed in *E. coli* to measure synthetic lethality (Butland et al. 2008; Typas et al. 2008). Synthetic lethality is defined as the combination of two non-lethal mutations that result in cell death.

The previously mentioned strategy is based on targeting crucial pathways for survival in bacteria. This strategy puts a heavy selection pressure on bacteria, resulting in the increased resistance of bacteria against current antibiotics. This directs us to also consider alternative drugs that put less survival pressure on bacteria. One example is to target bacterial virulence in order to prevent the pathogen from attacking the host by inhibiting mechanisms that are essential during infection (e.g., adhesion) or that cause disease symptoms (e.g., toxins). Some successes have been achieved in neutralizing bacterial toxins as at least six antitoxin candidates are in clinical trials (Cegelski et al. 2008). Systems biology can help us understand the underlying regulatory mechanisms of virulence and the consequences of pathogen-host interactions (Cegelski et al. 2008; Kaushik and Sehgal 2008). For instance, by performing a microarray experiment in which the pathogen-host interaction was measured, Eriksson et al. (2003) showed that more than half of the up-regulated genes were genes of unknown function. In the long term such fundamental knowledge provides the basis for the identification of possible drug targets (Table 5.2).

Vaccination is another strategy to prevent bacterial infections. Vaccine candidates are proteins that are present during infection and are located on the surface of the bacteria. The first vaccine candidates that were discovered by the use of microarrays were identified for *Neisseria meningitidis*. Grifantini et al. (2002) started by doing a microarray experiment assessing the expression of *N. meningitidis* during adhesion to epithelial cells. They found that 189 genes were up-regulated during adhesion. These genes and their corresponding gene products are thus important during infection. About 40% of these 189 up-regulated genes encode for a protein located on the surface of the bacterium, indicating that the cell membrane undergoes a reorganization during adhesion to the host. In addition, Grifantini et al. (2002) identified five adhesion-induced antigens that were capable of inducing bactericidal antibodies in mice and therefore, formed good vaccine candidates. Examples in other bacteria also exist (Yang et al. 2006; Merrell et al. 2002; Voyich et al. 2003).

Finally, the study of bacterial populations revealed that bacterial cells do not function in isolation, but rather as part of a large community. For instance, bacteria can communicate to secrete a particular protein or to differentiate and produce an

Table 5.2 An overview of expression studies that focus on microarrays dealing with the host–pathogen interaction

Species	Host	Study
Borrelia burgdorferi	Rat peritoneal cavities	11830671
Chlamidiae trachomatis	Human epithelial cervix HeLa 229 cells	12815105
Streptococcus pyogenes	Human polymorphonuclear leukocyte	12574517
Mycobacterium tuberculosis	Human and mouse macrophages	12953091 11576227
Neisseria meningitidis	Human serum and human epithelial and endothelial cells	12531357 12172557 12486052
Salmonella typhimurium	Murine macrophage-like cell line	12492857
Escherichia coli	Murine bladder	11744708
Vibrio cholerae	Human stool samples	12050664
Leptospira interrogans	Serial passages in guinea pigs for preservation of virulence	17109759

The species (Species) and the host (Host) on which the microarray experiment was performed; the host (Host) and the PubMed ID (Study) of the experiment are indicated.

extracellular matrix (Andre and Godelle 2005). Instead of developing drugs that target the cells themselves, it may therefore also be interesting to consider developing drugs against the cooperation mechanism between cells in a colony. Andre and Godelle (2005) developed a model that demonstrates that developing resistance to a drug against cooperative behavior is much slower and more difficult than evolving resistance to an antibiotic against individual cells. It is thus of importance to find out what is causing this cooperative behavior and how we can control it.

In addition, subpopulations with different phenotype may exist in a population of genetically identical organisms. From a clinical point of view it is especially important to find out how these bacterial subpopulations are able to survive an antibacterial treatment (Balaban et al. 2004) or to circumvent the host immune system (Ackermann et al. 2008). Ackermann et al. (2008), for instance, showed that for *S. typhimurium* a small part of the population triggers the host innate immune response by invading the host cell. This response will kill not only the invading subpopulation of Salmonella cells but also many of the competitor gut commensals. As a result of the latter, the infection process by the remaining subpopulations of Salmonella cells will be more successful. In this example it is critical to investigate the differences between the active networks of the two subpopulations (Dwyer et al. 2008).

5.5 Conclusions and Perspectives

Bacteria have been studied as model organisms in molecular biology for many decades. In systems biology their use has been lagging behind as compared to yeast. Thanks to the increasing availability of "omics" data in bacteria, however, bacteria have now gained a similar status to yeast as models for systems biology tool development. In comparison with eukaryotes, bacteria indeed offer some advantages. Firstly, because of their relatively simple regulatory networks, the inference problem is more tractable than it is in higher eukaryotes. Moreover, for the most common model organisms such as *E. coli* and *B. subtilis* the TRN is at least already partly known, allowing developers of network reconstruction methods to benchmark and test their algorithms. Finally, bacteria are easy to manipulate so that a rather straightforward validation of predictions is possible.

This chapter gives a description of different methods for the inference of TRNs that are available today and some applications of these methods on bacteria. The inference of networks is not a trivial task considering the complexity of most biological systems, the noisy character of the data, and the under-sampling of many biological systems. Each of the methods for the analysis of "omics" data deals with these problems in a different way, relying on different assumptions and simplifications of the biological reality. There exist supervised and non-supervised methods, methods dealing with expression data only and methods relying on the integration of several data sources, methods that reveal the interaction between a transcription factor and its target gene directly, and methods identifying transcriptional modules prior to the assignment of the regulatory program. In this chapter, we have outlined the main advantages and limitations of available network reconstruction tools and shown that no single best tool exists. As each of the methodologies highlights different aspects of the biology, the results obtained from different methods are often complementary (Michoel et al. 2009). The usefulness of a method of tackling a particular problem thus depends largely on the nature of the problem and the available data. Therefore, it is necessary to tune methods to the specificity of the problem and the design of experiments.

Currently, most methods for inferring TRNs make use of microarray, motif, and ChIP-chip data but will soon also benefit from novel high-throughput data. For instance, newly developed DNA sequencing technologies not only allow for rapid, less expensive sequencing of large and complex genomes but are also expected to take over from the array-based transcriptome (Brenner et al. 2000) and ChIP-chip analyses (Robertson et al. 2007; Johnson et al. 2007).

In the process of reverse-engineering the TRN, most methods account mainly for the direct regulator–target gene binding but disregard the role of a plethora of other regulatory mechanisms that have a direct or indirect effect on the TRN, such as sRNAs (Waters and Storz 2009), chromatin modeling (Blot et al. 2006), metabolite-based feedback (Shi and Shi 2004), etc. At this stage this oversimplification of the inference problem is partly due to the lack of sufficient amounts of high-throughput data at these additional levels of regulation. For prokaryotes, high-throughput data on other levels of the regulatory network, like protein-protein interaction data (Butland et al. 2005; Arifuzzaman et al. 2006), metabolic data, phenome data (Bochner et al. 2001; Zhou et al. 2003; Baba et al. 2006), and genetic interaction data (Butland et al. 2008), among others, are emerging. From each of these data sources the corresponding networks can be derived. But as cellular behavior results from the interplay between these distinct networks, novel methods need to either compare the responses triggered by these different networks or integrate the different network layers in a comprehensive way. Few examples of this integration exist at this stage and further improvement is thus possible (De Keersmaecker et al. 2006).

The novel field of high-throughput analysis is evolving from mere screening towards comprehensive network reconstruction and has already unveiled interesting biological findings. Systems biology offers a tremendous potential for drug and or vaccine development as well as for future research domains such as synthetic microbiology.

References

Ackermann M, Stecher B, Freed NE et al (2008) Self-destructive cooperation mediated by phenotypic noise. Nature 454:987–990

Andre J, Godelle B (2005) Multicellular organization in bacteria as a target for drug therapy. Ecol Lett 8:800–810

Arifuzzaman M, Maeda M, Itoh A et al (2006) Large-scale identification of protein–protein interaction of *Escherichia coli* K-12. Genome Res 16:686–691

Baba T, Ara T, Hasegawa M et al (2006) Construction of *Escherichia coli* K-12 in-frame, single-gene knockout mutants: the Keio collection. Mol Syst Biol 2:2006

Babu MM, Teichmann SA (2003) Evolution of transcription factors and the gene regulatory network in *Escherichia coli*. Nucleic Acids Res 31:1234–1244

Balaban NQ, Merrin J, Chait R et al (2004) Bacterial persistence as a phenotypic switch. Science 305:1622–1625

Bammler T, Beyer RP, Bhattacharya S et al (2005) Standardizing global gene expression analysis between laboratories and across platforms. Nat Methods 2:351–356

Barrett T, Troup DB, Wilhite SE et al (2007) NCBI GEO: mining tens of millions of expression profiles – database and tools update. Nucleic Acids Res 35:D760–D765

Ben Yehuda S, Fujita M, Liu XS et al (2005) Defining a centromere-like element in *Bacillus subtilis* by identifying the binding sites for the chromosome-anchoring protein RacA. Mol Cell 17:773–782

Bergmann S, Ihmels J, Barkai N (2003) Iterative signature algorithm for the analysis of large-scale gene expression data. Phys Rev E Stat Nonlinear Soft Matter Phys 67:031902

Blot N, Mavathur R, Geertz M et al (2006) Homeostatic regulation of supercoiling sensitivity coordinates transcription of the bacterial genome. EMBO Rep 7:710–715

Bochner BR, Gadzinski P, Panomitros E (2001) Phenotype microarrays for high-throughput phenotypic testing and assay of gene function. Genome Res 11:1246–1255

Bonneau R, Reiss DJ, Shannon P et al (2006) The Inferelator: an algorithm for learning parsimonious regulatory networks from systems-biology data sets de novo. Genome Biol 7:R36

Bonneau R, Facciotti MT, Reiss DJ et al (2007) A predictive model for transcriptional control of physiology in a free living cell. Cell 131:1354–1365

Brazma A, Hingamp P, Quackenbush J et al (2001) Minimum information about a microarray experiment (MIAME)-toward standards for microarray data. Nat Genet 29:365–371

Brenner S, Johnson M, Bridgham J et al (2000) Gene expression analysis by massively parallel signature sequencing (MPSS) on microbead arrays. Nat Biotechnol 18:630–634

Bruggeman FJ, Westerhoff HV (2007) The nature of systems biology. Trends Microbiol 15:45–50

Bumann D (2008) Has nature already identified all useful antibacterial targets? Curr Opin Microbiol 11:387–392

Butland G, Peregrin-Alvarez JM, Li J et al (2005) Interaction network containing conserved and essential protein complexes in *Escherichia coli*. Nature 433:531–537

Butland G, Babu M, Diaz-Mejia JJ et al (2008) eSGA: *E. coli* synthetic genetic array analysis. Nat Methods 5:789–795

Cegelski L, Marshall GR, Eldridge GR et al (2008) The biology and future prospects of antivirulence therapies. Nat Rev Microbiol 6:17–27

Cheng Y, Church GM (2000) Biclustering of expression data. Proc Int Conf Intell Syst Mol Biol 8:93–103

Cho BK, Knight EM, Barrett CL et al (2008a) Genome-wide analysis of Fis binding in *Escherichia coli* indicates a causative role for A-/AT-tracts. Genome Res 18:900–910

Cho BK, Barrett CL, Knight EM et al (2008b) Genome-scale reconstruction of the Lrp regulatory network in *Escherichia coli*. Proc Natl Acad Sci USA 105:19462–19467

de Hoon MJ, Makita Y, Imoto S et al (2004) Predicting gene regulation by sigma factors in *Bacillus subtilis* from genome-wide data. Bioinform 20(Suppl 1):i101–i108

De Keersmaecker SC, Thijs IM, Vanderleyden J et al (2006) Integration of omics data: how well does it work for bacteria? Mol Microbiol 62:1239–1250

Demeter J, Beauheim C, Gollub J et al (2007) The Stanford Microarray Database: implementation of new analysis tools and open source release of software. Nucleic Acids Res 35:D766–D770

Dhollander T, Sheng Q, Lemmens K et al (2007) Query-driven module discovery in microarray data. Bioinformatics 23:2573–2580

Dwyer DJ, Kohanski MA, Collins JJ (2008) Networking opportunities for bacteria. Cell 135:1153–1156

Eriksson S, Lucchini S, Thompson A et al (2003) Unravelling the biology of macrophage infection by gene expression profiling of intracellular *Salmonella enterica*. Mol Microbiol 47:103–118

Ernst J, Beg QK, Kay KA et al (2008) A semi-supervised method for predicting transcription factor–gene interactions in *Escherichia coli*. PLoS Comput Biol 4:e1000044

Faith JJ, Hayete B, Thaden JT et al (2007) Large-scale mapping and validation of Escherichia coli transcriptional regulation from a compendium of expression profiles. PLoS Biol 5:e8

Freiberg C, Brotz-Oesterhelt H (2005) Functional genomics in antibacterial drug discovery. Drug Discov Today 10:927–935

Freiberg C, Brotz-Oesterhelt H, Labischinski H (2004) The impact of transcriptome and proteome analyses on antibiotic drug discovery. Curr Opin Microbiol 7:451–459

Gama-Castro S, Jimenez-Jacinto V, Peralta-Gil M et al (2008) RegulonDB (version 6.0): gene regulation model of *Escherichia coli* K-12 beyond transcription, active (experimental) annotated promoters and Textpresso navigation. Nucleic Acids Res 36:D120–D124

Gao F, Foat BC, Bussemaker HJ (2004) Defining transcriptional networks through integrative modeling of mRNA expression and transcription factor binding data. BMC Bioinform 5:31

Getz G, Levine E, Domany E (2000) Coupled two-way clustering analysis of gene microarray data. Proc Natl Acad Sci USA 97:12079–12084

Grainger DC, Hurd D, Harrison M et al (2005) Studies of the distribution of *Escherichia coli* cAMP-receptor protein and RNA polymerase along the *E. coli* chromosome. Proc Natl Acad Sci USA 102:17693–17698

Grainger DC, Hurd D, Goldberg MD et al (2006) Association of nucleoid proteins with coding and non-coding segments of the *Escherichia coli* genome. Nucleic Acids Res 34:4642–4652

Grainger DC, Aiba H, Hurd D et al (2007) Transcription factor distribution in *Escherichia coli*: studies with FNR protein. Nucleic Acids Res 35:269–278

Grifantini R, Bartolini E, Muzzi A et al (2002) Previously unrecognized vaccine candidates against group B meningococcus identified by DNA microarrays. Nat Biotechnol 20:914–921

Grote A, Klein J, Retter I et al (2009) PRODORIC (release 2009): a database and tool platform for the analysis of gene regulation in prokaryotes. Nucleic Acids Res 37:D61--D65

Hartwell LH, Hopfield JJ, Leibler S et al (1999) From molecular to modular cell biology. Nature 402:C47–C52

Herrgard MJ, Covert MW, Palsson BO (2003) Reconciling gene expression data with known genome-scale regulatory network structures. Genome Res 13(11):2423–2434; Epub (14 Oct 12003) 13:2423–2434

Hertzberg L, Zuk O, Getz G et al (2005) Finding motifs in promoter regions. J Comput Biol 12:314–330

Hibbs MA, Hess DC, Myers CL et al (2007) Exploring the functional landscape of gene expression: directed search of large microarray compendia. Bioinformatics 23:2692–2699

Hutter B, Schaab C, Albrecht S et al (2004) Prediction of mechanisms of action of antibacterial compounds by gene expression profiling. Antimicrob Agents Chemother 48:2838–2844

Ihmels J, Bergmann S, Barkai N (2004) Defining transcription modules using large-scale gene expression data. Bioinformatics 20:1993–2003

Irizarry RA, Warren D, Spencer F et al (2005) Multiple-laboratory comparison of microarray platforms. Nat Methods 2:345–350

Johnson DS, Mortazavi A, Myers RM et al (2007) Genome-wide mapping of in vivo protein-DNA interactions. Science 316:1497–1502

Joshi A, De SR, Marchal K et al (2009) Module networks revisited: computational assessment and prioritization of model predictions. Bioinform 25:490–496

Kaern M, Elston TC, Blake WJ et al (2005) Stochasticity in gene expression: from theories to phenotypes. Nat Rev Genet 6:451–464

Kaushik DK, Sehgal D (2008) Developing antibacterial vaccines in genomics and proteomics era. Scand J Immunol 67:544–552

Keseler IM, Bonavides-Martinez C, Collado-Vides J et al (2009) EcoCyc: a comprehensive view of *Escherichia coli* biology. Nucleic Acids Res 37:D464–D470

Kitano H (2002) Computational systems biology. Nature 420:206–210

Laub MT, Chen SL, Shapiro L et al (2002) Genes directly controlled by CtrA, a master regulator of the Caulobacter cell cycle. Proc Natl Acad Sci USA 99:4632–4637

Lazzeroni L, Owen A (2002) Plaid models for gene expression data. Statist Sinica 2:61–86

Lemmens K, De Bie T, Dhollander T et al (2009) DISTILLER: a data integration framework to reveal condition dependency of complex regulons in *Escherichia coli*. Genome Biol 10:R27

Lucchini S, Rowley G, Goldberg MD et al (2006) H-NS mediates the silencing of laterally acquired genes in bacteria. PLoS Pathog 2:e81

Luscombe NM, Babu MM, Yu H et al (2004) Genomic analysis of regulatory network dynamics reveals large topological changes. Nature 431:308–312

Madeira SC, Oliveira AL (2004) Biclustering algorithms for biological data analysis: a survey. IEEE/ACM Trans Comput Biol Bioinform 1:24–45

Marchal K, De Keersmaecker S, Monsieurs P et al (2004) In silico identification and experimental validation of PmrAB targets in *Salmonella typhimurium* by regulatory motif detection. Genome Biol 5:R9

Margolin AA, Nemenman I, Basso K et al (2006) ARACNE: an algorithm for the reconstruction of gene regulatory networks in a mammalian cellular context. BMC Bioinform 7(1):S7

Matys V, Kel-Margoulis OV, Fricke E et al (2006) TRANSFAC and its module TRANSCompel: transcriptional gene regulation in eukaryotes. Nucleic Acids Res 34:D108–D110

Merrell DS, Butler SM, Qadri F et al (2002) Host-induced epidemic spread of the cholera bacterium. Nature 417:642–645

Michoel T, De Smet R, Joshi A et al (2009) Comparative analysis of module-based versus direct methods for reverse-engineering transcriptional regulatory networks BMC Syst Biol 3:49

Molle V, Fujita M, Jensen ST et al (2003a) The Spo0A regulon of *Bacillus subtilis*. Mol Microbiol 50:1683–1701

Molle V, Nakaura Y, Shivers RP et al (2003b) Additional targets of the *Bacillus subtilis* global regulator CodY identified by chromatin immunoprecipitation and genome-wide transcript analysis. J Bacteriol 185:1911–1922

Mordelet F, Vert JP (2008) SIRENE: supervised inference of regulatory networks. Bioinform 24:i76–i82

Murali TM, Kasif S (2003) Extracting conserved gene expression motifs from gene expression data. Pacific Symp Biocomput 8:77–88

Navarre WW, Porwollik S, Wang Y et al (2006) Selective silencing of foreign DNA with low GC content by the H-NS protein in Salmonella. Science 313:236–238

Orlando V (2000) Mapping chromosomal proteins in vivo by formaldehyde-crosslinked-chromatin immunoprecipitation. Trends Biochem Sci 25:99–104

Parkinson H, Kapushesky M, Shojatalab M et al (2007) ArrayExpress – a public database of microarray experiments and gene expression profiles. Nucleic Acids Res 35:D747–D750

Perez AG, Angarica VE, Vasconcelos AT et al (2007) Tractor_DB (version 2.0): a database of regulatory interactions in gamma-proteobacterial genomes. Nucleic Acids Res 35:D132–D136

Qi Y, Ge H (2006) Modularity and dynamics of cellular networks. PLoS Comput Biol 2:e174

Quackenbush J (2001) Computational analysis of microarray data. Nat Rev Genet 2:418–427

Reiss DJ, Baliga NS, Bonneau R (2006) Integrated biclustering of heterogeneous genome-wide datasets for the inference of global regulatory networks. BMC Bioinform 7:280

Ren B, Robert F, Wyrick JJ et al (2000) Genome-wide location and function of DNA binding proteins. Science 290:2306–2309

Robertson G, Hirst M, Bainbridge M et al (2007) Genome-wide profiles of STAT1 DNA association using chromatin immunoprecipitation and massively parallel sequencing. Nat Methods 4:651–657

Sasik R, Woelk CH, Corbeil J (2004) Microarray truths and consequences. J Mol Endocrinol 33:1–9

Segal E, Shapira M, Regev A et al (2003) Module networks: identifying regulatory modules and their condition-specific regulators from gene expression data. Nat Genet 34:166–176

Sheng Q, Moreau Y, De Moor B (2003) Biclustering microarray data by Gibbs sampling. Bioinform 19(Suppl 2):ii196–ii205

Shi Y, Shi Y (2004) Metabolic enzymes and coenzymes in transcription – a direct link between metabolism and transcription? Trends Genet 20:445–452

Sierro N, Makita Y, de Hoon M et al (2008) DBTBS: a database of transcriptional regulation in *Bacillus subtilis* containing upstream intergenic conservation information. Nucleic Acids Res 36:D93–D96

Stickler D (1999) Biofilms. Curr Opin Microbiol 2:270–275

Stolovitzky G, Monroe D, Califano A (2007) Dialogue on reverse-engineering assessment and methods: the DREAM of high-throughput pathway inference. Ann N Y Acad Sci 1115:1–22

Tanay A, Sharan R, Shamir R (2002) Discovering statistically significant biclusters in gene expression data. Bioinformatics 18(Suppl 1):S136–S144

Thieffry D, Huerta AM, Perez-Rueda E et al (1998) From specific gene regulation to genomic networks: a global analysis of transcriptional regulation in *Escherichia coli*. Bioessays 20:433–440

Thijs IM, De Keersmaecker SC, Fadda A et al (2007) Delineation of the *Salmonella enterica* serovar Typhimurium HilA regulon through genome-wide location and transcript analysis. J Bacteriol 189:4587–4596

Tompa M, Li N, Bailey TL et al (2005) Assessing computational tools for the discovery of transcription factor binding sites. Nat Biotechnol 23:137–144

Typas A, Nichols RJ, Siegele DA et al (2008) High-throughput, quantitative analyses of genetic interactions in *E. coli*. Nat Methods 5:781–787

Van den Bulcke T, Lemmens K, Van de Peer Y et al (2006a) Inferring transcriptional networks by mining omics data. Curr Bioinform 1:301–313

Van den Bulcke T, Van LK, Naudts B et al (2006b) SynTReN: a generator of synthetic gene expression data for design and analysis of structure learning algorithms. BMC Bioinform 7:43

Voyich JM, Sturdevant DE, Braughton KR et al (2003) Genome-wide protective response used by group A Streptococcus to evade destruction by human polymorphonuclear leukocytes. Proc Natl Acad Sci USA 100:1996–2001

Waters LS, Storz G (2009) Regulatory RNAs in bacteria. Cell 136:615–628

Yang HL, Zhu YZ, Qin JH et al (2006) In silico and microarray-based genomic approaches to identifying potential vaccine candidates against *Leptospira interrogans*. BMC Genom 7:293

Yue H, Eastman PS, Wang BB et al (2001) An evaluation of the performance of cDNA microarrays for detecting changes in global mRNA expression. Nucleic Acids Res 29:E41

Zaki MJ, Hsiao C (2002) CHARM: an efficient algorithm for closed itemset mining. In: Grossman R, Han J, Kumar V, Mannila H, Motwani R (eds) Proc Second SIAM International Conference on Data Mining (SDM '02)

Zhou L, Lei XH, Bochner BR et al (2003) Phenotype microarray analysis of *Escherichia coli* K-12 mutants with deletions of all two-component systems. J Bacteriol 185:4956–4972

Chapter 6
Host–Pathogen Systems Biology

Christian V. Forst

6.1 Introduction

Systems biology is an approach to study, analyze, and, finally, control biological systems. Unlike traditional research that typically focuses on single genes, systems biology, as defined by Leroy Hood, studies the complex interactions of all levels of biological information. Biological systems are particularly attractive for systems level exploration, as summarized by researchers at the Institute for Systems Biology (ISB, Seattle, WA, USA) (Aderem 2005). These systems possess emergent properties and are robust and modular. Although the success of systems biology lies in the integrative and iterative approach between experimental and computational/theoretical sciences, in this chapter we will focus on the latter aspect involving, for example, the construction of theoretical models, the conduction of in silico experiments, and their theoretical analysis as well as quantitative predictive modeling.

The research of host–pathogen interactions in its broadest definition is a very mature field. It is closely linked to our growing understanding of the immune system and immune responses (Box 6.1) (Goldsby et al. 1999). Host–pathogen interactions can be interpreted as the battle of two systems. For example, pathogens hijack host cells and render host cell capabilities to the pathogens' own advantage (Kahn et al. 2002), or they evolve so rapidly that their sheer diversity overwhelms the immune system, as is the case during HIV infection (Simon and Frost 2002).

The detailed mechanistic analysis of host–pathogen systems, encompassing all aspects of such a multi-level problem, from molecular interactions to organism responses, is still in its infancy. Components and sub-problems have been addressed by theoretical and experimental approaches, often focusing on either host-response

C.V. Forst
UT Southwestern Medical Center in Dallas, TX, USA

V. Sintchenko (ed.), *Infectious Disease Informatics*,
DOI 10.1007/978-1-4419-1327-2_6, © Springer Science+Business Media, LLC 2010

or pathogen interference by mimicking the missing "partner." The paradigm of integrating these different types of dynamic models into a multi-level host-pathogen system is host-pathogen systems biology. The ultimate goal for host-pathogen systems biology is not only the discovery and comprehension of the underlying biology but also the establishment of a robust framework for more efficient drug development and therapeutic intervention. Examples, approaches, and perspectives are given in this chapter.

Box 6.1 Immune system overview

Immune response is an essential defense mechanism against pathogens, and is available to most multi-cellular organisms. Even unicellular organisms, such as the well-known mold *Penicillium chrysogenum* produce chemical components to kill pathogens. Among other defense mechanisms, chemical agents are part of an innate immune response. The following list captures the defense mechanisms repertoire of both, the innate and adaptive immune system:

Innate Immune Response (plants, animals)

- Barriers to pathogen entry
- Mechanical responses to eliminate antigens
- Chemical agents

Adaptive Immune Response

- *Phagocytes*
- Fever; elevated temperature inhibits growth of microbes
- Inflammatory responses to attract white blood cells (*leukocytes*) to the infection site.
- *Natural Killer (NK) cells* to kill pathogen-infected and cancer cells.

Adaptive Immune Response (higher animals)

- Synthesis of antibodies to bind antigens and promote their elimination.
- T-cell killing of virus-infected cells.
- Activation of macrophages to destroy phagocytosed pathogens.

The innate and adaptive immune responses are complementary components of multi-cellular host defense. The innate immune response provides the initial defense against infections with responses occurring within hours after infection. In contrast, the adaptive immune response requires several days to develop after infection. Innate immunity relies on germline-encoded receptors and is limited to some extent in its diversity, although some diversification is achieved by heterodimerization of TLRs or the semi-invariant NKT-cells. NKT-cells, a special type of T-cells with properties of NK cells blur the distinction between innate and adaptive immunity by using the complex

Box 6.1 (continued)

machinery of somatic recombination to produce receptors recognizing a narrow range of antigenic diversity. On the other hand, the receptors of the adaptive response that are also produced by somatic recombinations of gene segments, experience a tremendous diversity. The adaptive immune system also produces memory cells to store receptor information for particular responses.

Innate Immune Response and Toll Like Receptor Pathways

The innate immune system is essential for host defense and is responsible for the early detection and containment of pathogens. The inflammatory response to pathogens is activated when the phagocyte recognizes the foreign invader using a battery of so called pattern recognition receptors (PRR) including toll like receptors (TLRs) (Akira and Takeda 2004), members of the C-type lectin receptor family (Gordon 2002), scavenger receptors (Platt et al. 2002), complement receptors (Ernst 1998), and integrins. Conserved pathogen-specific chemical motifs that are recognized by these receptors include carbohydrates, glycolipids, glycoproteins, nucleic acids (DNA and double-stranded RNA), proteolipids, and proteins. Stimulation of PRRs results in activation of a broad spectrum of interacting signaling pathways, revealing a system of extraordinary complexity. Additional receptors, such as cytokine, chemokine, or growth factor receptors, add to the specificity of the immune response (Lewis et al. 2001).

6.2 Systems Biology in Drug Discovery

Systems biology approaches offer novel strategies to shorten the cumbersome path from an identified target to an approved drug (Simon and Frost 2002); (Butcher 2005). Computational systems biology provides "in silico" models for cost-effective decision-making during multi-million-dollar drug development, moving away from a target-based reductionistic approach that severely fails, particularly in the case of chronic multifactorial diseases (Ho and Lieu 2008).

The term "systems biology" refers to many different techniques and models for probing and understanding biological complexity, spanning multiple levels of spatial and temporal scales (Fig. 6.1). Because biological complexity is an exponential function of the number of systems components and the interactions between them, such efforts are currently limited to simple organisms or to specific pathways in higher organisms. Limiting systems biology studies to specific functional sub-systems is even more pronounced in host–pathogen systems biology that focuses on more than one organism. Where systems biology is applied to drug discovery, three principal approaches can be identified (Butcher et al. 2004): (1) the bioinformatics integration of "omics" data (a bottom-up approach); (2) integrative mathematical cell models (an intermediate approach); and (3) computer models of disease or organ system physiology from cell and organ response information available in the literature (a top-down approach to target selection, clinical indication and clinical trial design).

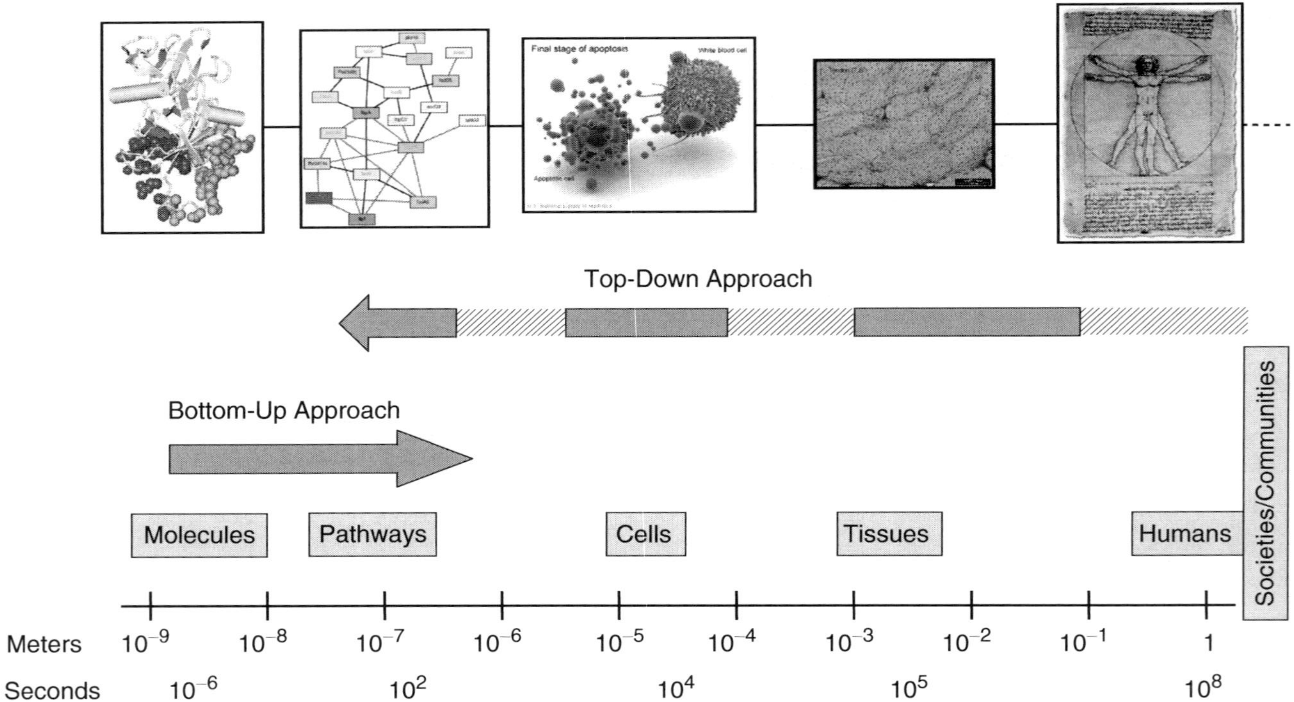

Fig. 6.1 Multiscale approaches in biological systems modeling, from molecules, pathways, cells and cell-cell interactions, and tissues to the whole organism. Life takes place on many different temporal and spatial scales. The spatial scale ranges orders of magnitudes from nanometers for chemicals and proteins to meters for the whole body. The temporal scale covers fast biochemical reactions happening in microseconds to the lifespan of an organism in years. The bottom-up approach ("omics") focuses on large-scale identification of molecular components. The top-down approach (modeling) attempts to form integrative (multi-level) models of human physiology and pathogen infection, which typically focus on relatively specific questions at particular scales, due to the limitations of current technologies (adapted from Butcher et al. 2004)

Pharmaceutical companies are known to work in all three areas. Initiated by Human Genome Sciences, organizations such as GeneGo (http://www.genego.com/), Ingenuity (http://www.ingenuity.com/), Jubilant Biosys (http://www.jubilantbiosys.com/), and PubGene (http://www.PubGene.org) offer services to access and visualize vast amounts of omics data in the context of biological networks. Gene expression patterns correlated with biochemical pathways or protein–protein interaction maps have yielded diagnostic insights into cancer (Alizadeh et al. 2000), and are expected to be useful for biomarker identification. Companies, such as BG Medicine (formerly Beyond Genomics), Biosystemix (resurrected from Molecular Mining), and Gene Logic (a spin-off into Ocimum Biosolutions), further use even more sophisticated algorithms to find biomarkers and diagnostic markers by scanning through omics data.

Other companies, such as Genomatica (http://www.genomatica.com/), use steady state flux balance analysis to model bacterial metabolic physiology. The modeling is made possible by the availability of the complete genome sequences and the fundamental assumption that bacteria will optimize their growth under various environmental conditions. Genomatica's models compute the resulting bacterial response constrained by nutrient availability (Schilling and Palsson 1998). These models can be considered as intermediate models in building biochemical networks by both bottom-up and top-down approaches incorporating all systemic interactions. The objective of these models with respect to drug targets is to find and induce growth-inhibiting conditions because it is safe to assume that non-growing bacteria are unlikely to cause human disease.

True top-down approaches in the pharmaceutical industry include multiscale modeling for particular syndromes and diseases with details down to the tissue level, and in some areas, down to enzymes and receptors. Such models also provide examples of computational simulation of disease to provide insights for researchers' decision-making processes (http://www.entelos.com/pubArchive/BANGS_A.pdf). For instance, Entelos offers disease models in asthma, atherosclerosis, type 1 diabetes, type 2 diabetes/obesity, and rheumatoid arthritis. In addition, customers can partner with Entelos to build new PhysioLab platforms in new disease areas, or to access technology through licensing agreements. Optimata has built models of cancer therapy and thrombopoiesis. Both Entelos and Optimata have shifted their business emphasis towards developing their own drugs by leveraging the risks in drug development with their internal modeling expertise (http://www.bio-itworld.com/issues/2007/oct/russell-transcript-optimata/). Gene Network Sciences started in the area of biosimuation using the bottom-up biochemical modeling of cancer cells. This company now has combined pattern recognition and reverse engineering with forward simulation. Reverse engineering is used to identify multiple models that explain cell physiological behavior. Forward simulation and optimization is conducted to select the best model from this set of possible models.

6.3 Computational Systems Biology Models, Methods and Tools

6.3.1 Scales and Models

The goal of host–pathogen systems biology is to understand physiology and infectious disease from the level of molecules, cellular networks (e.g., metabolic, regulatory, and signaling networks), cells (host cells as well as various viruses and bacterial pathogens), tissues, organs, and ultimately whole organisms. A comprehensive systems model may span about ten orders of magnitudes in scale and even more in time (Fig. 6.1). Two distinct strategies for modeling along many levels of description can be recognized: bottom-up and top-down approaches which can also be integrated in a third, hybrid strategy.

It could be argued that a full understanding of a host–pathogen system requires knowledge of all of its components. A *bottom-up approach* focuses on the measurement and description of complex systems utilizing building blocks, their interactions, and dynamic properties, such as kinetic parameters. With respect to molecular biology, bottom-up modeling has started during the post-genomic revolution, with a plethora of "omic" information available. For example, it can be used to investigate which genes, proteins, or phosphorylation states of proteins are expressed or upregulated in an infection process, leading to testable hypothesis that the regulated species are important to disease induction or progression. Through the integration of genomic, proteomic, and metabolomic data models have been developed to mechanistically describe intra- and inter-cellular processes, for example, during drug response or disease progression.

In contrast, modeling (or the *top-down approach*) attempts to develop integrative and predictive multi-scale models of biological processes. A long-term goal would be a model of in silico human–pathogen physiology and infection. However, with the current technology, such modeling focuses on relatively specific problems at particular scales, for example, at the pathway, immune cell-system, or organ level.

Bottom-up models are serving as scaffolds for top-down models by providing information of possible and potential interactions and sub-processes, how these sub-processes respond to drugs and infection, and how matter and information are passed between sub-processes and through different scales. Such hybrid approaches benefit from bottom-up molecular biological measurements and knowledge, as well as from top-down predictive modeling. A "post-genomic physiology" could span many different levels of biology, from molecules to whole organisms, moving away from "naïve reductionism" to a discipline that fosters integration and synthesis, as Strange envisioned in a review (Strange 2005).

6.3.2 Methods

Complementary to the biological hierarchy of host–pathogen systems, methodological descriptions and simulations of such systems have been performed on a different level of detail. Interaction networks and network models of biological systems have been studied at the levels of (1) topological connections, (2) qualitative connections, (3) quantitative connections, and (4) higher order interactions, as has been reviewed by Bower and Bolouri (2001).

Networks are assembled from interaction data, physical measurements, or computational predictions of protein–protein, protein–DNA, protein–small chemical, or other identified interactions, or by inference of correlations between cellular components. Undirected edges indicate interactions between components. Networks of this kind are often referred to as *interaction maps* or *networks*. *Network Biology* (Forst 2002; Barabási and Oltvai 2004) is a method that studies (inter/intra) cellular networks and their genomic, proteomic, and metabolomic foundations. Network biology assembles the basis of systems biology in providing information on biological components, their interactions, and their functional interplay in biological networks. One particular aspect of network biology is focused on the graph-structure of the underlying interaction map by providing quantifiable measures such as node degree distribution, mean path length, and clustering coefficients, as well as by identifying architectural features such as the existence of motifs and modules and their hierarchical structure (Strange 2005). These measures can be particularly interesting when linked to phenotypic properties of the biological system such as system survival. Jeong et al. (2001) have shown in a yeast protein interaction network, that proteins essential for survival are highly connected.

Qualitative connections and causality indicate how input nodes affect the output nodes. Directional edges reflect causal relationships as well as qualitative types of interactions (e.g., activating or inhibitory interactions). Qualitative models include metabolic flux-models assuming steady state conditions (Schilling et al. 1999; Schuster et al. 2000), some considering gene-regulation (Covert and Palsson 2002), or Boolean network models (Stuart and Kauffman 1969; Kauffman 1993). At the level of quantitative connections, such functions are assigned sets of interactions that describe the dynamic co-dependent behavior of inputs and an output. Methods of choice cover power law models, such as S-systems (Savageau 1969, 1970), reaction kinetics modeled by ordinary differential equations (Ackers et al. 1982; Novak and Tyson 1993), or stochastic simulations (Gillespie 1977; McAdams and Arkin 1998). For example, with respect to cell-signaling cascades (see Sect. 11.4.2), Goldstein et al. (2004) propose a three-part protocol for defining such a mathematical model:

- Selection of a set of components and identification of their interactions on the basis of what is known about the system
- Selection of parameters that quantify the cellular concentrations of the components and the strength of the interactions between components (known as "rate constants")
- Selection of a mathematical formulation of a method of simulation. Reaction-network models are based on the assumption that each species is uniformly distributed throughout the cell. Reaction-diffusion models allow for the variation of species concentrations in different cellular compartments

At the level of higher order interactions and reaction rules, higher-level nodes and connections represent abstract concepts that can be expanded into sub-level sub-graphs on the basis of reaction rules in a hierarchical fashion. Examples include signaling networks and metabolic reactions, with context dependent or rule-based interactions and different types of nodes (Faeder et al. 2003).

6.3.3 Static Networks

Although the pathway analysis experiences a renaissance with the establishment of systems biology, the field of research exploring biological pathways is more than 100 years old. Seminal contributions by Haldane (1924), Miller (1953), Oparin (1957), and Orgel (1968) are noteworthy in this context, especially with respect to the evolution of metabolic networks. The breakthrough of large-scale network analysis came with the post-genomic revolution and the utilization of the plethora of genomic data through publicly available databases. Early examples of such databases established in the 1990s are the Kyoto Encyclopedia of Genes and Genomes (KEGG) (Kanehisa et al. 2004), Karp's and Riley's EcoCyc (Karp et al. 2002), and the WIT-system (now commercialized as ERGO: http://ergo.integratedgenomics.com/) by Overbeek et al. (2000). Nowadays, a large selection of biological network databases is accessible, with a variety of genomic, interaction, and network information. PathGuide (http://www.pathguide.org/) lists almost 300 biological pathway resources on protein–protein interactions, metabolic pathways, signaling pathways, pathway diagrams, gene regulatory networks, protein–compound interactions, and genetic interaction networks on its web site.

Research areas, such as graph theory and statistical physics, contribute significantly to the understanding of biological networks. Generic graph properties and their scale-free nature have been described by the pioneering work of Barabasi and his coworkers (Barabasi and Oltvai 2004a, b). Following this network biology approach, further large-scale statistic and detailed graph-topological analyses have been performed on biological networks. For example, Barabasi et al. identified hierarchically organized sub-networks within large biological networks, Girvan and Newman identified "Community Structures" (Girvan and Newman 2002), and Alon and his group classified all possible network modules with six or less modules and identified particular functional circuits, such as feed-forward and feed-backward loops (Alon 2007).

6.3.4 Response Networks

With respect to molecular-biological networks, *response networks* were first mentioned by Magasanik (1995). Groundwork for a systematic, theoretical analysis of response networks was laid by Zien et al. (2000) and further developed independently by Ideker et al. (2002). The idea behind response network analysis is the analysis of experimental data, such as expression profiles, in the context of biological networks. Through a superposition of experimental data with network information, it has become possible to identify networks that best represent the system response according to the experimental conditions tested. In principle, graph comparison is a NP hard problem, which typically can only be addressed by exhaustive enumeration techniques. On the other hand, methods for comparative network analysis for biological systems have been developed in the past. Such methods have been proven powerful in a number of applications including metabolic (Dandekar et al. 1999; Forst and Schulten 1999, 2001; Ogata et al. 2000) and protein interaction networks (Kelley et al. 2003) as well as in the correlation of protein interaction networks with gene expression (Nakaya et al. 2001). Recently, a method has been developed to correlate and compare response networks for the identification of common and specific responses (Cabusora et al. 2005).

6.3.5 Modeling Techniques

Different modeling techniques are chosen depending on the dynamics under consideration at a particular level of description, particularly with respect to quantitative modeling (Sect. 6.3.2). Two major types of models can be distinguished – mathematical models using equations and computer algorithms applying a detailed set of rules. Mathematical models include deterministic and stochastic differential equations, discrete Boolean networks, and statistical methods. Computational algorithms encompass agent-based models (ABMs), such as cellular automata, or event-based simulations (e.g., Petri-Nets).

A plethora of tools and software on biological systems modeling have been developed and are available for download, often free for academic users. Many modeling tools are using the Systems Biology Markup Language (SBML) for portable model description, since SBML (http://www.systems-biology.org/) has been developed by Hucka and co–workers (Hucka et al. 2003). Because of the wealth of tools available, we refer the potential user to the Systems Biology website by "Kitano's Symbiotic Systems Project" (http://www.systems–biology.org/), which lists model editors (http://www.systems–biology.org/002/008.html) for graphic assisted model construction, simulation tools (http://www.systems–biology.org/002/005.html) for deterministic and stochastic simulations, analysis tools (http://www.systems–biology.org/002/006.html), and utilities (http://www.systems-biology.org/002/007.html). Physiology modeling software is not yet well integrated with molecular and cellular modeling tools. A list of physiological modeling groups and tools can be found at "The Federation of American Scientists" web site (http://www.fas.org/main/content.jsp?formAction = 297&contentId = 94).

6.4 Intracellular Models

Host–pathogen system models, which fall in the "omics" category, comprise interaction maps or interaction networks that represent components of a network and their interactions for further analysis.

6.4.1 Genomic Foundation of Host-Pathogen Interactions

One example of a true host–pathogen network model is the tryptophan (trp) biosynthesis network of a class of obligate intracellular pathogens, *Chlamydiae* (Fig. 6.2a, Forst 2002). Chlamydia primarily infect mucosal epithelial cells with consecutive infection of subepithelial tissue (Campbell et al. 1993). Chlamydia infections progress through a life-cycle with three distinct stages. The host is invaded by *elementary bodies* (EBs), which represent the extracellular infectious stage. After infection, EBs develop into intracellular *reticular bodies* (RBs), which replicate and further maturate into EBs, which then lyse the host-cell and initiate another round of pathogen infection. The cycle between EBs, RBs, and lysis of host-cells characterizes the acute disease state. A third state of development, the persistence state, is recognized, and describes the chronic disease progress. In tissue culture, the persistence state of Chlamydiae can be introduced by various factors, specifically interferon-γ (IFN-γ), nutrient limitations, or other environmental stress. For example, it is well documented that tryptophan levels in host cells decrease because of an effect of IFN-γ (Girvan and Newman 2002; Byrne et al. 1986). It has further been recognized that tryptophan depletion may play a role in the development of chronic disease conditions (Beatty et al. 1994).

Investigating the tryptophan biosynthesis pathway in a particular *Chlamydiae* species, *Chlamydia psittaci*, is interesting, because it shows the interdependence and interconnectivity between pathogen and host. Thus, it may help to explain the development of the chronic disease. This pathway assembles an almost complete biosynthetic unit. Interestingly, genes encoding the enzymatic subunits *trpA*α and *trpA*β, which are typically present in tryptophan operons and which are responsible for catalyzing the reaction from chorismate to anthranilate, are absent in the *C. psittaci* tryptophan operon and are not encoded elsewhere in the genome. Instead, the *C. psittaci* tryptophan operon includes two genes *kynU* and *kprS*, both of which are atypical components of the classic tryptophan operon (Fig. 6.2b). This can only be deciphered by systematic metabolic network analysis. *KynU* encodes kynureninase, an enzyme that converts kynurenine into anthranilate. *KprS* codes for 5-phospho-D-ribosyl-1-pyrophosphate (PRPP) synthetase, a component needed in the first steps of tryptophan biosynthesis (Fig. 6.2a). The complete tryptophan network, including the tryptophan-salvage pathway of the host, is shown in Fig. 6.2a (the pathway starts at IFN-γ with black arrows outside the oval). For tryptophan biosynthesis, *C. psittaci* obtains an alternative source of anthranilate by hijacking the host's tryptophan depletion pathway by intercepting the by-product kynurenine. At first, the tryptophan depletion pathway of the host is activated by inducing indoleamine-2,3-dioxygenase through IFN-γ (reaction with EC-number[1] 1.13.11.11 in Fig. 6.2a). Then, *C. psittaci* utilizes host kynurenine by kynU to produce its own tryptophan, enabling intracellular growth and causing

[1]The classification system for enzymes and biochemical reactions by the Nomenclature Committee of the International Union of Biochemistry and Molecular Biology (NC-IUBMB): http://www.chem.qmul.ac.uk/iubmb/enzyme/

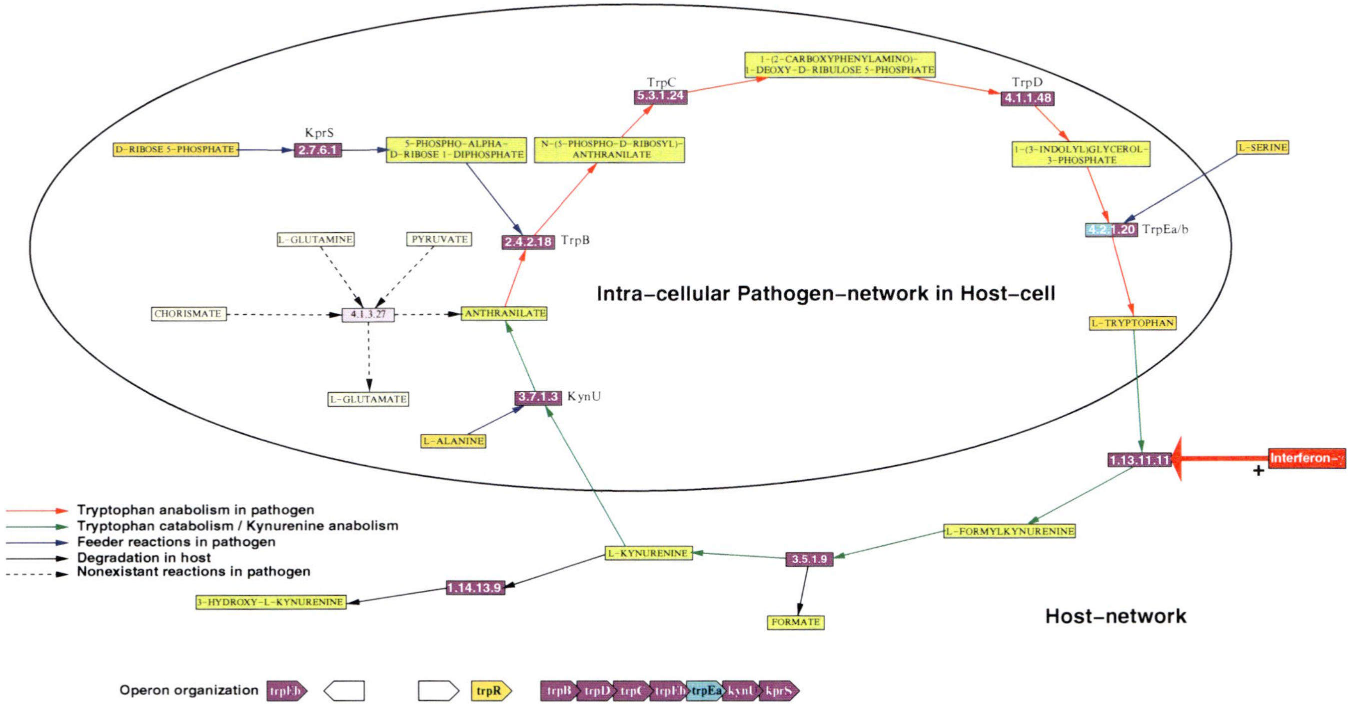

Fig. 6.2 (**a**) A shared trp metabolic network between the pathogen *C. psittaci* and a human host. The sub-network in the oval denotes the pathogen part of the network. *Red thin arrows* indicate anabolic trp reactions in the pathogen, *green arrows* are catabolic reactions, *blue arrows* refer to feeder reaction that provide required compounds (L-alanine (ala), ser and PRPP) for trp biosynthesis, and *black solid arrows* denote degradation reactions in the host. The *black dashed arrows* are reactions not found in the pathogen. The *red bold arrow* describes the activation of indoleamine-2,3-dioxygenase (EC 1.13.11.11) by IFN-γ. (**b**) The operon organization of the trp-operon in *C. psittaci* is shown below. Unusual for typical trp-operons is the presence of *kprS* and *kynU*. *Numbers* above the genes refer to intergenic distances [with kind permission of Springer Science and Business Media (Forst 2002)]

chronic infections. The knowledge of such a metabolic host-pathogen system accelerates drug development of successful antimicrobials against chronic chlamydial infection.

6.4.2 *Large-Scale Host Response Models*

The application of massive omics data analysis in the context of biological networks (Cabusora et al. 2005; Ideker et al. 2001) leads directly to large-scale studies of (human) host response. Particularly in the case of viral-host systems, the whole disease process is triggered by few viral components and the host cellular machinery is required for the viral life cycle. Thus, studying viral-host systems by analyzing host response is a well-justified and valid approach. One of the first comprehensive analyses of the global host response of viral infection was conducted by Kash and coworkers in the case of the 1918 influenza virus induced response in an animal model (Kash et al. 2006). A biological network of selected genes that were induced more than twofold ($P < 0.01$) in the lungs of mice infected with the recombinant 1918 influenza virus (r1918), as compared with uninfected controls, is used to depict the activation of cell-death responses during r1918 infection. Another example is a recent study of human response against the avian influenza virus H5N1 in an epithelial cell culture model. By combining the biological response network with the hierarchical knowledge representation from the gene ontology (GO), one can easily identify essential host response processes, such as "immune response" and "response to virus," as well as cell cycle activation (Tatebe et al. 2009). Figure 6.3 presents a combined network of host cell response 24 h after infection with an H5N1 influenza strain. Other success stories of large–scale network biology analysis include the identification of transcriptional regulatory networks governing the latency and early reactivation phases of HIV-1 (Bandyopadhyay et al. 2005), and the inference of virus-host protein interaction maps for two herpesviruses, Kaposi sarcoma-associated herpesvirus and varicella zoster virus, by comparative network proteomics using yeast two-hybrid network data (Uetz et al. 2006). Such gene-regulatory and protein interaction network information and large-scale virus-host interaction data boost our knowledge of the function of many still poorly understood viral proteins, as well as the large number of remaining "unknown" genes in host pathways. This will lead to a more detailed understanding of viral pathogenesis and will provide potential new targets for interfering with either the virus or the host at key points in the infection (Torigoe et al. 1998; Valitutti et al. 1995; Wofsy et al. 2001).

6.4.3 *Immune-Receptor Signaling*

Mathematical and computational network models have studied the process of signaling through receptors of the immune system (Magasanik 1995). Mathematical, dynamic models of immune-receptor signaling are essentially performed on two

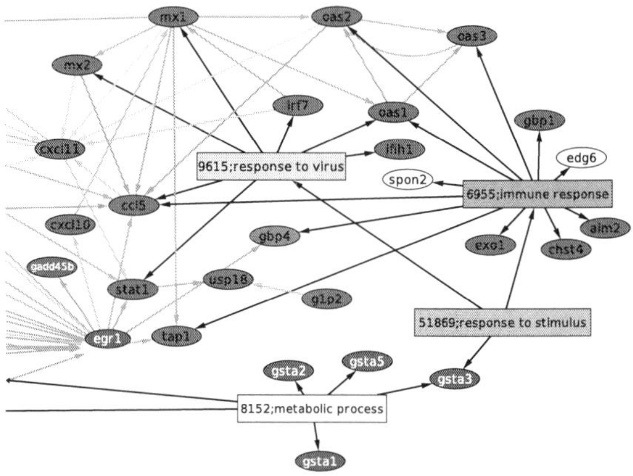

Fig. 6.3 Network of significant genes and GO nodes at 24 h: The gene ontology nodes found to be significant in the BiNGO analysis are shown as *rectangles*, with the *orange* nodes being more statistically significant. The genes associated with the GO nodes are listed in *ovals* connected by *black arrows* to the GO nodes. These genes are further connected to other genes in the Human Network via *yellow* and *orange arrows*. *Red ovals* indicate up-regulated genes while *blue ovals* indicate down-regulated genes

distinct levels of description (1) "simple models" and (2) "detailed models" (see Sect. 6.3 and Box 6.1).

Using the FcεRI receptor as an example, Faeder et al. (2003) have developed a detailed signaling model that takes into account downstream components affecting the signaling cascade (Fig. 6.4). Figure 6.4a shows the four components of the receptor, the ligand (IgE dimer), the receptor (FcεRI), and the two kinases Lyn and Syk. The nine basic interactions are shown in Fig. 6.4b, which include association and dissociation, transphosphorylation, i.e., catalysis of phosphorylation, and dephosphorylation. A surprising aspect of this model is that, because of combinatorial complexity, four components and nine interactions expand to a signaling network with 354 species and 3,680 reactions (one particular reaction "species" is depicted in Fig. 6.4c). The simulation results of the FcεRI signaling model show complex behavior of phosphorylation profiles because of competing behavior or two previously known but independently observed phenomena in cell signaling, i.e. kinetic proofreading and serial engagement. Kinetic proofreading is a process in signaling cascades where required preservation of initial receptor interaction during the subsequent time-dependent steps increases the fidelity of the response (Torigoe et al. 1998). Serial engagement (Valitutti et al. 1995; Wofsy et al. 2001) is observed in T cell signaling in which a single MHC peptide can stimulate a substantial number of TCRs. The complex phosphorylation profiles as a function of the ligand–receptor off-rate are induced by the change of the balance between kinetic proofreading and serial engagement changes, moving down the signaling cascade. Because serial engagement increases with the off-rate, an increase in phosphorylation with off-rate

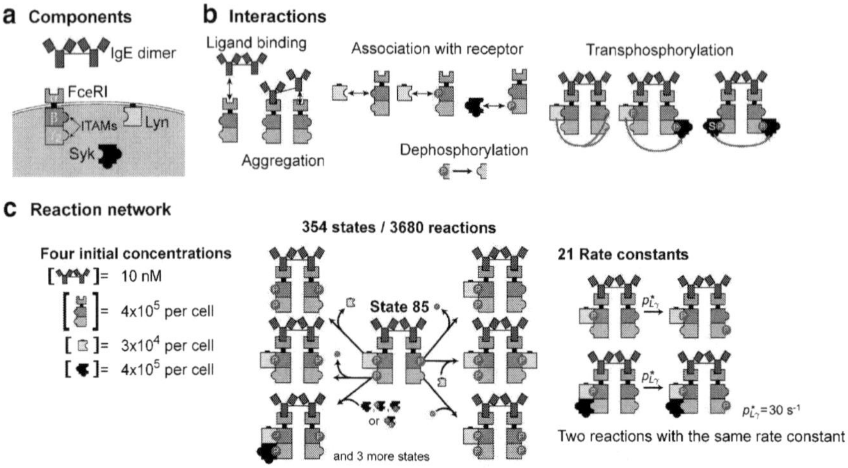

Fig. 6.4 A detailed model of early events in FcεRI signaling. (**a**) The four components in the model are the IgE dimer, the receptor (FcεRI), and the kinases Lyn and Syk. Of the two cytosolic domains of the receptor each contains an immunoreceptor tyrosine-based activation motif (ITAM). (**b**) There are nine basic interactions, five for association/dissociation between signaling components, three transphosphorylation reactions, and one for the spontaneous dephosphorylation of phosphorylated sites. (**c**) Considering all possible combinations between components, basic interactions yield 354 complexes and phosphorylation states, each of which is tracked as a separate species. The species are connected by 3,680 reactions assembling a large biological network that is defined by a small number of parameters (the initial conditions of 4 proteins and 21 rate constants). One typical species is illustrated along with nine different reactions, of which six are explicitly shown. Reactions seven to nine are generated by using different phosphorylation states of Syk (*gray square*) to form additional states from the complex in the center to the bottom-left complex (indicated by "and 3 more states"). The states are connected by a large biochemical reaction network (composed of 3,680 reactions). A small number of parameters, the initial concentrations of the 4 proteins and 21 rate constants, define this network because the same rate constant can be used for many similar reactions. The figure shows 2 of the 24 reactions in which Lyn transphosphorylates γ-ITAM. The $p^{*}_{L\gamma}$ indicates the reaction rate of these two reactions [reproduced with permission from Nature Reviews Immunology (Goldstein et al. 2004) copyright (2004) Macmillan Magazines Ltd]

indicates that serial engagement is the dominant effect, whereas a decrease indicates that kinetic proofreading is dominant. Depending on the timely occurrence of a particular phosphorylation event, either kinetic proofreading or serial engagement is dominant. For example, the phosphorylation profile of γ-ITAM (Immunoreceptor Tyrosine-based Activation Motif) passes through a maximum indicating a transition between control by kinetic proofreading and control by serial engagement. Thus, the detailed signaling model shows that kinetic proofreading and serial engagement are emergent properties and the interplay of these mechanisms gives rise to an optimal off-rate at which the highest response is achieved.

The ultimate goal of immune-receptor signaling models is to understand how the components of a signaling cascade work together to direct cellular responses to changes in the extracellular environment. Both simple and detailed mathematical models have contributed to the understanding of essential host-pathogen signaling events through immune receptors by identifying the fundamental mechanisms that are involved in determining, regulating, and therapeutically modifying immune responses.

6.5 Intercellular or Cell Host-Pathogen Interaction Models

Complementing the molecular biology modeling approach described above, the following models capture the dynamics of the immune response to infecting pathogens at the inter-cellular level in host–pathogen systems biology. A comprehensive review of the mathematic modeling techniques of immune systems has been presented by Perelson and Weisbuch (1997). Such modeling techniques are deeply rooted in Theoretical Biology, one of the foundations of systems biology. Below is the model of Anderson and May (Anderson and May 1980, 1981), which describes insect diseases. The host population consists of two portions: susceptible or healthy organisms, and infected individuals. The model captures changes in the density of susceptible (S), infected individuals (I) and pathogens (P):

$$\begin{aligned}\frac{\mathrm{d}S}{\mathrm{d}t} &= r*(S+I)-\gamma P*S,\\ \frac{\mathrm{d}I}{\mathrm{d}t} &= \gamma P*S-(\alpha+B)*I,\\ \frac{\mathrm{d}P}{\mathrm{d}t} &= \gamma*I-[\mu+\gamma(S+I)]*P.\end{aligned} \tag{6.1}$$

Parameters in this model are r, reproduction rate; γ, transmission coefficient; α, death rate of infected individuals; B, natural death rate; λ, pathogen emission rate; and μ, rate of natural pathogen decay.

The basic idea of viral infection models is simple and lead to the development of *viral dynamics* as a research field (Nowak and May 2000). Analogous to the model by Anderson and May (above), viral infection models consist of three types of cells, target cells T, infected cells I, and virus particles (virions) V. Infected cells produce new virus particles at a constant rate p and die at rate δ. Virions are cleared by the immune system at rate c. The rate in which a target cell is infected is k.

$$\begin{aligned}\frac{\mathrm{d}T}{\mathrm{d}t} &= \gamma-\mathrm{d}T-kVT,\\ \frac{\mathrm{d}I}{\mathrm{d}t} &= kVT-\delta I,\\ \frac{\mathrm{d}P}{\mathrm{d}t} &= pI-cV.\end{aligned} \tag{6.2}$$

The above equations, motivated by (6.1) and further developed by Perelson et al. (1996), describe the basic model of viral dynamics and have been used to study primary human immunodeficiency virus (HIV) infection. More complex models include specific components of the immune systems, such as the cytotoxic T lymphocytes (CTLs; DeBoer et al. 2001), other non-toxic lymphocytes, and cytokines/chemokines (Wodarz et al. 2000). These models essentially include specific expressions for resting, active, memory, and cytotoxic T-cells. Mathematical models of HIV infections have also proved to be useful in exploring the response of HIV to antiviral therapy. Specifically, the evolution and evasion of HIV under the selection pressure of the immune system and drug-treatments have been extensively studied. A review by Frost discusses the benefit of evolutionary dynamic HIV models for the understanding of HIV response to highly active antiretroviral therapy (Simon and Frost 2002). This review specifically discusses the role of such models in the design and analysis of structured treatment interruption studies, to reduce drug toxicities, to boost HIV-specific immune responses, and to allow drug resistance mutation to be reversed in highly drug-experienced patients.

Building upon these pharmacokinetic models of HIV evolution in human hosts, Dixit and Perelson developed a hybrid model of HIV dynamics under antiretroviral therapy that combines pharmacokinetics and intracellular delay, the time which is required for an infected cell to replicate virus (Dixit and Perelson 2004). This model helps to accurately determine the pharmacological delay and the time-dependent efficacy of drug action.

Particularly in the case of the tuberculosis causing pathogen *Mycobacterium tuberculosis*, tremendous advances have been made in the modeling of host-pathogen interactions on multiple scales. For example, Denise Kirschner and her group together with experimentalists have developed a suit of approaches and tools to study the interaction of the immune system with the intracellular pathogen *M. tuberculosis* at a number of biological and spatial levels. The starting point for these studies was a simple model of a mycobacteria-macrophage interaction, the preferred host cell for *M. tuberculosis*, in the lung (Wigginton and Kirschner 2001). This model, implemented as ODEs, tracked the time-evolution of the concentrations of three sub-populations of macrophages, two bacterial populations, three T-helper cell populations, and four key cytokines. It predicted that latency, active disease, and clearance could be observed under different host conditions. Kirschner and coworkers later extended this basic model to include CD8+ T cells (Sud et al. 2006) and the tumor-necrosis factor α (TNFα) (Marino et al. 2007).

A significant improvement towards multi-cell models and the analysis of infections on the tissue and organ level has been made by considering heterogeneous models of TB infections due to variation within cell populations. Structured models (Cushing 1998) describing, for example, different developmental stages of immune cells have been used to generalize age-structured population models known from theoretical ecology (Blasi et al. 1982) and epidemiology (Auranen et al. 2004). One example was a multi-scale model that combined mutual inhibition of two key transcription factors with a cell-population model to describe T_H-cell differentiation (Yates et al. 2004).

Using ABMs of spatially distributed cell population and tissues is an elegant approach to close the gap to physiological models discussed in Sect. 6.6. ABMs are the preferred modeling techniques in situations in which overall numbers are small or in which simulation of discrete individual behavior is desired. As such behavior becomes more sophisticated, ABMs offer a powerful approach to integrating individuals with the next scale above. ABMs have been used in medical setting in spatial models of wound healing (Drasdo 2003) and tumor growth (http://www.biomedtown.org/biomed_town/VPH/VPHnews/tumather) (Alber et al. 2003). Kirschner and coworkers applied ABMs to develop a two-dimensional model for simulating both the spatial and temporal events of granuloma formation and maintenance (Segovia-Juarez et al. 2004). Granulomas are characteristic multicellular structures within the lung tissue of infected individuals. Granuloma formation is a complex process involving interactions of bacteria, specific immune cells, including macrophages, CD4+ and CD8+ T cells, as well as immune effectors such as chemokines and cytokines. The formation and dynamics of these granulomas potentially play a central role in the pathogenesis of the disease. Two-dimensional lattice ABMs representing 2 × 2 mm of lung tissue were developed (Segovia-Juarez et al. 2004). As agents, T cells, macrophages in different stages (resting, activated, infected, chronically infected, or dead), bacteria, and tissue were included. The model predicted three distinct and robust infection outcomes :(1) a granuloma that was tightly packed, small, and showed little necrosis, and that was able to contain bacterial growth; (2) a larger and more diffuse granuloma that failed to restrict bacterial dissemination; and (3) a granuloma that cleared the bacteria load altogether and then dispersed.

6.6 Large Scale Models of Host–Pathogen Physiology

Physiological models go a step further. These models integrate molecular, cellular, and organ levels in a top-down approach to put in place an organ-level framework and to add increasing complexity in a modular format. With the beginning of the new millennium, a number of funding networks, consortia, and projects have been formed to target the multilevel modeling and simulation of the human physiology. Most noteworthy are two main physiology modeling approaches: the publicly funded and developed international *IUPS Physiome Project* at the University of Auckland in New Zealand (http://www.physiome.org.nz/), lead led by Peter Hunter, and the proprietary *PhysioLab*™ by Entelos, Foster City, CA led by Cynthia Stokes. Other efforts in the public domain include the *Virtual Soldier Research* at The University of Iowa with the objective of developing a new generation of digital humans by creating realistic human models related to anatomy, biomechanics, physiology, and real-time intelligence, as well as the *Digital Human Project*, which targets the representation of the body's processes from DNA molecules and proteins to cells, tissues, and gross anatomy. Another company that uses the multi-scale physiological model similar to Entelos is Rosa & Co., with PhysioPD (http://www.rosapharma.com/services/physio_pd) as their main R&D application. This company is focused on the rapid development of decision-focused models that are used to assess which uncertainties will have the greatest impact on outcomes. Depending on the decisions, these models range from population pharmacokinetic models to full-scale physiologically based population pharmacokinetic/pharmacodynamic models with clinical trial simulations. In order to provide more value, Rosa & Co. developed these models with the client in the belief that most insights arise from the model creation process.

Worthwhile to mention with respect to host–pathogen systems biology are efforts put forward by the European Commission, implementations of the Virtual Physiological Human Network of Excellence (http://www.vph-noe.eu/) within the EuroPhysiome (http://www.europhysiome.org/) roadmap: ImmunoGrid (http://igrid-ext.cryst.bbk.ac.uk/immunogrid/site), the European Virtual Human Immune System Project and TUMATHER (http://calvino.polito.it/~mcrtn/), the European Commission Research Training Network on modeling, mathematical methods, and computer simulation for tumor growth.

Any attempt to link molecular and cellular events with physiological function must deal with length scales that range from 1 nm (typical diameter of a protein) to the 1 m scale of a human body. Similarly, the range of timescales must encompass the 1 μs that is characteristic of Brownian motion and the 10^9 s of a human lifetime. It is clear that no single model can cover a factor of 10^9 in space and a factor of 10^{15} in time (see Fig. 6.1). A more reasonable approach is to develop models for a limited range of spatial and temporal scales and to develop techniques to link the parameters of this hierarchy of models. This means that, at any one level, there is a "black box" that groups all of the detail at the level below (in either spatial or temporal sense) into a mathematical expression. The parameters of this expression are determined directly from experiments, but can be related to another, more detailed, model at the finer spatial or temporal level. For example, according to a review by Peter Hunter and Thomas Borg (Hunter and Borg 2003), *"The Physiome Project will provide a framework for modeling the human body, using computational methods that incorporate biochemical, biophysical and anatomical information on cells, tissues and organs. The main project goals are to use computational modeling to analyze integrative biological function and to provide a system for hypothesis testing."*

The goals of the IUPS Physiome Project are the following:

- To develop and capture observations of physiological phenomenon and interpret theme in terms of mechanisms (a fundamentally reductionist goal)
- To integrate experimental information into quantitative descriptions of the functioning of humans and other organisms (modern integrative biology glued together via modeling)
- To disseminate experimental data and integrative models for teaching and research
- To foster collaboration among investigators worldwide, and thereby speed up the discovery of how biological systems work

- To determine the most effective targets (molecules or systems) for therapy, either pharmaceutical or genomic
- To provide information for the design of tissue-engineered, biocompatible implants

One example of the hierarchical modeling technique based on the IUPS Physiome approach is the integrative cardiac model (Winslow et al. 2000). The authors have developed a method for the risk stratification and treatment of Sudden Cardiac Death syndrome (SCD). Their approach is to collect data from the molecular to the organ level and to develop integrative models of the normal and failing heart. Gene expression profiles are used to build gene-regulatory network maps. Together with measurements on Ca^{2+}, K^{+} concentration and action potential time-series, Winslow et al. developed cellular models of the myocytes (Fig. 6.5a) which have been combined with 3D spatiotemporal models of the heart using Reaction-Diffusion Equations that include parameters from the ionic models as well as from the micro-anatomy of the ventricles (Fig. 6.5):

$$\frac{\partial v(\mathbf{x},t)}{\partial t} = \frac{1}{C_{\mathrm{m}}}\left[-I_{\mathrm{ion}}(\mathbf{x},t) - I_{\mathrm{app}}(\mathbf{x},t) + \frac{1}{\beta}\left(\frac{k}{k+1}\right)\nabla\cdot(M_i(\mathbf{x})\nabla v(\mathbf{x},t))\right],$$

where $\mathbf{x}$ is the spatial position, $v(\mathbf{x}, \mathrm{t})$ is the transmembrane voltage, C_{m} is the membrane capacitance per unit area, $I_{\mathrm{ion}}(\mathbf{x}, t)$ is the sum of the ionic currents per unit area through the membrane (from the ionic models), $I_{\mathrm{app}}(\mathbf{x}, t)$ is an applied stimulus current per unit area, β is the ration of membrane area to tissue volume, k is the anisotropy ration, and $M_i(\mathbf{x})$ is the intracellular 3×3 conductivity tensor at each point $\mathbf{x}$ (Musante et al. 2002). Entelos (Foster City, CA, USA) has developed complex simulations of disease physiology using a framework, PhysioLab (http://www.entelos.com/physiolabModeler.php), for determining differential equations based on empirical data in humans (Musante et al. 2002). In these models, cells or even tissues are represented as *black-boxes*, without explicit internal network models, that respond to inputs by providing specified dynamic outputs. Using such an organ level framework of a disease physiology, Stokes and colleagues have developed a

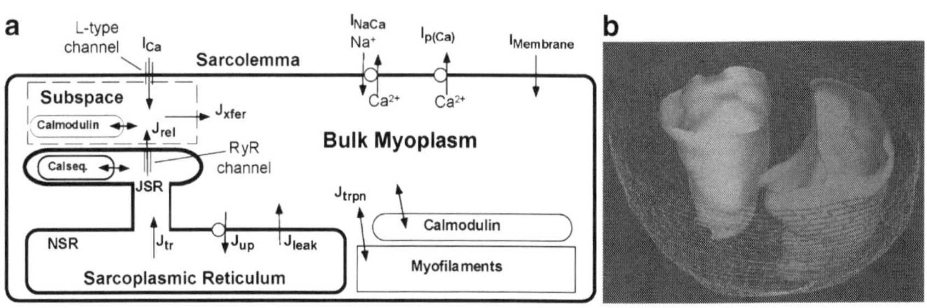

Fig. 6.5 (**a**) Schematic diagram of mechanisms involved in intracellular Ca^{2+} cycling in cardiac ventricular myocytes. (**b**) Reconstruction of epicardial (*blue wire mesh*) and endocardial surfaces (right ventricle endocardium - *red*; left ventricle endocardium - *gold*) (c.f. Winslow et al. 2000)

computational model of chronic asthma that includes interactions among cells, and their response to one another and their environment (Stokes et al. 1999). Different steady states of this disease, including chronic eosinophilic inflammation, chronic airway obstruction, airway hyper-responsiveness, and elevated IgE levels, can be induced in the model (Fig. 6.6). The in silico asthmatic model responds as expected to various drugs, such as β_2-agonists, glucocorticoids, and leukotriene antagonists. Furthermore, this model accurately predicts a decrease in airway eosinophils without much therapeutic improvement in airway conduction after reduction in the interleukin (IL)-5 protein as observed in clinical trials of an anti-IL-5 antibody in asthmatic patients.

6.7 Conclusion

Many aspects of host-pathogen systems have been addressed by both experimental discoveries and mathematical/computational models. However, a comprehensive analysis of entire host-pathogen interaction (including pathogen interference, host-response, pathogen-response, etc.) is a far fetched goal. Current models focus predominantly on host responses to infections or pathogen biology in a simulated host-environment. The size of these model systems depends essentially on the detail of description. They range from large interaction maps (bottom-up models) with thousands of components and interactions, through conceptual top-down cellular or organ models consisting of interacting *black-boxes* without detailed knowledge of internal processes within each individual *black-box*, to small mechanistic dynamic models describing few steps of a much larger system.

Efforts are under way to combine both pathogen action and host response in a comprehensive, multi-scale (hybrid) model, merging top-down approaches with "omic" bottom-up approaches (Kirschner and Marino 2005). Integrated with sophisticated experimental techniques, such as quantitative protein expression, tags by quantum dots for localization, and nano-biotechnological measurements on single cells, they promise new insights into the complexity of host–pathogen systems. Potential applications of host–pathogen systems biology range from biological target identification and drug discovery to bio-threat assessment and personalized health care. As with any modeling approach, theoretical models raise the challenge of experimental validation and the iterative cycle of improvement inherent to the modeling effort. With respect to drug discovery, success stories are still anecdotal. Until a given model shows a track record of successful predictions it will be risky to rely on it for drug development decisions. For the foreseeable future, modeling predictions will most likely be only one of many inputs into the decision making process in the pharmaceutical industry. A long-term goal for host-pathogen systems biology, as envisioned by an increasing number of national and international funding opportunities and seed projects, would include full scale in silico models of an individualized human fighting against pathogenic infections.

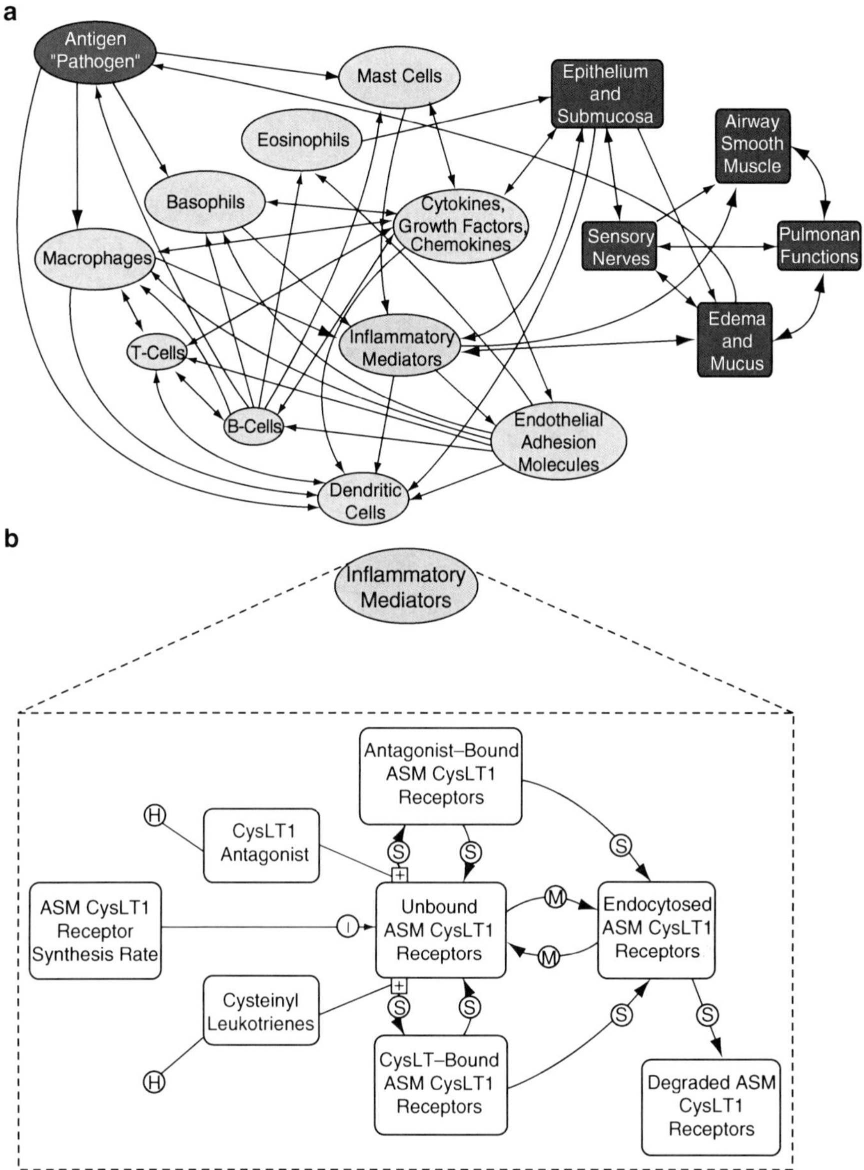

Fig. 6.6 Systems model of asthma physiology (cf. Lewis et al. 2001). (*Top*) The overall systems diagram shows the phenomenological interaction between the antigen trigger, cytokines, immune system cells, and organ response. The *bold pathway*, for example, indicates leukotrine production by macrophages and induced asthmatic response. (*Bottom*) The molecular-level details of the "Inflammatory Mediators" black-box (*top*) show pathways modulating the effect of cysteinyl leukotriene on Airway Smooth Muscle (ASM) via cysteinyl leukotriene receptor subtype 1 (CysLT1). *Arrow labels* correspond to following functions: H half-life, I increases, M moves, S change of state, + stimulates

References

Ackers GK, Johnson AD, Shea MA (1982) Quantitative model for gene regulation by lambda phage repressor. Proc Natl Acad Sci USA 79:1129–1133

Aderem A (2005) Systems biology: its practice and challenges. Cell 121:511–513

Akira S, Takeda K (2004) Toll-like receptor signaling. Nat Rev Immunol 4:499–511

Alber M, Kiskowski MA, Glazier JA, Jiang Y (2003) On cellular automaton approaches to modeling biological cells. In: Rosenthal J, Gilliam DS (eds) Mathematical systems theory in biology, communications, computation, and finance. Springer, Berlin, pp 1–40

Alizadeh AA, Eisen MB, Davis RE et al (2000) Distinct types of diffuse large B-cell lymphoma identified by gene expression profiling. Nature 403:503–511

Alon U (2007) Network motifs: theory and experimental approaches. Nat Rev Genet 8:450–461

Anderson RM, May RM (1980) Infection diseases and population cycles of forest insects. Science 210:658–661

Anderson RM, May RM (1981) The population dynamics of microparasites and their vertebrate hosts. Philos Trans R Soc Lond B 291:451–524

Auranen K, Eichner M, Leino T, Takala AK et al (2004) Modelling transmission, immunity and disease of *Haemophilus influenzae* type b in a structured population. Epidemiol Infect132:947–957

Bandyopadhyay S, Kelley R, Ideker T (2005) Discovering regulated networks during HIV-1 latency and reactivation. In: Proceedings of the Pacific Symposium on Biocomputing, pp 354–366

Barabasi A-L, Oltvai ZN (2004a) Network biology: understanding the cell's functional organization. Nat Rev Genet 5:101–113

Barabási A-L, Oltvai ZN (2004b) Network biology: understanding the cell's functional organization. Nat Rev Genet 5:101–113

Beatty WL, Morrison RP, Byrne GI (1994) Persistent Chlamydiae: from cell culture to a paradigm for chlamydial pathogenesis. Microbiol Rev 8:686–699

Blasi GD, Iannelh M, Sinestrari E (1982) Approach to equilibrium in age structured populations with an increasing recruitment process. J Math Biol 13:371–382

Bower JM, Bolouri H (2001) Computational modeling of genetic and biochemical networks. MIT, Cambridge, MA

Butcher EC (2005) Can cell systems biology rescue drug discovery?. Nat Rev Drug Discov 4:461–467

Butcher EC, Berg EL, Kunkel EJ (2004) Systems biology in drug discovery. Nat Biotechnol 22:1253–1259

Byrne GI, Oeyjahhkn LK, Landry GJ (1986) Induction of tryptophan catabolism is the mechanism for gamma-interferon-mediated inhibition of intracellular *Chlamydia psittaci* replication in T24 cells. Infect Immunol 53:347–351

Cabusora L, Sutton E, Fulmer A, Forst CV (2005) Differential network expression during drug and stress response. Bioinformatics 21:2898–2905

Campbell LA, Patton DL, Moore DE, Cappuccio AL et al (1993) Detection of *Chlamydia trachomatis* deoxyribonucleic acid in women with tubal infertility. Fertil Steril 59:45–50

Covert MW, Palsson BO (2002) Transcriptional regulation in constraint-based metabolic models of *Escherichia coli*. J Biol Chem 277:28058–28064

Cushing JM (1998) An introduction to structured population dynamics. Society for Industrial and Applied Mathematics, Philadelphia, PA

Dandekar T, Schuster S, Snel B, Huynen MA, Bork P (1999) Pathway alignment: application to the comparative analysis of glycolytic enzymes. Biochem J 343:115–124

DeBoer RJ, Oprea M, Antia R, Murali-Krishna K et al (2001) Recruitment times, proliferation, and apopotosis rates during the CD8+ T cell response to LCMV. J Virol 75:10663–10669

Dixit NM, Perelson AS (2004) Complex patterns of viral load decay under antiretroviral therapy: influence of pharmacokinetics and intracellular delay. J Theor Biol 226:95–109

Drasdo D (2003) On selected individual-based approaches to the dynamics in multicellular systems. In: Alt W, Chaplain M, Griebel M, Lenz J (eds) Polymer and cell dynamics. Birkhäuser, Switzerland, pp 169–204

Ernst JD (1998) Macrophage receptors for *Mycobacterium tuberculosis*. Infect Immunol 66:1277–1281

Faeder JR, Hlavacek WS, Reischl I, Blinov ML et al (2003) Investigation of early events in FcepsilonRI-mediated signaling using a detailed mathematical model. J Immunol 170:3769–3781

Forst CV (2002) Network genomics – A novel approach for the analysis of biological systems in the post-genomic era. Mol Biol Rep 29:265–280

Forst CV, Schulten K (1999) Evolution of metabolism: a new method for the comparison of metabolic pathways using genomic information. J Comput Biol 6:343–360

Forst CV, Schulten K (2001) Phylogenetic analysis of metabolic pathways. J Mol Evol 52:471–489

Gillespie D (1977) Exact stochastic simulation of coupled chemical reactions. J Chem Phys 81:2340–2361

Girvan M, Newman MEJ (2002) Community structure in social and biological networks. Proc Natl Acad Sci USA 99:7821–7826

Goldsby RA, Kindt TJ, Osborne BA (1999) Immunology. W. H. Freeman, San Fransisco, CA

Goldstein B, Faeder JR, Hlavacek WS (2004) Mathematical and computational models of immune-receptor signaling. Nat Rev Immunol 4:445–456

Gordon S (2002) Pattern recognition receptors: doubling up for the innate immune response. Cell 111:927–930

Haldane JBS (1924) A mathematical theory of natural and artificial selection, 1924

Ho RL, Lieu CA (2008) Systems biology: an evolving approach in drug discovery and development. Drugs R&D 9:203–216

Hucka M, Finney A, Sauro HM et al (2003) The systems biology markup language (SBML): a medium for representation and exchange of biochemical network models. Bioinformatics 19:524–531

Hunter PJ, Borg TK (2003) Integration from proteins to organs: The Physiome Project. Nat Rev Mol Cell Biol 4:237–243

Ideker T, Thorsson V, Ranish JA, Christmas R et al (2001) Integrated genomic and proteomic analyses of a systematically perturbed metabolic network. Science 292:929–934

Ideker T, Ozier O, Schwikowski B, Siegel AF (2002) Discovering regulatory and signalling circuits in molecular interaction networks. Bioinformatics 18(Suppl 1):S233–S240

Jeong H, Mason SP, Barabasi AL, Oltvai ZN (2001) Lethality and centrality in protein networks. Nature 411:41–42

Kahn RA, Fu H, Roy CR (2002) Cellular hijacking: a common strategy for microbial infection. Trends Biochem Sci 27:308–314

Kanehisa M, Goto S, Kawashima S, Okuno Y, Hattori M (2004) The KEGG resource for deciphering the genome. Nucleic Acids Res 32:D277–D280

Karp P, Riley M, Saier M, Paulsen IT et al (2002) The EcoCyc Database. Nucleic Acids Res 30:56–58

Kash JC, Tumpey TM, Proll SP et al (2006) Genomic analysis of increased host immune and cell death responses induced by 1918 influenza virus. Nature 443:578–581

Kauffman SA (1993) The origin of order. Oxford University Press, New York

Kelley BP, Sharan R, Karp RM, Sittler T et al (2003) Conserved pathways within bacteria and yeast as revealed by global protein network alignment. Proc Natl Acad Sci USA 100:11394–11399

Kirschner D, Marino S (2005) *Mycobacterium tuberculosis* as viewed through a computer. Trends Microbiol 13:206–211

Lewis AK, Paterson T, Leong CC, Defranoux N et al (2001) The roles of cells and mediators in a computer model of chronic asthma. Int Arch Allergy Immunol 124:282–286

Magasanik B (1995) Nitrogen response networks of yeast and bacteria. J Cell Biol 19A:326

Marino S, Sud D, Plessner H, Lin PL et al (2007) Differences in reactivation of tuberculosis induced from anti-TNF treatments are based on bioavailability in granulomatous tissue. PLoS Comput Biol 3:e194
McAdams HH, Arkin A (1998) Simulation of prokaryotic genetic circuits. Annu Rev Biophys Biomol Struct 27:199–224
Miller SL (1953) A production of amino acids under possible primitive earth conditions. Science 117:528–529
Musante CJ, Lewis AK, Hall K (2002) Small- and large-scale biosimulation applied to drug discovery and development. Drug Discov Today 7:S192–S196
Nakaya A, Goto S, Kanehisa M (2001) Extraction of correlated gene clusters by multiple graph comparison. Genome Inform 12:44–53
Novak B, Tyson JJ (1993) Modeling the cell division cycle: M-phase trigger, oscillations, and size control. J Theor Biol 165:101–134
Nowak MA, May RM (2000) Virus dynamics: mathematical principles of immunology and virology. Oxford University Press, Oxford
Ogata H, Fujibuchi W, Goto S, Kanehisa M (2000) A heuristic graph comparison algorithm and its application to detect functionally related enzyme clusters. Nucleic Acids Res 28:4021–4028
Oparin OJ (1957) The origin of life on earth. Academic, New York
Orgel LE (1968) Evolution of the genetic apparatus. J Mol Biol 38:381–383
Overbeek R, Larsen N, Pusch GD, D'Souza M et al (2000) WIT: integrated system for high-throughput genome sequence analysis and metabolic reconstruction. Nucleic Acids Res 28:123–125
Perelson A, Weisbuch G (1997) Immunology for physicists. Rev Mod Phys 69:1219–1267
Perelson AS, Neumann AU, Markowitz M, Leonard JM, Ho DD (1996) HIV-1 dynamics in vivo: virion clearance rate, infrected cell life-span, and viral generation time. Science 271:1582–1586
Platt N, Haworth R, Darley L, Gordon S (2002) The many roles of the class A macrophage scavenger receptor. Int Rev Cytol 212:1–40
Savageau MA (1969) Biological systems analysis, II. The steady-state solutions for an n-pool system using a power-law approximation. J Theor Biol 25:370–379
Savageau MA (1970) Biological systems analysis, II. Dynamic solutions using a power-law approximation. J Theor Biol 26:215–226
Schilling CH, Palsson BO (1998) The underlying pathway structure of biochemical reaction networks. Proc Natl Acad Sci USA 95:4193–4198
Schilling CH, Edwards JS, Palsson BO (1999) Toward metabolic phenomics: analysis of genomic data using flux balances. Biotechnol Prog 15:288–295
Schuster S, Fell DA, Dandekar T (2000) A general definition of metabolic pathways useful for systematic organization and analysis of complex metabolic networks. Nat Biotechnol 18:326–332
Segovia-Juarez JL, Ganguli S, Kirschner D (2004) Identifying control mechanisms of granuloma formation during *M. tuberculosis* infection using an agent-based model. J Theor Biol 231:357–376
Simon D, Frost W (2002) Dynamics and evolution of HIV-1 during structured treatment interruptions. AIDS Rev 4:119–127
Stokes CL, Lewis AK, Paterson T, Leong CC et al (1999) Asthma PhysioLab: a dynamic, computer-based mathematical model of acute and chronic asthma. In: Serving Humanity, Advancing Technology, Annual International Conference of the IEEE Engineering in Medicine and Biology, p 1208
Strange K (2005) The end of "naïve reductionism": rise of systems biology or renaissance of physiology?. Am J Physiol Cell Physiol 288:C968–C974
Stuart A, Kauffman (1969) Metabolic stability and epigenesis in randomly constructed genetic nets. J Theor Biol 22:437

Sud D, Bigbee C, Flynn JL, Kirschner DE (2006) Contribution of CD8+ T cells to control of *Mycobacterium tuberculosis* infection. J Immunol 176:4296–4314

Tatebe K, Zeytun A, Harrod KS, Hoffmann R, Ribeiro R, Forst CV (2009) Response network analysis of differential gene expression in human epithelial lung cell during avian influenza infections. BMC Bioinform (under review)

Torigoe C, Inman JK, Metzger H (1998) An unusual mechanism for ligand antagonism. Science 281:568–572

Uetz P, Dong Y-A, Zeretzke C et al (2006) Herpesviral protein networks and their interaction with the human proteome. Science 311:239–242

Valitutti S, Muller S, Cella M, Padovan E, Lanzavecchia A (1995) Serial triggering of many T-cell receptors by a few peptide-MHC complexes. Nature 375:148–151

Wigginton JE, Kirschner D (2001) A model to predictcell-mediated immune regulatory mechanismsduring human infection with *Mycobacterium tuberculosis*. J Immunol 166:1951–1967

Winslow RL, Scollan DF, Holmes A, Yung CK et al (2000) Electrophysiological modeling of cardiac ventricular function: From cell to organ. Ann Rev Biomed Eng 2:119–155

Wodarz D, Page KM, Arnaout RA, Thomsen AR et al (2000) A new theory of cytotoxic t-lymphocyte memory: implications for HIV treatment. Philos Trans R Soc Lond B Biol Sci 355:329–343

Wofsy C, Coombs D, Goldstein B (2001) Calculations show substantial serial engagement of T cell receptors. Biophys J 80:606–612

Yates A, Callard R, Stark J (2004) Combining cytokine signalling with T-bet and GATA-3 regulation in Th1 and Th2 differentiation: a model for cellular decision-making. J Theor Biol 231:181–196

Zien A, Küffner R, Zimmer R, Lengauer T (2000) Analysis of gene expression data with pathway scores. In: Proceedings of ISMB'00. American Association for Artificial Intelligence, Menlo Park, CA, pp 407–417

Chapter 7
Text Mining for Discovery of Host–Pathogen Interactions

Stephen Anthony, Vitali Sintchenko, and Enrico Coiera

7.1 Introduction

The vast body of published literature in biomedicine represents an immensely valuable source of knowledge. Text mining tools and techniques aid in the organization and navigation of this tremendously rich body of knowledge. Traditional approaches to text mining have a strong basis in finite-state pattern extraction (Appelt et al. 1993; Roche and Schabes 1997). With relationship extraction, a great majority of relations are encoded in free text using common linguistic patterns, with the remainder being encoded in tabular or graphical form. Extraction methods predominantly incorporate cue-phrase-based approaches, regular expression pattern mining, or grammar induction. These approaches belong to a class of structural content-based extraction methods. Although the structure of language is infinitely complex, these approaches can produce reasonably reliable results. While the majority of relations may be mined using methods based on the structure of language alone, there is in reality a long tail of less common structures that will not be captured. The distribution is analogous to that of word usage where a small number of words are used most often, with a large number of words making up the remainder (Zipf 1932). The less frequently occurring relations will generally not be captured through the use of syntax alone because they are irregular and difficult to profile given their sparsity.

One obvious and common case that exemplifies the weakness of structural approaches is cross-sentence anaphoric reference. The following passage contains a relationship between a pathogen and syndromes that grammar-based solutions will not easily capture on their own.

> The MRSA clones were mainly isolated from children (overall median age, 3 years). They caused a variety of clinical syndromes, including toxic shock syndrome and suppurative infections.

S. Anthony (✉)
Centre for Health Informatics, University of New South Wales, Sydney, NSW, Australia

V. Sintchenko (ed.), *Infectious Disease Informatics*,
DOI 10.1007/978-1-4419-1327-2_7,

Another linguistic class of problems not readily captured by purely syntactic means is that of semantic modification. Examples include negation, speculation, varying degrees of association, and context-dependent relationships. The first sentence below reflects a speculative association between a genotype and syndrome that is not easily distinguished from a legitimate or actual relationship. The second sentence provides evidence of a relationship between a pathogen and two syndromes, although only within a specifically prescribed context.

> This is the first report of a Beijing genotype association with HIV status, which may be an association unique to tuberculous meningitis.

> Group B Streptococcus (GBS) causes severe infections in very young infants and invasive disease in pregnant women and adults with underlying medical conditions.

Semantic considerations such as these must be addressed for automatically extracted relationships to be useful. In the remainder of this chapter, we explore existing technologies that begin to incorporate the semantic layer.

7.2 Corpus Construction

The desired output of a text processing system dictates the type of data that are collected for training and evaluation. In many cases, there will be no existing corpus that completely satisfies the requirements. In some cases, it is possible to augment existing data sets with additional annotations. At the other end of the spectrum, corpus construction may involve the recruitment of participants in order to acquire raw data upon which domain experts could then apply annotations. Once the application is determined and a source for the data is identified, an annotation scheme and an annotation guideline need to be developed (Wilbur et al. 2006).

It is useful to collect a small sample and create a pilot corpus as the annotation guidelines are being developed. The corpora used for text mining purposes can range from a collection of speaker turns to large document sets that comprise a range of sources and modalities. The document selection process is an important stage of corpus construction, particularly when the documents are being sourced from a larger collection, such as the Internet, or large literature databases, such as PubMed. Search terms and queries must be carefully crafted in order to capture a balanced and representative sample of the target domain. Any necessary refinement of the collection or annotation process will emerge as the guidelines are being developed and the annotations applied to the test corpus. The quality of the annotations and any further refinements to the annotation guidelines are easily revealed through the application of inter-annotator agreement measures. Inter-annotator agreement gauges the consistency of annotations across annotators and should be measured often during the annotation process. The most commonly used measure for inter-annotator agreement is Cohen's kappa coefficient (Cohen 1960).

The next consideration is the system development methodology itself. Generally, a well-balanced mixture of examples is desirable. However, the data should reflect the expected distribution that will be observed in practice. Annotation is an expensive step, and if there exists a minority class that is difficult or costly to capture, annotating a proportional amount of cases to that found in real world scenarios will be more efficient.

Annotation can be performed in a number of ways, either by hand or using computational tools to assist in the process. Commonly used tools that aid annotation include Callisto (Day et al. 2004), GATE (Cunningham et al. 2002), and Knowtator (Ogren 2006). The employment of annotation tools aims to improve productivity. The use of these tools implies the adoption of a standard format and the possibility of automatic data validation. For some tasks, annotation can be semi-automated. For example, words may be automatically pre-tagged using a statistically derived heuristic and subsequently corrected by human annotators. Heuristics can be based on observations, such as most common tag, most frequent sense, surrounding context, or any other empirically derived statistic. Often, correcting existing annotations is faster than creating new ones.

7.3 Biomedical Corpora

There exist a small, yet rapidly expanding number of corpora for the biomedical domain. Most of these corpora are related to the detection of genes and proteins. Currently, very few corpora target relationships that exist between biomolecular entities. Even fewer corpora exist that target relationships between genes or pathogens and diseases or syndromes. Some of the most notable collections include the GENIA corpus, the NLPBA corpus, GENETAG, PennBioIE, and the TREC Genomics Track data. Specifically, the GENIA corpus (Kim et al. 2003) contains a collection of 2,000 PubMed abstracts annotated for entities that are involved in biochemical reactions (e.g., amino acid, DNA, nucleotide, peptide, protein, RNA). The text collection is restricted to abstracts that match the MeSH search terms *human*, *blood cells*, and *transcription factors*. The NLPBA corpus (Kim et al. 2004) is a modified version of the GENIA corpus labelled for five entities across 18,546 training and 3,856 evaluation sentences. GENETAG (Tanabe et al. 2005) is a corpus of 20,000 PubMed sentences labelled for genes and proteins. PennBioIE (Liberman and Mandel 2008) comprises 2,257 PubMed abstracts annotated for paragraphs, sentences, tokens, parts of speech, entities, and Penn Treebank structure. Recent TREC Genomic Track (Hersh et al. 2006) collections have focused on the extraction of passages relevant to topical questions. The 2007 collection was organized around a set of 36 topics such as Antibodies, Diseases, Drugs, Genes, and Strains. Below is a set of example questions from the corpus. Relevant passages from the literature are associated with each question as part of the corpus itself.

What [ANTIBODIES] have been used to detect protein TLR4?
What [DISEASES] are associated with lysosomal abnormalities in the nervous system?
What [DRUGS] have been tested in mouse models of Alzheimer's disease?

7.4 Named Entity Recognition

Automatic recognition of relatively well-defined entities, such as gene symbols, protein names, and syndromes, can achieve sufficient performance levels to warrant their use for relationship extraction. A common problem is that of lexical ambiguity such as that between *Alzheimer's* in reference to a type of mental condition or the possessive form of a person's name.

There also exist numerous issues related to the determination of beginning and end entity spans. An example from the biomedical domain is that of "*human T-cell leukemia lymphotropic virus type 1 tax protein*." It is extremely difficult to identify the beginning and end of such entities in running text without appropriate dictionaries and some appreciation of the context of use. Other issues include the general lack of naming conventions, excessive use of abbreviations, the frequent use of synonyms, and disagreement among the experts (Leser and Hakenberg 2005).

Selecting the appropriate semantic types and granularity of entities is an important consideration for many text-processing tasks. General-purpose parsers will tend to recognize a generic set of entities such as Person, Location, Organization, and Temporal expressions. Biomedical Named Entity Recognition (BNER) systems such as ABNER (Settles 2005) or the GENIA tagger (Tsuruoka and Tsujii 2005) recognize biomolecular entities such as Cell, DNA, Gene, Protein, and Virus. The MetaMap program has a much wider repertoire of 135 semantic types that includes Antibiotic, Bacterium, Cell Function, Disease or Syndrome, Event, Gene or Genome, Molecular Function, and Organism.

Common approaches to entity recognition include the use of gazetteers, rule-based grammars and data-driven machine learning methods. Statistical machine learning methods are used in many publicly available NER systems and reduce the recognition problem to one of feature engineering. The bulk of the effort is concentrated on the selection of a set of features that provides the learning algorithm with optimal discriminatory power. However, the feature space itself is static and is not updated once it is defined by the human expert. This often implies that a change in the domain results in a large drop in performance and is one weakness of modern machine learning approaches. The way in which the features are utilized does vary, and learning algorithms are designed to deal with the variations and noise that exist within the training data. A good feature for one instance may introduce errors in other instances because of interdependencies between features. Such dependencies are accounted for in different ways, depending on the class of learning algorithm that is deployed.

Examples of machine learning algorithms that are used extensively in the biomedical domain include Conditional Random Fields (CRF) (Sutton and McCallum 2007; Sutton et al. 2007) and Support Vector Machines (SVM) (Cristianini and Shawe-Taylor 2000; Steinwart et al. 2008). A CRF is a type of graphical model that weighs feature functions according to the values of the input sequence. CRFs employ an arbitrary number of feature functions that are conditioned across the positions of hidden states and the input sequence. This approach can be contrasted to Hidden Markov Models (HMM) that use constant probabilities to model state transitions and emissions. The approach is an effective compromise between dynamic features and a fixed feature space used for training classifiers. The SVM algorithm is based on the notion of the maximum separation of margins and is typically used for classification and regression. Conceptually, the algorithm casts input data into two sets of vectors in an n-dimensional space. Mapping to higher dimensions assists in the construction of maximal separating hyperplanes.

7.5 Syntactic Parsing

The syntactic analysis of natural language aims to identify grammatical structure with respect to a formal grammar. Selecting an appropriately suitable parsing formalism is an important consideration. Commonly used parsing systems employ either dependency grammars or phrase structure grammars.

Dependency grammars are defined by the functional dependencies that exist between headwords and their dependants. Word order is not defined in dependency grammar making it a useful candidate for free word order languages. Other closely related theories of syntax include Link Grammar and Operator Grammar.

Phrase structure grammars are defined by a series of rewrite rules or context-sensitive transformations. For this reason, they are also commonly referred to as constituency grammars, or transformational grammars. They include theories such as Government Biding (GB), Head-Driven Phrase Structure Grammar (HPSG), Tree Adjoining Grammar (TAG), Lexical Functional Grammar (LFG), and Combinatory Categorical Grammar (CCG).

The use of syntactic analyses in biomedical text mining tasks is commonplace (Daraselia et al. 2004; Pyysalo et al. 2006; Smith and Wilbur 2009). Recent advances have seen the combination of multiple syntactic analyses (Miwa et al. 2008). Identifying the contributions of different grammatical formalisms is an important area of investigation. A number of studies have been conducted that explore the differences between parsers and their output representations when used in biomedical applications (Clegg and Shepherd 2007; Miyao et al. 2009). Once entities and syntactic structure have been identified, it is possible to determine the relationships that exist between entities.

7.6 Relationship Extraction

The extraction of pertinent relationships between biological entities aids our understanding of the organisms, processes, and interactions that underlie and control biological functions. As an example, knowledge of the relationships that exist between genes and diseases is required for an appreciation of disease models such as phenotype disease networks. In the same fashion, it is the knowledge of the regulatory and physical interactions between genes and proteins that provides insights into the mechanics of regulatory and metabolic pathways. Relationship extraction provides a means of transforming the vast amount of fragmented knowledge that is scattered across millions of unstructured texts into structured forms. There are a few examples of structured resources. These include the EMBL-Bank and GenBank for gene sequences, SwissProt and UniProt for protein sequences, and KEGG and BioCyc for metabolic pathways. The complexity of these data makes the maintenance of such resources costly. Automatic relationship extraction can aid in the construction and unification of such resources. Examples of existing systems include Arrowsmith (Smalheiser and Swanson 1998; Smalheiser et al. 2006) and BITOLA (Hristovski et al. 2006).

The great majority of existing relationship extraction systems have been targeted at the genomic level that encompasses gene-protein or protein-protein interactions (Hunter et al. 2008; Manine et al. 2009). The protein-protein interaction extraction system of Kim et al. (2008) transforms the problem from one of pattern matching to one of kernel construction. Kernels are developed incrementally from features at the lexical level to incorporate structural features derived from dependency graphs. Other relationship extraction systems have been developed for the identification of disease models such as phenotype disease networks (Hidalgo et al. 2009) and gene–disease interactions (Chun et al. 2006).

7.7 Case Study: Pathogen–Host Relationship Extraction

This section details the steps involved in extracting relationships between a predetermined set of entities. The entities chosen for the case study are Genotype, Pathogen, and Syndrome. The corpus is a collection of abstracts sourced from PubMed. For the relationship extraction task we extend our definition of genotype to include genes, and the definition of syndrome to include diseases. Examples of each category are listed in the Table 7.1.

The corpus is a collection of 43 PubMed abstracts from articles related to infectious diseases. Document selection and annotation are performed by a domain expert-microbiologist. The corpus contains 367 sentences, 59 pathogens, 101 genotypes, and 91 syndromes. An example sentence manually annotated for entity types, as well as automatically annotated using the MetaMap program, is displayed in Table 7.2. The foundations of relationship extraction are based on the results of entity recognition. The following sections introduce approaches that may be

Table 7.1 Examples of genotype, pathogen, and syndrome entities

Genotype	Pathogen	Syndrome
sea-seh gene	*Escherichia coli*	Liver abscess
enterotoxin gene	*Streptococcus agalactiae*	Community-acquired MRSA
egc	*Streptococcus pyogenes*	Toxic-shock-like syndrome
rmpA gene	*Mycobacterium tuberculosis*	GBS pathogenesis
nuoG gene	*Klebsiella pneumoniae*	Invasive disease

Table 7.2 Sentential analysis with semantic annotation

Word	Lemma	MetaMap semantic type	Named entity
Our	Our	–	–
Data	Datum	IDEA_OR_CONCEPT	–
Support	Support	CLINICAL_ATTRIBUTE	–
a	a	–	–
Statistical	Statistical	INTELLECTUAL_ PRODUCT-RESEARCH_ ACTIVITY	–
Correlation	Correlation	INTELLECTUAL_ PRODUCT-RESEARCH_ ACTIVITY	–
Between	Between	–	–
The	The	–	–
rmpA	rmpA	–	Genotype
gene	gene	GENE_OR_GENOME	Genotype
And	And	–	–
Virulence	Virulence	QUALITATIVE_CONCEPT	–
In	In	–	–
Terms	Term	IDEA_OR_CONCEPT-TEMPORAL_CONCEPT	–
Of	Of	–	–
Abscess	Abscess	–	Syndrome
Formation	Formation	FUNCTIONAL_CONCEPT	–
For	For	–	–
These	These	–	–
Hypermucoviscous	Hypermucoviscous	–	–
K.	K.	–	Pathogen
Pneumoniae	Pneumonia	DISEASE_OR_SYNDROME	Pathogen
Strains	Strain	INTELLECTUAL_PRODUCT	–
.	.	–	–

employed for the construction of genotype, pathogen, and syndrome recognizers that approximate the performance of human experts.

7.7.1 Gene and Genotype Recognition

Gene nomenclature standardization efforts have been under consideration for many decades. The International Committee on Genetic Symbols and Nomenclature published recommendations for naming genes in 1957 (ICGSN 1957). For the human genome a complete set of naming guidelines was published in 1979 as a result of the Edinburgh Human Genome Meeting (EHGM) (Shows et al. 1979). One of the largest ongoing efforts in this domain initiated by the Human Genome Organisation (HUGO), is that of the HUGO Gene Nomenclature Committee (HGNC) (Eyre et al. 2006; Bruford et al. 2008). Despite these and similar efforts many scientists continue to adopt alternative nomenclatures that reflect the functional characteristics and abbreviations that signify the research history of particular genes (Tamames and Valencia 2006; Lacroix 2009). These alternate names hinder the consolidation and exchange of biological knowledge. Even though gene names are not entirely uniform or systematic, current nomenclatures still offer utility for automatic gene identification.

A lexical ambiguity issue that makes automatic gene recognition difficult is that of overlapping terms. Many gene symbols for example correspond to common English words such as "*end*" (Rv0670c), and "*folD*" (Rv3356). The issue is particularly problematic if orthography is disregarded. Although there is some overlap between gene symbols and common English words the variation and ambiguity between terms employed to represent genes posits a much larger problem. An example of variation and ambiguity is that of the gene symbol ACT. The variety of genes for which the symbol ACT is associated follow with the HGNC approved symbol for each gene in parentheses: acyl-CoA thioesterase 7 (ACOT7), or four and a half LIM domains 5 (FHL5), or alpha-1-antichymotrypsin (SERPINA3). The most ambiguous gene recorded in the HGNC database is the epithelial cell adhesion molecule (EPCAM), with an astounding 20 aliases on record.

One approach to automatic gene recognition is via the combination of gazetteers. Most gene symbols do not overlap with English words and therefore an English lexicon may be used as a blacklist to preclude common words. In other words, terms that appear in this list are labeled as unlikely gene symbol candidates. This approach can be used in conjunction with a list of known gene symbols, essentially forming a whitelist. When a term is identified as a member of both lists, these items are greylisted and must be handled as special cases.

The degree of overlap between gene databases and English words is directly related to the terminologies that are employed, as well as to the method of comparison. For example, the HGNC database currently contains 28,110 approved gene symbols. Overlap can be measured against independent lexicons, either with or without the consideration of orthography. When compared against the 233,615 words from the Webster's Second International Dictionary in a case-insensitive manner, there is an overlap of 163 terms. When compared against the smaller set of 147,306 terms from WordNet 3.0 (Fellbaum 1998), there is a slightly larger overlap of 166 terms. This increase can be explained by the fact that WordNet addresses a broader domain coverage, and is actively updated. When compared against a smaller set of 56,647 words from the Debian dictionaries-common package the overlap is reduced to 82 terms, as would be expected.

A similar proportion of overlap is observed for bacterial gene symbols. Consider, for example, a list of 1,476 gene symbols for *Mycobacterium tuberculosis* H37Rv (virulent strain), sourced from the Comprehensive Microbial Resource (CMR, http://cmr.jcvi.org/tigr-scripts/CMR/CmrHomePage.cgi). When compared against WordNet, there are 50 terms that overlap. When compared against the smaller set of Debian dictionaries-common words there are 27 overlapping terms, listed below in Table 7.3.

Table 7.3 Gene symbols that overlap with common words

add	alaS	amiD	apt	ask	citE
cobS	cysT	deaD	end	far	folD
folK	gap	hisS	lipS	map	mas
menD	metE	proS	proW	purE	sec
serA	sigH	sodA			

If a term is identified as being ambiguous, a number of heuristics may be deployed. Heuristics may be introduced at any stage of processing. Heuristics may be as simple as pattern matching. Patterns can be basic regular expressions that capture a variety of gene symbols such as ACOT7, FLJ30846, IL-28R1, R33729_1, and Za11. However, these types of patterns grossly overgenerate and require more stringent constraints. Wider contextual cues can be used to constrain the problem of overgeneration. These cues often take the form of lexical context words or any other type of language model. Obvious cue words include *gene* and *genotype* that follow uncommon dictionary words or unrecognized terms.

Heuristics are not the only way to disambiguate problematic terms. Language models are easily transformed into features that can be used to develop classification models. Features such as a dictionary or non-dictionary word, a word on either side, the proportion of uppercase letters, and the proportion of numbers can be used effectively as features for the classification of gene symbols. Other approaches to gene recognition incorporate language modeling that involves the generation of profiles extracted for each gene mention from their context of occurrence in publications (Xu et al. 2007). One of many potential applications of such profiles is the discovery of related genes via similarity measures.

Although the identification of gene names is fraught with difficulty many of these problems can be overcome through the combination of gazetteers and machine learning.

7.7.2 Pathogen Recognition

Pathogens do not suffer from many of the same problems as gene names and syndromes, although there are a number of idiosyncratic naming conventions and behaviors to account for when automatically recognizing this class of entity. Aliases are some of the more common problems that afflict the recognition of entities representing pathogens. An example is the use of the terms *methicillin-resistant Staphylococcus aureus*, or *oxacillin-resistant Staphylococcus aureus* to refer to the same pathogenic strain of *Staphylococcus aureus* that is often resistant to different classes of antibiotics. Abbreviations are also problematic as they affect the recognition of pathogen entities in text.

As an example *Group B Streptococcus* is commonly abbreviated to GBS and *methicillin-resistant Staphylococcus aureus* to MRSA. A commonly used convention in the life sciences literature in relation to pathogens is the abbreviation of the genus name on subsequent uses. For example, *Staphylococcus aureus* will be abbreviated to *S. aureus* and *Klebsiella pneumoniae* abbreviated to *K. pneumoniae* on subsequent mentions. As part of the normalization process, an association must be made between full forms and their respective abbreviated forms.

7.7.3 Disease and Syndrome Recognition

Diseases and syndromes can be identified using techniques similar to those introduced for gene recognition. A number of systems have already been developed for the recognition of diseases using related methods and demonstrate reasonable performance (Chun et al. 2006; Bundschus et al. 2008). These systems employ terminological resources that provide disease terms that include Medical Subject Headings (MeSH), the National Cancer Institute (NCI) Thesaurus, the Systematized Nomenclature of Medicine-Clinical Terms (SNOMED CT), as well as the overarching Unified Medical Language System (UMLS) Metathesaurus. Positive evidence exists for the contribution of terminological resources with respect to the recognition of disease entities (Jimeno et al. 2008). An assessment of disease entity extraction is then conducted on a corpus of manually re-annotated sentences. Competitive results are achieved, suggesting that disease terminology is relatively standardized both within the literature and within the resources themselves.

Syndromes, however, are merely indicative of specific etiological disease and there are no known terminological resources that define a comprehensive set of syndromes for text mining purposes. The list of terms employed to represent syndromes for our case study is sourced from the Medical Dictionary for Regulatory Authorities (MedDRA). This list contains 18,483 adverse event terms that are employed as a representative sample of syndromes.

7.7.4 Association Mining

The previous sections outlined possible approaches and resources that can be used to recognize specific sets of entities. Of interest are the relationships that exist between individual syndromes and specific microorganisms or pathogens, pathogens and their genotypes, and genotypes and syndromes. A pathogen will be related to any number of genotypes, and any single genotype will be related to any number of syndromes. Take, for example,the following sentences for which entities are identified.

> [K. pneumoniae]$_{\text{Pathogen}}$ [genotype K1]$_{\text{Genotype}}$ is an emerging pathogen capable of causing catastrophic [septic ocular or central nervous system complications]$_{\text{Syndrome}}$ from pyogenic [liver abscess]$_{\text{Syndrome}}$ independent of underlying diseases in the host. (PMID: 17599305)

> We report a patient transferred from Alaska to Washington State with a [magA (+)]$_{\text{Genotype}}$ [K. pneumoniae]$_{\text{Pathogen}}$ [liver abscess]$_{\text{Syndrome}}$ and describe a simple approach for recognition of these hypervirulent strains. (PMID: 15695726)

> The association of the [magA gene]$_{\text{Genotype}}$ with the hypermucoviscosity phenotype relevant to the pathogenesis of [Klebsiella pneumoniae]$_{\text{Pathogen}}$ [liver abscess]$_{\text{Syndrome}}$ has been reported in Taiwan. (PMID: 16619144)

> Our data support a statistical correlation between the [rmpA gene]$_{\text{Genotype}}$ and virulence in terms of [abscess]$_{\text{Syndrome}}$ formation for these hypermucoviscous [K. pneumoniae]$_{\text{Pathogen}}$ strains. (PMID: 16619144)

A straightforward method for relationship extraction is one based on co-occurrence. Entities found in the same span are proposed as candidates that are potentially related. The span may be defined at any level within a document, including paragraph, sentence, or clause level. Such an approach introduces a great deal of noise through spurious relationships. The relationships at this level are in fact only loose associations. With simple normalization the associations can be generated on the basis of common entities. In this example, the pathogen *Klebsiella pneumoniae* (*K. pneumoniae*) is common among all sentences and is thereby able to produce the following network (Fig. 7.1).

The network is generated from four different sentences identified across three distinct documents. Automatically digesting this information into a graphical form has a number of advantages. Firstly, it is now possible to easily identify the genes and syndromes that are associated with the pathogen. Secondly, it also becomes possible to infer relationships between genes and syndromes themselves on the basis of their relationship to pathogens. When combined with networks generated for other pathogens, it becomes possible to cluster pathogens together based on network topology (Bales et al. 2007).

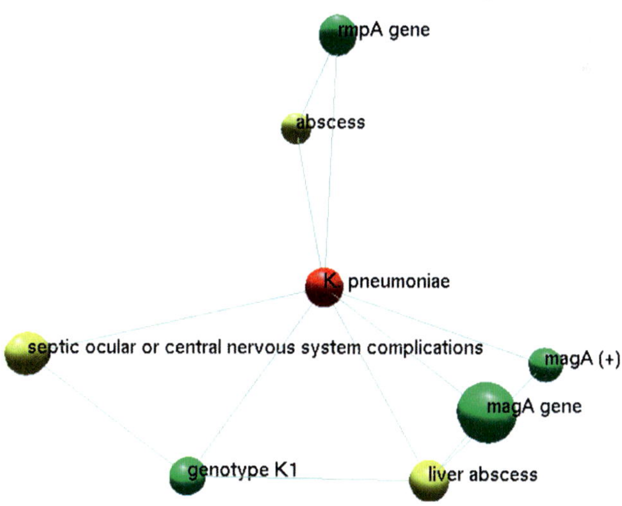

Fig. 7.1 Literature-mined pathogen–host network

The relationships extracted thus far are on the basis of sentences that contain all of the entity types. A more comprehensive approach to relationship extraction involves the use of entity recognition, syntactic analysis, predicate-argument structure, anaphora resolution, and semantic interpretation. Such an approach reduces the level of ambiguity and noise as relationships are semantically constrained. Section 7.7.5 introduces a potential solution in the pursuit of these goals.

7.7.5 Potential Directions for Relationship Extraction

As with syntactic parsing, approaches to semantic parsing vary widely. A common semantic representation is Predicate Argument Structure (PAS), where each sentential predicate is associated with each of its arguments. For example, the main verb in the sentence below is *support*, and could be represented as follows:

> [Our data]$_{\text{Subject}}$ support [a statistical correlation between the [rmpA gene]$_{\text{Gene}}$ and virulence in terms of [abscess]$_{\text{Syndrome}}$ formation]$_{\text{Object}}$ [for these hypermucoviscous [K. pneumoniae]$_{\text{Pathogen}}$ strains]$_{\text{Indirect object}}$.

The PAS for this sentence identifies the subject, object, and indirect object arguments of the main verb *support*. It is *data* that supports the *statistical correlation* between the *rmpA gene* and *abscess*. When further deconstructed, *correlation* predicates a number of arguments, a *statistical* modifier, as well as the two noun phrase objects, *rmpA gene* and *virulence in terms of abscess formation*. The following relationships can be inferred for this sentence:

correlation(abscess, *rpmA*)

association(*K. pneumoniae*, abscess)

association(*K. pneumoniae*, *rmpA*)

There are numerous representations that capture the meaning of a sentence, with the most common representation based on either PropBank (Palmer et al. 2005) or FrameNet (Fillmore et al. 2003). The question to ask is whether the representation provides enough meaningful "hooks" to accommodate the task at hand. In other words, does the representation afford sufficient expression to be useful to the application? In the current setting the application is relationship extraction and, for example, one has to make sure that PAS can extract relationships between genes and syndromes, such as the relationship between *rpmA* and *abscess, that* is modulated through the notion of *virulence*. This is a detail that should be captured by the semantic representation. Moreover, the representation should readily allow access to such modifications. Other phenomena that should be addressed by any semantic formalism include the resolution of pronouns and ellipses, quantifier scope ambiguities, analyses of tense and aspect, and the ability to distinguish between distributive and collective readings of plural noun phrases.

It is known that the identification of semantic arguments aids information extraction to some extent (Surdeanu et al. 2003). Realistically semantic types are generalizations that overlay syntactic constituents. The approach is little more informative than language models based on syntax. It is the pragmatic aspects of meaning that can better guide interpretation rather than compositional sentence level representations alone. Understanding the intention behind a statement allows for a more directly accessible and potentially more usable representation. The characterization of the intention behind statements in a restricted domain whose general purpose is well defined is far simpler than identifying the intention of sentences in free text. Once the intention of a statement has been identified, at least putatively, the ability to refine the interpretation improves.

Abstracts from scientific publications by definition and convention present a summary of research aims, methods, and key findings. By assuming this to be common ground, then it is possible to extract these types of facts from abstracts. The intention of each statement in the abstract will be in some way constructed in pursuit of these goals. That is, the author crafts each sentence to support the goal of informing the reader of the aims, methods, and key findings concluded by the research. It has been shown that such discourse level labels can be automatically assigned for sentences in biomedical abstracts with high accuracy (Chung 2009).

Ascertaining the purpose of a given statement to some extent elicits the relationships embedded within the sentence. The notion of discourse coherence relationships differs from that of discourse segment purpose. Theories of discourse coherence, such as Rhetorical Structure Theory (RST) (Mann and Thompson 1987), propose relationships such as Explanation, Result, Contrast, and Generalization. A definition of discourse segment purpose must allow for a more robust assignment of purposeful labels and their consolidation into a wider meaning base. The following passage comprises the first three sentences of an abstract. The text is analyzed for discourse segment purpose.

> [Multidrug-resistant tuberculous meningitis]$_{NP}$ is [fatal]$_{ADJP}$ [without rapid diagnosis and use of second-line therapy]$_{PP}$.
>
> [It]$_{NP}$ is [more common]$_{ADJP}$ [in human immunodeficiency virus (HIV)-positive patients]$_{PP}$.
>
> [Beijing genotype strains of Mycobacterium tuberculosis]$_{NP}$ are associated [with drug resistance, particularly multidrug resistance]$_{PP}$, and [their prevalence]$_{NP}$ is increasing [worldwide]$_{NP}$.

The initial concept that is introduced relates to the potential consequences of tuberculosis meningitis. It is framed as a conditional assertion where it is suggested that, without some early intervention, tuberculosis meningitis is usually fatal. The statement evokes certain expectations that can be viewed as recommendations for the prevention of disease. It is these actions that one expects to become the primary focus of the discussion. The reader is provoked to ask certain questions, including those related to the methods by which the proposed solutions of *rapid diagnosis* and *second-line therapy* may be instantiated. These types of questions and expectations guide the reader's interpretation and understanding.

The second sentence is an elaboration of the main concept, providing information about the distribution of the disease among the affected population. The sentence provides support for the claims that follow and refines the expectations of the reader.

The background information provided to the reader allows for a more felicitous reading. This type of communication act is referred to by theories of discourse as common ground or mutually recognized intention (Grice 1989; Stalnaker 2002).

The third sentence provides new information in linking a particular strain of the virus with drug resistance. The claim that prevalence is increasing worldwide is an elaboration in reference to this strain. The sentence serves a number of functions. Firstly, the statement alerts the readers and attracts their attention. Secondly, the new information introduces further questions that require explanation. It is these questions that help elicit the relationships embedded within the text. A simple question that may be invoked by this sentence concerns the relationship between the *Beijing genotype* and *Mycobacterium tuberculosis*. The answer to this question is explicitly stated in the text. Questions such as how the *Beijing genotype* strains are associated with *drug resistance* embed other relationships.

The proposal to formally encode expectations and explanations as a mechanism for the extraction of relationships is novel and remains untested. It is clear that syntax and semantics alone cannot capture the wide array of relationships that are embedded within text. Pragmatic phenomena must be harnessed in conjunction with domain knowledge in order to extract reliable relationships.

For details related to the field of text mining in general and approaches for transforming unstructured data into structured forms see Feldman and Sanger (2007) and Kao and Poteet (2007). A comprehensive review of text mining for the biosciences that includes details on corpora and corpus annotation, terminological resources, and biomedical named entity recognition can be found in Ananiadou and McNaught (2006). For further details on theories of discourse and pragmatics the reader is referred to Tannen et al. (2001) and Horn and Ward (2004). Also see Ginzburg (1996) and Roberts (1996) for more discussion on formalisms that can be used for the representation of discourse goals and intentions.

7.8 Concluding Remarks

A number of techniques and resources for the extraction of relationships have been outlined in this chapter. Issues related to the automatic recognition of host-pathogen named entities are addressed. It is clear that accurate and robust entity recognition is essential for relationship extraction. Clustering is one of many applications of the relationship extraction process, and can elucidate knowledge embedded within text. As shown in the case study, the relationships that are extracted on the basis of the recognition of genotype, pathogen, and syndrome entities provide a basis for clustering. The syndromic relationships can be used as features for grouping genotypes and pathogens with latent similarity.

However, there are many semantic phenomena that need to be addressed for more reliable relationships to be identified and, in particular, for new discoveries to be made. The example of negation has been discussed as a semantic phenomenon that must be addressed for successful relationship extraction. A related issue is that of absence as opposed to negation. As shown in the sentence below, evidence of the absence of a gene in relation to a pathogen is itself a relationship worth extracting.

> Both MRSA strains were multiresistant and lacked the Panton–Valentine leukocidin-encoding gene.

It is suggested that discourse purpose can aid in the relationship extraction process by addressing these and similar problems. The approach to discourse purpose recognition advocated here is based on the notion of expectations and explanations. The questions that are evoked by statements can, in the given context, be evaluated with respect to their explanations. The explanations that are generated for each of these expectations isolate the relationships between entities of interest.

Relationship extraction enables many interesting and useful applications in infectious disease research. It is particularly useful in the biosciences where the interpretation of complex interactions has become a major bottleneck. The study of host–pathogen interactions on the host, organ, tissue, and molecular levels offers new opportunities for text-mining applications. Incentives for the pursuit of relationship extraction include the ability to integrate data from disparate sources, the provision of information sharing, and the potential for breakthrough discoveries.

References

Ananiadou S, McNaught J (2006) Text mining for biology and biomedicine. Artech House, Boston, MA

Appelt DE, Hobbs JR et al (1993). FASTUS: a finite-state processor for information extraction from real-world text. In: The 13th international joint conference on artificial intelligence (IJCAI-93). Chambéry, France

Bales ME, Lussier YA et al (2007). Topological analysis of large-scale biomedical terminology structures. J Am Med Inform Assoc 14(6):788–797

Bruford, EA, Lush MJ et al (2008). The HGNC Database in 2008: a resource for the human genome. Nucleic Acids Res 36(Database Issue):D445–D448

Bundschus M, Dejori M et al (2008) Extraction of semantic biomedical relations from text using conditional random fields. BMC Bioinform 9:207

Chun HW, Tsuruoka Y et al (2006) Extraction of gene-disease relations from Medline using domain dictionaries and machine learning. Pac Symp Biocomput 4–15

Chung GY (2009) Sentence retrieval for abstracts of randomized controlled trials. BMC Med Inform Decis Making 9:10

Clegg AB, Shepherd JI (2007) Benchmarking natural-language parsers for biological applications using dependency graphs. BMC Bioinform 8:24

Cohen J (1960) A coefficient of agreement for nominal scales. Educ Psychol Meas 20(1):37–46

Cristianini N, Shawe-Taylor J (2000) An introduction to support vector machines and other kernel-based learning methods. Cambridge University Press, Cambridge.

Cunningham H, Maynard D et al (2002) GATE: A framework and graphical development environment for robust NLP tools and applications. In: The 40th anniversary meeting of the association for computational linguistics, Philadelphia

Daraselia N, Yuryev A et al (2004) Extracting human protein interactions from MEDLINE using a full-sentence parser. Bioinformatics 20(5):604–611

Day D, Kozierok R et al (2004) Callisto: a configurable annotation workbench. In: The fourth international conference on language resources and evaluation (LREC 2004). Lisbon, Portugal

Eyre TA, Ducluzeau F et al (2006) The HUGO Gene Nomenclature Database, 2006 updates. Nucleic Acids Res 34(Databases Issue):D319–D321

Feldman R, Sanger J (2007) The text mining handbook: advanced approaches in analyzing unstructured data. Cambridge University Press, New York

Fellbaum C (1998) WordNet: an electronic lexical database. MIT, Cambridge, MA

Fillmore CJ, Johnson CR et al (2003) Background to framenet. Int J Lexicogr 16(3):235–250

Ginzburg J (1996) Interrogatives: questions, facts, and dialogue. The handbook of contemporary semantic theory. Blackwell, Oxford, pp 385–422

Grice H (1989) Studies in the way of words. Harvard University Press, Cambridge, MA

Hersh WR, Cohen A et al (2006) TREC 2006 genomics track overview. In: The 15th text retrieval conference (TREC 2006), Gaithersburg, MD, pp 52–78

Hidalgo CA, Blumm N et al. (2009) A dynamic network approach for the study of human phenotypes. PLoS Comput Biol 5(4):e1000353

Horn L, Ward G (eds) (2004) The handbook of pragmatics. Blackwell Handbooks in Linguistics. Blackwell, Oxford

Hristovski D, Friedman C et al (2006) Exploiting semantic relations for literature-based discovery. In: AMIA annual symposium proceedings, pp 349–353

Hunter L, Lu Z et al (2008) OpenDMAP: an open source, ontology-driven concept analysis engine, with applications to capturing knowledge regarding protein transport, protein interactions and cell-type-specific gene expression. BMC Bioinform 9:78

ICGSN (1957) Report of the International Committee on Genetic Symbols and Nomenclature. Union of International Sci Biol Ser B, Colloquia No. 30

Jimeno A, Jimenez-Ruiz E et al (2008) Assessment of disease named entity recognition on a corpus of annotated sentences. BMC Bioinformatics 9(Suppl 3):S3

Kao A, Poteet SR (2007) Natural language processing and text mining. Springer, London

Kim JD, Ohta T et al (2003) GENIA corpus – semantically annotated corpus for bio-textmining. Bioinformatics 19(Suppl 1):i180–i182

Kim J, Ohta T et al (2004) Introduction to the bio-entity recognition task at JNLPBA. In: The international joint workshop on natural language processing in biomedicine and its applications (NLPBA), Geneva, Switzerland, pp 70–75

Kim S, Yoon J et al (2008) Kernel approaches for genic interaction extraction. Bioinformatics 24(1):118–126

Lacroix M (2009) Poor usage of HUGO standard gene nomenclature in breast cancer studies. Breast Cancer Res Treat 114(2):385–386

Leser U, Hakenberg J (2005) What makes a gene name? Named entity recognition in the biomedical literature. Brief Bioinform 6(4):357–369

Liberman, M., Mandel M (2008). PennBioIE, Linguistic Data Consortium, Philadelphia

Manine AP, Alphonse E et al (2009) Learning ontological rules to extract multiple relations of genic interactions from text. Int J Med Inform. Epub ahead of print 22 apr. PMID: 19398370

Mann WC, Thompson SA (1987) Rhetorical structure theory: a theory of text organization. Information Sciences Institute, Marina del Rey, CA

Miwa M, Sætre R et al (2008) Combining multiple layers of syntactic information for protein–protein interaction extraction. In: The 3rd international symposium on semantic mining biomed (SMBM), Turku, Finland, pp 101–108

Miyao Y, Sagae K et al (2009) Evaluating contributions of natural language parsers to protein–protein interaction extraction. Bioinformatics 25(3):394–400

Ogren PV (2006) Knowtator: a plug-in for creating training and evaluation data sets for Biomedical Natural Language systems. In: The 9th International Protégé Conference, Stanford, CA

Palmer M, Gildea D et al (2005) The proposition bank: a annotated corpus semantic roles. Comput Linguistics 31(1):71–106

Pyysalo S, Ginter F et al (2006) Evaluation of two dependency parsers on biomedical corpus targeted at protein-protein interactions. Int J Med Inform 75(6):430–442

Roberts C (1996) Information structure: towards an integrated theory of formal pragmatics. OSU Working Papers in Linguistics, vol. 49, pp 91–136

Roche E, Schabes Y (1997) Finite-state language processing. MIT, Cambridge, MA

Settles B (2005) ABNER: an open source tool for automatically tagging genes, proteins and other entity names in text. Bioinformatics 21(14):3191–3192

Shows TB, Alper CA et al (1979) International system for human gene nomenclature (1999) ISGN (1979). Cytogenet Cell Genet 25(1–4):96–116

Smalheiser NR, Swanson DR (1998) Using ARROWSMITH: a computer-assisted approach to formulating and assessing scientific hypotheses. Comput Methods Programs Biomed 57(3):149–153

Smalheiser NR, Torvik VI et al (2006) Collaborative development of the Arrowsmith two node search interface designed for laboratory investigators. J Biomed Discov Collab 1:8

Smith LH, Wilbur WJ (2009) The value of parsing as feature generation for gene mention recognitionJ Biomed Inform. Epub ahead of print PMID: 19345281

Stalnaker RC (2002) Common ground. Linguistics Philos 24(5–6):701–721

Steinwart I, Christmann A et al (2008) Support vector machines. Springer, Dordrecht

Surdeanu M, Sanda H et al (2003) Using predicate-argument structures for information extraction. In: The 41st annual meeting of the association for computational linguistics, Sapporo, Japan, pp 8–15

Sutton C, McCallum A (2007) An introduction to conditional random fields for relational learning. In: Getoor L, Taskar B (eds) Introduction to statistical relational learning. MIT, Cambridge, MA, pp 93–127

Sutton C, McCallum A et al (2007) Dynamic conditional random fields: factorized probabilistic models for labeling and segmenting sequence dData. J Machine Learn Res 8:693–723

Tamames J, Valencia A (2006) The success (or not) of HUGO nomenclature. Genome Biol 7(5):402

Tanabe L, Xie N et al (2005) GENETAG: a tagged corpus for gene/protein named entity recognition. BMC Bioinform 6(Suppl 1):S3

Tannen D, Schiffrin D et al (2001) The handbook of discourse analysis. Blackwell, Malden, MA

Tsuruoka Y, Tsujii JI (2005) Bidirectional inference with the easiest-first strategy for tagging sequence data. HLT/EMNLP 2005, Vancouver, BC, Canada, pp 467–474

Wilbur WJ, Rzhetsky A et al (2006) New directions in biomedical text annotation: definitions, guidelines and corpus construction. BMC Bioinform 7:356

Xu H, Fan JW et al (2007) Gene symbol disambiguation using knowledge-based profiles. Bioinformatics 23(8):1015–1022

Zipf GK (1932) Selective studies and the principle of relative frequency in language. MIT, Cambridge, MA

Chapter 8
A Network Approach to Understanding Pathogen Population Structure

Caroline O. Buckee and Sunetra Gupta

8.1 Introduction

Epidemiological models have been used effectively to understand patterns of disease transmission, to estimate important epidemiological parameters, and ultimately, to design and assess public health policies. The simplest of these mathematical formulations partitions the host population into those that are susceptible to an infection and those that are infected or immune. The proportion of hosts within each compartment is then tracked over time, with the density of infected hosts contributing to the force of infection. These frameworks assume that hosts mix randomly within the population, so that each susceptible host is equally likely to become infected, and infected hosts contribute equally to the force of infection. This "mean-field" assumption allows for analytically tractable equations describing disease dynamics, and has led to valuable insights into the relative significance of different epidemiological parameters and the importance of the basic reproductive number, R_0, for defining the likelihood of disease persistence. R_0 translates as the average number of secondary cases of a disease caused by a single infection within a wholly susceptible population. In these models, a threshold condition exists such that a disease can persist if $R_0 > 1$. This threshold behavior facilitates analyses of which epidemiological factors can be changed through public health policies to reduce the R_0 of a disease to a value below 1.

However, host populations are generally not homogeneous, with individuals exhibiting heterogeneity due to a range of factors including their age, varying levels of susceptibility, or behavior. Different types of heterogeneities are taken into account within compartmental models by partitioning the host population further, for example, into different age groups. Decisions must then be made about mixing patterns between different groups of hosts, and the simplest assumption is random mixing between all hosts. Populations are also spatially heterogeneous, grouped into towns or cities for example, and individuals interact with a particular network of friends,

C.O. Buckee (✉)
Department of Zoology, University of Oxford, Oxford, United Kingdom

V. Sintchenko (ed.), *Infectious Disease Informatics*,
DOI 10.1007/978-1-4419-1327-2_8, © Springer Science+Business Media, LLC 2010

co-workers, and family members. These interactions and spatial heterogeneities also underlie the spread of infectious disease across the population. Depending on the mode of transmission of the pathogen (for example, airborne versus sexually transmitted), the host contact network may have profound consequences not only for transmission, but also for the evolution of the pathogen population. We will discuss the implications of different types of host contact network for the transmission of pathogens, with specific reference to the importance of understanding the network topologies underlying the spread of sexually transmitted diseases such as HIV.

Another assumption made by the simple epidemiological models described above is that the pathogen population is also homogeneous. However, for many important pathogens of humans, including those responsible for HIV, meningitis and malaria, genetic loci encoding immunogenic proteins are highly polymorphic across the pathogen population. Genetic diversity at these antigenic loci allows pathogens to invade the same hosts multiple times, since an immune response against one variant or strain may not be protective against infection by a pathogen strain with different antigenic determinants. We will briefly discuss the evolutionary forces that shape the population structure of these loci, and then focus on how host contact networks may affect the evolution of antigenically diverse pathogen populations.

Understanding the evolution of genetically diverse pathogen species is often hampered by high rates of recombination between pathogen genomes. Recombination causes otherwise unrelated genomes to share genetic material, disrupting phylogenetic signals within gene sequences and complicating sequence analysis. Standard phylogenetic techniques cannot accommodate high rates of recombination, since different positions within an alignment, upon which these techniques rely, may have different underlying trees. In the absence of appropriate phylogenetic tools for recombining antigenic loci, a new approach to understanding the evolution of malaria parasite antigens has been developed, in which relationships between sequences are visualized as a network (Bull et al. 2008). The structure of antigen networks reflects the extent to which different genes recombine with one another. The identification of a recombination hierarchy within malaria parasite antigens, in which particular groups of genes recombine more frequently than others, provides insights into the epidemiological patterns of infection and disease observed in endemic regions.

We conclude with a discussion of the implications of antigen networks, and networks of host immune responses, for the evolution and epidemiology of the flu virus. Annual epidemics of influenza are caused primarily by distinct viruses that are sufficiently different from previous outbreak strains to evade the immune responses of most hosts. Mathematical models of influenza outbreaks often require strong constraints on viral evolution, in order to reproduce this single-strain outbreak dynamic. We will present an alternative model of influenza dynamics, in which the range of antigenic possibilities for the flu virus is limited because of functional constraints (Recker et al. 2007). Within this framework, a network of host immune responses drives successive outbreaks of individual strains.

8.2 Contact Networks and Disease Transmission

Disease transmission depends upon contact between hosts, and since most individuals within a population have a very few social or sexual contacts relative to the overall population size, assumptions of random mixing may be inappropriate for particular research questions. Within mathematical models of disease transmission that include a host contact network structure, each individual makes contact with specific people, and these generally remain fixed. The analysis of network structure and function has a long history within the mathematical literature on graph theory and percolation theory, and this provides a wealth of useful tools for theoretical epidemiologists. However, the structures of real contact networks are often hard to elucidate. We will first discuss the implications for disease transmission within sexual contact networks, before exploring the role of contact networks for directly transmitted infections.

8.2.1 Sexually Transmitted Diseases and Host Contact Networks

Large-scale surveys of sexual behavior have uncovered an extremely high variability in the number of sexual contacts per individual (Johnson et al. 2001). For sexually transmitted infections (STIs), this highly skewed distribution of contacts, with most individuals having a few sexual partners and a small number of individuals having many, has a profound impact on disease transmission. Early mathematical frameworks approximated the sexual contact network by partitioning the population into subgroups with different levels of sexual activity (Gupta et al. 1989; Koopman et al. 1988; Yorke et al. 1978). These models showed that the prevalence among the general population may be relatively low, and the maintenance of STIs within the population relies on a highly infected core group of individuals with many sexual contacts and a high frequency of sexual activity. Attempts to provide a finer grained examination of host population structures in the context of STI transmission led to the development of pair-formation models, in which men and women form relatively long-term sexual contacts in pairs (Dietz and Hadeler 1988), while explicit network formulations are required for the inclusion of concurrent sexual contacts (i.e., contact with more than one individual). For example, a stochastic individual-based model with explicit network structure was used to illustrate how concurrency causes a dramatic increase in the size and variability of an epidemic, confirming the importance of the distribution of contacts among the population, not just the mean behavior (Morris and Kretzschmar 1997).

In fact, the distributions of real sexual contact networks often appear to follow a power law that is scale-free; having an exponent between 2 and 3 (Schneeberger et al. 2004). As discussed in the introduction, traditional epidemiological models exhibit a threshold condition where R_0 must be greater than one for a disease to persist. In heterogeneous networks, this epidemic threshold decreases with the standard deviation of the connectivity distribution of the host population (Anderson and May 1992). This effect is amplified in scale-free networks, and May and Lloyd have shown that there is no threshold behavior at all for epidemics upon infinite scale-free networks (May and Lloyd 2001). Real sexual networks are of course not infinite, however, which means that the variance in the partnership distribution is finite and an epidemic threshold does exist (Jones and Handcock 2003), even if it is extremely small (Pastor-Satorras and Vespignani 2002). Furthermore, when two types of nodes are present within a scale-free network, representing males and females in a heterosexual population (Gomez-Gardenes et al. 2008), the epidemic threshold is higher than if all nodes are the same; the transmissibility of the disease must be much higher to cause an epidemic. This observation, and the discovery that scale-free networks exhibit extreme fragility to the removal of high-degree nodes, suggests that targeted public health strategies (vaccinating particular individuals, for example) may be an effective strategy (Callaway et al. 2000).

8.2.2 Directly Transmitted Diseases and Host Contact Networks

For directly transmitted diseases, our knowledge of the structure of the host contact networks underlying transmission is less clear. Early social science experiments emphasized the importance of "hubs" within social networks and the unexpectedly short path length through the network between any two individuals (Milgram 1967). This phenomenon of an unexpectedly connected network, or a "small-world" network, has implications for the spread of disease since long-range connections can allow the pathogen to reach different parts of the host population rapidly. Empirical estimates of social network structure have generally come from surveys of particular groups of individuals like school children (http://www.cpc.unc.edu/projects/addhealth), which only cover a small fraction of individuals and do not measure casual acquaintances. A more comprehensive recent survey in several European countries showed that children and young adults tend to exhibit more assortative mixing - that is, the extent to which "like mixes with like" (Mossong et al. 2008). They also have a higher frequency of contacts and more long-duration contacts than adults do, suggesting that some of the results pertaining to sexual core groups discussed above may also be applicable to directly transmitted diseases. In an age-structured model of measles among school children, for example, Schenzle (1984) showed that the timing of school terms and holidays, coupled with strong age structure and close contacts between children within schools, could maintain biennial outbreaks in the absence of seasonal forcing. In a detailed stochastic simulation model, Eubank et al. (2004) used census and land-use data to develop an agent-based model of human movement, constrained by transportation infrastructure, to explore how a disease may spread on a relatively realistic network. They showed that the critical factor determining whether a disease would spread or not was the speed at which infected individuals quarantined themselves by going home, either due to symptoms or encouragement by public health officials.

In the absence of large empirical data sets on the structure of host networks, theoretical studies have generally focused on simplified hypothetical contact networks. Keeling and Eames, for example, reviewed the numerous models of disease spread on lattice networks, random networks, and small-world networks (Keeling and Eames 2005). Different types of network structure have different effects on disease transmission, yet a few findings are general to a range of structures. Clustering of hosts and localized contacts, for example, tends to dramatically reduce the spread of an infection because a local depletion of susceptible hosts occurs following its introduction (Diekmann et al. 1998). In lattice models, a traveling wave of infection is observed, and these capture spatial aspects of the spread of diseases such as measles (Grenfell et al. 2001). As the number of long-range links increases, as it does within small-world networks, infection can rapidly reach all parts of the host network (Watts and Strogatz 1998). Using percolation theory to explore the effects of small-world structure on disease transmission, Moore and Newman (2000) showed that these long-range links can dramatically reduce the epidemic threshold and increase the probability of a disease outbreak. The generality and mathematical tractability of these theoretical studies provide an important basis for our understanding of the effects of heterogeneity in contact patterns for disease spread. As larger data sets on human behavior become available from sources such as mobile phones, it will be possible to start addressing how they relate to reality.

8.3 Host Contact Networks and Pathogen Evolution

In addition to contributing to the rate and spatial dynamics of disease spread, host contact networks have important implications for the evolution of a pathogen species being transmitted across the population. For example, pathogens that recombine frequently may be restricted in the rate and type of genetic exchange possible due to the non-random mixing of hosts, and due to local differences in strain composition of the pathogen population. In addition, the evolutionary forces placed on the pathogen by the host, such as drug pressure for example, may be heterogeneous across the host network.

8.3.1 Evolution of Pathogen Traits and Host Contact Networks

In classical models, the relationship between a pathogen's transmissibility and virulence is linked, with virulence being an unavoidable consequence of high transmissibility (Anderson and May 1992). These frameworks have shown that the trade-off between virulence and transmissibility can lead to intermediate levels of virulence and transmissibility being selected. However, Rand et al. (1995) showed that in a spatial setting with an explicit lattice contact network, pathogens will not evolve too high a transmission rate, even in the absence of this trade-off. Pathogens with extremely high transmission rates will rapidly exhaust local clusters of hosts, leading to a threshold effect not found in meanfield models: above a critical intrinsic transmissibility pathogens cannot survive. When pathogen transmissibility was allowed to evolve in this model, it evolved to this critical level and remained stable. Once immunity is included in spatially explicit lattice models of pathogen evolution, interesting dynamics result from the "blocking" behavior of immune hosts, which tend to surround infectious hosts (Boots et al. 2004). In this case, an evolutionary bistability exists wherein virulent strains are favored among sparsely distributed hosts and avirulent strains are favored among dense, locally mixing hosts. Large shifts in virulence were observed within the model framework due to this bistability, when strains encountered a newly susceptible population for example. These models illustrate the importance of local contacts on the evolution of virulence.

The observation that hosts within a sexual contact network may be divided into a small core group and a peripheral, less active group also has implications for the evolution of pathogen traits. For example, it has been shown (Eames and Keeling 2006) that the level of assortative mixing determines whether pathogen strains with different characteristics can coexist or not. They explored the coexistence of two strains within a serially monogamous host network: one strain that replicates slowly, but produces long infectious periods ("slow") and one that replicates rapidly, but produces short infections ("fast"). Within a mean-field model of competing strains, the "fast" strain would outcompete the "slow" strain at equilibrium which would be driven to extinction. The heterogeneity within the host network allowed the two strains to coexist in different parts of the host population, with the fast strain dominating the core group and the slow strain dominating the less sexually active hosts. Among these individuals, hosts generally recovered from the "fast" strain, before they could transmit it. Figure 8.1 shows the results of the model, with the region of coexistence maximized when assortative mixing is strong (Fig. 8.1a). Under these conditions, hosts with the most potential contacts, or large "neighborhoods," will be dominated by the "fast" strain, whereas the "slow" strain will dominate among hosts with few potential contacts (Fig. 8.1b). Thus, heterogeneity within the pathogen population may be maintained by heterogeneities in human contact patterns.

8.3.2 Pathogen Population Structure

Many pathogens are structured into distinct lineages or strains, defined by different antigenic determinants that stimulate a specific immune response in the host. Antigenically diverse pathogen species can re-invade hosts multiple times, because immunity to one strain may not provide protection against a different strain. For example, there are approximately 90 different variants of the capsular polysaccharide

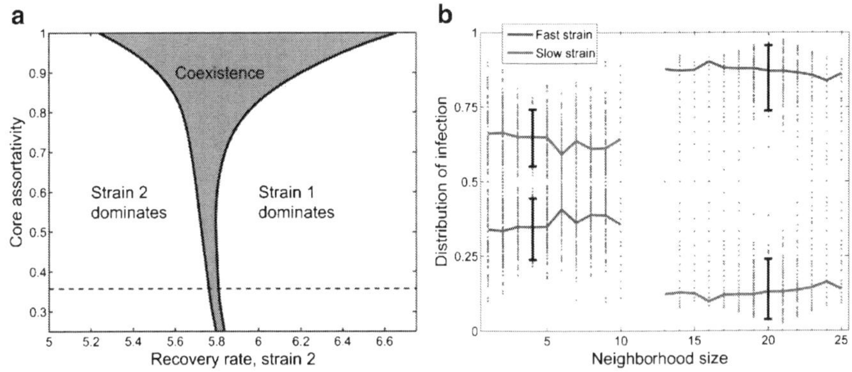

Fig. 8.1 The coexistence of "fast" and "slow" strains within a heterogeneous host population, from Eames and Keeling (2006). (**a**) The relationship between the level of assortative mixing and strain coexistence. The coexistence region is plotted for a second strain with a fixed value of transmission rate for a range of population mixing patterns, where assortativity is defined as the proportion of the core group that mix only with other members of the core group (thus when assortativity = 1, there is no mixing between the core group and the rest of the population). The dashed line shows the level of assortativity corresponding to random mixing. (**b**) The results of a stochastic simulation, with the proportion of hosts infected by each strain shown in red and blue for individuals belonging to different behavioral groups, represented as belonging to different size neighborhoods. 90% error bars are shown around two representative points (Reprinted with permission of University of Chicago Press)

surrounding the bacteria *Streptococcus pneumoniae*, and each generates a specific immune response that will not fully protect the host against infection with a different variant. Here, one locus is responsible for generating immunity, and different capsular polysaccharide variants circulate essentially independently. For pathogens with multiple immunodominant loci, theoretical frameworks have shown that host immune responses can organize the pathogen population into discrete strains characterized by non-overlapping combinations of antigenic determinants (Fig. 8.2). Dominant strains will not be competing for hosts, since immune responses will be directed against these non-overlapping combinations. Most hosts in the population will therefore be exposed to and protected against infection by one or both of the dominant strains, preventing the emergence of recombinant combinations of antigenic determinants.

However, localized host contact network structure can disrupt the strong herd immunity effect described above, and cause the discrete strain structure of the pathogen population to break down. Using host networks that varied between a mean-field approximation (wherein contacts between hosts were randomly changed each time step), a local, regular mixing pattern, and a "small-world" contact structure, Buckee et al. (2004) used a stochastic individual-based model to explore these effects. As in the mean-field model described above, pathogens were defined by two antigenic loci, each with two alleles to which hosts gained specific immunity following infection. Here, two metrics of pathogen population structure, diversity and discordance, were used to determine the structure of the pathogen

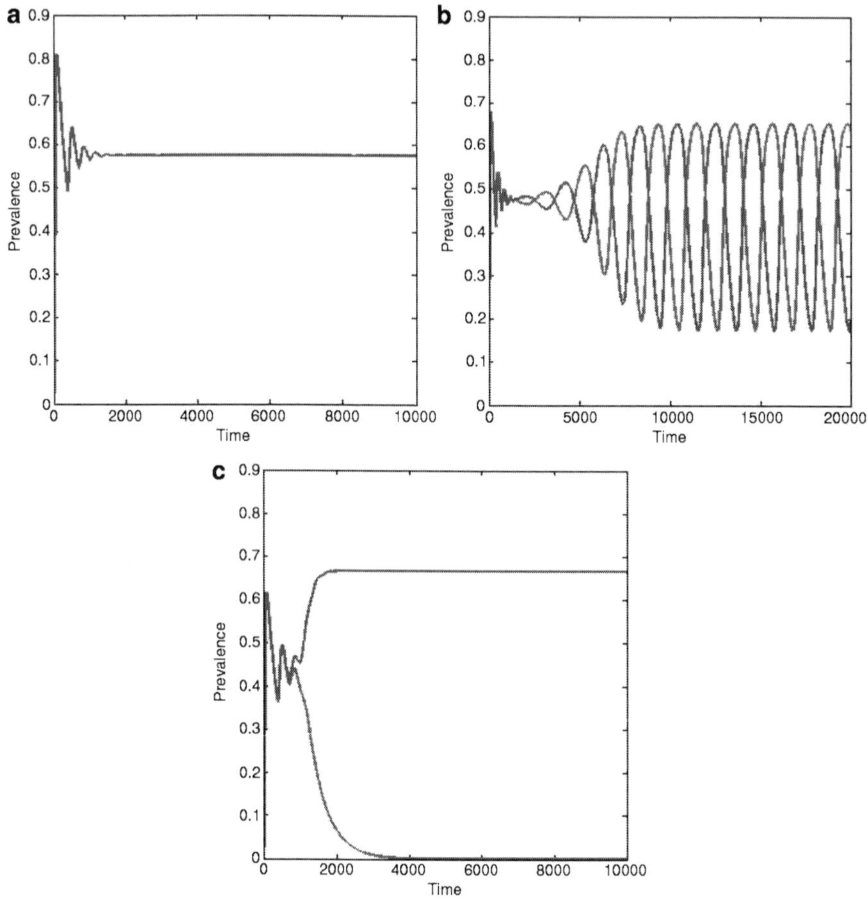

Fig. 8.2 Model results showing the effect of cross-immunity on the population structure of antigenically variable pathogens, from Gupta et al. (1996). Strains are defined by two antigenic determinants, represented as a circle and a triangle, and all strains are assumed to have equal intrinsic transmissibility. Each antigenic determinant has two alleles, either white or yellow. When cross-immunity between strains sharing antigenic alleles is low (**a**), all strains coexist at the same prevalence at equilibrium. When cross-immunity between strains antigenic alleles is high (**c**), however, the pathogen population is dominated by two strains with non-overlapping combinations of antigenic alleles, since they will not be competing for hosts. Here, recombinants will be suppressed because most hosts in the population have been exposed to antigenic alleles from parent strains. At intermediate levels of cross-immunity (**b**), an oscillatory dynamic is observed, with sets of strains with non-overlapping antigenic alleles alternately dominating the population

population, once equilibrium had been reached. Diversity measures the evenness with which a pathogen population is partitioned into all of its possible different strains, in a calculation based on the Shannon-Weaver diversity index (Shannon and Weaver 1949):

$$D = \frac{\sum_{i=1}^{Ns} p_i \log(1/p_i)}{\log(N_s)} \tag{8.1}$$

where p_i is the frequency of strain i in the population, and N_s, the number of strains, $= 2^n$. D ranges between 0 and 1, with $D = 1$ indicating that all of the possible strain types in the pathogen population are equally represented. Discordance measures the extent of non-overlapping structure between strains, the signature of strong immune selection. We use a taxonomic distinctness measure, previously used to calculate the average phylogenetic distance between species within a community. Here, we use weights to quantify allelic differences between strains, measuring the Hamming distance as the number of loci at which strains differ. Since the maximum Hamming distance possible in a pathogen population is known, we divide by the maximum Hamming distance (the number of loci), to get a discordance (H) measure between 0 and 1:

$$H = \frac{1}{n}\left(\frac{\sum\sum_{i<j} w_{ij} p_i p_j}{\sum\sum_{i<j} p_i p_j}\right) \tag{8.2}$$

where w_{ij} is the number of loci with different alleles for strains i and j; p_i and p_j are the frequencies of strain i and j in the pathogen population, respectively, and n is the number of loci.

As the network structure was varied between an essentially random mixing network, a small-world network, and a locally mixing host network, the size of the clusters of pathogen strains decreased (Fig. 8.3). Although non-overlapping strains dominated within local clusters, when the contacts between hosts were primarily local, different subsets of strains dominated different clusters. This caused the diversity of the overall pathogen population to increase and the discordance to decrease, as the clustering coefficient of the network increased. This result is analogous to ecological and other network models in which local dynamics equilibrate more rapidly than global dynamics. In ecological models, for example, the patchiness of resources can increase the diversity of interacting species.

8.3.3 Community Structure in Host Networks and Pathogen Population Structure

Community structure has been identified as an important characteristic of many real-world networks, including those representing social interactions. However, the analysis of community structure in networks to date has focused on understanding the structure rather than the consequences of the dynamics of an infection. The extent to which connected communities of hosts can affect the composition of the pathogen

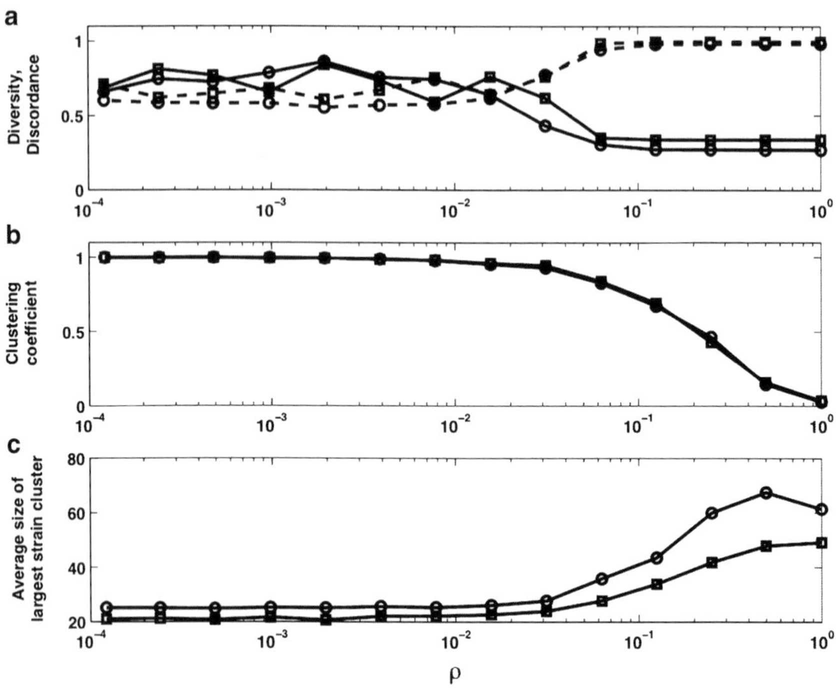

Fig. 8.3 The effect of host contact network patterns on pathogen population structure, from Buckee et al. (2004). The parameter ρ determines the degree of mixing within the host network, such that when $\rho = 1$, hosts mix essentially randomly, and when ρ is small, hosts mix only locally. The top panel shows the effects of ρ on discordance (*dashed line*) and diversity (*solid line*) for two simulations. The middle panel shows the effects of ρ on the clustering coefficient, calculated as in Watts and Strogatz (1998). The bottom panel illustrates how mixing affects the average size of the largest strain cluster within the host network

population was recently explored using an explicit network formulation (Buckee et al. 2007). Implementing a similar model as described above, pathogen strains were defined by two loci each with two possible antigenic alleles. Two communities of hosts were loosely linked to each other, with hosts having the majority of their links within their own community, and fewer with individuals in the other community. Each community was seeded with a different subset of strains, in order to distinguish the dynamics of the system.

To compare the antigenic profiles in different communities, the same diversity metric discussed earlier was used, based on the Shannon-Weaver diversity index (Shannon and Weaver 1949). Here, when cross-immunity was high the within-community diversities were expected to be relatively low. If both communities were dominated by the same subsets of strains, then the overall diversity would also be low. Conversely, when different dominant strains were maintained in each community, the overall population showed high diversity, since all strains existed at more or less the same prevalence overall. Thus, the diversity of individual communities could be compared to that of the overall population, with the difference between local and global diversities being analogous to the difference in prevalence of the same strains in different communities for the deterministic model.

Interestingly, the extent to which the host population remained structured into identifiable communities did not equate to the differentiation of the pathogen population. Figure 8.4 illustrates how the maintenance of different pathogen profiles within different communities changed as the connectivity between them increased (note the logarithmic scale of the x axis). The pathogen population was dominated by one subset of non-overlapping strains at very low levels of connectivity between the two communities. In contrast, community structure was still easily detected in the host network at connectivities that were orders of magnitude higher. Identifiable community structure in host networks may not reflect the differentiation of the processes occurring upon them, and conversely a lack of genetic differentiation between pathogens from different host communities may not reflect strong mixing between them.

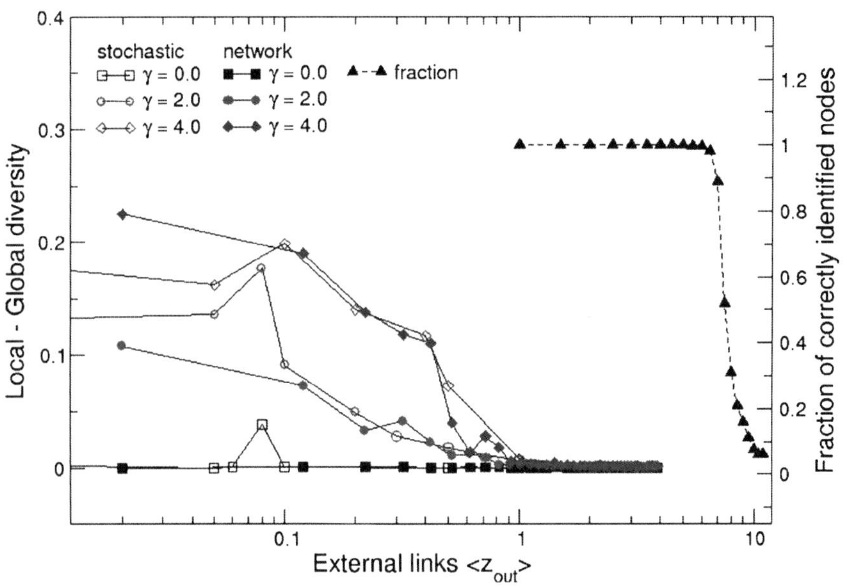

Fig. 8.4 A comparison of pathogen dynamics and host community structure, from Buckee et al. (2007). Black triangles represent the host population, illustrating how the differentiation of the two host communities changes as the average number of links to hosts in a different community $<z_{\text{out}}>$ increases (detection of host differentiation relies on an algorithm correctly identifying nodes as belonging to one community or another - here, host communities remain well defined until $z_{\text{out}} = 3$). The pathogen population structure is shown for different levels of cross-immunity, γ, with two types of host population taken into account, one "stochastic" and one "network." The stochastic model is individual-based, but without explicit network structure. In both cases, the pathogen population is behaving essentially as a homogeneous population when the host communities are still very well differentiated

We have shown how host contact networks can affect disease transmission in terms of the rate and extent of epidemic outbreaks and threshold effects for disease persistence. We have also discussed how these network structures differ for STIs and directly transmitted diseases, and how assumptions about host networks can affect the evolution of virulence and pathogen population structures. We now introduce the use of networks as a conceptual tool for understanding the population structure of highly diverse pathogens species.

8.4 Antigen Networks

8.4.1 Malaria Antigen Networks

The malaria parasite, *Plasmodium falciparum*, exhibits high levels of genetic diversity among parasite antigen proteins expressed on the surface of infected red blood cells (Marsh and Howard 1986; Su et al. 1995). To evade the host immune system, the parasite has evolved a system of antigenic variation in which different variants of this family of proteins, called PfEMP1 (*P. falciparum* erythrocyte membrane protein 1), are expressed sequentially in a mutually exclusive manner during infection. Each parasite has approximately 60 *var* genes encoding PfEMP1. As host antibodies are generated to one variant, the parasite switches expression to another, prolonging infection and increasing the chances of transmission to another host (Scherf et al. 1998). PfEMP1 is the primary target of the protective immune response and an important virulence factor through its role in the binding of red blood cells to the host endothelium. It makes understanding the diversity and expression of these antigens critical to public health efforts such as vaccination (Bull et al. 1998; Newbold et al. 1997; Rowe et al. 1995). Unfortunately, high rates of recombination lead to enormous diversity on a population level and within individual parasite antigen repertoires, complicating studies of the parasite population structure at these loci.

The full genome sequences of three *P. falciparum* genomes have now been completed, and different groups of *var* genes have been identified based on their upstream promoter regions, position in the chromosome, and direction of transcription. Three main groups – UpsA, UpsB, and Ups C – have been named for the upstream promoter region of the gene (Lavstsen et al. 2003). Studies of the expression of *var* genes in different hosts have shown that the UpsA group is associated with expression in young hosts and hosts with severe disease (Jensen et al. 2004; Kyriacou et al. 2006; Magistrado et al. 2008). However, understanding the evolutionary relationships between *var* genes and *var* gene groups remains a major challenge. Traditional phylogenetic techniques rely on an initial alignment of sequences leading to a distance matrix, which is then transformed into a phylogenetic tree. Most *var* genes contain multiple mosaic fragments of sequences from a variety of different recombination events with unrelated genomes, however, so a distance matrix may not represent evolutionary relationships adequately. Recently, a technique which is not reliant on an alignment has been developed by Bull et al. (2008) for visualizing the relationships between *var* sequence fragments. Sequence tags are represented as nodes in a network, with the edges between nodes representing exact sequence matches at variable regions termed "position specific polymorphic blocks" (PSPBs). These regions show considerable mosaicism, such that otherwise unrelated sequences may share one or more PSPB. Figure 8.5 illustrates a network of over 1,000 *var* gene fragments from clinical and lab isolates, as well as several *P. reichenowi* (the primate malaria parasite) homologues.

The network has two distinct lobes with dense links within them and loose links between them. These lobes correspond relatively well to different groups of *var* genes based on the classification systems described above, with all UpsA genes (using the system reported by Lavstsen et al. 2003) found within the smaller lobe (Fig. 8.5). Thus, the separation of groups seems to represent a recombination hierarchy, with certain types of *var* gene recombining with each other more frequently than others. This hierarchical recombination is thought to maintain functional differences between the groups of *var* genes, which exhibit different binding specificities.

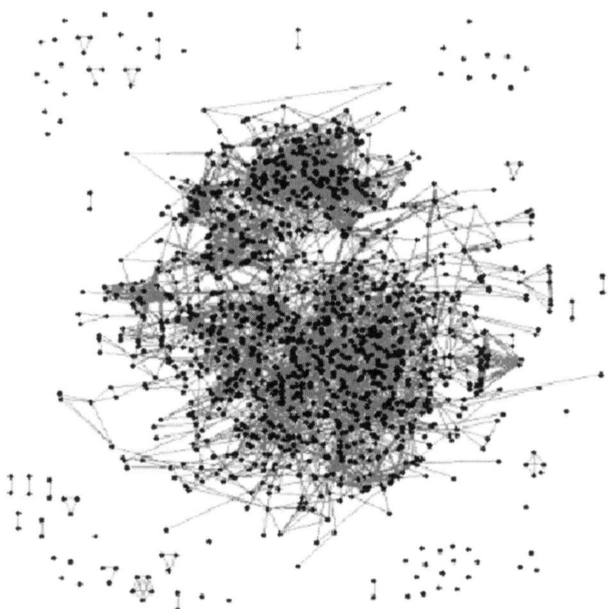

Fig. 8.5 A network of *Plasmodium falciparum var* gene fragments, from Bull et al. (2008). Here, each node in the network represents a sequence tag from the DBLα domain of *var* genes from wild isolates, laboratory strains, and chimp malaria parasite homologues. Nodes in the network are connected if they share an exact match at the amino acid level at one of four PSPBs (see main text), at positions chosen to represent intermediate polymorphism. Exact matches are assumed to represent recombination events

Interestingly, *P. reichenowi* genes fall within the network, indicating that many of these polymorphic regions are relatively ancient. The diversity of the malaria parasite is therefore primarily generated by recombination, which shuffles variable regions between *var* genes. A major field of research in the next few years will be the search for antigenic epitopes within *var* gene sequences and the investigation of how the expression of different *var* epitopes relates to parasite phenotype and disease patterns in the host. This network approach to visualizing the relationships between sequences represents a transparent and simple tool for assessing recombination in the absence of appropriate phylogenetic tools.

8.4.2 Conceptual Antigen Networks and Influenza Dynamics

The technique described above provides a useful way to analyze sequence data while avoiding assumptions implicit within phylogenetic frameworks. A network approach to understanding pathogen population structure can also be useful on a more conceptual level. Pathogen antigens are usually functional in some way, for example, they may mediate binding to host receptors. Therefore, the extent of potential phenotypic diversity among antigenic loci is generally constrained, in spite of strong diversifying selection imposed by host immunity. The structure of "antigen space" will determine how a pathogen population responds to selection, and networks have recently been used to conceptualize this structure for antigenic determinants of the influenza virus.

The influenza virus, though less antigenically complex than the malaria parasite, also shows genetic and immunological diversity on a population level. Two main antigenic determinants, the surface glycoproteins haemaglutinnin (HA) and neuraminidase (NA), generate strain specific responses in humans. Although periodically the reassortment of viral segments can lead to global flu pandemics, annual outbreaks are thought to result from a gradual process of mutation or "antigenic drift." Each outbreak is dominated by a genetically restricted viral population with antigenic characteristics sufficiently different from previous strains that it can evade the host population's immune responses. This observation leads to a major conceptual problem: why should only a restricted set of viruses evade host immune responses at one time? For most mathematical models of influenza, in which the viral population mutates irreversibly in a particular direction in "antigen" space, either structured constraints on mutation or the inclusion of short-term cross-immunity are required for the appearance of single strain epidemic behavior. Without these assumptions, several variants can emerge within the host population at the same time.

Koelle et al. (2006) assume that "neutral networks" underlie influenza evolution. Here, mutation allows different strains to explore neutral networks that do not change the antigenic properties of the virus within sequence space clusters, extending the period of cross-immunity between strains. Occasionally, a small sequence change from a particular position within these networks will lead to a large phenotypic change or "cluster transition", allowing the strain to escape herd immunity and cause an outbreak. This framework provides a detailed description of the sequence-level evolution of the viral population, but requires many assumptions about the mutation process itself. Another model by Ferguson et al. (2003) recreates observed phylogenetic patterns, but requires the assumption of short-lived, strain-transcending immunity to reduce the diversity of the influenza population sufficiently between outbreaks. Recker et al. (2007) introduced an alternative model of influenza outbreaks that brings together concepts of both host networks and antigen networks within a model of single-strain epidemics. Here, a network of epitope-specific host immune responses against a limited network of antigenic epitopes can drive single-strain outbreaks without explicit assumptions about the direction of mutation or transient immunity. The model assumes a number antigenic epitopes, each with a varying number of possible alleles, representing various levels of functional constraints. Thus "antigen" space is not infinite, but is constrained within a defined network of possible epitope combinations. The order in which variants emerge in successive outbreaks then depends on the network of host immune responses present at each point of time in the population.

Hemaglutination inhibition (HI) assays have been used to explore the immunological relationships between different influenza strains over a 40-year period (Smith et al. 2004). HI assays measure the cross-reactivity or antigenic distance of different influenza strains by exposing ferrets to particular strains, and testing the ability of the exposed sera to recognize different strains. The results of these comparisons show that flu strains between 1968 and 2003 show a zigzagging pattern across "antigen space", when the HI assays are subjected to multi-dimensional scaling (Fig. 8.6). However, since ferret sera have a limited range of cross-recognition, many elements in the matrix of responses to different strains are absent. This creates a false sense of antigenic distance between viral samples that cannot be compared using this technique. Figure 8.6 shows the output from the model with a restricted antigenic space defined by five loci each with two alleles. Although the antigenic network in this case is a 5-dimensional hypercube, when subjected to the same analysis as the data from Smith et al. (i.e., a projection onto lower dimensional space using MDS), a similar zigzag pattern occurs. Figure 8.7 illustrates this concept, suggesting that a "true" antigenic map would in fact, cycle through the network of epitope combinations in a manner that reflects the immune structure of the host population.

8.5 Conclusion

The representation of data and ideas in the form of networks has been successful, both because intuitively the interactions between components of a system are easily understood within a network framework and because a body of mathematical theory provides the tools necessary to analyze networks in a quantitative way.

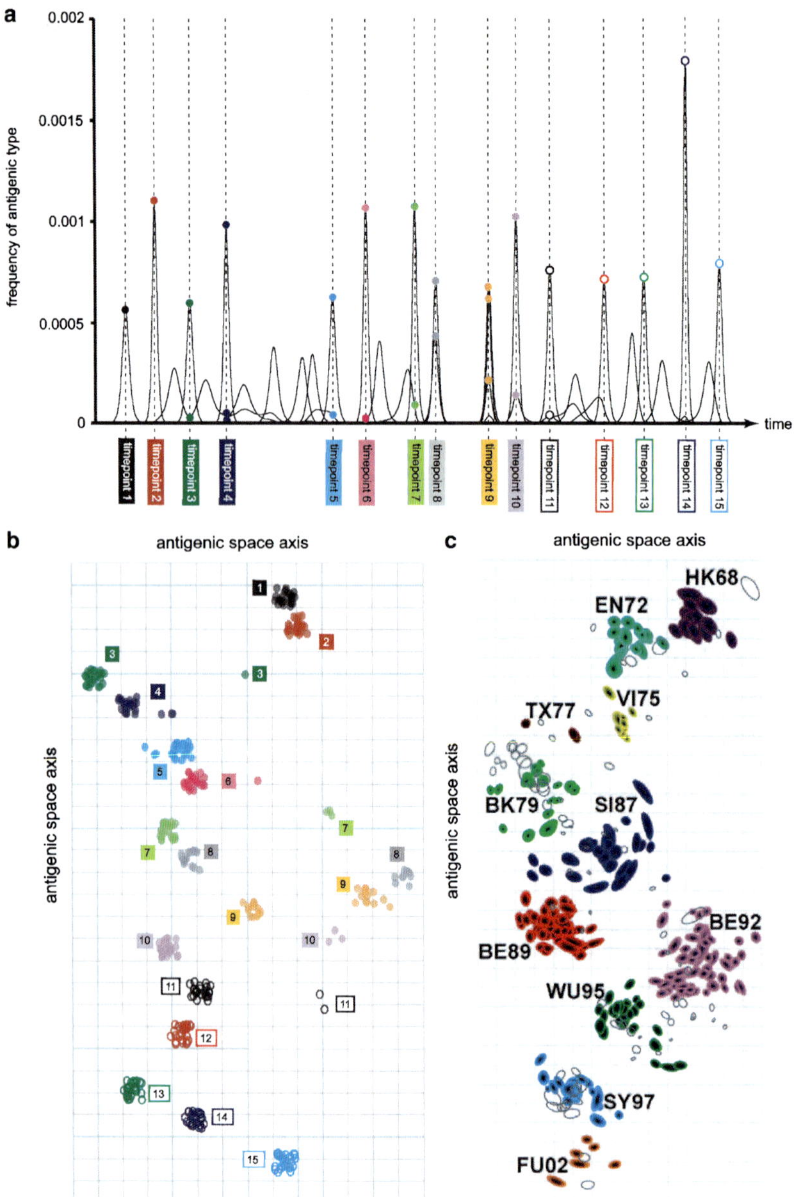

Fig. 8.6 Model output from Recker et al (2008) compared to data. (**a**) The dynamics of the model, showing the proportions of hosts infectious over time. (**b**) Hypothetical samples from individuals with influenza at different time points (the *dotted lines* in (**a**)) were analyzed using "antigenic cartography," or MDS, as in Smith et al (2004). (**c**) The observed antigenic map from Smith et al (2004)

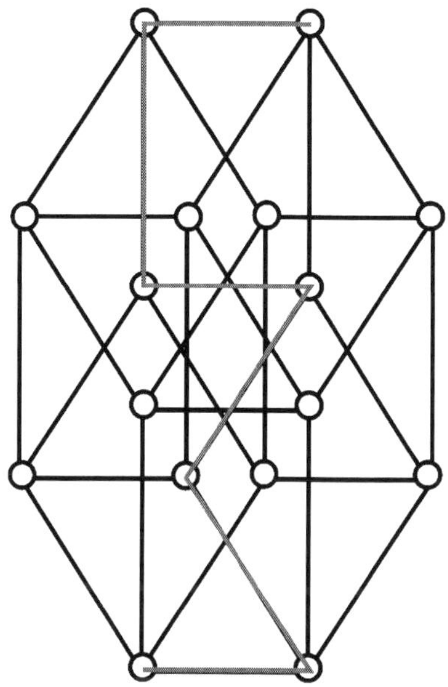

Fig. 8.7 A schematic showing the effects of MDS analysis (see main text) on a multi-dimensional network of antigenic epitopes. By projecting a multidimensional network onto two-dimensional space, the path between epitope combinations in successive outbreaks (shown in red) in the model appears to follow a zigzag pattern in one direction. In this model, however, epitope combinations are constrained, such that the path through the network is eventually recycled

We have discussed how networks of contacts between hosts can dramatically alter the epidemiology and evolution of infectious diseases, and how networks of pathogen antigens can inform our understanding of pathogen biology. Although we still have no reliable empirical observations of social networks in the real world, hypothetical host contact networks have provided a solid theoretical basis for our understanding of how disease transmission may be affected by localized social interactions. Generally, the analysis of these networks has shown a reduced rate of spread when contacts are local, and an increase in the heterogeneity of pathogen traits such as virulence when host contacts are heterogeneous. Thus, pathogen transmission and evolution often reflects structural constraints imposed by the host population.

We believe a network approach has utility for understanding the evolution of frequently recombining pathogens. Both phylogenetic and population genetic techniques have yet to deal thoroughly with high rates of recombination, complicating the analysis of relationships between pathogen strains and species. Visualizing relationships between antigenic loci as networks can therefore provide a framework for data analysis, as discussed in relation to the malaria parasite. In addition, a more conceptual application may also be valuable for framing assumptions about antigenic relationships within abstract mathematical models, as explored above for the influenza virus. For most antigens that serve a specific functional purpose, such as binding to host receptors, the limits of antigenic space are likely to be finite and structured. In this context, an explicit representation of the network of possible evolutionary trajectories can have important implications for disease dynamics. Furthermore, a mapping of this network onto the network of immune responses of the host population can provide insight into the dynamical relationships between a pathogen and its host.

References

Anderson RM, May R (1992) Infectious diseases of humans: dynamics and control. Oxford University Press, New York

Boots M, Hudson PJ, Sasaki A (2004) Large shifts in pathogen virulence relate to host population structure. Science 303(5659):842–844

Buckee C, Danon L, Gupta S (2007) Host community structure and the maintenance of pathogen diversity. Proc Biol Sci 274(1619):1715–1721

Buckee CO, Koelle K, Mustard MJ, Gupta S (2004) The effects of host contact network structure on pathogen diversity and strain structure. Proc Natl Acad Sci U S A 101(29):10839–10844

Callaway DS, Newman ME, Strogatz SH, Watts DJ (2000) Network robustness and fragility: percolation on random graphs. Phys Rev Lett 85(25):5468–5471

Diekmann O, Heesterbeek JAP, Metz JAJ (1998) A deterministic epidemic model taking account of repeated contacts between the same individuals. J Appl Prob 35:462–468

Dietz K, Hadeler KP (1988) Epidemiological models for sexually transmitted diseases. J Math Biol 26(1):1–25

Eames KT, Keeling MJ (2006) Coexistence and specialization of pathogen strains on contact networks. Am Nat 168(2):230–241

Eubank S, Guclu H, Kumar VS, Marathe MV, Srinivasan A et al. (2004) Modelling disease outbreaks in realistic urban social networks. Nature 429(6988):180–184

Gomez-Gardenes J, Latora V, Moreno Y, Profumo E (2008) Spreading of sexually transmitted diseases in heterosexual populations. Proc Natl Acad Sci U S A 105(5):1399–1404

Grenfell BT, Bjornstad ON, Kappey J (2001) Travelling waves and spatial hierarchies in measles epidemics. Nature 414(6865):716–723

Jones JH, Handcock MS (2003) Social networks: sexual contacts and epidemic thresholds. Nature 423(6940):605–606

Keeling MJ, Eames KT (2005) Networks and epidemic models. J R Soc Interface 2(4):295–307

Lavstsen T, Salanti A, Jensen AT, Arnot DE, Theander TG (2003) Sub-grouping of *Plasmodium falciparum* 3D7 var genes based on sequence analysis of coding and non-coding regions. Malar J 2:27

May RM, Lloyd AL (2001) Infection dynamics on scale-free networks. Phys Rev E Stat Nonlin Soft Matter Phys 64(6 Pt 2):066112

Milgram S (1967) The small world problem. Psychol Today 1(1):1–67

Moore C, Newman ME (2000) Exact solution of site and bond percolation on small-world networks. Phys Rev E Stat Phys Plasmas Fluids Relat Interdiscip Topics 62(5 Pt B):7059–7064

Morris M, Kretzschmar M (1997) Concurrent partnerships and the spread of HIV. AIDS 11(5):641–648

Mossong J, Hens N, Jit M, Beutels P, Auranen K et al. (2008) Social contacts and mixing patterns relevant to the spread of infectious diseases. PLoS Med 5(3):e74

Pastor-Satorras R, Vespignani A (2002) Epidemic dynamics in finite size scale-free networks. Phys Rev E Stat Nonlin Soft Matter Phys 65(3 Pt 2A):035108

Rand DA, Keeling M, Wilson HB (1995) Invasion, stability and evolution to criticality in spatially extended host-pathogen systems. Proc R Soc Lond B: Biol Sci 259(1354):55–63

Recker M, Pybus OG, Nee S, Gupta S (2007) The generation of influenza outbreaks by a network of host immune responses against a limited set of antigenic types. Proc Natl Acad Sci U S A 104(18):711–716

Schenzle D (1984) An age-structured model of pre- and post-vaccination measles transmission. IMA J Math Appl Med Biol 1(2):169–191

Schneeberger A, Mercer CH, Gregson SA, Ferguson NM et al (2004) Scale-free networks and sexually transmitted diseases: a description of observed patterns of sexual contacts in Britain and Zimbabwe. Sex Transm Dis 31(6):380–387

Shannon C, Weaver W (1949) The mathematical theory of communication. University of Illionois Press, Urbana

Smith DJ, Lapedes AS, de Jong JC, Bestebroer TM, Rimmelzwaan GF et al (2004) Mapping the antigenic and genetic evolution of influenza virus. Science 305(5682):371–376

Watts DJ, Strogatz SH (1998) Collective dynamics of 'small-world' networks. Nature 393(6684):440–442

Chapter 9
Computational Epitope Mapping

Matthew N. Davies and Darren R. Flower

9.1 Introduction

> As the complexity of a system increases, [the decision maker's] ability to make precise and significant statements about its behavior diminishes until a threshold is reached beyond which precision and significance (or relevance) become almost mutually exclusive characteristics.... A corollary principle may be stated succinctly as: The closer one looks at a real-world problem, the fuzzier becomes its solution.
>
> L.A. Zadeh (1973) *Outline of a new approach to the analysis of complex systems and decision processes*

Different dictionaries contain different definitions of *epitope*, one of the most used, abused, and certainly oft-cited pieces of modern biomedical terminology. To paraphrase the Oxford English Dictionary (OED), an epitope is a surface region of an antigen to which an antibody may bind with high specificity, or, more generally, an epitope is an antigenic determinant. The OED traces the use of the word "epitope" to a 1960 review by Niels Kaj Jerne (1911–1994) (Jerne 1960). This work acknoledges the lack of a strict definition of *epitope*, linking it to many meanings, including surface configurations, dingle, determinants, structural themes, immunogenic elements, haptenic groups, and antigenic patterns. The termes "epitope" is intimately linked with the concept of self and non-self. Likewise, self – the word and concept embodied by it – has a variety of meanings, some harmonious, and some in conflict. It is important to note that *the self* is an idea able to encompass, or at least encapsulate, both the physical and psychological manifestations of identity.

In this chapter, we attempt to explore the notion of molecular self, as exemplified by the epitope. We examine the diverse and synergistic combination of molecular patterns that comprise the language of immunological recognition. We discuss how the words "self" and "epitope" relate to the ability of the immune system to identify molecules, cells, and organs as belonging to the host and to differentiate itself from nonself: molecules, cells, and organs of exogenous and potentially pathogenic origin. However, rather than saying "definition," in the singular, we

---

M.N. Davies (✉)
SGDP Centre, Institute of Psychiatry, London, United Kingdom

V. Sintchenko (ed.), *Infectious Disease Informatics*,
DOI 10.1007/978-1-4419-1327-2_9,

should rather say "definitions," in the plural, for within immunology there are many definitions of epitope.

One attempt to escape semantic constraints is provided by Polly Matzinger's danger model (Matzinger 2002). This proposes that the immune system reacts to danger signals, be they of external origin or from injured cells. Thus, the danger model effaces the immune self, replacing it with the idea of danger signals. Any molecular signal of whatever origin that is itself dangerous or can act as a flag for the presence of other dangerous substances could act in this way. Such models are uncomplicated yet compelling.

Self or nonself could encode recognition signals. Self is thus encoded as being part of the host and nonself encoded as being part of some identified non-host, a particular bacterium or set of bacteria, for example. Alternatively, theories can be formulated wherein either, but not both, the self or the non-self act as empty placeholders. For example, "self" could be identified as something possessing one or more signals of being part of the host; and "non-self" as being anything that does not. An entity is thus seen as non-self if it lacks a self-signal. The reverse could hold. Non-self could be identified as something possessing one or more signals of being part of a specific non-host organism; self would then be anything else. An entity is thus seen as "self" if it lacks a non-self signal.

Thus, put simply, we have three alternatives: A double-positive model, a self-positive model, or a non-self-positive model. There are many logical and practical problems with third alternative. It would necessitate the existence of generic signals across all non-host organisms or an effectively infinite capacity to store knowledge. At least as far as we know, neither of these can be realized in the context of a finite immune system. Obviously, a double-positive model allows for substances that are neither self nor non-self. The real immune self, however, is a composite exhibiting several features derived from all clear alternatives. Janeway provided a potential clarification, arguing that immunity discriminates between non-infectious self and infectious non-self (Janeway and Medzhitov 2002). The somatic component is provided by the life history of the host organism. Mammals evolve more slowly than microorganisms, and so pathogens would always be at an advantage in the immunological arms race.

The self is both molecules – peptides and proteins – and signals. The self and the non-self are merely, yet not solely, molecules that are recognized – or more properly bound – by other molecules. And that, as they say, is all that self and non-self ultimately are: molecules and their recognition by the host. All the rest is just waffle, confusion, and obfuscation. The signal in the self is the recognition event – Major Histocompatibility Complex (MHC) or antibody mediated – which triggers the immune system to respond.

It was once felt that the MHC would provide a simple, straightforward, unambiguous, and unequivocal criterion capable of discriminating self from non-self. Class I MHC molecules are expressed by almost every nucleated cell and are able to act to identify self. This is realized in the ternary complex of peptide-MHC and the T-cell Receptor (TCR), which is the necessary preliminary to the activation of the T-cell and thus the initiation of concomitant immune responses.

The self, and thus the non-self, are dynamic. Under certain constraints and in response to certain stimuli, the immune system is capable of attacking host constituents. Autoimmunity is immunity turning against the self that it is intended to defend. An autoimmune response is usually associated with disease, but can also be a normative function of homeostatic maintenance and control. Thus, the immune self is not a stable and unchanging entity but is clearly context dependent and prone to environmental influence.

The fundamental molecular mechanisms underlying cellular and humoral immunity are quite different. T-cell immunity is mediated by the molecular recognition of peptides bound to MHC molecules, essentially short denatured fragments excised from proteins via proteolytic degradation. B-cell- mediated immunity is made manifest by the antibody recognition of a protein antigen's three-dimensional structure. The molecular recognition events at the heart of cellular immunity are essentially conformation-independent: instead, they are mediated by the recognition of amino acid side chains within the context of a peptide-MHC complex. Humoral immunity is, by contrast, highly dependent on the conformation of a folded protein.

9.2 The Principal Molecular Varieties of Epitope

In a cellular context, the discrimination of "self" vs. "non-self" by the immune system has largely focused on the recognition of fragments derived by proteolysis from the host and pathogen proteins presented by classical MHC molecules. The innate system is also pivotal to the sensing of non-self, while the innate and adaptive immune systems are intimately connected and cooperate highly. Protective immunity results from the interplay of the antigen-specific adaptive immune system with the more generic, less specific innate response. The recognition properties of the innate system do not exhibit any optimization of specificity or selectivity. However, those of the adaptive immune system employ receptors that can undergo a refinement process that significantly enhances their recognition of whole antigens or derived peptides.

Innate immunity has a pivotal role in regulating adaptive immunity, in generating strong adaptive responses, and in the development of immune memory. Most of the operation of the innate immune system is preprogrammed and uses widely distributed receptors capable of recognizing generic targets: conserved molecular patterns characteristic of microbial life. It does this through the recognition, by "pattern recognition receptors" or PRRs, of evolutionarily conserved epitopes or so-called "pathogen associated molecular patterns" or PAMPs (de Diego et al. 2007). PRRs react to molecular structures found in or on pathogenic, but not normal vertebrate, cells. Each PRR has its own binding properties and cellular expression, and engages with different signaling pathways. This diversity within innate immunity protects the host from the diverse plethora of pathogens present in the environment. PRRs detect disturbances to the immune microenvironment (including the discrimination of "non-self") and initiate appropriate innate responses (Areschoug and Gordon 2008).

PRRs bind multiple ligands by recognizing common PAMPs rather than binding to unique epitopes. However, the range of peptides bound by antibodies, and particularly by the MHC-TCR system, is much, much larger than is generally supposed. The PRR engagement of PAMPs elicits a response that is typically proinflammatory, involving cytokine generation that activates immune cells. Such reactions are crucial to disease management, but must be tightly regulated as an excessive immune response is pernicious.

PRRs are encoded by germ line genes. Since the structures of such receptors are inherited and resulted entirely from conventionally understood evolutionary processes, their specificity is fixed. They evolve relatively slowly by the mechanisms of natural selection through standard processes of point mutation, gene duplication, and so on. The germ line nature of these receptors necessarily limits the eventual repertoire of recognition specificity exhibited by the innate immune system; it does not permit the recognition of previously unknown antigens, yet except by chance. Yet over long periods, it can evolve to ignore self-molecules and thus manifest robust discrimination between noninfectious self and infectious non-self.

Several distinct families of PRRs are known, including the following long list (Areschoug and Gordon 2008; Kornbluth and Stone 2006). Arguably the most important, or at least the most prominent, are the so-called toll or toll-like receptors (TLRs). Humans have ten TLRs; they sense both intracellular pathogens (viruses) and extracellular pathogens (bacteria and fungi). Some bind particular patterns contained in microbial DNA that are absent from vertebrate DNA. More specifically, ssRNA is recognized by TLR7 and TLR8, and dsRNA is recognized by TLR3. TLR4 recognizes LPS, Taxol, bacterial HSP60, F protein, and fibronectin. TLR5 binds flagellin. TLR2 and TLR6 recognize many ligands, such as bacterial lipoproteins or peptidoglycan. TLR9, found on DCs and B-cells, detects CpG motifs in DNA. An activated TLR-dependent signaling cascade ultimately induces the expression of a variety of response molecules.

dsRNA is also recognized within the cytoplasm by another PRR – RNA helicases such as RIG-I. These are important PRRs, as are the cytosolic NOD-like receptors (NLRs), which undertake key activities in innate immunity as intracellular sensors of cell damage and pathogens. While TLRs signal from the cell surface or early endosome, bacterial molecules activate NLRs intracellularly. Models of bacterial infection suggest a pro-inflammatory role for NLRs, including the regulation of the "inflammasome." Other PRRs include scavenger receptors and C-type lectin-like receptors such as mannose receptors, which detect mannosylated lipoarabinomannans, β-glucan receptors, and DC-SIGN, which detects carbohydrate moieties on a variety of proteins.

Whole classes of organisms usually share pAMPs. Lipopolysaccharide (LPS) is, for example, found as a common component of gram- negative bacteria. PAMP-stimulated PRRs induce the maturation and migration of APCs, the upregulation of antigen-loaded Class I and Class II molecules, the cell surface expression of co-stimulatory molecules, and the production of cytokines and chemokines.

Costimulatory molecules flag the microbial origin of a presented antigen, help to activate antigen-specific T-cells, and create an inflammatory environment that amplifies adaptive immune responses. Activation of different PRRs can alter the form taken by subsequent immune responses.

Traditionally, when attempting to analyze and predict properties of the cellular response, immunoinformaticians have centred their attention and their efforts solely on the specificity of MHC molecules (Flower 2003; Flower and Doytchinova 2002). In more recent and more enlightened times, attention has turned to the richer, deeper, more challenging world of antigen presentation (Vivona et al. 2008; Davies and Flower 2007). Our understanding of the manifold mechanisms underlying antigen presentation, and thus the manifestation of the molecular components of the immune self, is as yet incomplete and partial. These mechanisms, as we currently picture them, are by no means simple. As with all exciting branches of science, many important aspects of these complex processes remain controversial.

MHCs bind peptides produced by the proteolytic degradation of proteins. There are many alternative processing pathways, but the two best-understood are the classical Class I and classical Class II. MHCs are not indiscriminate binders, but importantly exhibit a finely tuned yet complex specificity for particular peptide sequences composed of the 20 commonly occurring amino acids. MHCs also display a wider specificity that is in itself quite catholic in terms of the molecules they can bind; MHCs are not restricted solely to peptides; they also bind a variety of other molecules. A wide range of Post-Translational Modifications (PTMs) and synthetically modified peptides are also bound by MHCs and recognized by T-cells. Many natural and synthetic small molecules also bind to MHCs. In addition, small molecule drug-like compounds can form complexes with MHCs. This can mediate pathological effects and has important implications for behavior-modifying odor recognition.

Cell-surface antigen presentation in the context of Class I and Class II MHC molecules demonstrates distinct differences. This arises in at least three different ways: one from the physical differences in peptide binding by the different classes; one from the TCR-mediated differences in the recognition of the two classes; and another from the significant differences in the complex machinery of antigen processing and transport that affects the conversion of whole proteins into fragmentary epitopes.

Class I MHC molecules survey important intracellular changes. These include viral infection, the presence of intracellular bacteria, or malignant cellular transformation as seen in tumor cells. The flagging or signaling of such profound cellular events ensures the induction of an appropriate immune response by circulating CD8 + T-cells. Class II MHC molecules proffer to circulating CD4 + T-cell-mediated immune surveillance markers sampling extracellular events.

Cellular antigen presentation is affected significantly by the innate response. PAMPs and the PRRs that bind them affect both Class I and Class II antigen processing and presentation pathways. They regulate and orchestrate the spatio-temporal dynamics of MHC biosynthesis, antigen sequestration, and the reordering of the cytoskeleton.

The natural repertoire of Class I MHC-presented peptides has a greater breadth than is widely supposed (Vyas et al. 2008; Lin et al. 2008; Loureiro and Ploegh 2006). MHC Class I ligands are derived primarily from degraded endogenously expressed intracellular proteins. Intracellular peptide fragments arise from two sources: self-peptides derived from the host genome and proteins from external sources such as pathogenic microbes, principally those originating from viral infection. This seeming simplicity obfuscates several layers of complexity.

Intracellular proteins, including newly synthesized proteins, are degraded quickly, producing large amounts of short peptides. Nonfunctional proteins, or defective ribosomal products (DRiP), result from errors in translation and processing. They form a significant proportion of newly synthesized proteins, which are rapidly digested by the proteasome. Viruses can invade host cells and generate viral proteins and bacteria can inject protein into the host cell via type III secretion systems; both are degraded by the host.

A multiprotein complex called the proteasome mediates intracellular protein degradation. The proteasome is a multimeric proteinase composed of a core of proteolytic enzymes flanked by a complex arrangement of regulatory elements capable of recognizing, among other things, an ubiquitin label. A whole variety of proteins including heat-denatured proteins, incorrectly assembled, mis-translated or mis-folded proteins, as well as regulatory proteins with limited half-lives, are targeted by the proteasome by being tagged with ubiquitin. The proteasome then proteolytically digests proteins in a stochastic manner, producing a population of relatively short peptides.

After peptides are degraded by the proteasome, they are transferred into the lumen of the endoplasmic reticulum (ER). The translocation process from cytosol to ER is ATP dependent. The so-called transporter associated with antigen processing, or TAP, effects peptide transit, and is also able to interact with peptide-free Class I HLA molecules in the ER. Newly synthesized Class I MHC molecules are understood to be unstable in a peptide-free state and are retained in the ER in a partially folded form. Formation of a MHC-peptide complex is quite intricate and complicated, and it is facilitated by a variety of proteins, including tapasin, calreticulin, and ERp57. Once complexed to peptide and β_2-microglobulin, the MHC protein leaves the ER and is transported to the cell surface. The peptide binding process is considered as the rate- limiting step of MHC protein assembly, as only a fraction of the peptides are able to bind to MHC.

Class II MHC expression is believed to be restricted primarily to professional antigen-presenting cells (APCs), including macrophages and dendritic cells (DCs). In the MHC Class II processing pathway, following the receptor-mediated endocytosis of exogenous antigens by APCs, presented proteins are targeted to the multi-compartment lysosomal-endosomal apparatus, passing first into endosomes, then into late endosomes, ending up in lysosomes. While in transit, antigens are proteolytically fragmented into peptides by cathepsins. Before final cell surface presentation, peptides are bound by Class II MHCs. MHC Class II ligands have a more variable length of 9–25 amino acids and are derived mainly from exogenous proteins. Peptide-bound Class II MHC molecules are ultimately translocated to the cell surface, where they are available for immune surveillance by $CD4^+$ T-cells.

Another entity mediating the recognition of self is the so-called B-cell epitope. These are regions of the surface of a protein, or other biomacromolecule, recognized by soluble or membrane-bound antibody molecules. The protection offered by all vaccines is mediated completely or predominantly through the induction of antibodies, which act mostly in infection at the bacteremic or viremic stage. Humoral immunogenicity, as mediated by soluble or membrane-bound cell surface antibodies through their binding of B-cell epitopes, is of prime importance for almost all existing vaccines, except BCG.

B-cell epitopes can be linear (also called continuous) or discontinuous. Linear epitopes are single, short, continuous subsequences within an antigen. Discontinuous epitopes are groups of individual, isolated residues forming patches on the surface of the antigen. The verity and exegesis of an epitope depends on the nature of their experimental determination. Linear epitopes are typically identified using an experimental screening procedure, i.e., PEPSCAN, where by overlapping sequences are assayed against pre-existing ex vivo antibodies. Discontinuous epitopes are usually identified from the structure of an antigen, typically one derived experimentally by X-ray crystallography or multidimensional NMR. Discontinuous epitopes are also identified by making site-directed mutants of the antigen and testing them for their effect on antibody binding.

Taken together, all of these mechanisms greatly increase the potential size and diversity of the immunogenic repertoire of reactive peptides. Thus, one may argue that, in the face of such complexity, the only realistic way to address this potential enormity of the peptide repertoire is via computational analysis and prediction. For the reasons adumbrated earlier, we shall concentrate on T-cell mapping.

9.3 T-cell and B-cell Epitope Prediction In Silico

Informatics, in the form of immunoinformatics, offers a considerable diversity of tools and techniques for undertaking epitope mapping in silico. With an ever-increasing number of pathogen genomes now available, the mapping of B-cell and T-cell epitopes, both computationally and experimentally, is becoming a central issue in vaccine discovery (De Groot 2006; De Groot and Berzofsky 2004). By using such approaches, computer-based prediction methods can greatly increase the celerity of T-cell and B-cell epitope discovery.

Experimentally determined IC50 and BL50 affinity data have been used to develop a variety of peptide sequence-based MHC binding prediction algorithms, which can distinguish binders from non-binders. Many different algorithms, mostly developed by the data-mining community and derived from research into artificial intelligence, have now been applied to immunoinformatic problems. MHC binding motifs are a straightforward and easily comprehended method of epitope detection, yet produce many false-positive and many false-negative results

(Rammensee et al. 1999). Support Vector Machines (SVMs) are machine-learning algorithms based on statistical theory that seeks to separate data into two distinct classes (in this case binders and non-binders) (Jardetzky et al. 1996; Donnes and Elofsson 2002). Hidden Markov Models (HMMs) are statistical tools where the system being modeled is assumed to be a Markov process with unknown parameters (Noguchi et al. 2002). In a HMM, the internal state is not visible directly, but variables influenced by the state are. HMMs aim to determine the hidden parameters from observable ones. A HMM profile can be used to determine those sequences with "binder-like" qualities. Bayesian Neural networks can also be applied to the problem, as they are better suited to recognizing complicated peptide patterns than more straightforward algorithms (Burden and Winkler 2005). Bayesian neural networks in particular have the advantages that they are robust, resistant, but not immune, to overtraining, capable of minimize the risk of overfitting, tolerate noisy or missing data, and can find the least complex model capable of explaining the data automatically.

Of the existing immunoinformatic prediction techniques, by far the most successful has been that of data-driven prediction of T-cell epitopes, at least for well-studied Class I MHC alleles (Peters et al. 2006). In a pivotal retrospective analysis, Deavin et al. (1996) compared several early direct T-cell epitope prediction methods without finding a single method with a high-enough accuracy to be useful. Today, work concentrates instead on predicting Class I MHC-peptide binding affinity. Where data are sufficiently abundant, such methods work well (Peters et al. 2006; Flower 2008).

Compared with Class I predictions, Predicting Class II epitopes is much more problematic. Such difficulties arise for several reasons. Chief among these is the unrestricted length of Class II epitopes. The structure of the open-ended Class II binding site does not constrain peptide lengths, allowing the binding of the full range of peptide lengths – 11–25 + amino acids. X-ray structures of Class II MHCs indicate that the binding site is typically occupied by a nine-residue subsequence, with the rest of the peptide extending out at one or both ends. Thus, immunoinformatic algorithms for Class II need to identify the central 9mer when attempting prediction, and to then develop predictive models for the bound nonameric sequence.

This search is complicated, conceptually at least, by the ability of MHCs to bind in a degenerative manner. Long peptides, in particular, might exhibit a hierarchy of multiple binding modes. However, relatively little is known concerning the explicit degeneracy of the binding process. Nonetheless, the fact that the binding groove is open at both ends in Class II molecules is consistent with the possibility. Whether this phenomenon actually occurs in reality seems unlikely on theoretical grounds, except in the case of repetitive sequences.

Moreover, our attempts to account for a possible multiplicity of binding modes, i.e., 2 or more 9mer subsequences, have not yet yielded a stable solution or workable algorithm. Another important issue is the influence of "flanking" residues on affinity and recognition: Arnold et al. identified residues at +2 or −2, relative to the core nonamer, as important for effective recognition by T-cells (Arnold et al. 2002).

We have sought to address this by increasing the core peptide region identified in our model by 2 in both directions, but again this did not yield a stable solution, perhaps suggesting that this phenomenon is a subsidiary one, at least statistically.

Recently, an attempt has been made to incorporate components of the Class I antigen presentation pathway, such as Proteasome cleavage (Saxova et al. 2003) and TAP binding (Doytchinova et al. 2004), into composite approaches to T-cell epitope prediction (Doytchinova et al. 2006; Peters and Sette 2005; Larsen et al. 2005; Dönnes and Kohlbacher 2005). Likewise, we have recently explored a CoMSIA method (Hattotuwagama et al. 2006) for distinguishing true epitopes from non-epitopes that bind MHCs with high affinity (Doytchinova and Flower 2006). These methods, which show encouraging improvements, compared to MHC-only approaches, use subsidiary stages, such as TAP binding, as additional filters to reduce the number of possible epitopes.

However, for the prediction of all immunological epitope data other than Class I MHC peptide binding, results have been unsatisfactory and inadequate. Over the last few years, several comparative studies have shown that the prediction of Class II T-cell epitopes is usually poor (El-Manzalawy et al. 2008a, b; Lin et al. 2008; Gowthaman and Agrewala 2008). Results are similar for structure-driven prediction of Class I and Class II T-cell epitopes (Knapp et al. 2009). Likewise, both structure- (Ponomarenko and Bourne 2007) and data-driven (Blythe and Flower 2005) prediction of antibody-mediated epitopes is known to be poor. Moreover, irrespective of the poor reported predictivity, there are several other problems, albeit different, for T-cell and B-cell epitope prediction.

For T-cell prediction, the major issue is the quality and availability of data. It has recently been shown that that T-cell epitopes, which were previously thought to be short peptides of 8–10 amino acids, can consist of up to 16 amino acids or perhaps even more. The existence of these longmer epitopes has greatly expanded the repertoire of peptides open to inspection by T-cells (Flower 2008). Similarly, over 3,000 different MHC alleles are known to exist in the global human population, indicating the potential for distinct peptide specificities among patients. Problematic as this seems, the situation is made worse by the fundamental logistic constraints of sampling within even a single allele specificity. A nonameric peptide has a 20^9 data space equating to 512 billion combinations of amino acid; considering that a single model is built from a few hundred peptides at most, the sampling ratio is infinitesimally small. While some work addresses the allele diversity issue – for example, pan-MHC methods (Zhang et al. 2009; Nielsen et al. 2008) and supertypes (Doytchinova and Flower 2005; Doytchinova et al. 2004) – little has been or can be done to circumvent the sampling issue.

Table 9.1 summarizes a fair cross-section of currently available servers for T-cell epitope prediction. Within the limits imposed by the quality and sufficiency of binding data then these methods work, and unequivocally so. Yet, as we have said, these limits can be very limiting indeed. For many alleles, the construction of useful and meaningful training and testing sets is highly problematic. Efforts are, ultimately, limited by the data themselves. A properly designed training set will resolve most issues.

Table 9.1 T-cell epitope prediction servers

Prediction server	URL	Class	Ref
CTLPred	http://www.imtech.res.in/raghava/ctlpred	I	Bhasin and Raghava (2004)
MMBPred	http://www.imtech.res.in/raghava/mmbpred	I	Bhasin and Raghava (2003)
NetMHC	http://www.cbs.dtu.dk/services/NetMHC	I	Buus et al. (2003)
BIMAS	http://thr.cit.nih.gov/molbio/hla_bind	I	Bhasin and Raghava (2006)
NetCTL	http://www.cbs.dtu.dk/services/NetCTL	I	Larsen et al. (2005)
ProPred-I	http://www.imtech.res.in/raghava/propred1	I	Singh and Raghava (2003)
NHLApred	http://www.imtech.res.in/raghava/nhlapred	I	Bhasin and Raghava (2006)
MHC-Thread	http://www.csd.abdn.ac.uk/gjlk/MHC-Thread/	II	Swain et al. (2001)
NetMHCII	http://www.cbs.dtu.dk/services/NetMHCII	II	Nielsen et al. (2007)
ProPred	http://www.imtech.res.in/raghava/propred	II	Singh and Raghava (2001)
SYFPEITHI	http://www.syfpeithi.de/	Both	Rammensee et al. (1999)
MHCPred	http://www.jenner.ac.uk/MHCPred	Both	Guan et al. (2006)
IEDB	http://tools.immuneepitope.org/	Both	Nielsen et al. (2003)
SVMHC	http://www-bs.informatik.uni-tuebingen.de/SVMHC	Both	Donnes and Elofsson (2002)
RankPep	http://bio.dfci.harvard.edu/RANKPEP/	Both	Reche et al. (2002)
ELF	http://www.hiv.lanl.gov/content/hiv-db/ELF/	Both	Korber et al. (2005)
EpiPredict	EpiPredict http://www.epipredict.de/Prediction/prediction.html	Both	Jung et al. (2001)

Data quantity, in particular, has important implications for the selection of appropriate prediction techniques. Guidelines that help us to address these issues are:

- In the absence of binding data, speculative molecular modeling is the only option. Here, supertype analysis can prove useful.
- When peptide number is below 50, binding motifs offer a pragmatic answer.
- With 50–100 peptides, quantitative matrices, SVMs, or QSAR are usable.
- With over 100 peptides, HMMs, artificial neural networks, Bayesian networks, or robust multivariate statistical models are useable.
- With very large data sets of many hundred peptides, most modern methods provide high-quality predictions, albeit within their own interpolative boundaries.

However, data diversity as well as quantity is an issue. As diversity in peptide sequence and affinity increases, so does the generality of the generated models. Highly degenerate data or data with a very narrow affinity range often prove difficult.

Predictive models should be tested before using via cross-validation, test sets, and randomization. However, the ideal testing strategy is to use experimental validation involving the blind prediction and testing of novel peptides.

Sequence-based B-cell epitope prediction methods are limited to the identification of linear epitopes. If we look back a decade or two, most predictors of either T-cells or B-cell epitopes were based on identifying maximally valued regions of sequences – essentially looking for peaks, or troughs, in some form of a propensity plot. This was long ago shown to be inappropriate for T-cell epitopes and consequently many advanced methods for T-cell epitopes prediction have arisen. However, many – most, if not actually all – B-cell epitope prediction methods continue to rely, wholly or in part, on finding such peaks. No single property is known to predict linear or discontinuous epitope location with any reliability or accuracy. Most prediction methods use properties related to surface exposure – such as accessibility, hydrophilicity, flexibility/mobility and loop and turn structures – since it is believed that epitopes, at least for non-denatured proteins, must be solvent-accessible if antibody binding is to occur.

Early approaches used the sliding-window method, adapting standard hydropathy scales to identify maximal property peaks. A correctly predicted epitope equates to any peak close to an antigenic residue. Short windows, which reduced erratic peak values, outperformed larger window sizes. Using datasets representing the most stringent examples of peer-reviewed publications describing linear epitope-mapped protein sequences, Blythe and Flower (2005) have explored the validity of B-cell epitope prediction using sequence profiles of amino acid scales. Using 484 amino acid scales and 50 epitope-mapped protein sequences, as defined using polyclonal antibodies, the analysis of both single sequence and combined profiles indicated that this approach is of limited value: the best sets of methods generated predictions only fractionally better than random.

The poor performance demonstrated by BCE prediction algorithms is troubling. No explanation seems overly convincing. It is unlikely that the available methodology is to blame, as data-mining techniques have proved much more successful in other areas. The explanation favored here again targets the experimental data as the source of the problem. The most widely available data derives from PEPSCAN, and there are reasons to suspect that this is not what it seems or what people believe it to be. Experimentally derived epitopes are identified by assayed against pre-existing antibodies with affinity for whole antigens. However, when such "epitopes" are mapped back onto antigen structures, their locations are scattered randomly through the protein. They would not form discrete patches as one would expect if they were simple mimics of crystallographically identified discontinuous epitopes. These in situ epitopes can be exposed or completely buried, and thus are inaccessible to antibody binding, and also in every state in between. If we compare the conformation adopted by antibody bound peptides with those in situ in the intact antigen, we see that they are typically very different. However, if we compare antibodies in intact antigen and in whole antigen-antibody complexes, they are very similar. Thus, how epitopes in a PEPSCAN analysis are recognized requires an explanation other than that of a simple one-to-one correspondence. One such explanation might be that denatured antigen is recognized in vivo by the preformed antibody. Another is that the isolated peptide has a conformation capable of imitating some or all of the surface features exhibited by a discontinuous epitope.

Nonetheless, many servers are now available that implement one or another of the many published algorithms and methods purported to predict B-cell epitopes. We sample these in Table 9.2. The inclusion of one approach at the expense of another should not be taken as any form of vindication nor as any kind of condemnation. Instead, we should bear in mind the significant provisos enumerated in the preceding paragraph when we seek to select B-cell epitope prediction methods.

An alternative to epitope mapping is the computational identification of whole antigens as opposed to the epitopes they contain. Antigens, as prospective subunit vaccines, must be immunogenic, but many facets of immunogenicity are as yet under-explored experimentally. Some intriguing anecdotal evidence has gathered over time and has suggested that large complex proteins or proteins distant in sequence from the host proteome are more likely to be immunogenic – while others – that aggregated protein is immanently more immunogenic – are less open to prediction. The manifestation of immunogenicity at the protein level arises from a complex process that combines both intrinsic and extrinsic factors and that operates at different scales and rates. Properties intrinsic to the host immune system (such as the possession of appropriate B- or T-cell epitopes) interact with properties intrinsic to the pathogen (such as its expression level, the time course of expression and secretion, and its location within the host cell) and with properties intrinsic to the protein itself (such as the presence of post-translational danger signals) to determine whether there will be an immunogenic response. The recognition of such signals is the job of the innate immune system, which provides a rapid yet non-specific response. Some efforts have been made to predict antigens directly (Doytchinova and Flower 2007a, b) and to develop alternative strategies for the identification of vaccine candidates with computer-aided reverse vaccinology (Serruto and Rappuoli 2006).

Table 9.2 B-cell epitope prediction severs

Server	URL	Ref
ABCpred	http://www.imtech.res.in/raghava/abcpred	Saha and Raghava GPS (2006)
Bepipred	http://www.cbs.dtu.dk/services/BepiPred	Larsen et al. (2006)
CEP	http://bioinfo.ernet.in/cep.htm	Kulkarni-Kale et al. (2005)
DiscoTope	http://www.cbs.dtu.dk/services/DiscoTope	Haste Andersen et al. (2006)
PEPPOP	http://diagtools.sysdiag.cnrs.fr/PEPOP/	Moreau et al. (2008)
Epitopia	http://epitopia.tau.ac.il/	Rubinstein et al. (2008)
Pep-3D-Search	http://kyc.nenu.edu.cn/Pep3DSearch/	Huang et al. (2009)
ElliPro	http://tools.immuneepitope.org/tools/ElliPro	Ponomarenko et al (2008)
LEPD	http://biotools.cs.ntou.edu.tw/lepd_antigenicity. php	Chang et al. (2008)
BCPREDS	http://ailab.cs.iastate.edu/bcpreds/	El-Manzalawy et al. (2008)
MIMOP	http://diagtools.sysdiag.cnrs.fr/MIMOP/	Moreau et al. (2006)
AAPRED	http://www.bioinf.ru/aappred/	Ya et al. (2009)

9.4 Conclusion

Computational support for vaccinology is clearly a discipline in transition. Sequence analysis and genome annotation have long been in use and are now being supplemented by many potent prediction techniques, principally those for T-cell epitopes. The computational identification of B-cells or T-cells or epitopes hardly closes the door on immunological prediction. Rather, it is the key that opens that door, or at least the crow-bar that prizes that door open. When there are enough data to build a good model and where the prediction method is sophisticated enough, it is more efficient to use prediction than to perform exhaustive experiments. That such approaches are not much more widely deployed says as much about the mind-set of immunologists and biotechnologists as it does about the reliability and poor exposure of prediction tools and techniques.

Yet we seek to provide a word of caution. Firstly, good data are essential. Despite the overenthusiasm of some in the field, much work remains to be done, or at least done well. T-cell epitope software is not as reliable as many claim, while B-cell epitope software has not developed to the point where it is of any practical use. Conversely, reverse vaccinology shows considerable promise. If reverse vaccinology is applied astutely as a tool in vaccine design and discovery, it can save enormous amounts of money, time, and wasted labor. Nevertheless, the development of accurate, robust, and reliable epitope mapping remains a key objective. Progress is being made but not quite as fast as its proponents maintain. The potential is huge, but only if researchers are willing to take up the technology and use it appropriately.

People's expectations of computational work are typically biased and thus unrealistic. Some expect perfection and are usually disappointed; others are highly critical and are almost impossible to reconcile with informatics methods. Neither of these two stances is wholly or completely correct. However, one is ever minded to sympathize and condole with the expression of such feelings. Informatics does not supplant, or even seek to supplant, experiment; instead, it helps to rationalize the increasingly complex, confusing, and confounding world of post-genomic research. It exists, at least in part, to save labor, time, and resources. Informatics requires intellectual effort comparable in scale but not in kind to experimental science. The two disciplines – informatics and experimental science – are complementary and distinct. Ultimately, informatics will find its place, although that may take some time yet.

References

Areschoug T, Gordon S (2008) Pattern recognition receptors and their role in innate immunity: focus on microbial protein ligands. Contrib Microbiol 15:45–60

Arnold PY, La Gruta NL, Miller T, Vignali KM et al (2002) The majority of immunogenic epitopes generate CD4 + T cells that are dependent on MHC class II-bound peptide-flanking residues. J Immunol 169:739–749

Bhasin M, Raghava GPS (2003) Prediction of promiscuous and high-affinity mutated MHC binders. Hybridomics 22:229–234

Bhasin M, Raghava GPS (2004) Prediction of CTL epitopes using QM, SVM and ANN techniques. Vaccine 22:3195–3201

Bhasin M, Raghava GPS (2006) A hybrid approach for predicting promiscuous MHC Class I restricted T-cell epitopes. J Biosci 32:31–42

Bhasin M, Raghava GPS (2006) A hybrid approach for predicting promiscuous MHC Class I restricted T-cell epitopes. J Biosci 32:31–42

Blythe MJ, Flower DR (2005) Benchmarking B-cell epitope prediction: underperformance of existing methods. Protein Sci 14:246–248

Burden FR, Winkler DA (2005) Predictive Bayesian neural network models of MHC Class II peptide binding. J Mol Graph Model 23:481–489

Buus S et al (2003) Sensitive quantitative predictions of peptide- MHC binding by a 'Query by Committee' artificial neural network approach. Tissue Antigens 62:378–384

Chang HT, Liu CH, Pai TW (2008) Estimation and extraction of B-cell linear epitopes predicted by mathematical morphology approaches. J Mol Recognit 21:431–441

Davies MN, Flower DR (2007) Harnessing bioinformatics to discover new vaccines. Drug Discov Today 12:389–395

Deavin AJ, Auton TR, Greaney PJ (1996) Statistical comparison of established T-cell epitope predictors against a large database of human and murine antigens. Mol Immunol 33:145–155

de Diego JL, Gerold G, Zychlinsky A (2007) Sensing, presenting, and regulating PAMPs. Ernst Schering Found Symp Proc 3:83–95

De Groot AS (2006) Immunomics: discovering new targets for vaccines and therapeutics. Drug Discov Today 11:203–209

De Groot AS, Berzofsky JA (2004) From genome to vaccine – new immunoinformatics tools for vaccine design. Methods 34:425–428

Donnes P, Elofsson A (2002) Prediction of MHC Class I binding peptides using SVMHC. BMC Bioinform 3:25–38

Dönnes P, Kohlbacher O (2005) Integrated modeling of the major events in the MHC Class I antigen processing pathway. Protein Sci 14:2132–2140

Doytchinova IA, Flower DR (2005) In silico identification of supertypes for Class II MHCs. J Immunol 174:7085–7095

Doytchinova IA, Flower DR (2006) Modeling the peptide-T-cell receptor interaction by the comparative molecular similarity indices analysis-soft independent modeling of class analogy technique. J Med Chem 49(7):2193–2199

Doytchinova IA, Flower DR (2007a) VaxiJen: a server for prediction of protective antigens, tumour antigens and subunit vaccines. BMC Bioinform 8:4

Doytchinova IA, Flower DR (2007b) Identifying candidate subunit vaccines using an alignment-independent method based on principal amino acid properties. Vaccine 25:856–866

Doytchinova IA, Guan P, Flower DR (2004) Identifiying human MHC supertypes using bioinformatic methods. J Immunol 172:4314–4323

Doytchinova I, Hemsley S, Flower DR (2004) Transporter associated with antigen processing preselection of peptides binding to the MHC: a bioinformatic evaluation. J Immunol 173(11):6813–6819

Doytchinova IA, Guan P, Flower DR (2006) EpiJen: a server for multistep T-cell epitope prediction. BMC Bioinform 7:131

El-Manzalawy Y, Dobbs D, Honavar V (2008a) On evaluating MHC-II binding peptide prediction methods. PLoS ONE 3:e3268

El-Manzalawy Y, Dobbs D, Honavar V (2008b) Predicting linear B-cell epitopes using string kernels. J Mol Recogn 21:243–255

Flower DR (2003) Towards in silico prediction of immunogenic epitopes. Trends Immunol 24:667–674

Flower DR (2008) Bioinformatics for vaccinology. Wiley

Flower DR, Doytchinova IA (2002) Immunoinformatics and the prediction of immunogenicity. Appl Bioinform 1:167–176

Gowthaman U, Agrewala JN (2008) In silico tools for predicting peptides binding to HLA-Class II molecules: more confusion than conclusion. J Proteome Res 7:154–163

Guan P et al (2006) MHCPred 2.0: an updated quantitative T-cell epitope prediction server. Appl Bioinform 5:55–61
Haste Andersen P, Nielsen M, Lund O (2006) Prediction of residues in discontinuous B-cell epitopes using protein 3D structures. Protein Sci 15:2558–2567
Hattotuwagama CK, Toseland CP, Guan P, Taylor DJ, Hemsley SL, Doytchinova IA, Flower DR (2006) Toward prediction of Class II mouse major histocompatibility complex peptide binding affinity: in silico bioinformatic evaluation using partial least squares, a robust multivariate statistical technique. J Chem Inf Model 46(3):1491–1502
Huang YX, Bao YL, Guo SY, Wang Y et al (2008) Pep-3D-Search: a method for B-cell epitope prediction based on mimotope analysis. BMC Bioinform 9:538
Janeway CA Jr, Medzhitov R (2002) Innate immune recognition. Annu Rev Immunol 20:197–216
Jardetzky TS et al (1996) Crystallographic analysis of endogenous peptides associated with HLADR1 suggests a common, polyproline II-like conformation for bound peptides. Proc Natl Acad Sci USA 93:734–738
Jerne NK (1960) Immunological speculations. Annu Rev Microbiol 14:341–358
Jung G et al (2001) From combinatorial libraries to MHC ligand motifs, T-cell superagonists and antagonists. Biologicals 29:179–181
Knapp B, Omasits U, Frantal S, Schreiner W (2009) A critical cross-validation of high throughput structural binding prediction methods for pMHC. J Comput Aided Mol Des 23:301–307
Korber BT et al (2005) HIV Molecular Immunology 2005. Los Alamos National Laboratory, Theoretical Biology and Biophysics
Kornbluth RS, Stone GW (2006) Immunostimulatory combinations: designing the next generation of vaccine adjuvants. J Leukoc Biol 80:1084–1102
Kulkarni-Kale U, Bhosle S, Kolaskar AS (2005) CEP: a conformational epitope prediction server. Nucleic Acids Res 33(Web Server issue):W168–171
Larsen JE, Lund O, Nielsen M (2006) Improved method for predicting linear B-cell epitopes. Immunome Res 2:2
Larsen MV et al (2005) An integrative approach to CTL epitope prediction. A combined algorithm integrating MHC-I binding, TAP transport efficiency, and proteasomal cleavage predictions. Eur J Immunol 35:2295–2303
Larsen MV, Lundegaard C, Lamberth K, Buus S, Brunak S, Lund O, Nielsen M (2005) An integrative approach to CTL epitope prediction: a combined algorithm integrating MHC Class I binding, TAP transport efficiency, and proteasomal cleavage predictions. Eur J Immunol 35:2295–2303
Lin HH, Zhang GL, Tongchusak S, Reinherz EL, Brusic V (2008) Evaluation of MHC-II peptide binding prediction servers: applications for vaccine research. BMC Bioinform 9(Suppl 12):S22
Lin ML, Zhan Y, Villadangos JA, Lew AM (2008) The cell biology of cross-presentation and the role of dendritic cell subsets. Immunol Cell Biol 86:353–362
Loureiro J, Ploegh HL (2006) Antigen presentation and the ubiquitin-proteasome system in host-pathogen interactions. Adv Immunol 92:225–305
Matzinger P (2002) An innate sense of danger. Ann NY Acad Sci 961:341–342
Moreau V, Granier C, Villard S, Laune D, Molina F (2006) Discontinuous epitope prediction based on mimotope analysis. Bioinform 22:1088–1095
Moreau V, Fleury C, Piquer D, Nguyen C et al (2008) PEPOP: computational design of immunogenic peptides. BMC Bioinform 9:71
Nielsen M et al (2003) Reliable prediction of T-cell epitopes using neural networks with novel sequence representations. Prot Sci 12:1007–1017
Nielsen M et al (2007) Prediction of MHC Class II binding affinity using SMM-align, a novel stabilization matrix alignment method. BMC Bioinform 8:238
Nielsen M, Lundegaard C, Blicher T, Peters B et al (2008) Quantitative predictions of peptide binding to any HLA-DR molecule of known sequence: NetMHCIIpan. PLoS Comput Biol 4:e1000107
Noguchi H et al (2002) Hidden Markov model-based prediction of antigenic peptides that interact with MHC Class II molecules. J Biosci Bioeng 94:264–270
Peters B, Sette A (2005) Generating quantitative models describing the sequence specificity of biological process with the stabilized matrix method. BMC Bioinform 6:132

Peters B, Bui HH, Frankild S, Nielson M et al (2006) A community resource benchmarking predictions of peptide binding to MHC-I molecules. PLoS Comput Biol 2:e65

Ponomarenko J, Bui HH, Li W, Fusseder N et al (2008) ElliPro: a new structure-based tool for the prediction of antibody epitopes. BMC Bioinform 9:514

Ponomarenko JV, Bourne PE (2007) Antibody-protein interactions: benchmark datasets and prediction tools evaluation. BMC Struct Biol 7:64

Rammensee H et al (1999) SYFPEITHI: database for MHC ligands and peptide motifs. Immunogenet 50:213–219

Reche PA et al (2002) Prediction of MHC Class I binding peptides using profile motifs. Hum Immunol 63:701–709

Rubinstein ND, Mayrose I, Pupko T. (2009) A machine-learning approach for predicting B-cell epitopes. Mol. Immunol 46:840–847

Saha S, Raghava GPS (2006) Prediction of continuous B-cell epitopes in an antigen using recurrent neural network. Proteins 65:40–48

Saxova P, Buus S, Brunak S, Kesmir C (2003) Predicting proteasomal cleavage sites: a comparison of available methods. Int Immunol 15(7):781–787

Serruto D, Rappuoli R. (2006) Post-genomic vaccine development. FEBS Lett 580:2985–2992

Singh H, Raghava GPS (2001) ProPred: prediction of HLA-DR binding sites. Bioinform 17:1236–1237

Singh H, Raghava GPS (2003) ProPred1: prediction of promiscuous MHC Class-I binding sites. Bioinform 19:1009–1014

Swain MT et al (2001) An automated approach to modelling Class II MHC alleles and predicting peptide binding. In: Bourbakis NS (ed) Proc 2nd IEEE Int Symp Biol.-Inform Biomed Engin. IEEE Computer Society Press, pp. 81–88

Vivona S, Gardy JL, Ramachandran S, Brinkman FS et al (2008) Computer-aided biotechnology: from immuno-informatics to reverse vaccinology. Trends Biotechnol 26:190–200

Vyas JM, Van der Veen AG, Ploegh HL (2008) The known unknowns of antigen processing and presentation. Nat Rev Immunol 8:607–618

Ya L, Davydov I, Tonevitsky AG (2009) published in Molekulyarnaya Biologiya 43(1):166–174

Zhang H, Lundegaard C, Nielsen M (2009) Pan-specific MHC Class I predictors: a benchmark of HLA Class I pan-specific prediction methods. Bioinform 25:83–89

Chapter 10
Pangenomic Reverse Vaccinology

Claudio Donati, Duccio Medini, and Rino Rappuoli

10.1 Introduction

The possibility of quickly obtaining the complete genome sequence of bacterial pathogens produced a dramatic change in the process leading to the development of protein-based vaccines. While the traditional approach is focused on the identification and purification of subunits such as toxins or capsular polysaccharides from the organism, it has become possible to obtain the complete set of potentially expressed proteins from the DNA sequence and, using a combination of computational and experimental approaches, to select a list of the potential antigens to be tested in animal models. This innovative process, termed "Reverse Vaccinology" and first applied to the case of *Neisseria meningitides* serogroup B (MenB), has dramatically increased the efficiency of vaccine development (Serruto and Rappuoli 2006). For MenB, the traditional approach is not feasible because its capsule polysaccharide is structurally identical to a self-antigen, while a protein-based vaccine had eluded researchers for decades due to the high allelic variation of known antigens. For this reason, the complete genome of a virulent isolate was sequenced (Tettelin et al. 2000), 28 novel protective antigens were identified by a combination of computational and experimental screening of the proteins predicted to be encoded in the genome (Pizza et al. 2000), and a potentially universal vaccine based on five of these antigens was developed (Giuliani et al. 2006). Later on, the original process based on the genomic sequence of a single isolate was extended into a Pangenomic Reverse Vaccinology (PRV) approach, for the design of a vaccine against *Streptococcus agalactiae* (Group B Streptococcus, GBS) (Maione et al. 2005; Tettelin et al. 2005). The whole pangenome of the species (see Chap. 2 for the introduction), rather than the genetic repertoire of a single isolate, was mined. This allowed the identification of a large number of previously unknown potential vaccine targets, including proteins encoded by genes that are not shared by the whole population of the pathogen under investigation.

R. Rappuoli (✉)
Novartis Vaccines and Diagnostics, Siena, Italy

V. Sintchenko (ed.), *Infectious Disease Informatics*,
DOI 10.1007/978-1-4419-1327-2_10,

The Pangenomic Reverse Vaccinology approach entails a succession of steps, in which a shorter list of the best candidates is obtained by successive filtering and prioritization steps starting from all the proteins predicted to be encoded within the pathogen pangenome. Although the details of the procedure depend on the biology of the pathogen, a general scheme can be given. Once the genomic sequence from an appropriate number of isolates has been determined (see Hogg et al. 2007 and Tettelin et al. 2008) for a discussion on how to establish the right number of genomes to be sequenced), the steps required for a complete PRV include: (i) the prediction of open reading frames (ORFs) from all genomes, along with the function and cellular localization of the predicted proteins, (ii) the clustering of the predicted ORFs into Clusters of Orthologous Genes (COGs), (iii) the bioinformatics screening of COGs to select and prioritize candidate antigens, (iv) the experimental validation of bioinformatics predictions, and (v) the study of the selected antigen distribution within the population structure of the species.

In this process, starting from the overall number of predicted ORFs and COGs, gene clusters are sequentially removed from the candidate list when the localization of their products is certainly cytoplasmic (based on both localization prediction and functional annotation), or when they encode for known antigens whose immunological characteristics have already been investigated. Since the goal of the in silico screening is to collect the largest number of candidates, with the only limit of making the further experimental investigation feasible and reasonably prioritized, the first selection steps are designed to favor sensitivity over specificity.

One of the most important goals of the in silico analysis is the improvement of the annotation of the candidates throughout the filtering steps, in order to have a prioritized list to be passed on for experimental validation. Features that affect the priority to be given to candidates include: (i) the presence/absence distribution profile in the whole pathogen population, or in compartments of the population that are known to be of major clinical relevance, (ii) the degree of sequence variability, (iii) the number of predicted trans-membrane domains that affect the viability of the candidate in the following experimental process, (iv) the presence of leader peptides or lipoprotein signatures, outer membrane anchoring motives and host-cell binding domains such as RGD (Brennan and Shahin 1996), (v) anomalous G + C content of the chromosomal region, which is often a signature of acquisition by horizontal transfer, (vi) the presence of tandem repeats in or at the 5′ ends of genes that characterize certain virulence genes (Saunders et al. 1998; Hood et al. 1996; Bentley et al. 2007), and (vii) the homology matches of the candidate genes to human genes in order to avoid potential autoimmune reactions. Also, databases of protein families such as Pfam (Finn et al. 2008), TIGRFams (Haft et al. 2003), and PHN-Families (Medini et al. 2006) are scanned to find distant homologies that can contribute to the assignment of a hypothetical subcellular localization or functional annotation to each predicted ORF or COG. Searches for orthologous proteins and virulence factors previously characterized in other organisms are also conducted.

In the earliest application of Reverse Vaccinology, the screening of vaccine candidates was typically performed using the algorithms described in Sect. 10.2. It was performed on the genome of an identified "type" strain, historically the only one or the first one for which the sequence was available. As a secondary step, conservation and variability of the candidates in other genomes was investigated through the methods described in Sect. 10.3, to add upon one-genome analysis. This approach, although conceptually simple, has the obvious limitation of being severely biased toward the type strain selected, which is particularly relevant in species with an "open" pangenome.

Conversely, if a sequence of more than one strain is available, a PRV approach can be taken. After the identification of the ORFs in all the available genomes, clusters of orthologous genes (COGs) are constructed as described in Sect. 10.3.1, and the list of candidates is built in terms of COGs rather than single antigens. As a consequence, if the number of genomes is large enough and the sampling is epidemiologically meaningful, each candidate is already intrinsically characterized in terms of its distribution within the species population and its sequence variability.

10.2 Single Genome Analysis

10.2.1 The Annotation Procedure

The first step of the procedure is the determination of the genes encoded by a genome sequence. Compared to the efforts for the prediction of genes in eukaryotes, there is relatively little work done for the prediction of genes in prokaryotes. Computational methods to identify protein-encoding genes specific to prokaryotic genomes have been developed in the late nineties, and are based on methods borrowed from artificial intelligence, aiming to statistically define the functional role of each site in a given DNA sequence using local patterns of nucleotide composition.

The software packages most widely used for this purpose are GeneMark (Lukashin and Borodovsky 1998) and GLIMMER (Delcher et al. 1999). GeneMark predicts the position of genes by identifying the sequence of switches from coding to noncoding state and vice-versa, and maximizes the probability of the observed DNA sequence using model parameters defined on the *Escherichia coli* genome. Since the statistical model of GeneMark does not allow the overlap of two genes, the prediction of the algorithm is then refined using a probabilistic model of Ribosomal Binding Sites (RBS) also derived from the annotated *E. coli* genome. Recently, a gene prediction method based on an improved algorithm for the identification of RBS and on a novel method for the identification of polycistronic operons have been proposed (Nishi et al. 2005).

Beside these model-based methods, an alternative approach to genome-scale gene prediction that leverages on the large body of knowledge on gene sequences available in sequence databases has been developed (Frishman et al. 1998). A probabilistic model of protein coding genes and of RBS has been derived from a combination of homology searches against a database of known genes. This model has demonstrated a high degree of sensitivity and a good accuracy in the prediction of the starting sites of genes.

The next step in the selection pipeline consists of the prediction of a biological function for each of the predicted genes using a combination of bioinformatics tools. At the end of this step, the raw DNA sequence is annotated by a list of the regions that are potentially translated into proteins. A putative function for these proteins, giving a preliminary picture of the biology of the organism, and a first selection of the candidates for development as vaccine targets is then defined.

Traditionally, characterizing directly the function of a single protein requires a large body of careful experimental work and is therefore unfeasible on a genomic scale. However, information on protein functions in a newly sequenced organism can be obtained using the large body of knowledge stored in protein sequences databases. The most widely used method for function prediction is the homology transfer of annotation. The idea behind this approach is that proteins that differentiated through speciation from a common ancestor usually maintain similar functions. However, this method is based on a number of assumptions that should be evaluated with caution, case by case.

In practice, homology is usually inferred from sequence conservation, and the homology-based annotation of protein sequences is accomplished by a combination of sequence similarity searches against annotated databases, such as UniProt (The UniProt Consortium 2008), GenBank (Benson et al. 2008), and Ensembl (Flicek et al. 2008), using one of the popular sequence alignment methods, such as, BLAST or PSI-BLAST (Altschul et al. 1997). Although it is widely accepted that the higher the level of sequence conservation, the more likely it is that the query and target proteins share a similar function, it is difficult to identify a safe threshold for the conservation of function, and verification of the sequence alignment by experts is recommended.

An often-overlooked difficulty in performing the functional annotation step on a genomic scale is inherent to the simplification necessarily introduced when condensing information about a complex concept such as protein function into a summary annotation. Proteins often perform more than one function, and the details depend on the biology of the organism. To help the researchers to describe protein function in a nonarbitrary way, several large interdisciplinary teams have attempted to build ontologies of biological functions (Bard and Rhee 2004). The model for these projects has been the Enzyme Commission, which defined a four-digit system (EC number) for the classification of enzymes. The Gene Ontology (GO) project also extends this approach to nonenzymatic functions by providing a controlled vocabulary to describe the function of any gene product. Thus, the GO has become the standard for the classification of protein function in biological systems (see Chap. 19 for the more detailed discussion about ontologies).

Another source of errors in proteins annotation is the fact that many proteins are multidomain, and often the query and the annotated target sequence share only a portion of the sequence, corresponding to one conserved domain. In these cases, the transfer of annotation from the target to the query sequence is not justified. Many databases of protein domains and domain families, such as TIGRFAMs (Haft et al. 2003) and Pfam (Finn et al. 2008), have been compiled and manually curated. However, the coverage of these databases is limited. To reduce the noise introduced by multidomain proteins and to bypass the bottleneck of manual curation, unsupervised methods of protein classification into families have been proposed [PHN-Fams (Medini et al. 2006)]. They allow the assignment of biological functions to single proteins and to protein complexes, improving the signal-to-noise ratio in sequence-base homology detection and homology transfer.

10.2.2 Review of the Methods for Protein Localization Prediction

Despite the growing size of sequence databases, 30–40% of the proteins encoded into a typical bacterial genome have no homology to proteins of known function. To circumvent this problem, a large set of computational tools has been developed to predict protein function directly from a sequence. For a recent review on the main methods, see (Punta and Ofran 2008). In this section, we concentrate on the prediction of the cellular localization of proteins from sequence, which is of fundamental importance for the selection of possible vaccine candidates, based on the concept that, in order to elicit antibody response from the host, proteins must be visible to the immune response and must therefore be present outside of the bacterial cell.

In bacterial cells, surface proteins are synthesized in the cytoplasm and are then transported to the extracellular space by one of several transport systems. One of these is the type I secretion system (Holland et al. 2005), which shuttles proteins directly outside the bacterial cell. In type II secretion systems (Pugsley 1993), the secretion is a two-step process, where the protein is first inserted into or translocated across the cytoplasmic membrane, and then, in Gram-negative bacteria, inserted into or translocated across the outer membrane. Other transport systems include type III (Hueck 1998) and type IV (Christie et al. 2005) secretion systems, which directly inject effector proteins into host cells through specialized, contact dependent secretion systems, and type V (Thanassi et al. 2005) secretion systems, or autotransporters, where the exported proteins contains a self-transporter domain that is secreted through a pore formed by a pore-forming domain encoded in the C-terminus of the protein.

Proteins are directed to one of these export system by signals encoded in their sequence, like the N-teminal signal peptides that direct the post-translational export out of the cytoplasm, or the differences in amino-acid-composition which are characteristic of proteins destined to engage to different compartments (Cedano et al. 1997). Although the complete set of sequence-encoded features directing the localization of the nascent protein is far from being characterized, these signals can be identified using computational methods. These methods are able to predict the localization of a protein from the amino-acid sequence of the protein alone, can be used in the high-throughput screening of large sets of proteins and can therefore be employed in the selection of candidate antigens from complete genome sequences.

The methods for the prediction of the localization of proteins in bacterial cells can be classified into three families: (I) computational methods based on statistical properties of the protein amino acid sequence (Gardy et al. 2003, 2005; Gardy et al. 2003; Yu et al. 2004); (II) feature-based methods (Bendtsen et al. 2004a, b, 2005; Juncker et al. 2003; Kall et al. 2007); (III) homology based methods. For a recent review of localization prediction methods in bacteria, see (Gardy and Brinkman 2006).

Computational methods employ a wide range of unsupervised classification techniques based on statistical properties of the sequence, such as its n-peptide composition (Yu et al. 2004), the homology to characterized proteins, the presence of signal peptides, trans-membrane helices, or specific motifs (Gardy et al. 2005). The different lines of evidences are often integrated into a single prediction by machine-learning tools, like PSORTb (Gardy et al. 2005). These methods, although having a precision as high as 96% in the identification of surface exposed proteins, usually have values of recall in the range of 70–80% against annotated datasets, due to the fact that prediction is performed only when enough evidence is accumulated (Gardy and Brinkman 2006). This drawback is especially important for extracellular proteins, for which the training datasets, which must be composed of experimentally characterized proteins, are usually smaller.

Feature-based methods rely on the identification of sequence features that strongly support specific localization, such as the presence of signal peptides (short N-terminal stretches of sequence that cause a protein to be exported out of the cytoplasm of bacterial cells). While the presence of one of these features can support or exclude a localization prediction, its absence does not have any predictive relevance on localization. Distinct prediction methods exist for each feature, and the choice of the best methods and their integration into a single prediction is left to the user.

Homology methods of protein localization are based on the assumption that homologous proteins share the same cellular localization. Therefore, through homology searches against annotated sequence databases, the annotation of well-characterized proteins can be transferred onto unknown proteins. The target databases can comprise specialized databases composed only of proteins of known localization, like PSORTdb (Rey et al. 2005) or PA-GOSUB (Lu et al. 2005), or more general databases like SwissProt (Boeckmann et al. 2003) or GenBank (Benson et al. 2008), where information on localization can be gained from text extraction from the annotation field. However, using a nonspecialized dataset poses additional difficulties due to the heterogeneity of annotations, and the possibility of imprecise or ambiguous annotations.

Recently, experimental protocols for the large-scale identification of surfaces exposed in gram-positive pathogenic bacteria have been developed and applied to the human pathogen Group A Streptococcus (GAS), responsible for an estimated 600 million cases of pharyngitis worldwide each year (Rodriguez-Ortega et al. 2006). In this work, bacteria are treated either with trypsin or with proteinase K and the released surface protein fragments are identified with mass spectrometry and comparisons to public databases of annotated proteins. As a result, 72 surface-exposed proteins, 68 of which also predicted by PSORT, were identified and the testing of a subset of these proteins in an animal model allowed the identification of a previously unknown protective antigen.

10.3 Pangenomic Analysis

Today, as the number of fully sequenced microbial genomes exceeds 800, it is clear that microbial diversity has been vastly underestimated and we are just "scratching the surface." In many species, there is extensive genomic plasticity; for example, the completion of the genome sequence of *Escherichia coli* O157:H7 revealed that this strain possesses >1,300 strain-specific genes compared with *E. coli* K12. These genes encode proteins that are involved in virulence and metabolic capabilities (Perna et al. 2001). Moreover, when the genomes of three *E. coli* strains (K12, O157:H7 and the uropathogenic strain CFT073) were compared, only 39.2% of genes could be found in all three strains (Welch et al. 2002). Other reports have also revealed an extensive amount of genomic diversity among strains of a single species (Tettelin et al. 2005; Brochet et al. 2006; Brzuszkiewicz et al. 2006). From these studies, it is evident that it is not possible to characterize a species from a single genome sequence. Recently, a new mathematical model was proposed to describe the pangenome of several species (Tettelin et al. 2008), showing that an unbounded number of genes for the species (an open pangenome) can be realistically obtained with a vanishing number of new genes found upon sequencing of very large number of new genomes (Fig. 10.1).

The pan-genome (see also Chap. 2) can be divided into three elements: a core genome that is shared by all strains; a set of dispensable genes that are shared by some but not all isolates; and a set of strain-specific genes that are unique to each isolate. Conversely, the dispensable and strain-specific genes, which are largely composed of hypothetical, phage-related and transposon-related genes (Tettelin et al. 2006), contribute to its genetic diversity. The concept of the pan-genome has practical applications in vaccine research. In fact, while the ideal vaccine candidate is a conserved protein encoded by a gene present in every isolate of the species, in several real cases the identification of such a candidate turned out to be impossible. Recently, it was shown that the design of a universal protein-based vaccine against GBS was only possible using dispensable genes (Maione et al. 2005). In addition, the sequencing of multiple genomes was instrumental in discovering the presence of pili in Group A and Group B Streptococci and in *Streptococcus pneumoniae*, an essential virulence factor that had been missed by conventional technologies for a century (Telford et al. 2006).

10.3.1 Methods for Ortholog Identification

The introduction of the concept of the pangenome has lead to the understanding that, in order to select antigens that are protective against the largest possible fraction of circulating strains of a given pathogen, it is important to compare several unrelated genomes from the same species and to identify the homologous genes in the different strains.

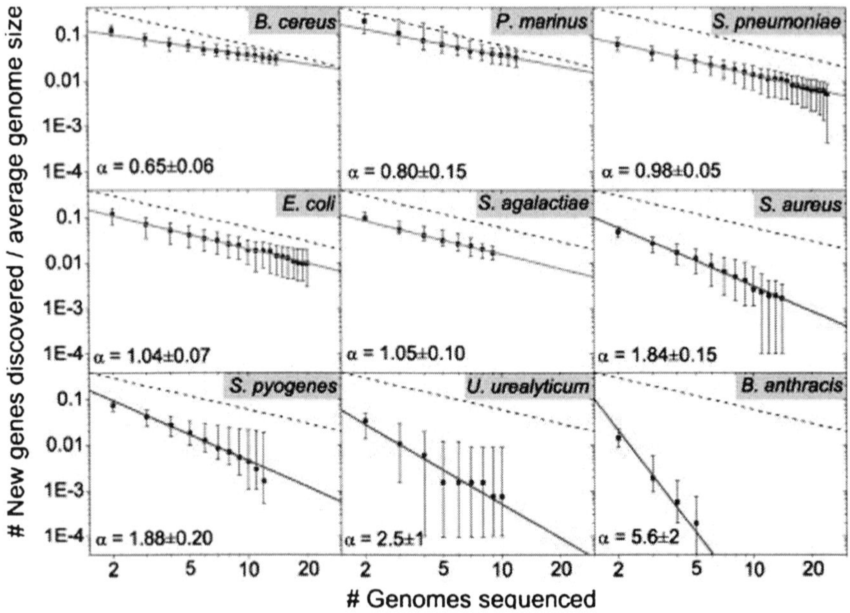

Fig. 10.1 Open and closed pangenomes for bacterial species. *Red curves* indicate closed pangenomes; *green curves* indicate open ones. Shown are the number n of new genes discovered for an increasing number N of genomes sequenced, normalized to the average genome size of the species, along with their 25–75 percentile intervals. *Solid curves* show the power law $n = \geq N^{-\alpha}$ least squares fit to data for $N \geq 3$, weighted for the 25–75 percentile interval. In each box a dashed guide to the eye shows the borderline power law $n \sim N^{-1}$ to facilitate the comparison of slopes (adapted from Tettelin et al. 2008)

Homologous proteins are proteins that share a common ancestry and can be characterized as orthologs, i.e., genes that originated through clonal inheritance, and paralogs, i.e., genes that originated through duplication followed by differentiation (Fitch 1970; Koonin 2005). Although both types of relationship identify homologous genes, it is important to distinguish between these two classes, since paralogs are likely to have adapted to perform a different function. Several strategies have been defined to identify orthologous and paralogous genes in different species, all of which have specific strengths and weaknesses (Chen et al. 2007). The most widely used of these methods, and those that are most easily adapted to the specific task of identifying homologs in different strains of a single species, are the BLAST-based methods, like the Reverse Best Hit (RBH) technique or the OrthoMCL algorithm (Li et al. 2003). The RBH identifies orthologs in different genomes as the bi-directional BLAST best hit in an all-versus-all comparison. However, this approach only compares pairs of genomes, and its generalizations to more than two genomes can lead to inconsistencies. OrthoMCL, although originally developed for identifying orthologs in eukaryotic genomes, can be fruitfully applied for many genomes of the same species, since it is able to compare more than two genomes at the same time.

A second approach consists of exploiting the fact that, although variable with regards to gene content, genomes from the same bacterial species often show a high level of synteny, i.e., the ordering of genes along the chromosome is mostly conserved and only a relatively small number of rearrangements are present. In this approach, groups of orthologous genes can be identified from RBH screens by selecting only those that share the same neighboring genes, directly identified using pair-wise or multiple whole-genome alignments obtained by programs such as MUMmer (Kurtz et al. 2004) or Mauve (Darling et al. 2004), or by using specialized software like DAGChainer (Haas et al. 2004), which identifies chains of gene pairs sharing conserved order in different genomes.

10.3.2 Allelic Variation in Candidate Antigens

From the pathogen perspective, exposing a core antigen to interaction with the host immune system is a big risk. As a consequence, core antigens are often highly variable, while dispensable antigens tend to be more conserved (a typical example being the porA and NadA antigens in *N. meningitides*, respectively).

Recently, *S. pneumoniae* was shown to express a multiproteic pilus able to elicit protection both by active and passive immunization in a mouse model of infection (Barocchi et al. 2006; LeMieux et al. 2006). These long surface exposed structures are encoded by a chromosomal element defined as the *rlrA* pathogenicity islet (Barocchi et al. 2006). The function of pneumococcal pili is currently an area of investigation. To date, pili are known to be involved in adherence to lung epithelial cells in vitro, as well as in colonization in a murine model of infection (Barocchi et al. 2006; Nelson et al. 2007). In order to determine the feasibility of a protein vaccine that includes pilus components, the distribution and sequence variability of the *rlrA* islet among a defined *S. pneumoniae* collection of clinical isolates was investigated in 386 clinical isolates from diverse geographic regions, selected for epidemiological relevance and to capture the genetic diversity within *S. pneumoniae* (Moschioni et al. 2008). It was found that approximately 30% of the isolates contained the *rlrA* islet, and that the presence of the *rlrA* islet correlated with the genetic background (defined by Multi Locus Sequence Typing, MLST), suggesting that the islet was probably acquired prior to the formation of the clonal complexes and steadily maintained during clonal diversification, even in CCs that show evidence of a complex evolutionary history.

The genetic variability of the *rlrA* pilus islet can be organized into 3 clades. Single protein alignments highlight RrgA and RrgC (84 and 98% of sequence identity between the most divergent clades, respectively) as the most promising components for a serotype-independent vaccine, while the variability of RrgB (49% sequence identity between the most divergent clades) makes this protein less attractive. Additionally, the *rlrA* clade type within a CC was determined to be identical, strongly suggesting clonal inheritance of this islet. In contrast, the serotype frequently varies within a CC. These findings clearly reveal that the association between the *rlrA* islet and serotype depends on the genetic link between serotype and genotype. In fact, there is a significant correlation between the serotype and the presence of the operon only for those serotypes that correspond to a restricted number of CCs, such as serotype 9V (CC 162) and serotype 3 (CC 180).

10.4 Experimental Validation

The in silico candidate selection procedure, outlined in Chap. 1 and performed according to the methods detailed in Chaps. 2 and 3, usually results in the selection of a large number of genes, covering as much as one fourth of the total number of ORFs in a genome, or one fifth of the COGs in a pangenome. In general, this amounts to several hundred genes. The priority rank assigned to candidates is aimed at screening the most promising antigens upfront, and the adherence of bioinformatics predictions with the actual features of candidate antigens is an open and theoretically relevant field of investigation.

However, from a practical standpoint, the conclusive selection of candidate antigens needs to be based on experimental evidence. Therefore, simple procedures that allow researchers to clone and express large numbers of genes are necessary. Fortunately, PCR and robotics development make this possible. In this case the availability of the genomic sequence for multiple isolates is also of paramount importance, in that they make it possible to design oligonucleotides for the PCR primers. These primers can be universal or specific for the different allelic versions of the candidate antigens.

10.4.1 Experimental Validation Procedure

The product of each PCR reaction is cloned and screened for expression in a heterologous system. Successful expression depends on the predicted localization of the protein. Integral membrane proteins have proven to be particularly difficult to produce by recombinant techniques in *Escherichia coli*. Once purified, the recombinant proteins are used to immunize mice and the post-immunization sera are analyzed to verify the computer-predicted surface localization of each polypeptide and its ability to elicit a quantitative and qualitative immune response. First, the immune sera are tested using western blot analysis of the recombinant proteins, outer membrane vesicles (OMVs) and total extracts of the bacterium to determine if the antibodies are able to recognize both the recombinant and the bacterial protein, and to confirm the predicted localization of the protein. A limitation of immunoblotting is that it requires the boiling of the samples, which results in a disruption to the native structure of antigens, preventing antibodies from binding to conformational epitopes. To further confirm the presence of the proteins on the bacterial surface and to assess their immunogenicity, sera are analyzed by ELISA and fluorescence-activated cell sorter analysis (FACS), in order to measure antibody titers and to determine the ability of antisera to bind to the surface of live bacteria.

The direct means to study the protective efficacy of candidate antigens is to test the immune sera in an animal model in which protection is dependent on the same effector mechanisms as in humans. The lack of reliable animal models has often hampered the development of vaccines, and alternative in vitro assays that are known to correlate with vaccine efficacy in humans have to be developed [e.g., assays that measure bactericidal activity (Goldschneider et al. 1969) and opsonophagocytosis (Ross et al. 1987)].

10.4.2 Reverse Vaccinology Case Studies

Serogroup B meningococcus (MenB) represents the first example of the application of reverse vaccinology and the demonstration of the power of genomic approaches for target antigen identification (Rappuoli 2000). *N. meningitidis* is a human pathogen that, despite available antibiotic therapy, is still a major cause of mortality as a result of sepsis and meningitis. Using traditional approaches, vaccines have been developed against serogroups A, C, Y, and W135, but for MenB, an efficacious vaccine is not yet available, due to the sequence variation of surface-exposed proteins and the cross-reactivity of the serogroup B capsular polysaccharide with human tissues.

While the *N. meningitidis* sequencing project was in progress, the incompletely assembled DNA fragments were screened by computer analysis to select proteins predicted to be on the bacterial surface or those with homologies to known bacterial factors involved in pathogenesis and virulence. After discarding cytoplasmic proteins and known *Neisseria* antigens, 570 genes predicted to code for surface-exposed or membrane-associated proteins were identified. Successful cloning and expression was achieved for 350 proteins, which were then purified and tested for localization, immunogenicity, and protective efficacy. Of the 85 proteins found to be surface-exposed, 22 were able to induce complement-mediated bactericidal antibody response, providing a strong indication of proteins capable of inducing protective immunity. In addition, to test the suitability of these proteins as candidate antigens for conferring protection against heterologous strains, the proteins were evaluated for gene presence, phase variation and sequence conservation in a panel of genetically diverse strains representative of the global diversity of the natural *N. meningitidis* population (Maiden et al. 1998). Most of the selected antigens were able to induce cross-protection against heterologous strains, demonstrating that the antigens, identified by in silico analysis, are good candidates for the clinical development of a vaccine against MenB (Pizza et al. 2000; Jodar et al. 2002). Five were finally selected, and formulated with different adjuvants to immunize mice. Mice sera were found to be bactericidal against a large and diverse panel of invasive, circulating strains, obtaining a potentially universal vaccine against the MenB population (Giuliani et al. 2006), which is currently in advanced clinical trials in humans.

The availability of an increasing number of bacterial genome sequences, together with the MenB example, has prompted the application of the reverse vaccinology approach to other pathogens, such as GBS (Maione et al. 2005). Use of computational algorithms applied to the genomic sequences of eight GBS genomes allowed the prediction that GBS contains 589 surface-associated proteins, of which 396 were core genes and the remaining 193 were absent in at least one strain. Each of these proteins was tested for protection against GBS, and four antigens were able to elicit protective immunity in an animal model. The important novelty of this study is that none of these antigens could be classified as universal, because three of them where absent in a fraction of the tested strains, and the fourth core gene showed negligible surface accessibility in some strains. The use of multigenome sequence information for vaccine design represented a major conceptual step from the common concept that a single genome sequence is sufficient to identify surface associated proteins to be tested as potential vaccine candidates. Multiple sequences may be needed to identify a vaccine formulation that is effective in the case of a highly differentiated species, and this situation is likely to be common to many important bacterial pathogens.

The use of bioinformatics tools in combination with molecular biology techniques enables the systematic investigation of the utility of potential genomic sequences to act as antigens for vaccine production. It is now possible to conceive the development of new vaccines against a wide variety of pathogens for which classical vaccinology has failed so far and, in theory, this approach could be extended to parasites and viruses.

10.5 Bacterial Population Genetics and Vaccine Design

The experimental validation step discussed in Chap. 4 completes a PRV approach to vaccine development, and results in one or more vaccine formulations that are introduced into early clinical trials for evaluation in humans. However, recent pangenomic and metagenomic studies increased our awareness of the degree of variability present in natural microbial populations, and its potential impact on vaccine design (Medini et al. 2008). In fact, it is now clear that an appropriate population genomic investigation is the essential companion of an effective PRV.

Strictly speaking, understanding the population structure of the species under investigation would be an important pre-requisite for any pangenomic approach, because it would allow an epidemiologically based selection of the genomes to be sequenced. However, the construction of epidemiologically sound collections of strains is often a difficult and time-consuming task, especially when dealing with occasional pathogens that are commonly carried by healthy individuals, and in cases where both the carriage and the invasive populations need to be investigated. As a consequence, the study of population structure and antigenic distribution is often carried in parallel with the PRV procedure. A continuous cross-talk between the two processes is indispensable for the development of an effective vaccine.

10.5.1 Genetic Variability Between Subpopulations

Very recently, it has been shown that the genetic variability within single naturally occurring, seemingly homogenous populations of bacteria could be much higher than expected (Thompson et al. 2005). Analyzing 12 randomly chosen clones, the authors demonstrated extensive allelic heterogeneity with no spatial or temporal substructure in the population and with genomes varying over 20% in size. These results could be relevant in the design of vaccines against mainly commensal pathogens that occasionally become pathogenic, such as nonpathogenic and pathogenic types of *E. coli* (Welch et al. 2002). For uropathogenic strains of *E. coli*, island acquisition resulted in the capability to infect the urinary tract and bloodstream and evade host defenses without compromising the ability to harmlessly colonize the intestine. If the genotypic variability of the colonizing population shares a degree of heterogeneity comparable to that found in environmental samples, formulates against these pathogens should be designed to cover a wide panel of circulating strains.

These findings demonstrate the importance of characterizing the structure of the populations of bacterial pathogens, in order to be able to predict the coverage obtained by vaccinating using proteins, which are not expressed by all bacterial strains, or which are present with different, noncross-protective alleles. Traditionally, bacterial strains are grouped into serotypes according to the ability of specific antisera to recognize them. However, due to the lack of general correlation between serotype designation and genetic background, Multi Locus Enzyme Electrophoresis (MLEE), a technique that classifies bacteria using the isoforms of about 15 metabolic enzymes, was later developed and successfully applied to define the population structure of several bacterial species (Selander et al. 1986). With the advancement of DNA sequencing techniques, it has become feasible to develop typing methods that are based on the sequencing of selected loci in the bacterial genomes. Using this approach, the multilocus sequence typing (http://www.mlst.net) schema, which is based on the sequencing of fragments of ~450 bp in seven housekeeping genes, has recently been developed and applied to a growing number of bacteria (Maiden et al. 1998) (Fig. 10.2).

For some bacterial taxa, often defined as "genetically monomorphic species," the level of diversification is so low that even more sensitive typing methods are required (Achtman 2008). For these organisms, the most promising method is the study of Single Nucleotide Polymorphisms (SNP). Originally developed for use in humans, and then applied to bacteria for the analysis of single genes (Moorhead et al. 2003; Robertson et al. 2004; Weissman et al. 2003), SNPs have recently been used to differentiate *B. anthracis* clinical samples that were collected from a disease outbreak (Read et al. 2002), to resolve the population structure of different pathogens (Alland et al. 2003; Gutacker et al. 2002; Roumagnac et al. 2006; Achtman et al. 2004; Smith et al. 1993) and to propose an *M. tuberculosis* typing scheme (Filliol et al. 2006).

Genetic typing methods constitute a useful tool for the identification of bacterial population structure and for the study of the distribution of genes of interest

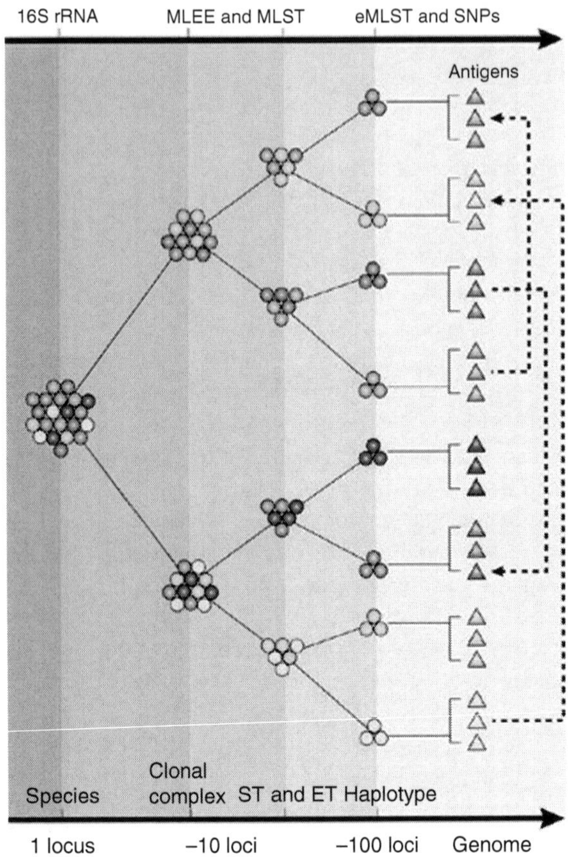

Fig. 10.2 Genetic markers and deviations from population structure. Schematic representation of different resolution levels within a typical population structure as identified by various typing schemes. Ribosomal RNA (rRNA) typing is the gold standard to differentiate species from other members of the same genus, class or even kingdom but, being based on a single locus, frequently lacks intra-species resolution. Multilocus typing schemes that are based on ~10 loci, either via enzyme electrophoresis (MLEE) or housekeeping-gene sequencing (MLST), provide fine intra-species resolution by defining electrophoretic and sequence types (ETs and STs, respectively) and clusters of types that group into clonal complexes. By measuring single-nucleotide polymorphisms (SNPs) at ~100 loci or applying an extended MLST (eMLST) schema that includes dispensable gene sequences, it is possible to further increase the typing resolution and define species-specific haplotypes. However, various genes that encode protein antigens have allelic distributions that do not correlate with MLST classification and, in principle, only complete genome coverage will be able to detect all of the non-clonal genetic variations that shape the fine structure of a bacterial population (modified from Medini et al. 2008)

amongst distinct subpopulations, as in the case of the *rlrA* islet distribution in *S. pneumoniae* strains. Although the rate of recombination is known to be high (Feil et al. 2001; Muzzi et al. 2008) for this organism, the three distinct forms of this operon present in the circulating strains correlate well with the clonal structure of *S. pneumoniae* reconstructed by MLST typing (Moschioni et al. 2008). Therefore, even for those organisms for which HGT is known to be important, typing methods can still yield useful information for specific antigens.

10.5.2 Vaccine-Oriented Antigenic Typing

When HGT has a nonnegligible impact on the antigenic profile of pathogenic strains in a given species, the relationship between conventional or newly proposed typing schemes and important antigens needs to be clarified. A typing scheme for a pathogenic species assumes critical importance when a new vaccine is developed and, as we have seen in previous sections, it is based on a combination of proteic antigens that are noncore, variable or both (see Fig. 10.2).

Although SNPs can be extremely powerful, owing to their provision of greater genomic coverage compared with other classification methods, their potential for vaccine-oriented population genetics is limited unless specific polymorphisms that perfectly associate with antigenic profiles are identified. In these cases, and when the level of expression for an antigen is substantially stable despite allelic variation or is clearly associated with different variants, serological tests can be successfully substituted with sequence-based typing methods. Conversely, phenotypic typing is required when the HGT of antigens is substantial, and/or when the level of expression of an antigen, or the overall degree of its accessibility to antibodies, is found to vary in the population.

10.6 Conclusion

In the pre-genomic era, nonproteic antigens such as capsular polysaccharides were known to vary within the pathogen population, and serological typing methods that directly addressed the variability of the polysaccharides had been available for decades. Despite the profound differences in the immunological properties of proteic and polysaccharidic antigens, from a typing perspective they are both surface exposed moieties of the pathogen that can be recognized by vaccine-induced antibodies. From a practical standpoint, the typing exercise to be performed is substantially analogous: it determines the portion of the pathogen population that carries that moiety in a sufficient amount to elicit immunity in vaccinated hosts. While effective, such an antigen-based approach is only capable of providing a static snapshot of the population. Bacterial populations are known to evolve very rapidly and genomic typing is an instrument capable of providing some hint into the population scale evolutionary dynamics of the pathogen. A combination of genetic typing (MLST, SNPs or, in the near future, rapid sequencing) and antigen-based serological typing is probably the optimal strategy to support the vaccine development phase as well as the continued clinical surveillance needed upon the introduction of a new vaccine into the field.

References

Achtman M (2008) Evolution, population structure, and phylogeography of genetically monomorphic bacterial pathogens. Annu Rev Microbiol 62:53–70

Achtman M et al (2004) Microevolution and history of the plague bacillus, *Yersinia pestis*. Proc Natl Acad Sci USA 101(51):17837–17842

Alland D et al (2003) Modeling bacterial evolution with comparative-genome-based marker systems: application to *Mycobacterium tuberculosis* evolution and pathogenesis. J Bacteriol 185(11):3392–3399

Altschul SF et al (1997) Gapped BLAST and PSI-BLAST: a new generation of protein database search programs. Nucleic Acids Res 25(17):3389–3402

Bard JB, Rhee SY (2004) Ontologies in biology: design, applications and future challenges. Nat Rev Genet 5(3):213–222

Barocchi MA et al (2006) A pneumococcal pilus influences virulence and host inflammatory responses. Proc Natl Acad Sci USA 103(8):2857–2862

Bendtsen JD et al (2004a) Feature-based prediction of non-classical and leaderless protein secretion. Protein Eng Des Sel 17(4):349–356

Bendtsen JD et al (2004b) Improved prediction of signal peptides: SignalP 3.0. J Mol Biol 340(4):783–795

Bendtsen JD et al (2005) Prediction of twin-arginine signal peptides. BMC Bioinform 6:167

Benson DA et al (2008) GenBank. Nucleic Acids Res 36(Database Issue):D25–D30

Bentley SD et al (2007) Meningococcal genetic variation mechanisms viewed through comparative analysis of serogroup C strain FAM18. PLoS Genet 3(2):e23

Boeckmann B et al (2003) The SWISS-PROT protein knowledgebase and its supplement TrEMBL in 2003. Nucleic Acids Res 31(1):365–370

Brennan MJ, Shahin RD (1996) Pertussis antigens that abrogate bacterial adherence and elicit immunity. Am J Respir Crit Care Med 154(4 Pt 2):S145–S149

Brochet M et al. (2006) Genomic diversity and evolution within the species *Streptococcus agalactiae*. Microbes Infect 8(5):1227–1243

Brzuszkiewicz E et al (2006) How to become a uropathogen: comparative genomic analysis of extraintestinal pathogenic *Escherichia coli* strains. Proc Natl Acad Sci USA 103(34): 12879–12884

Cedano J et al (1997) Relation between amino acid composition and cellular location of proteins. J Mol Biol 266(3):594–600

Chen F et al (2007) Assessing performance of orthology detection strategies applied to eukaryotic genomes. PLoS ONE 2(4):e383

Christie PJ et al (2005) Biogenesis, architecture, and function of bacterial type IV secretion systems. Annu Rev Microbiol 59:451–485

Darling AC et al (2004) Mauve: multiple alignment of conserved genomic sequence with rearrangements. Genome Res 14(7):1394–1403

Delcher AL et al (1999) Improved microbial gene identification with GLIMMER. Nucleic Acids Res 27(23):4636–4641

Feil EJ et al (2001) Recombination within natural populations of pathogenic bacteria: short-term empirical estimates and long-term phylogenetic consequences. Proc Natl Acad Sci USA 98(1):182–187

Filliol I et al (2006) Global phylogeny of *Mycobacterium tuberculosis* based on single nucleotide polymorphism (SNP) analysis: insights into tuberculosis evolution, phylogenetic accuracy of other DNA fingerprinting systems, and recommendations for a minimal standard SNP set. J Bacteriol 188(2):759–772

Finn RD et al (2008) The Pfam protein families database. Nucleic Acids Res 36(Database issue):D281–D288

Fitch WM (1970) Distinguishing homologous from analogous proteins. Syst Zool 19(2):99–113

Flicek P et al (2008) Ensembl 2008. Nucleic Acids Res 36(Database issue):D707–D714

Frishman D et al (1998) Combining diverse evidence for gene recognition in completely sequenced bacterial genomes. Nucleic Acids Res 26(12):2941–2947

Gardy JL, Brinkman FS (2006) Methods for predicting bacterial protein subcellular localization. Nat Rev Microbiol 4(10):741–751

Gardy JL et al (2003) PSORT-B: Improving protein subcellular localization prediction for Gram-negative bacteria. Nucleic Acids Res 31(13):3613–3617

Gardy JL et al (2005) PSORTb v.2.0: expanded prediction of bacterial protein subcellular localization and insights gained from comparative proteome analysis. Bioinform 21(5):617–623

Giuliani MM et al (2006) A universal vaccine for serogroup B meningococcus. Proc Natl Acad Sci USA 103(29):10834–10839

Goldschneider I, Gotschlich EC, Artenstein MS (1969) Human immunity to the meningococcus. I. The role of humoral antibodies. J Exp Med 129(6):1307–1326

Gutacker MM et al (2002) Genome-wide analysis of synonymous single nucleotide polymorphisms in *Mycobacterium tuberculosis* complex organisms: resolution of genetic relationships among closely related microbial strains. Genetics 162(4):1533–1543

Haas BJ et al (2004) DAGchainer: a tool for mining segmental genome duplications and synteny. Bioinform 20(18):3643–3646

Haft DH, Selengut JD, White O (2003) The TIGRFAMs database of protein families. Nucleic Acids Res 31(1):371–373

Hogg JS et al (2007) Characterization and modeling of the *Haemophilus influenzae* core and supragenomes based on the complete genomic sequences of Rd and 12 clinical nontypeable strains. Genome Biol 8(6):R103

Holland IB, Schmitt L, Young J (2005) Type 1 protein secretion in bacteria, the ABC-transporter dependent pathway (review). Mol Membr Biol 22(1–2):29–39

Hood DW et al (1996) DNA repeats identify novel virulence genes in *Haemophilus influenzae*. Proc Natl Acad Sci USA 93(20):11121–11125

Hueck CJ (1998) Type III protein secretion systems in bacterial pathogens of animals and plants. Microbiol Mol Biol Rev 62(2):379–433

Jodar L et al (2002) Development of vaccines against meningococcal disease. Lancet 359(9316):1499–1508

Juncker AS et al. (2003) Prediction of lipoprotein signal peptides in Gram-negative bacteria. Protein Sci 12(8):1652–1662

Kall L, Krogh A, Sonnhammer EL (2007) Advantages of combined transmembrane topology and signal peptide prediction-the Phobius web server. Nucleic Acids Res 35(Web Server issue):W429–W432

Koonin EV (2005) Orthologs, paralogs, and evolutionary genomics. Annu Rev Genet 39:309–338

Kurtz S et al (2004) Versatile and open software for comparing large genomes. Genome Biol 5(2):R12

LeMieux J et al (2006) RrgA and RrgB are components of a multisubunit pilus encoded by the *Streptococcus pneumoniae* rlrA pathogenicity islet. Infect Immun 74(4):2453–2456

Li L, Stoeckert CJ, Jr, Roos DS (2003) OrthoMCL: identification of ortholog groups for eukaryotic genomes. Genome Res 13(9):2178–2189

Lukashin AV, Borodovsky M (1998) GeneMark.hmm: new solutions for gene finding. Nucleic Acids Res 26(4):1107–1115

Lu P et al (2005) PA-GOSUB: a searchable database of model organism protein sequences with their predicted Gene Ontology molecular function and subcellular localization. Nucleic Acids Res 33(Database issue):D147–D153

Maiden MC et al (1998) Multilocus sequence typing: a portable approach to the identification of clones within populations of pathogenic microorganisms. Proc Natl Acad Sci USA 95(6): 3140–3145

Maione D et al (2005) Identification of a universal Group B streptococcus vaccine by multiple genome screen. Science 309(5731):148–150

Medini D, Covacci A, Donati C (2006) Protein homology network families reveal step-wise diversification of Type III and Type IV secretion systems. PLoS Comput Biol 2(12):e173

Medini D et al (2008) Microbiology in the post-genomic era. Nat Rev Microbiol 6(6):419–430
Moorhead SM, Dykes GA, Cursons RT (2003) An SNP-based PCR assay to differentiate between *Listeria monocytogenes* lineages derived from phylogenetic analysis of the *sigB* gene. J Microbiol Methods 55(2):425–432
Moschioni M et al (2008) *Streptococcus pneumoniae* contains 3 rlrA pilus variants that are clonally related. J Infect Dis 197(6):888–896
Muzzi A et al (2008) Pilus operon evolution in *Streptococcus pneumoniae* is driven by positive selection and recombination. PLoS ONE 3(11):e3660.
Nelson AL et al (2007) RrgA is a pilus-associated adhesin in *Streptococcus pneumoniae*. Mol Microbiol 66(2):329–340
Nishi T, Ikemura T, Kanaya S (2005) GeneLook: a novel ab initio gene identification system suitable for automated annotation of prokaryotic sequences. Gene 346:115–125
Perna NT et al (2001) Genome sequence of enterohaemorrhagic *Escherichia coli* O157:H7. Nature 409(6819):529–533
Pizza M et al (2000) Identification of vaccine candidates against serogroup B meningococcus by whole-genome sequencing. Science 287(5459):1816–1820
Pugsley AP (1993) The complete general secretory pathway in gram-negative bacteria. Microbiol Rev 57(1):50–108
Punta M, Ofran Y (2008) The rough guide to in silico function prediction, or how to use sequence and structure information to predict protein function. PLoS Comput Biol 4(10):e1000160
Rappuoli R (2000) Reverse vaccinology. Curr Opin Microbiol 3(5):445–450
Read TD et al (2002) Comparative genome sequencing for discovery of novel polymorphisms in *Bacillus anthracis*. Science 296(5575):2028–2033
Rey S et al (2005) PSORTdb: a protein subcellular localization database for bacteria. Nucleic Acids Res 33(Database issue):D164–D168
Robertson GA et al (2004) Identification and interrogation of highly informative single nucleotide polymorphism sets defined by bacterial multilocus sequence typing databases. J Med Microbiol 53(Pt 1):35–45
Rodriguez-Ortega MJ et al (2006) Characterization and identification of vaccine candidate proteins through analysis of the group A Streptococcus surface proteome. Nat Biotechnol 24(2):191–197
Ross SC et al (1987) Killing of *Neisseria meningitidis* by human neutrophils: implications for normal and complement-deficient individuals. J Infect Dis 155(6):1266–1275
Roumagnac P et al (2006) Evolutionary history of *Salmonella typhi*. Science 314(5803): 1301–1304
Saunders NJ et al (1998) Simple sequence repeats in the *Helicobacter pylori* genome. Mol Microbiol 27(6):1091–1098
Selander RK et al (1986) Methods of multilocus enzyme electrophoresis for bacterial population genetics and systematics. Appl Environ Microbiol 51(5):873–884
Serruto D, Rappuoli R (2006) Post-genomic vaccine development. FEBS Lett 580(12):2985–2992
Smith JM et al (1993) How clonal are bacteria? Proc Natl Acad Sci USA 90(10):4384–4388
Telford JL et al (2006) Pili in gram-positive pathogens. Nat Rev Microbiol 4(7):509–519
Tettelin H et al (2000) Complete genome sequence of *Neisseria meningitidis* serogroup B strain MC58. Science 287(5459):1809–1815
Tettelin H et al (2005) Genome analysis of multiple pathogenic isolates of *Streptococcus agalactiae*: implications for the microbial "pan-genome". Proc Natl Acad Sci USA 102(39):13950–13955
Tettelin H et al (2006) Towards a universal group B Streptococcus vaccine using multistrain genome analysis. Expert Rev Vaccines 5(5):687–694
Tettelin H et al (2008) Comparative genomics: the bacterial pan-genome. Curr Opin Microbiol 11(5):472–477
Thanassi DG et al (2005) Protein secretion in the absence of ATP: the autotransporter, two-partner secretion and chaperone/usher pathways of gram-negative bacteria (review). Mol Membr Biol 22(1–2):63–72

The UniProt Consortium (2008) The universal protein resource (UniProt). Nucleic Acids Res 36(Database issue):D190–D195

Thompson JR et al (2005) Genotypic diversity within a natural coastal bacterioplankton population. Science 307(5713):1311–1313

Weissman SJ et al (2003) Enterobacterial adhesins and the case for studying SNPs in bacteria. Trends Microbiol 11(3):115–117

Welch RA et al (2002) Extensive mosaic structure revealed by the complete genome sequence of uropathogenic *Escherichia coli*. Proc Natl Acad Sci USA 99(26):17020–17024

Yu CS, Lin CJ, Hwang JK (2004) Predicting subcellular localization of proteins for Gram-negative bacteria by support vector machines based on n-peptide compositions. Protein Sci 13(5):1402–1406

Chapter 11
Immunoinformatics: The Next Step in Vaccine Design

Tobias Cohen, Lenny Moise, William Martin, and Anne S. De Groot

11.1 Introduction

T cells have come to be recognized as critical mediators of competent and lasting humoral and cytotoxic immune responses elicited by vaccines (Gillespie et al. 2000; Gianfrani et al. 2000). This recognition has catalyzed the development of computer-driven (*immunoinformatics*) methods for defining T-cell epitopes directly from protein sequences, allowing investigations into the role of T cells to leapfrog directly to the leading edge of immunology and vaccine research. A number of laboratories are currently pursuing *T-cell directed vaccination*, hypothesizing that this approach may provide the solution to the development of vaccines against human pathogens for which no vaccine has yet been developed.

Critical to the development of T-cell epitope-driven vaccines has been the elucidation of the correlates of "immunity" for a wide range of important human pathogens. The link between epitope-specific responses, the establishment of T-cell memory, and protection from disease has been confirmed for human immunodeficiency virus (HIV), hepatitis B virus (HBV), hepatitis C virus (HCV), malaria (Blattman et al. 2000; Harrer et al. 1996; Doolan et al. 1997), and other infectious diseases. Several laboratories have published data supporting the hypothesis that a protective immune response to a number of pathogens requires the development of broad T-cell responses to an ensemble of different epitopes (Gianfrani et al. 2000; Gillespie et al. 2000). Following exposure to a pathogen, epitope-specific memory T-cell clones are established (Blattman et al. 2000). These clones respond rapidly and efficiently upon any subsequent infection, secreting cytokines, killing infected host cells, and marshalling other cellular defenses against the pathogen.

Despite these proven linkages between T cells and protection against infectious disease, and despite the availability of immunoinformatics tools, new and effective epitope-driven vaccines have been slow in coming. The first T-cell epitope-driven vaccines for a globally relevant infectious disease, HIV, failed in the clinic, raising

---

T. Cohen (✉)
Brown University, School of Medicine, Providence, RI, USA

V. Sintchenko (ed.), *Infectious Disease Informatics*,
DOI 10.1007/978-1-4419-1327-2_11,

questions about T-cell epitope-driven vaccine efficacy for human pathogens (Wilson et al. 2008). To better understand the reason for the failure of previous *epitope-driven vaccines*, it is important to remember that vaccine design, delivery, and the quality of the resulting immune response all play critical roles in vaccine efficacy. These factors can be summarized as follows:

Immunogen + Adjuvant + Delivery vehicle = Vaccine

It is the authors' view that without the proper payload of effective epitopes (in terms of quality and quantity), the right adjuvant (triggering the right type of immune response) and the right delivery vehicle (focusing on the right immune compartment), a T-cell epitope-driven vaccine is bound to fail. Therefore, rather than abandoning all hope on the basis of the failure of two of the best known "T-cell-directed" HIV vaccines (Wilson et al. 2008; McElrath et al. 2008), we and others believe that T-cell-directed vaccines will work if properly constructed (Hanke 2008). These failed HIV vaccines were developed prior to the era of the availability of higher-quality vaccine design tools that have now enabled the effective selection of highly cross-conserved HIV epitopes, and the delivery of those epitopes in properly designed constructs (whether as a polytope or an immunogenic consensus protein as described by our laboratory and by Hanke et al. 2008).

Although commonly overlooked by vaccine developers, both T-help and cytotoxic T-cell response are also required for effective vaccines against viruses. In fact, nowhere is the lack of effective T-cell-directed vaccine design more obvious than for influenza vaccines. Currently licensed conventional inactivated vaccines (CIV) for influenza induce a hemagglutination inhibition (HI) response that is primarily strain-specific, and is relatively short-lived due to modifications of the B-cell antigenic components in the year-to-year variation of the influenza virus. Evidence supporting the development of T-cell directed vaccines for influenza includes the following: (1) T-cell help is required for high specific IgG antibody titers (Kamperschroer et al. 2006), (2) vaccine efficacy is improved when cross-reactive helper-T-cell populations are present from prior infection and/or vaccination (Rasmussen et al. 2001), (3) the rate of viral clearance depends on the presence of CD4 + T cells, (Belz et al. 2002), (4) cytotoxic T cells are required for viral clearance (Rasmussen et al. 2001), and (5) memory T-helper (TH) cells specific to a previous influenza strain lead to distinct cross-strain antibody responses (Marshall et al. 1999). It will not be long before influenza vaccine developers realize that much is to be gained from the addition of T-cell directed immune response to highly conserved influenza epitopes (McMurry et al. 2008). Perhaps a more universal influenza vaccine will then be developed.

Cell-mediated immunity is critical for the control of chronic bacterial infections and for protection against disease. This is particularly true for vaccines against bacterial diseases caused by *Mycobacterium tuberculosis* (Mtb) and *Listeria monocytogenes*, which prefer an intracellular lifestyle as a means of escaping the humoral (antibody) defense. Mtb is an example of an intra-cellular pathogen (it also persists in granulomas in the extracellular state). It is well known that antibodies to Mtb do not protect against tuberculosis. In contrast to antibody response, the adoptive transfer of T cells does confer protection from tuberculosis (TB) in the mouse model (Lefford 1975). In humans, the induction of broad T-cell mediated immunity to Mtb requires the involvement of Type 1 cytokines including interleukin 2 (IL-2), interferon γ (IFN-γ), and tumor necrosis factor α (TNF-α) (Kaufmann and Hess 1999). In addition, it is clear that CD4 + T cells are involved since TH1-mediated immune responses have been demonstrated in mouse protection studies (Cooper et al. 1993; Flynn et al. 1993; Scanga et al. 2000). TH1-biased T-cell responses to Mtb epitopes not contained in Bacillus Calmette-Guérin (BCG) may help control immune response in latently-infected individuals. Although TH1 cytokines are essential for protection, their levels of production do not strictly correlate with disease state. CD8 + T cells also play a significant role in TB immunity (Serbina and Flynn 2001). Importantly, studies of human CD8 + restricted responses to Mtb antigens have revealed that these cells play an important role in the control of Mtb replication in the alveolar macrophage (Canaday et al. 2001; Lewinsohn et al. 2007; Cho et al. 2000). Taken together, these data suggest that the stimulation of TH1-biased CD4 + and CD8 + T cell responses may be critical to the control (and eradication) of latent bacterial infection in chronically infected individuals.

11.2 Technological Advances

Several technical advances have enabled vaccines to be better designed for T-cell dependent immune responses. These are (1) improved immunoinformatics tools for vaccine design, (2) improved delivery vehicles, and (3) improved vaccine adjuvants.

11.2.1 Immunoinformatics for Vaccine Design

One limitation of conventional vaccination, and to a lesser extent natural infection, is that the immune system often focuses strongly on a surface antigen, which can be the most mutable immunogen of the pathogen. This is clearly the case in the context of influenza infection, in which the immune response focuses on hemaglutinin (HA), a major surface glycoprotein. In the case of HIV and other viruses, vaccination with more conserved, subdominant epitopes has been shown to circumvent this hierarchy and potentiate cross strain protection (Ostrowski et al. 2002; Nara and Lin 2005). Similarly, a conserved TH-directed vaccine may stimulate a more "democratic" immune response, increasing the number targets for T-cell recognition, thereby providing T-cell help to antibody response, despite potential viral variability (Santra et al. 2002; Subbramanian et al. 2003; Scherle and Gerhard 1986, 1988; Russell and Liew 1979; Johansson et al. 1987). In addition, broadening the T-cell repertoire might make it possible to impair viral immune-escape mechanisms and decrease viral loads sufficiently to disrupt transmission.

To identify T-cell epitopes for vaccine development, it is necessary to determine which peptides of a pathogen's proteome will bind to the human major histocompatability complex (MHC). Only about 2% of peptides have the ability to bind to MHC, the critical first step required for T-cell response. Another critical determinant of T-cell epitope immunogenicity is the strength of epitope binding to MHC molecules (Lazarski et al. 2005). It is this peptide-MHC interaction that is modeled by immunoinformatics tools [reviewed in Brusic et al; Petrovsky and Brusic 2006 and De Groot et al. (De Groot and Berzofsky 2004)].

EpiVax, Inc., has developed a suite of computer algorithms that can be applied to the development of epitope-based vaccines; this suite includes EpiMatrix, ClustiMer, Conservatrix, BlastiMer, Aggregatrix, Optimatrix, and VaccineCAD. The *EpiMatrix* algorithm, which rates the MHC binding capability for every 9 mer in a protein sequence, has been benchmarked using a set of "gold standard" epitopes published by the IEDB (Immune Epitope Database) (Zhang et al. 2008). Using this set of epitopes as an objective standard, EpiVax assessed the predictive accuracy of the EpiMatrix algorithm relative to eight well-known epitope-mapping tools (such as SYFPEITHI and BIMAS). The comparisons confirm that the EpiMatrix algorithm is the most accurate predictive tool currently available: http://www.EpiVax.com/comps/(Username: guest, Password: welcome) (Ardito 2009). In addition to the EpiMatrix algorithm for T-cell epitope identification, the EpiMatrix toolset also includes a set of analysis and design tools directly applicable to the vaccine design process. *ClustiMer*, an ancillary algorithm used with EpiMatrix, maps MHC motif matches along the length of a protein and calculates the density of motifs for eight common class II HLA alleles: DRB1*0101, DRB1*0301, DRB1*0401, DRB1*0701, DRB1*0801, DRB1*1101, DRB1*1301, and DRB1*1501. Typical T-cell epitope "clusters" range from 9 to roughly 25 amino acids in length, and considering their affinity to multiple alleles and across multiple frames, they can contain anywhere from 4 to 40 binding motifs, also known as promiscuous epitopes. The *Conservatrix* algorithm identifies conserved segments from among any given set of variable protein isolates. Pairing EpiMatrix with Conservatrix allows users to identify peptides, which are both potentially antigenic and conserved in circulating disease strains. *BlastiMer* compares the peptides' sequence to the human proteome to ensure that the peptides do not contain too much homology to any human protein. The *Aggregatrix* algorithm addresses the classical "set cover" problem by guiding the selection of a portfolio of epitopes that collectively "cover" a wide variety of both the known circulating strain variants of a given pathogen and the majority of common human HLA types. *OptiMatrix* is an algorithm that is used for designing altered peptide ligands that optimize the "aggretope." Specifically, OptiMatrix guides strategic substitutions of the MHC-contact residues such that the peptide binds more strongly; the TCR-facing residues are free to interact as they would in the unaltered peptide. The *VaccineCAD* algorithm arranges putative T-cell epitopes to create optimized "string-of-beads" vaccine immunogens. The *EpiAssembler* algorithm was developed, especially for use with highly variable RNA viruses. It is used to analyze the universe of viral isolates and to create composite epitope sequences in which constituent overlapping

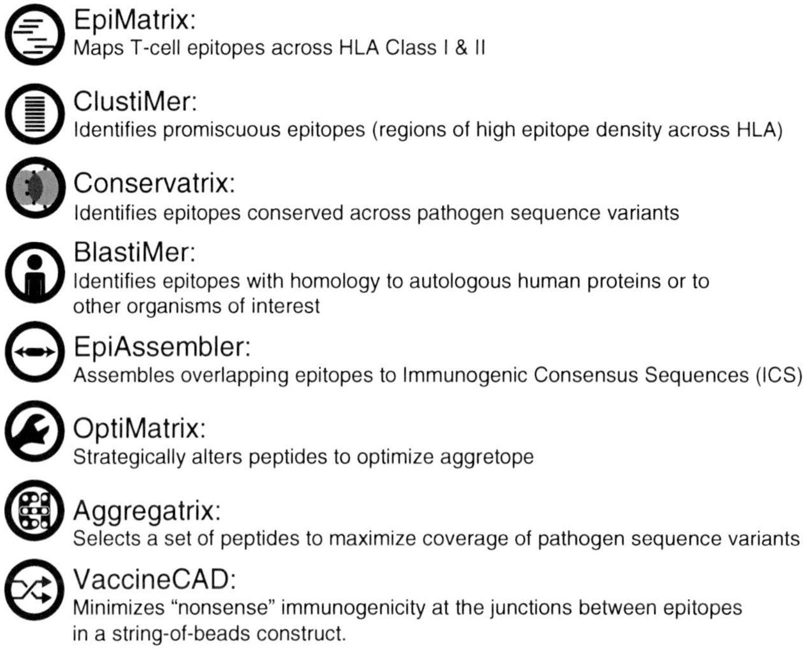

Fig. 11.1 EpiVax vaccine development toolkit

epitopes are both highly conserved and highly immunogenic (Koita et al. 2006). Taken collectively, these tools allow researchers to quickly and effectively identify T-cell epitopes and design new antigens for experimental study.

EpiVax's immunoinformatics toolkit, developed by De Groot & Martin and summarized in Fig. 11.1, has been used to rank proteins for potential immunogenicity (Koren et al. 2007; De Groot et al. 2007) and to design and evaluate vaccines (De Groot, Knopf, et al. 2007; De Groot et al. 2001; Bond et al. 2001). Over the course of a decade of research, these tools have been validated in vitro and in vivo. The tools are currently in use for both vaccine and protein therapeutics design (Koita et al. 2006; Koren et al. 2007; De Groot et al. 1997; Bond et al. 2001; McMurry et al. 2005; Dong et al. 2004; Tatarewicz et al. 2007). We note that Korber et al. have recently implemented very similar tools for the design of HIV-1 vaccines, although these tools have yet to be tested in animal models (Thurmond et al. 2008). For published descriptions of these tools, see previously published chapters in the Springer Immunomics series (De Groot et al. 2007; De Groot et al. 2008).

11.2.2 Improved Delivery Vehicles

The same range of delivery vehicles that exist for conventional vaccines can be used for the development of T-cell epitope-driven vaccines. For example, immunome derived vaccines (IDV) and epitope-based vaccines IDV can be formulated and delivered as string-of-beads multi-epitope proteins or as peptides in a carrier vehicle such as a liposome or viral-like protein (VLP). Alternatively, the sequences of IDV antigens or epitope strings can be inserted into a DNA vector, or a viral or bacterial vector such as adenovirus or *Salmonella*. *Dendritic cells* are the desired targets of such delivery vehicles. However, as with conventional vaccines, the efficiency of this targeting varies.

11.2.2.1 Targeting Dendritic Cells

As critical intermediaries between antigens and lymphocytes (Steinman 2001), dendritic cells are a logical site for vaccine targeting. In addition to their role as mediators of immune responses, dendritic cells play a critical role in sensing environmental cues that can lead to induction of tolerance (Mahnke and Enk 2005). Dendritic cell function is differentiation-dependent (see Fig. 11.2): Antigen capture takes place in the immature state, whereas potent immune stimulation follows after a complex maturation program resulting from stimulation by microbial toll-like receptor (TLR) ligands, innate lymphocytes, and CD40 ligation (Janeway and Medzhitov 2002; Takeda et al. 2003; Bendelac and Medzhitov 2002; Caux et al. 1994) (see section on adjuvants). In this view, immature dendritic cells induce tolerance

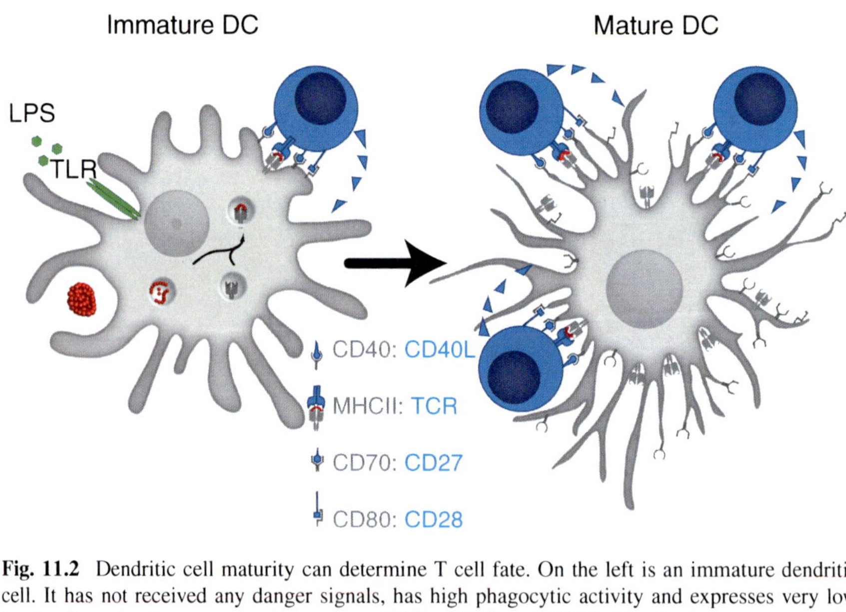

Fig. 11.2 Dendritic cell maturity can determine T cell fate. On the left is an immature dendritic cell. It has not received any danger signals, has high phagocytic activity and expresses very low levels of costimulatory molecules such as CD40, CD80, and CD70. T cells that are reactive for antigens presented by a dendritic cell in this state will be anergized. On the right is a mature dendritic cell. It has received several danger signals such as TLR stimulation or TNF-α mediated signaling from macrophages and NK cells and now expresses high levels of CD40, CD80, CD70 and MHC class II in addition to secreting IL-12 which facilitates the binding and activation of T cells

to self proteins, and mature dendritic cells stimulate immunity to pathogens. The term "mature" has also been used to phenotypically characterize dendritic cells expressing high surface levels of MHC class II, CD40, CD80, and CD86, all markers associated with T-cell priming ability.

Despite the discrimination between mature and immature dendritic cells in the immunological literature, recent studies show that phenotypically mature dendritic cells do not always stimulate immunity and, in some cases, induce tolerance (Spörri and Reis e Sousa 2005; Albert et al. 2001; Menges et al. 2002; Fujii et al. 2004). Moreover, antigen presentation by dendritic cells in the steady-state does not necessarily result in T-cell inactivation (Scheinecker et al. 2002) and, in some cases, stimulates immunity (Mayerova et al. 2004; Shibaki et al. 2004). Indeed, a new paradigm is emerging from these apparent contradictions: immature dendritic cells may give rise to a range of "effector" dendritic cells that lead to different T-cell fates (Reis e Sousa 2006). The type of "effector dendritic cell" appears to be dependent on a number of cytokine and chemokine signals, which can be replicated in the vaccine context through the use of the appropriate adjuvants.

A further innovation in vaccine design has been the use of targeting molecules to gain entry into dendritic cells. Dendritic cells express *DEC-205*, an endocytic receptor that enhances antigen uptake and presentation. DEC-205 is a member of the multi-lectin receptor family and contains a cysteine-rich domain in its amino-terminus, a fibronectin type II domain, and multiple C-type lectin domains, which are important for the binding and uptake of carbohydrate antigens (Jiang et al. 1995). Upon endocytosis, DEC-205 enters MHC class II compartments (MIICs), late endosome/lysosome compartments that are rich in class II MHC, mediating enhanced antigen presentation (Jiang et al. 1995). MIICs are the sites where peptides, formed by lysosomal proteolysis, bind to MHC class II molecules just before they are transported to the cell surface for presentation to CD4 + T cells (Mahnke et al. 2000). In addition, DEC-205 mediates the presentation of protein antigens through the exogenous TAP-dependent MHC class I pathway, leading to the activation of CD8 + T cells (Bonifaz et al. 2002).

The trafficking properties of DEC-205 have made it an excellent marker for dendritic cells through targeted antigen delivery via monoclonal αDEC-205. Ovalbumin (OVA), HIV-Gag and hen egg lysozyme peptide 46–61 are examples of antigens that have been conjugated to αDEC-205. It is not surprising, however, that these hybrid antibodies elicit either immunogenic or immuno-suppressive responses in vivo, depending on immunization conditions such as the presence of adjuvants. Strong inflammatory T-cell responses are generated by the coadministration of agonistic αCD40 or other dendritic cells maturation stimuli such as lipopolysaccharide or poly I:C (Bonifaz et al. 2002; Hawiger et al. 2001; Trumpfheller et al. 2006; Boscardin et al. 2006). Without αCD40 agonist, dendritic cells in the steady state give rise to sustained levels of CD4 +/CD25 + Tregs and to deletion of antigen-specific T cells (Bonifaz et al. 2002; Hawiger et al. 2001; Kretschmer et al. 2005; Mahnke et al. 2003; Bruder et al. 2005). The mechanism of tolerance induction has not been fully explained.

11.2.2.2 Mucosal Delivery

Mucosal vaccination has received more attention in the TB vaccine field recently (Dietrich et al. 2006; Santosuosso et al. 2006). Mucosal surfaces are important for priming immune responses, especially for a pathogen like Mtb that infects a host via the mucosal surface of the lung (Fig. 11.3). A number of studies have shown signs of protection against Mtb by mucosal vaccination (Wang et al. 2004; Chen et al. 2004; Giri et al. 2005).

Larger complexes are more avidly taken up by antigen presenting cells (APCs) in mucosal linings. This has provided impetus for the selection of a range of delivery vehicles that presents as complexes to APCs. For example, we have used cationic liposomes for epitope delivery. These stable structures are prepared from three lipid components: dioleylphosphatidylethanolamine, dimethylaminoethanecarbamol-cholesterol, and polyethylene glycol 2000-phosphatidyl-ethanolamine. The lipids are mixed and exposed to a range of stresses that allow them to form unilamellar liposomes. Unilamellar liposomes are then mixed with adjuvant (such as CpG oligodeoxynucleotide) and the vaccine immunogen (proteins or peptides). Vesicles that are consistently less than 150 nm in diameter, the proper size for APC uptake, are produced by extruding the liposomes through polycarbonate filters. Liposomes can be stored up to a few weeks before use, and when used in combination with our DNA-prime vaccines, they have been shown to be very effective at inducing high quality immune responses that are protective against challenge in murine models (Fig. 11.4).

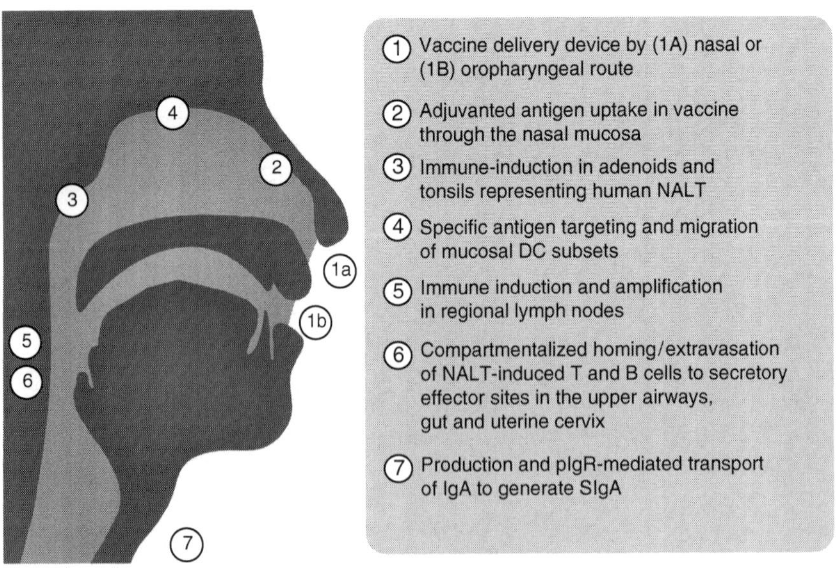

Fig. 11.3 Dynamics of intranasal delivery. Adapted from: Stefan B. Svenson, Sixth World Congress on Vaccines, Immunization & Immunotherapy Milan, Italy 23–25 September 2008, Karolinska Institutet/SLU, Stockholm, Sweden

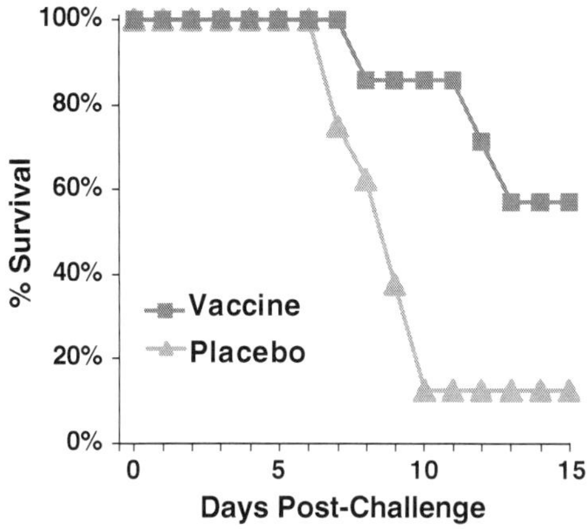

Fig. 11.4 Epitope-based prime-boost vaccine induces protective immunity to tularemia in HLA DRB1*0101 (class II) transgenic mice. Groups of 7 vaccinated and 8 control mice were challenged intratracheally with 5XLD50 *F. tularensis* LVS and survival was monitored over a 3-week period. The vaccine conferred 56% protection compared to 0% with pVax 1 vector control. Three additional ex-periments conducted under similar conditions yielded comparable results. Adapted From: Stephen H. Gregory, Stephanie Mott, Jennifer Phung, JinHee Lee, Leonard Moise, Julie A. McMurry, William Martin, Anne S. De Groot. Epi-tope-based Vaccination against Pneumonic Tularemia. Manuscript submitted to Vaccine. November 2008

11.2.2.3 Improved Adjuvants

Factors extrinsic to processing, such as the cytokine milieu induced in response to a particular component of a vaccine (Krieg et al. 1998) or pathogen (Ghosh et al. 1998), also play a role in the conditioning of the immune response. Thus, while T-cell epitopes may be necessary to drive immune response, they are not sufficient. Co-stimulatory molecules that provide T cells with a second activating signal, the right cytokine milieu, and other factors directing the nature of the immune response (TH1 vs. TH2) are also crucial (Shahinian et al. 1993; Kuchroo et al. 1995). Adjuvants provide this added "boost" in the context of whole protein and epitope-based vaccines. The choice of adjuvants for use in humans is relatively extensive, and each adjuvant has advantages and disadvantages, as reviewed elsewhere (Fraser et al. 2007).

When dendritic cells are exposed to various stimuli, such as lipopolysaccharide (LPS), monophosphoryl lipid (MPL) A, or poly-I:C, their ability to stimulate B and T cells increases. LPS, MPL-A, and poly-I:C stimulate dendritic cells by binding to a family of pattern-recognition receptors known as toll-like receptors (TLR). In addition, TLR ligand interactions in NK cells produce an array of cytokines (TNF-α, IL-1β, TNF-α, IL-6, and prostaglandin (PG) E2 (Munz et al. 2005). It is not clear which maturation stimulus is best for the induction of effector T cells in vivo. However, dendritic cells that are matured using the array of cytokines listed above are homogenous, have a high viability, migrate well to chemotactic stimuli, and induce effector T cells both in vitro and in vivo. Cytokines alone or those produced by TLR signaling can also induce TH1 responses in vivo. Interestingly, it has recently become evident that PGE2 has to be part of the maturation stimulus in order to obtain functional CC chemokine receptor 7 (CCR7) expression. CCR7 probably guides DCs into the lymph node in response to CCL19 and CCL21 (Luft et al. 2002; Scandella et al. 2002).

Unmethylated CpG motifs in oligodeoxynucleotides (CpG ODN) have emerged as potent vaccine adjuvants that activate TLR9 on antigen presenting cells (Vollmer 2005). We typically formulate CpG ODN in liposomes because this preparation improves their uptake and immunostimulatory activity (Gursel et al. 2001).

11.2.2.4 Multi-functional T Cells

T-cell immune responses are mediated by multiple mechanisms, making their characterization highly complex. Typically, a T-cell response is characterized by its magnitude, as defined by the frequency of antigen-specific T cells or the expression of a particular effector function, including cytokine secretion, cytotoxic activity, or proliferation. T-cell function is defined by a complex set of parameters, and the full potential of a functional T cell cannot be described by any one of these. It is the specific combination of functions carried out by T cells in response to infection or vaccination that uniquely describe the quality of the T cell response. A T cell that produces only one cytokine is a poor quality one; one that produces multiple cytokines is a high quality T cell. Recent reports have shown that pathogen-clearing protective immune responses are carried out by multi-functional (i.e., "high quality") T cells (Seder et al. 2008). A vaccine that can elicit this kind of immune response holds promise as a protective agent. Flow cytometry is the method of choice to quantify individual functions of T cells independently and simultaneously on the single-cell level. Multi-functional T cell quantification requires at least six-color technology. One color is needed for viability measurement to remove unwanted cell populations. T cell lineage is identified by surface staining for CD3, CD4, and CD8. Effector functions are determined by staining for multiple cytokines, such as IFN-γ, IL-4, TNF-α, and IL-2 for TH1-focused vaccines, or for cytotoxic activity markers, such as perforin, granzymes, and CD107.

11.3 Advantages and Disadvantages of T-cell Directed Vaccines

One reason for the relative paucity of IDV in clinical development is that the immunoinformatics tools for developing these vaccines have really only evolved in the last 10 to 15 years, while the average length of time to develop a vaccine is typically 20 years or more. Currently, immunoinformatics tools are more commonly used for antigen discovery and for testing vaccines in animal models. For example, Duraswaimy et al. developed an adenovirus-vectored vaccine expressing Class I Epstein-Barr virus epitopes; this vaccine successfully prevented and treated tumors associated with Hodgkin's disease and nasopharyngeal carcinoma in transgenic mice (Duraiswamy et al. 2003). There is even evidence in murine models that epitopes from wild type p53 can be used therapeutically to fight p53-associated tumors (DeLeo and Whiteside 2008). Tian et al. created a vaccine for infectious bronchitis virus (IBV) from MHC I, MHC II, and B-cell epitopes that proved to be successful in chickens (Tian et al. 2008). In mice, peptides containing epitopes from sperm were used as a vaccine against fertilization, with a single injection causing a 75% reduction in fertility for 9–10 months (Naz 2009). Since developing clinical trials is such a lengthy process, it is likely that IDV and epitope-based IDV will begin to enter clinical trials and emerge on the market in greater numbers in 5 to 10 years.

Another reason for the delayed implementation of IDV is that many researchers have had limited access to validated tools beyond epitope mapping tools. Tools such as ClustiMer, VaccineCAD, EpiAssembler, and Aggregatrix are validated algorithms that are not yet widely used. In addition, researchers, having not had an opportunity to test them in their own work, are unfamiliar with their application, and are often skeptical about their accuracy. Acceptance of immunoinformatics tools for vaccine design tools is gradually improving, as can be measured by the number and size of NIH grants awarded (recent US$24M contract for the Immune Epitope Data Base, for example) and by the number of "computational immunology" papers listed on PubMed (38 published papers claiming "computational immunology" as a key phrase in 2001; 232 such publications in 2007).

Despite the slow start, epitope-based IDVs have a lot of benefits compared with the traditional methods of vaccination that have been used for the past 50 years. Epitope-based vaccines are much safer than traditional vaccines because they operate under the principle of exposing the body to the minimal activating unit of the immune system: epitopes. Moreover, epitopes cannot reactivate like an attenuated virus vaccine. T cell epitopes, in contrast to their parent proteins, are generally too small to be biologically active; this makes the epitope-based vaccine approach particularly advantageous in cases wherein the whole antigen, though immunologically important, has a deleterious effect on the host (Wang et al. 1985). Another distinct advantage of epitope-based vaccines is the ability to efficiently exclude from consideration any sequence that has significant homology with human proteins. Sequences that do bear high degrees of homology with human proteins are poor candidates for vaccines for two reasons. First, the vaccine could be actively or passively tolerated. Secondly, if the vaccine is not tolerated, it has the potential to engender an autoimmune response.

A number of IDV have been tested in clinical trials (Elliott et al. 2008; Gahery et al. 2006; Asjö et al. 2002; Kran et al. 2004). Because epitope-based IDV are generally considered to be safe when compared with other vectored or attenuated live vaccines, most have progressed more rapidly than traditional vaccine candidates from preclinical concept into the clinic. Epitope-based IDV may also provide essential T-cell help for antibody-directed vaccines; this concept has been exploited to improve existing polysaccharide vaccines such as *Haemophilus influenza* type B (HiB) (Falugi et al. 2001) and Pneumococcal vaccines (Sen et al. 2006).

In the cancer vaccine field, in which epitope-based vaccines are well established, many such vaccines are currently in Phase I/II clinical trials (Pietersz et al. 2006). Immunoinformatics tools can also be used to improve vaccines already in the clinic; this may be one application of the tools that will progress more rapidly from concept to implementation.

11.4 Examples of T-cell Epitope-Driven Vaccines

11.4.1 TulyVax

TulyVax provides evidence that epitope-based vaccines are effective at generating protective immunity. Like TB, tularemia is a disease caused by an intracellular bacterium, *Francisella tularensis*, that is transmitted via the aerosol route. A moderately effective live vaccine strain (LVS) is available, which is thought to protect via cell-mediated immunity, but suffers from safety issues. Starting with the *F. tularensis* genome of the Schu4 isolate, we have employed a computational approach to design and test an epitope-based vaccine to provide protection against tularemia in a preclinical model.

Intracellular *F. tularensis* alters its environment by secreting proteins that interact with the host cell (Gil et al. 2006). Our strategy for the selection of epitopes within the *F. tularensis* genome was therefore focused on proteins that were predicted to be secreted. Each selected protein was parsed into all possible 9 mer peptides; EpiMatrix was then used to assess each 9 mer for putative binding to an array of Class II HLA alleles. The top scoring 40 promiscuous class II epitopes were synthesized and screened in vitro using our in-house HLA Class II competition-binding assay. Peptides that bound with high affinity to the selected alleles were then tested *ex vivo* in ELISpot assays with blood obtained from individuals that had prior *F. tularensis* infection. Positive responses were observed to 21 of 25 individual peptides; the pool of peptides was recognized in 95% of subjects, and an average response of 1,000 spots over the background were elicited. Vaccination with the multi-epitope vaccine in a *DNA prime, peptide in liposome boost* method resulted in protection (Gregory et al. 2008). Prototype vaccines containing the immunogenic epitopes were then assembled via Vaccine CAD for in vivo studies.

We immunized DRB1*0101 transgenic mice intratracheally with a 14-epitope DNA vaccine construct and boosted with peptide epitopes formulated in liposomes with CpG ODN. Splenocytes from the vaccinated mice were shown to respond in vitro to the peptides encoded in the vaccine (by ELISpot and intracellular cytokine staining, data not shown). Cytokine levels were found to be elevated in mice receiving DNA prime in addition to peptide when compared with those receiving peptide alone. Following intratracheal challenge with 6200 colony forming units (5 times the 50% lethal dose) of LVS, 57% (4/7 mice) of immunized mice survived compared with 13% (1/8 mice) of nonimmunized mice (as shown in Fig. 11.4). Considering that this SchuS4 vaccine contained only 7 epitopes that were conserved in the LVS challenge strain, this study was an unqualified success.

11.4.2 HelicoVax

In the context of chronic infections, pathogens may be able to survive by subverting the inflammatory response and inducing tolerance instead. Modification of the immune response by vaccination could result in clearance. *H. pylori* is one of the most common human pathogens, infecting approximately 50% of the world population, leading to a gastric cancer incidence of approximately 18 to 32 per 100,000 cases (Rupnow et al. 2001).

Putative T-cell epitope clusters were identified by computational analysis from highly conserved J99 and 26695 open reading frames (ORFs) using the EpiMatrix algorithm. About 1,152 epitope clusters were identified from 1107 conserved ORFs. Of these, 150 epitope clusters, with no more than 25.9% human homology (determined via BlastiMer), were selected as vaccine candidates. These 150 clusters represent 130 distinct ORFs, of which only two were previously published antigens. In a soluble DR1 competition binding assay, 57 peptides (73% of those successfully synthesized) demonstrated moderate affinity (i.e., 50% inhibition of competitor (IC_{50}) < 100 μM); 28 demonstrated high affinity (i.e., IC_{50} < 10 μM).

Fifty epitopes were incorporated into "HelicoVax," a DNA-prime/peptide-boost vaccine, and tested in p27 −/− mice preinfected with the murine-adapted *H. pylori* SS1 strain (Kuzushita et al. 2005). This transgenic strain is susceptible to developing gastric cancer following *H. pylori* infection. Two groups of mice (20 per group) were primed with a plasmid DNA vaccine, either intranasally or intramuscularly, and then boosted intranasally with peptides formulated in liposomes. Control groups (20 per group) neither received SS1 lysate vaccine nor were vaccinated. Immunogenicity was measured 45 weeks postinfection. IFN-γ ELISpot assays of epitope-stimulated splenocytes demonstrated that 47/50 peptides (94%) were immunogenic following intranasal (IN) or intramuscular (IM) DNA immunization with the multi-epitope vaccine (Fig. 11.5, filled bars), compared with only 4/50 epitopes that were recognized in SS1 lysate-immunized animals (Fig. 11.5, open bars). There is no consistent difference between IN and IM DNA immunization on the single-epitope level, but intranasal vaccination generally correlates with stronger responses, particularly among the most immunogenic epitopes (Fig. 11.5). However, despite the fact that the vaccine stimulated strong and sustained immune responses, we cannot definitively conclude that it had a significant effect on bacterial burden as measured by real time (RT)-PCR, because bacterial loads at 45 weeks postinfection are too low in this model for there to be a significant difference between the vaccinated and the control mice. IN immunized animals eradicated infection, whereas IM-immunized, placebo-immunized, and SS1-immunized animals did not. The bacterial burden was assessed by standard quantitative PCR

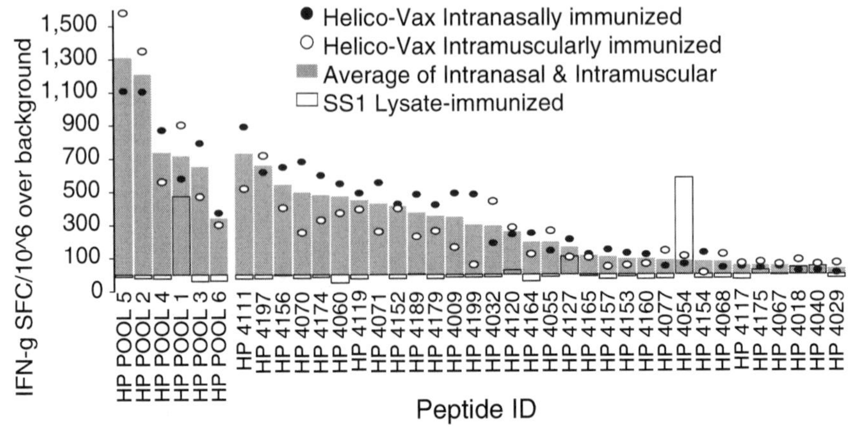

Fig. 11.5 IFN-gamma ELISpot responses to HelicoVax peptides (*right*) and peptide pools (*left*). The responses among mice intranasally administered vaccine are represented by filled circles; the responses among mice intramuscularly immunized HelicoVax are represented by open circles. The average of these HelicoVax responses is represented by the filled bars. Responses among SS1 Lysate-immunized mice are represented by the open bars. Adapted from Moise L, McMurry JA, Moss S, Martin WD, De Groot AS. Therapeutic epitope-based vaccine clears Helico-bacter pylori infection in p27 knockout mice. Manuscript in preparation.

methods (Ozpolat et al. 2000). The ratio of SS1/glyceraldehyde 3-phosphate dehydrogenase (GAPDH) PCR cycles to reach a product threshold was the criterion for measurement of *H. pylori* levels (Fig. 11.6). The goal of our current HelicoVax vaccine development program is to reduce chronic infection and to modulate metaplastic disease in this well-established model of gastric carcinogenesis in *H. pylori*-infected mice.

Notably, for this vaccine development program, we did not screen epitopes in *H. pylori*-infected humans. The rationale for skipping this step was primarily scientific: the literature shows that people who have *H. pylori* infection have weak responses to *H pylori* epitopes because of the immunomodulatory effects of the bacteria (Fan et al. 1994; Quiding-Järbrink et al.2001). Thus, it is not thought that even detectible responses would be particularly helpful in identifying the "protective" epitopes.

11.4.3 VennVax

We proposed to develop a safe, new smallpox vaccine based on epitopes conserved between the vaccinia virus (VV) and Variola (VAR) "immunomes." Such a vaccine could be used as both a prophylactic and a therapeutic intervention in the event of a bioterrorist attack. To do this, we first identified peptide sequences that were

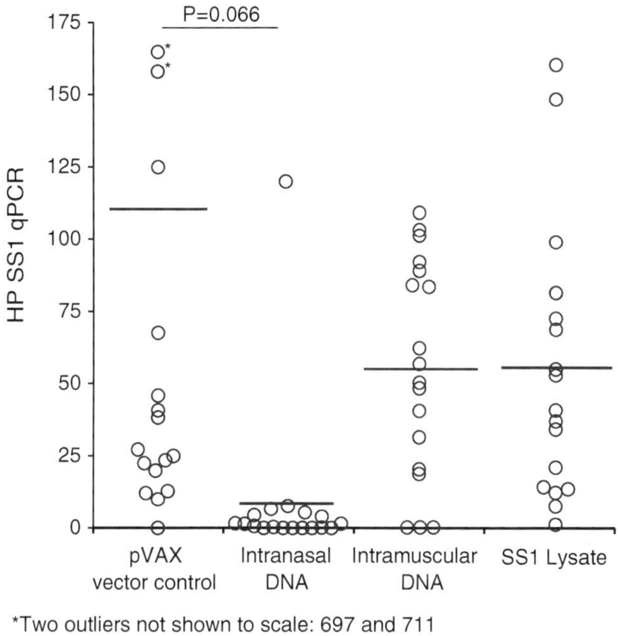

Fig. 11.6 Bacterial burden. Quantitative PCR was used to assess bacterial burden. 19/20 mice intranasally immunized with HelicoVax DNA vaccine cleared the bacteria, compared with only 2/20 mice given vector control and 3/20 mice intramuscularly immunized with HelicoVax DNA. The greater degree of protection afforded by intranasally administered HelicoVax DNA, compared to intramuscularly administered HelicoVax DNA, is concordant with the higher degree of immunogenicity afforded by the vaccine delivered by the IN route. Adapted from Moise L, McMurry JA, Moss S, Martin WD, De Groot AS. Therapeutic epitope-based vaccine clears Helico-bacter pylori infection in p27 knockout mice. Manuscript in preparation.

conserved between the VV and Var genomes. We then used EpiMatrix to score these conserved sequences to identify highly promiscuous T cell epitopes. The highest scoring candidates were synthesized as peptides and validated in both soluble MHC binding assays and in T cell assays using blood from individuals that had been Vaccinia-immunized. In vitro binding assays showed that 13 out of 14 (93%) peptides bound with high affinity to the human MHC class I HLA-A*0201 molecule and 73 of 90 (81%) bound with high affinity to HLA-B7. Ninety-one percent of the variola/vaccinia epitopes identified were confirmed in ELISpot assays. We then developed a DNA vaccine based on these epitopes and tested it for immunogenicity in *HLA transgenic mice*. In our first study, the immunization of DRB1*0101 transgenic mice stimulated significant T-cell responses to 6 of 25 epitopes (24%). In comparison (2007–08), DRB1*0301 mice immunized with the same 25-epitope set responded to 10 (40%) of epitopes, of which two were also reactive in DRB1*0101 mice. A vaccine encoding a second set of 25 epitopes stimulated significant responses for 8 (32%) epitopes in DRB1*0301 mice.

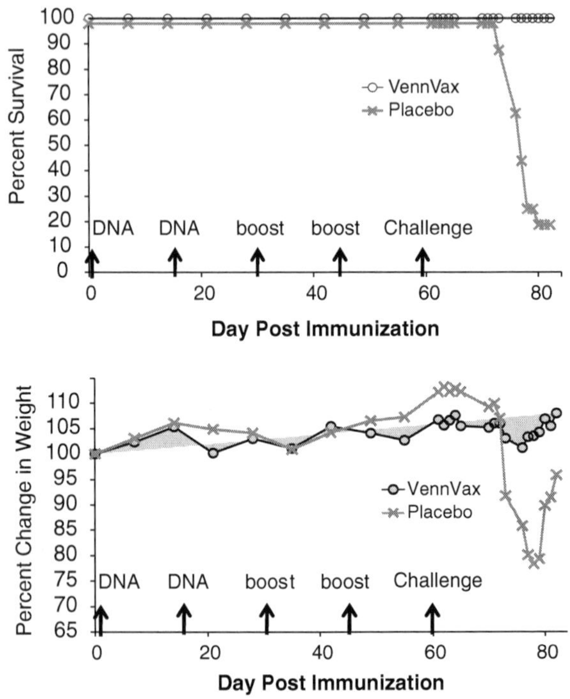

Fig. 11.7 Epitope-based prime-boost vaccine induces protective immunity to lethal vaccinia challenge in HLA DRB1*0301 (class II) transgenic mice. Groups of 18 vaccinated and 16 control mice were challenged intranasally with 10 LD50 Vaccinia Moyer strain and survival and weight were monitored over a 40 day period. The vaccine conferred 100% protection compared to 19% with pVax 1 vector control prime, empty liposome boost. Adapted from: Leonard Moise, Julie A. McMurry, William Martin, Mark Buller, Jill Schrewier, and Anne S. De Groot. Manuscript in preparation

In lethal challenge studies, vaccinated mice maintained 100% survival, and their body weights remained normal. By comparison, the placebo mice demonstrated only 17% survival, and the mice that did survive experienced dramatic weight loss (Fig. 11.7).

11.5 Concluding Remarks

Future vaccine approaches may need to move away from "whole" protein or pathogen vaccines for a wide range of reasons. Multiple antigen or epitope vaccinations such as the approach illustrated here could be one way to elicit the sort of strong TH1 response necessary to pathogens following infection, in the context of a therapeutic vaccine. And although often surmised, the linkage between immune responses to whole antigen vaccines and potential adverse effects is recently becoming evident. For example, the one Lyme disease vaccine tested in clinical trials contained a single sequence that was cross-reactive with human myelin protein (self), leading to possible postvaccination side effects and contributing to the withdrawal of the vaccine from the market. Further, there is considerable evidence that in some individuals, chronic infections (such as EBV) result in autoimmune disorders (such as multiple sclerosis, or reactive arthritis). Since specific HLA haplotypes are associated with this adverse response to infection (DR2 and DR4), it follows that immune epitopes could be the root cause.

Using immunoinformatics tools, we have begun to discover whether or not these observations are true, and we are developing vaccines that limit the immune target to epitopes that are not cross-reactive with self. This approach could also be useful for a wide range of pathogens for which genomes have been partially or completely mapped. As described in this article, our group is actively pursuing the development of epitope-driven vaccines for *F. tularensis* and *H. pylori*. We have progressed from genome-derived epitope mapping to challenge studies in less than 1 year for some of our vaccine development programs. Thus, the question is not *whether* to begin making immunome-derived vaccines, but *when* to begin. New approaches to vaccine development are required, and there is no better time to implement these new technologies than now.

References

Albert ML, Jegathesan M, Darnell RB (2001) Dendritic cell maturation is required for the cross-tolerization of CD8 + T cells. Nat Immunol 2(11):1010–1017

Ardito M (2009) Manuscript in preparation (PDF of poster, available at www.EpiVax.com)

Asjö B, Stavang H, Sørensen B, Baksaas I et al (2002) Phase I trial of a therapeutic HIV type 1 vaccine, Vacc-4x, in HIV type 1-infected individuals with or without antiretroviral therapy. AIDS Res Hum Retroviruses 18:1357–1365

Belz GT, Wodarz D, Diaz G et al (2002) Compromised influenza virus-specific CD8(+)-T-cell memory in CD4(+)-T-cell-deficient mice. J Virol 76(23):12388–12393

Bendelac A, Medzhitov R (2002) Adjuvants of immunity: harnessing innate immunity to promote adaptive immunity. J Exp Med 195:F19–F23

Blattman JN, Sourdive DJ, Murali-Krishna K et al (2000) Evolution of the T-cell repertoire during primary, memory, and recall responses to viral infection. J Immunol 165(11):6081–6090

Bond KB, Sriwanthana B, Hodge TW et al (2001) An HLA-directed molecular and bioinformatics approach identifies new HLA-A11 HIV-1 subtype E cytotoxic T lymphocyte epitopes in HIV-1-infected Thais. AIDS Res Hum Retroviruses 20:703–717

Bonifaz LC, Bonnyay DP, Mahnke K et al (2002) Efficient targeting of protein antigen to the dendritic cell receptor DEC-205 in the steady state leads to antigen presentation on major histocompatibility complex class I products and peripheral CD8 + T cell tolerance. J Exp Med 196:1627–1638

Boscardin SB, Hafalla JC, Masilamani RF et al (2006) Antigen targeting to dendritic cells elicits long-lived T cell help for antibody responses. J Exp Med 203(3):599–606

Bruder D, Westendorf AM, Hansen W, Prettin S et al (2005) On the edge of autoimmunity: T-cell stimulation by steady-state dendritic cells prevents autoimmune diabetes. Diabetes 54(12):3395–3401

Canaday DH, Wilkinson RJ, Li Q et al (2001) CD4(+) and CD8(+) T cells kill intracellular Mycobacterium tuberculosis by a perforin and Fas/Fas ligand-independent mechanism. J Immunol 167:2734–2742

Caux C, Massacrier C, Vanbervliet B et al (1994) Activation of human dendritic cells through CD40 cross linking. J Exp Med 180:1263–1272

Chen L, Wang J, Zganiacz A, Xing Z (2004) Single intranasal mucosal *Mycobacterium bovis* BCG vaccination confers improved protection compared to subcutaneous vaccination against pulmonary tuberculosis. Infect Immun 72(1):238–246

Cho S, Mehra V, Thoma-Uszynski S et al (2000) Antimicrobial activity of MHC Class I-restricted CD8 + T cells in human tuberculosis. P Natl Acad Sci USA 97:12210–12215

Cooper AM, Dalton DK, Stewart TA et al (1993) Disseminated tuberculosis in interferon gamma gene-disrupted mice. J Exp Med 178:2243–2247

De Groot AS, Berzofsky JA (2004) From genome to vaccine – new immunoinformatics tools for vaccine design. Methods 34:425–428

De Groot AS, Jesdale BM, Szu E et al (1997) An interactive Web site providing major histocompatibility ligand predictions: application to HIV research. AIDS Res Hum Retroviruses 13:529–531

De Groot AS, Bosma A, Chinai N et al (2001) From genome to vaccine: in silico predictions, ex vivo verification. Vaccine 19(31):4385–4395

De Groot AS et al (2007) Immunomics Reviews, vol 1. Springer, NY

De Groot AS, Knopf PM, Rivera D, Martin W (2007) Immunoinformatics applied to modifying and improving biological therapeutics. In: Schönbach C, Ranganathan S, Brusic V (eds) Immunoinformatics (Immunoinfomatics Reviews), Kluwer publications 1:109–132. ISBN: 978-0-387-72967-1

De Groot AS, McMurry J, Moise L, Martin B (2008) Immunome-derived vaccines. In: Falus Falus A (ed) Springer immunomics series, Series: 2, vol. 1. Kluwer publications. Immunomics Reviews, Submitted, 2008

DeLeo AB, Whiteside TL (2008) Development of multi-epitope vaccines targeting wild-type sequence p53 peptides. Expert Rev Vaccines 7(7):1031–1040

Dietrich J, Andersen C, Rappuoli R, Doherty TM, Jensen CG, Andersen P (2006) Mucosal administration of Ag85B-ESAT-6 protects against infection with Mycobacterium tuberculosis and boosts prior bacillus Calmette-Guerin immunity. J Immunol 177(9):6353–6360

Dong Y, Demaria S, Sun X et al (2004) HLA-A2-restricted CD8 + -cytotoxic- T cell responses to novel epitopes in *Mycobacterium tuberculosis* superoxide dismutase, alanine dehydrogenase, and glutamine synthetase. Infect Immun 72:2412–2415

Doolan DL, Hoffman SL, Southwood S et al (1997) Degenerate cytotoxic T-cell epitopes from P. Falciparum restricted by multiple HLA-A and HLA-B supertype alleles. Immunity 7(1):97–112

Duraiswamy et al (2003) Therapeutic LMP1 Polyepitope Vaccine for EBV-associated Hodgkin Disease and Nasopharyngeal Carcinoma. Blood 101(8):3150–3156

Elliott SL, Suhrbier A, Miles JJ, Lawrence G et al (2008) Phase I trial of a CD8 + T-cell peptide epitope-based vaccine for infectious mononucleosis. J Virol 82:1448–1457

Falugi F, Petracca R, Mariani M, Luzzi E et al (2001) Rationally designed strings of promiscuous CD4(+) T-cell epitopes provide help to Haemophilus influenzae type b oligosaccharide: a model for new conjugate vaccines. Eur J Immunol 31(12):3816–3824

Fan XJ, Chua A, Shahi CN, McDevitt J, Keeling PW, Kelleher D (1994) Gastric T lymphocyte responses to *Helicobacter pylori* in patients with *H.pylori* colonisation. Gut 35(10):1379–1384

Flynn JL, Chan J, Triebold KJ et al (1993) An essential role for interferon gamma in resistance to *Mycobacterium tuberculosis* infection. J Exp Med 178:2249–2254

Fraser CK, Diener KR, Brown MP, Hayball JD (2007) Improving vaccines by incorporating immunological coadjuvants. Expert Rev Vaccines 6:559–578

Fujii S, Liu K, Smith C et al (2004) The linkage of innate to adaptive immunity via maturing dendritic cells in vivo requires CD40 ligation in addition to antigen presentation and CD80/86 costimulation. J Exp Med 199(12):1607–1618

Gahery H, Daniel N, Charmeteau B, Ourth L et al (2006) New CD4 + and CD8 + T-cell responses induced in chronically HIV type-1-infected patients after immunizations with an HIV type 1 lipopeptide vaccine. AIDS Res Hum Retroviruses 22:684–694

Ghosh S, Pal S, Das S, Dasgupta SK, Majumdar S (1998) Lipoarabinomannan induced cytotoxic effects in human mononuclear cells. FEMS Immunol Med Microbiol 21:181–188

Gianfrani C, Oseroff C, Sidney J et al (2000) Human memory CTL response specific for influenza A virus is broad and multispecific. Hum Immunol 61:438–452

Gil H, Platz GJ, Forestal CA, Monfett M et al (2006) Deletion of TolC orthologs in *Francisella tularensis* identifies roles in multidrug resistance and virulence. Proc Natl Acad Sci USA 103(34):12897–12902

Gillespie GM, Wills MR, Appay V et al (2000) Functional heterogeneity and high frequencies of cytomegalovirus-specific CD8(+) T lymphocytes in healthy seropositive donors. J Virol 74:8140–8150

Giri PK, Sable SB, Verma I, Khuller GK (2005) Comparative evaluation of intranasal and subcutaneous route of immunization for development of mucosal vaccine against experimental tuberculosis. FEMS Immunol Med Microbiol 45(1):87–93

Gregory SH, Mott S, Phung J, Lee J, Moise L et al (2008) Epitope-based vaccination against pneumonic tularemia (Manuscript submitted to Vaccine)

Gursel I, Gursel M, Ishii KJ, Klinman DM (2001) Sterically stabilized cationic liposomes improve the uptake and immunostimulatory activity of CpG oligonucleotides. J Immunol 167(6):3324–3328

Hanke T (2008) STEP trial and HIV-1 vaccines inducing T-cell responses. Expert Rev Vaccines 7(3):303–309

Harrer T, Harrer E, Kalams SA et al (1996) Cytotoxic T-lymphocytes in asymptomatic long-term nonprogressing HIV-1 infection. Breadth and specificity of the response and relation to in vivo viral quasispecies in a person with prolonged infection and low viral load. J Immunol 156(7):2616–2623

Hawiger D, Inaba K, Dorsett Y et al (2001) Dendritic cells induce peripheral T cell unresponsiveness under steady state conditions in vivo. J Exp Med 194(6):769–779

Janeway CA Jr, Medzhitov R (2002) Innate immune recognition. Annu Rev Immunol 20:197–216

Jiang W, Swiggard WJ, Heufler C et al (1995) The receptor DEC-205 expressed by dendritic cells and thymic epithelial cells is involved in antigen processing. Nature 375:151–155

Johansson BE, Moran TM, Kilbourne ED (1987) Antigen-presenting B cells and helper T cells cooperatively mediate intravirionic antigenic competition between influenza A virus surface glycoproteins. Proc Natl Acad Sci USA 84(19):6869–6873

Kamperschroer C, Dibble JP, Meents DL et al (2006) SAP is required for TH cell function and for immunity to influenza. J Immunol 177:5317–5327

Kaufmann SH, Hess J (1999) Impact of intracellular location of and antigen display by intracellular bacteria, implications for vaccine development. Immunol Lett 65:81–84

Koita OA, Dabitao D, Mahamadou I et al (2006) Confirmation of immunogenic consensus sequence HIV-1 T-cell epitopes in Bamako, Mali and Providence, RI. Hum Vaccin 2(3):119–128

Koren E, De Groot AS, Jawa V et al (2007) Clinical validation of the "in silico" prediction of immunogenicity of a human recombinant therapeutic protein. Clin Immunol 124(1):26–32

Kran AM, Sørensen B, Nyhus J, Sommerfelt MA et al (2004) HLA- and dose-dependent immunogenicity of a peptide-based HIV-1 immunotherapy candidate (Vacc-4x). AIDS 18:1875–1883

Kretschmer K, Apostolou I, Hawiger D, Khazaie K et al (2005) Inducing and expanding regulatory T cell populations by foreign antigen. Nat Immunol 6(12):1219–1227

Krieg AM, Yi AK, Schorr J, Davis HL (1998) The role of CpG dinucleotides in DNA vaccines. Trends Microbiol 6:23–27

Kuchroo VK, Das MP, Brown JA, Ranger AM et al (1995) B7–1 and B7–2 costimulatory molecule activate differentially the TH1/TH2 developmental pathways: application to autoimmune disease therapy. Cell 80:707–718

Kuzushita N, Rogers AB, Monti NA, Whary MT et al (2005) p27kip1 deficiency confers susceptibility to gastric carcinogenesis in Helicobacter pylori-infected mice. Gastroenterol 129(5):1544–1556

Lazarski CA, Chaves FA, Jenks SA et al (2005) The kinetic stability of MHC class II:peptide complexes is a key parameter that dictates immunodominance. Immun 23(1):29–40

Lefford MJ (1975) Transfer of adoptive immunity to tuberculosis in mice. Infect Immun 11:1174–1181

Lewinsohn DA, Winata E, Swarbrick GM et al (2007) Immunodominant tuberculosis CD8 antigens preferentially restricted by HLA-B. PLoS Pathog 3(9):1240–1249

Luft T, Jefford M, Luetjens P, Toy T et al (2002) Functionally distinct dendritic cell (DC) populations induced by physiologic stimuli: prostaglandin E(2) regulates the migratory capacity of specific DC subsets. Blood 100:1362–1372

Mahnke K, Enk AH (2005) Dendritic cells: key cells for the induction of regulatory T cells? Curr Top Microbiol Immunol 293:133–150

Mahnke K, Guo M, Lee S et al (2000) The dendritic cell receptor for endocytosis, DEC-205, can recycle and enhance antigen presentation via major histocompatibility complex class II – positive lysosomal compartments. J Cell Biol 151:673–683

Mahnke K, Qian Y, Knop J, Enk AH (2003) Induction of CD4 +/CD25 + regulatory T cells by targeting of antigens to immature dendritic cells. Blood 101(12):4862–4869

Marshall D, Sealy R, Sangster M, et al (1999) TH cells primed during influenza virus infection provide help for qualitatively distinct antibody responses to subsequent immunization. J Immunol 163:4673–4682

Mayerova D, Parke EA, Bursch LS et al (2004) Langerhans cells activate naive self-antigen-specific CD8 T cells in the steady state. Immun 21(3):391–400

McElrath MJ, De Rosa SC, Moodie Z et al (2008) HIV-1 vaccine-induced immunity in the test-of-concept Step Study: a case-cohort analysis. Lancet 372(9653):1894–1905

McMurry J, Sbai H, Gennaro ML et al (2005) Analyzing *Mycobacterium tuberculosis* proteomes for candidate vaccine epitopes. Tuberculosis 85:95–105

McMurry JA, Johansson BE, De Groot AS (2008) A call to cellular & humoral arms: enlisting cognate T cell help to develop broad-spectrum vaccines against influenza. A Hum Vaccin 4(2):148–157

Menges M, Rossner S, Voigtlander C et al (2002) Repetitive injections of dendritic cells matured with tumor necrosis factor alpha induce antigen-specific protection of mice from autoimmunity. J Exp Med 195(1):15–21

Munz C, Steinman RM, Fujii S (2005) Dendritic cell maturation by innate lymphocytes: coordinated stimulation of innate and adaptive immunity. J Exp Med 202(2):203–207

Nara PL, Lin G. (2005) HIV-1: the confounding variables of virus neutralization. Curr Drug Targets Infect Disord 5(2):157–170

Naz RK. Status of contraceptive vaccines. Am J Reprod Immunol. 2009 61(1):11–18

Ostrowski M, Galeota JA, Jar AM et al (2002) Identification of neutralizing and nonneutralizing epitopes in the porcine reproductive and respiratory syndrome virus GP5 ectodomain. J Virol 76(9):4241–4250 Erratum in 2002 J Virol 76(13):6863

Ozpolat B, Actor JK, Rao XM, Lee S et al (2000) Quantitation of *Helicobacter pylori* in the stomach using quantitative polymerase chain reaction assays. Helicobacter 5(1):13–21

Petrovsky N, Brusic V (2006) Bioinformatics for study of autoimmunity. Autoimmunity 39(8):635–643

Pietersz GA, Pouniotis DS, Apostolopoulos V (2006) Design of peptide-based vaccines for cancer. Curr Med Chem 13:1591–1607

Quiding-Järbrink M, Lundin BS, Lönroth H, Svennerholm AM (2001) CD4 + and CD8 + T cell responses in Helicobacter pylori-infected individuals. Clin Exp Immunol 123(1):81–87

Rasmussen IB, Lunde E, Michaelsen TE et al (2001) The principle of delivery of T cell epitopes to antigen-presenting cells applied to peptides from influenza virus, ovalbumin, and hen egg lysozyme: implications for peptide vaccination. Proc Natl Acad Sci USA 98:10296–10301

Reis e Sousa C (2006) Dendritic cells in a mature age. Nat Rev Immunol 6(6):476–483

Rupnow MF, Shachter RD, Owens DK, Parsonnet J (2001) Quantifying the population impact of a prophylactic *Helicobacter pylori* vaccine. Vaccine 20(5–6):879–885

Russell SM, Liew FY (1979) T cells primed by influenza virion internal components can cooperate in the antibody response to haemagglutinin. Nature 280(5718):147–148

Santosuosso M, McCormick S, Zhang X, Zganiacz A, Xing Z (2006) Intranasal boosting with an adenovirus-vectored vaccine markedly enhances protection by parenteral *Mycobacterium bovis* BCG immunization against pulmonary tuberculosis. Infect Immun 74(8):4634–4643

Santra S, Barouch DH, Kuroda MJ et al (2002) Prior vaccination increases the epitopic breadth of the cytotoxic T-lymphocyte response that evolves in rhesus monkeys following a simian-human immunodeficiency virus infection. J Virol 76(12):6376–6381

Scandella E, Men Y, Gillessen S, Forster R, Groettrup M (2002) Prostaglandin E2 is a key factor for CCR7 surface expression and migration of monocyte-derived dendritic cells. Blood 100:1354–1361

Scanga CA, Mohan VP, Yu K et al (2000) Depletion of CD4(+)T cells causes reactivation of murine persistent tuberculosis despite continued expression of interferon gamma and nitric oxide synthase 2. J Exp Med 192:347–358

Scheinecker C, McHugh R, Shevach EM et al (2002) Constitutive presentation of a natural tissue autoantigen exclusively by dendritic cells in the draining lymph node. J Exp Med 196(8):1079–1090

Scherle PA, Gerhard W (1986) Functional analysis of influenza-specific helper T cell clones in vivo. T cells specific for internal viral proteins provide cognate help for B cell responses to hemagglutinin. J Exp Med 164(4):1114–1128

Scherle PA, Gerhard W (1988) Differential ability of B cells specific for external vs. internal influenza virus proteins to respond to help from influenza virus-specific T-cell clones in vivo. Proc Natl Acad Sci USA 85(12):4446–4450

Seder RA, Darrah PA, Roederer M (2008) T-cell quality in memory and protection: implications for vaccine design. Nat Rev Immunol 8(4):247–258

Sen G, Chen Q, Snapper CM (2006) Immunization of aged mice with a pneumococcal conjugate vaccine combined with an unmethylated CpG-containing oligodeoxynucleotide restores defective immunoglobulin G antipolysaccharide responses and specific CD4 + -T-cell priming to young adult levels. Infect Immun 74(4):2177–2186

Serbina NV, Flynn JL (2001) CD8(+).T cells participate in the memory immune response to Mycobacterium tuberculosis. Infect Immun 69:4320–4328

Shahinian A, Pfeffer K, Lee KP, Kundig TM et al (1993) Differential T-cell costimulatory requirements in CD28-deficient mice. Science 261:609–612

Shibaki A, Sato A, Vogel JC et al (2004) Induction of GVHD-like skin disease by passively transferred CD8(+) T-cell receptor transgenic T cells into keratin 14-ovalbumin transgenic mice. J Invest Dermatol 123(1):109–115

Spörri R, Reis e Sousa C (2005) Inflammatory mediators are insufficient for full dendritic cell activation and promote expansion of CD4 + T cell populations lacking helper function. Nat Immunol 6(2):163–170

Steinman RM (2001) Dendritic cells and the control of immunity: enhancing the efficiency of antigen presentation. Mt Sinai J Med 68:160–166

Subbramanian RA, Kuroda MJ, Charini WA et al (2003) Magnitude and diversity of cytotoxic-T-lymphocyte responses elicited by multiepitope DNA vaccination in rhesus monkeys. J Virol 77(18):10113–10118

Takeda K, Kaisho T, Akira S (2003) Toll-like receptors. Ann Rev Immunol 21:335–376

Tatarewicz SM, Wei X, Gupta S et al (2007) Development of a maturing T-cell-mediated immune response in patients with idiopathic Parkinson's disease receiving r-metHuGDNF via continuous intraputaminal infusion. J Clin Immunol 27(6):620–627

Thurmond J, Yoon H, Kuiken C et al (2008) Web-based design and evaluation of T-cell vaccine candidates. Bioinform 24(14):1639–1640

Tian L, Wang HN, Lu D, Zhang YF, Wang T, Kang RM (2008) The immunoreactivity of a chimeric multi-epitope DNA vaccine against IBV in chickens. Biochem Biophys Res Commun 2008 377(1) 221–225

Trumpfheller C, Finke JS, Lopez CB et al (2006) Intensified and protective CD4 + T cell immunity in mice with anti-dendritic cell HIV gag fusion antibody vaccine. J Exp Med 203(3):607–617

Vollmer J (2005) Progress in drug development of immunostimulatory CpG oligodeoxynucleotide ligands for TLR9.Expert Opin Biol Ther 5(5):673–682

Wang D, Liebowitz D, Kieff E. (1985) An Epstein-Barr virus membrane protein expressed in immortalized lymphocytes transforms established rodent cells. Cell 43:831–840

Wang J, Thorson L, Stokes RW, Santosuosso M et al (2004) Single mucosal, but not parenteral, immunization with recombinant adenoviral-based vaccine provides potent protection from pulmonary tuberculosis. J Immunol 173(10):6357–6365

Wilson CC, Newman MJ, Livingston BD et al (2008) Clinical phase 1 testing of the safety and immunogenicity of an epitope-based DNA vaccine in human immunodeficiency virus type 1-infected subjects receiving highly active antiretroviral therapy. Clin Vaccine Immunol 15(6):986–994

Zhang Q, Wang P, Kim Y et al (2008) Immune epitope database analysis resource (IEDB-AR). Nucleic Acids Res 36:W513–W518

Chapter 12
Understanding the Shared Bacterial Genome

Jonathan R Iredell and Sally R. Partridge

12.1 Introduction

Bacteria have been adapting to change since long before eukaryotic organisms developed. In many ways, the limited success of antibiotics due to such adaptation is much less surprising than the apparent unwillingness of humans to accept and accommodate this fact. This chapter deals with some of the natural adaptive capacities of eubacterial genomes, with particular focus on the shared genome of the Gram-negative bacteria, especially antibiotic resistance in the *Enterobacteriaceae*, as a model.

The Gram stain divides bacteria into those that retain significant amounts of stain in their cell wall ("Gram-positive") and those that do not ("Gram-negative") (Gram 1884). This simple distinction remains relevant today because it distinguishes bacteria with double-membraned cell envelopes, including a lipid-rich outer membrane, from those with single peptidoglycan-rich cell walls. The presence or absence of the periplasmic space, which separates the two membranes of the Gram-negative bacteria, is probably the most important biological divide among the eubacteria. Failure to recognize this fact may lead to inappropriate extrapolations from the simpler and better-explored genetic paradigms of the Gram-positive bacteria, to the differently adapted Gram-negative bacteria.

A variety of terms have been applied to the nonessential parts of the genome, including the "dispensible" genes of the "pangenome" (see also Chap. 2), the "floating genome" (Liebert et al. 1999), the "metagenome" (Chung et al. 2008; Holmes et al. 2003; Robinson et al. 2008), and the "mobile genome" (Li et al. 2009). Lack of clarity in all of these terms has rendered them less useful, although the terms "metagenome" and "metagenomics" have come in to the general lexicon. These terms refer to the unsorted genes found by high-volume sequencing approaches which may be, at one extreme, "fixed" in multiple unrelated chromosomes or, at the other extreme, a pool of truly promiscuous genes found in any one of a number of genetic contexts.

J.R. Iredell (✉)
Centre for Infectious Diseases and Microbiology, Westmead Hospital, Sydney West Area Health Service, University of Sydney, NSW, Sydney, Australia

V. Sintchenko (ed.), *Infectious Disease Informatics*,
DOI 10.1007/978-1-4419-1327-2_12,

In the opening chapters, the concept of the "pangenome" is introduced. In this model, comparative genomics allows the bioinformatician to define a "core" genome (always present in a given species), a "dispensible" genome (present in some but not all members of a species), and a "strain-specific" genome (unique to a given isolate or strain). All this borrows heavily from our traditional concept of an organized genome and begs the definition of the eubacterial species, a concept of uncertain value in organisms for which participation in a mobile gene pool is the key adaptation. A large highly organized eukaryotic genome is intolerant of mutations, translocations, insertions, deletions, etc., relying instead on carefully regulated genomic systems. Such a genome has great power and sophistication but is relatively inflexible and slow to adapt. By contrast, a very small genome (e.g., HIV, ~10 kb) must rely on cooperative populations to complement defects in individual members. This is the least sophisticated genomic strategy but one that is characterized by enormous flexibility with the capacity to rapidly generate novel subpopulations with high spontaneous mutation rates.

The eubacterial genome (typically ~2–6 Mb) is quite distinct from both of these examples, relying on genomic plasticity, a key part of which is the ability to share genetic information. The notion of the eubacterial genome as a work in progress (Chaps. 1 and 2) helps to remind us that not every gene is a "core" gene or even a "functional" gene. However, this tends to deny one of the key functions of the genome itself: to generate adaptive capacity. Recombinatory evolution is a primary driver in rapid eubacterial adaptation, such as is required to resist antimicrobials. Rare gene capture events mobilize useful genetic elements into a common gene pool, which in turn becomes progressively more efficient in the capture and transmission of such material. This pool can be seen as evolving independently of the bacterial hosts that both enrich it and benefit from it. Bacteria with less access to a high-flux genomic pool (i.e., those whose biological niche is unique and/or isolated) are less likely to successfully exploit this adaptive strategy and will be less dependent on it. Differences in participation in the shared gene pool are important in the approach to the bacterial genome and are essential to understanding it.

12.2 Ecological Niche and Adaptive Capacity

The outer lipid-rich layer of Gram-negative bacteria provides a gated hydrophobic barrier that allows efficient exploitation of aqueous environments. The protected space (the periplasm) between this and the inner cell wall gives Gram-negative bacteria a sheltered environment in which to elaborate delicate sensor systems to selectively sample the outside world. Transmembrane sensor/regulator machinery assembled on the inner (cytoplasmic) membrane has access to the cell's energy stores and protein production systems and can directly interact with cytoplasmic DNA to regulate a variety of cellular processes (Zhu et al. 2002). It is similarly clear that bacteria intercommunicate effectively to co-ordinate group activity ("quorum sensing") (Bassler 2002) and that bidirectional signaling between host and pathogen is important in determining the process of human infection by many pathogens.

In the human gut, bacteria have access to rich and diverse sources of carbon and nitrogen and have only to compete with each other. Exposure to multiple populations of bacteria, which use a range of strategies to compete, means that the ability to efficiently acquire new strategies is in itself a major biological advantage. One well-studied example of such adaptation is the acquisition of bacteriophage-transduced adhesins (Karaolis et al. 1999) and toxins (Waldor and Mekalanos 1996) by *Vibrio cholerae*. This acquisition allows it to invade and disseminate massively using a new life cycle which involves human infection. To us, this is epidemic cholera. Genetic changes, which allowed avoidance of host responses, were at the basis of the feared eighth (O:139 or "Bengal" strain) cholera pandemic (Faruque et al. 2003; Stroeher et al. 1995).

An important adaptation now required by organisms already adapted to human commensalism or parasitism is the ability to withstand antibacterial drugs. These drugs have been developed only in the last several decades but have been met by an efficient adaptive response on each occasion (Fig. 12.1). There is no reason to be surprised by this, nor any reason to suppose that this will change. However, we argue that such adaptations may be anticipated, and even managed, if we understand their bases.

The mechanism of bacterial adaptation varies with the natural adaptive capacity of the organism. The adaptation style can be divided for the sake of discussion into those organisms with existing systems which can be readily employed for new purposes, with or without modification (improvisation), and those organisms which rely on their ability to acquire completely new characteristics (innovation). Improvisation is an efficient strategy since, by definition, material from which to develop new capacity is already available within the organism. Sometimes there is a cost, which comes from sacrificing one function for another, but often it is an adaptation which is well tolerated or can be compensated for.

P. aeruginosa is one of the best studied models of bacterial pathogenesis. Isolates infecting patients with Cystic Fibrosis (CF) are genetically distinct from environmental isolates and non-CF clinical isolates (Finnan et al. 2004; Reik et al. 2005; Wood and Smyth 2006), and are more likely to contain known virulence-associated genes (Finnan et al. 2004). Importantly, the CF lung is a unique environment in which hypermutable *Pseudomonas* populations with specific defects in their mismatch-repair systems are not uncommon. This adaptation, reviewed in (Hall and Henderson-Begg 2006), makes the organism more likely to undergo further spontaneous changes, including development of antibiotic resistance. The genome of *P. aeruginosa* encodes several efflux pumps important in resistance to antibiotics. The ability to combine various responses to external toxins means that *P. aeruginosa* is less dependent on the mobile genome, in which it also participates (McGowan 2006).

The capacity to innovate is a function of the ability to acquire new material and to effectively integrate it. The opportunity comes from access to genetic potential, and is most available in a diverse polymicrobial community in which organisms have developed alternative adaptations to the same environment. The relative importance of each of the mechanisms widely available for DNA acquisition

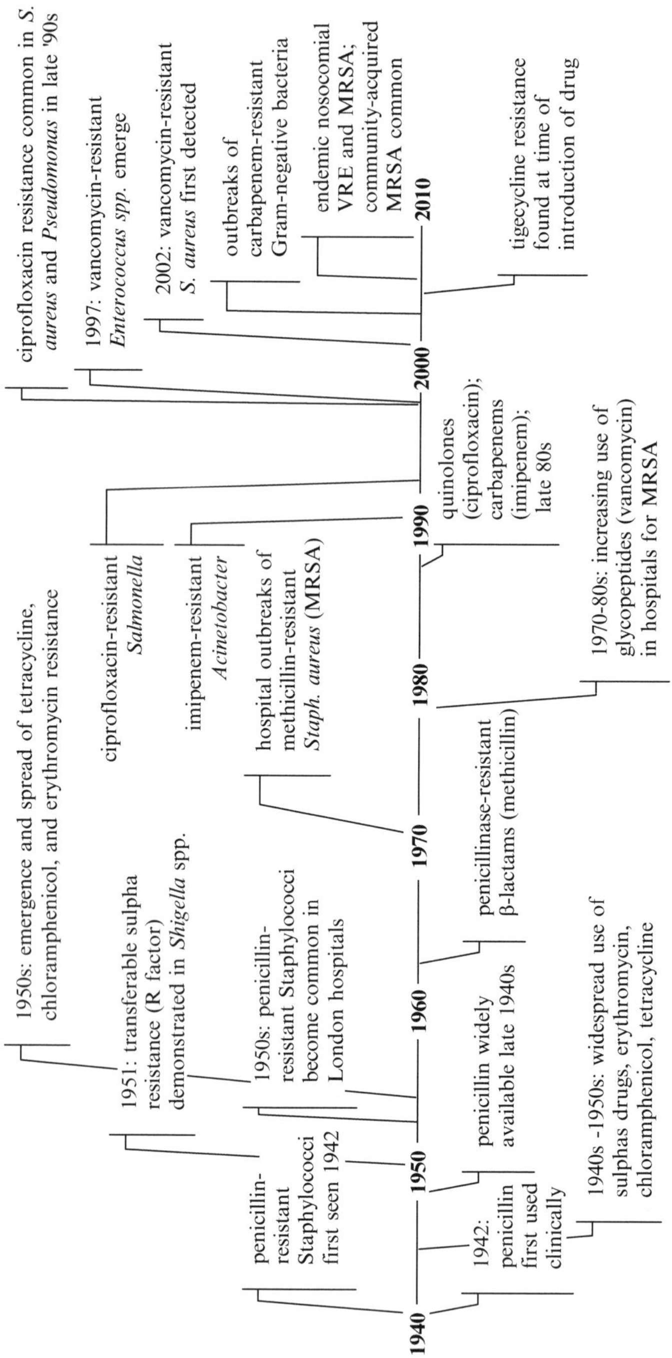

Fig. 12.1 Bacterial adaptation to antibiotics

(bacteriophage transduction, direct uptake from the environment, and conjugal transfer between cells) is logically a function of the niche/s occupied by the organism. In the case of aquatic organisms such as the *Pseudomonadaceae* and *Vibrionaceae*, adapted to aqueous environments scarce in resources, small protected packages of DNA must be able to shuttle freely between sparsely distributed cells: the bacteriophages are specialized viruses which mediate DNA transfer in the course of symbiotic or parasitic relationships with bacteria. In the case of organisms such as the staphylococci, adaptation to survive a dry environment means the clonal expansion of successful organisms is an important epidemiological feature. In these organisms, a significant element of genomic diversity is expected to be generated by mutation in relatively less mobile components of the genome.

In the case of the medically important *Enterobacteriaceae* (e.g., *E. coli* and *Klebsiella* spp.), adapted to close living in an aqueous nutrient-rich environment (the gut), direct cell-cell communication and genetic transfer are the most important adaptations. This has special implications for the rapid acquisition of varied and multiple resistance mechanisms in these types of bacteria. For organisms adapted to sharing genetic material, it is generally most efficient to acquire (and possibly then adapt) what is already available elsewhere. Mechanisms for the integration of genes into the genome are important for this, and necessary to profit from access to bacteriophage-packaged or conjugally transferred DNA. Genes of value to the wider pool include those conferring resistance to antibiotics, which may be mobilized from, say, a chromosomal position, by any one of a number of specialized processes ("gene capture") and thus brought into the wider gene pool. It follows that the actual gene "capture" events are rare and that associations with individual gene capture systems are relatively stable. Once a useful gene is captured in such a manner, it would be most efficient if the capture process also enhances genetic mobility. Natural selection will tend to favor events that enhance the mobility and flexibility of a useful genetic package within the gene pool.

The shared genome is therefore characterized; at least in the Gram-negative bacteria where it has been well studied, by common themes in the genetic relationships between the genes and gene capture elements. We will discuss antibiotic resistance genes in the medically important *Enterobacteriaceae* to illustrate this. The mosaic patterns observed reflect a process of recombinatory evolution, which is likely to be most prominent in those organisms which are most adapted to gene sharing. The extent to which an organism participates in gene sharing, and therefore the extent to which its genome is recombinatory and mosaic, and shared with other species, is a function of its ecological niche.

12.3 The Shared Genome

Organisms such as *E. coli* are naturally rich in conjugative plasmids, which provide highly efficient vehicles for DNA exchange. These plasmids themselves occupy an intracellular niche in which gene sharing is advantageous to the host bacterium and to the plasmid itself, and it is therefore unsurprising that the conjugative plasmids of *E. coli* found in humans are rich in antibiotic resistance genes and gene capture elements of various types.

12.3.1 Gene Capture and Transfer

Genes which confer resistance to medically important antibiotics are associated with a few types of mobile elements (ME), which have some features in common but capture and move resistance genes in different ways. Large transposons (Tn; Fig. 12.2a, f) carry both "transposition" functions required for movement and resistance genes (Grindley 2002). Classical insertion sequences (IS; e.g., IS*26*) contain only genes which encode their own mobility (Chandler and Mahillon 2002), but a pair inserted either side of an antibiotic resistance gene can capture it as part of a composite transposon (C-Tn; Fig. 12.2b) that subsequently moves as a unit. ISEcp*1* is unusual in that it is able to capture and move adjacent resistance genes (Fig. 12.2c) (Poirel et al. 2005), IS*CR* elements (Fig. 12.2d) achieve the same result but use a distinct mechanism (Toleman et al. 2006). Class 1 integrons (In) capture one or more gene cassettes (Fig. 12.2e) by site-specific recombination, creating arrays (of typically less than nine cassettes) inserted between two highly conserved elements (5′- and 3′-CS) (Hall et al. 1999).

12.3.2 Associations Between R Genes and ME

As indicated above, gene capture is a rare event, but such an event may only need to happen once for a gene to enter the pool, from which it can be efficiently acquired by different organisms. Thus, each resistance gene is typically associated

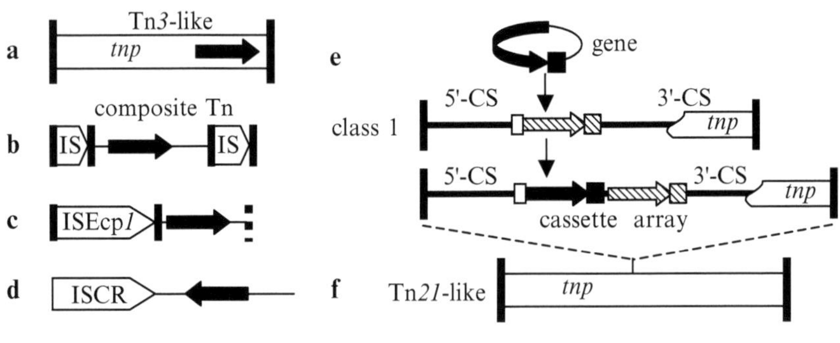

Fig. 12.2 Mobile gene-capture elements. *Black arrow*, resistance gene; Tn, transposon; *tnp*, transposition functions; *vertical bar*, terminal inverted repeat (IR); IS, insertion sequence; *broken vertical bar*, alternative IR used by ISEcp*1* to mobilise adjacent genes; ISCR elements have no IR and mobilise adjacent genes by a different ("rolling circle") mechanism; CS, (integron) conserved sequence; *white box*, *attI1* recombination site; *black box*, *attC* recombination site

Table 12.1 Associations between resistance genes and mobile elements

Gene type[a]	Mobile element[b]				
	Tn	C-Tn	Gene cassette	IS*CR*	ISEcp*1*
"older"	*strB*	*tetD, aph*	*dfrA,B, catB, aadA,*	*dfrA, sul2, catA2*	
ESBL	TEM	SHV	OXA, VEB	CTX-M-9,-14b	CTX-M-3,-14a,-15
ampC				DHA	CMY-2-like
CPMase	KPC		IMP, VIM, SIM, GIM	SPM	
AG-R		*npmA*	*aadB, aacC1, aac(6')-Ib*	*armA, rmtD*	*rmtC*
FQ-R			*aac(6')-Ib*-cr	*qnrA, qnrB2*	

Note:[a]ESBL, extended-spectrum β-lactamase; CPMase, carbapenemase; G-R, aminoglycoside (e.g., gentamicin)-resistance; FQ-R, fluoroquinolone (e.g., ciprofloxacin)-resistance
[b]Tn, transposon; C-Tn, composite transposon; IS, insertion sequence. Gene names in uppercase (e.g., TEM) are simplifications of full *bla* β-lactamase gene names (e.g., bla_{TEM})

with a particular ME, and the types of ME originally identified some time ago in association with "older" resistance genes (conferring resistance to antibiotics that are now less important) also remain important in the capture of "new" resistance genes (Table 12.1).

Association with a particular type of ME may dictate where a resistance gene can be incorporated after it has been captured. For example, gene cassettes are usually inserted into integrons, and integrons are generally found in specific sites found in Tn21-like transposons and in related sites in plasmids (Minakhina et al. 1999), IS*CR1* usually lies between partial duplications of the 3′-CS of class 1 integrons (Toleman et al. 2006) while most IS, large transposons and, probably, ISEcp1 exhibit no particular insertion-site preferences. The initial spread of a resistance gene may thus depend on which ME it is captured by and the availability of appropriate sites for insertion of that ME-resistance gene combination. However, the subsequent success of genes associated with the same ME may be more dependent on their wider genetic contexts (Walsh 2006). This is discussed in more detail below, and we will use a common and important antibiotic resistance phenotype as an example.

12.3.3 β-Lactamases Conferring Resistance to Cephalosporins

β-lactamases are the most important transferable mechanisms of resistance to the most important class of modern antibiotics and are encoded by variants of a few *bla* gene families (http://www.lahey.org/Studies/). The third-generation cephalosporins (3GC) are potent broad-spectrum antibiotics that were developed to combat rising resistance to β-lactam antibiotics, but were soon confronted by the emergence of "extended spectrum" β-lactamases in bacteria. The "classical" extended-spectrum β-lactamases (ESBL) confer resistance to 3GC antibiotics, but two more β-lactamase

Table 12.2 Summary of β-lactamases conferring resistance to 3GC

β-Lactamase type[a]	β-Lactamase (*bla*-) genes	β-Lactam resistance[b]
ESBL	CTX-M; SHV; TEM; VEB; PER, GES	3-GC
AmpC	CMY; DHA; FOX; MOX; ACC;	3-GC; APP-β
Carbapenemases		
Metallo-β-lactamases	IMP; VIM; GIM; SIM; SPM	3-GC; APP-β; CPM[c]
Serine-mediated	KPC; (GES)[d]	3-GC; APP-β; CPM[c]

Note:
[a]ESBL, extended-spectrum β-lactamase
[b]3-GC, third-generation cephalosporin (e.g., ceftrixaone, ceftazidime); APP-β, antipseudomonal penicillin-β-lactamase inhibitor combination (e.g., ticarcillin-clavulanate; piperacillin-tazobactam); CPM, carbapenem (e.g., imipenem, meropenem)
[c]Full resistance to carbapenems usually requires additional permeability defects
[d]Only some GES variants can confer carbapenem resistance

groups have overlapping antibiotic-hydrolyzing activities (Table 12.2). Several *bla* genes may be present in one organism at the same time and the frequent co-occurrence of multiple β-lactamases makes it difficult to predict which genes are present from the phenotype, greatly complicating screening and diagnosis (Babic et al. 2006).

For the clinical microbiologist, the crucial question is whether the genetic basis for a given resistance phenotype is sufficiently predictable to enable the development of surveillance and/or diagnostic tools. There are many genes associated with the common and important antibiotic resistance phenotypes but there are very few data to address the local prevalence of a given resistance gene, the extent of natural diversity, or the way in which this changes over time and place.

Significantly, however, available data suggest that the gene pool may be relatively limited (Rossolini et al. 2008) – that is, there tends to be a limited set of genes and genetic elements responsible for the local antibiotic resistance phenotypes and these are commonly shared between unrelated organisms (Baquero 2004). This implies that surveillance and monitoring of the genome has the potential to be a vital tool for both predicting and containing the spread of resistance and for development of diagnostic applications.

12.3.4 Genetic Disequilibrium Within the Mobile Gene Pool; the Multi-(Antibiotic) Resistance Region

Studies of the wider contexts of ME-resistance gene combinations indicate that they are often found in large multiresistance regions (MRR; e.g., Fig. 12.3). Other resistance genes have yet to be detected inside MRR, but only limited contextual data is available for many of these.

The insertion of a resistance gene into an MRR will result in links to other resistance genes, potentially allowing coselection by unrelated antibiotics, and/or may enable

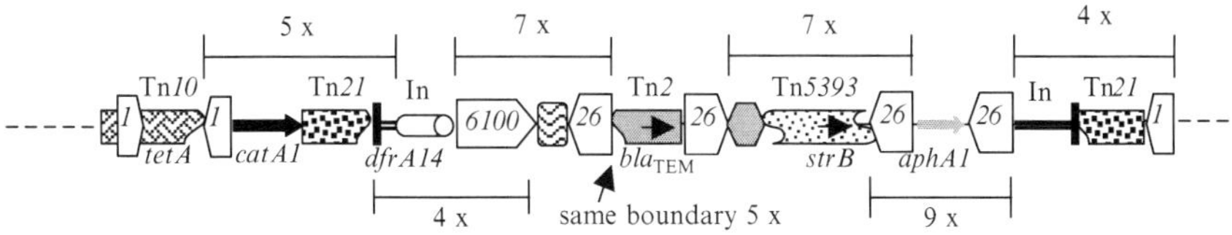

Fig. 12.3 Recombination and mosaicism in multiresistance regions. The MRR in pRSB107 (GenBank accession no. AJ851089; not to scale), showing some common combinations of components and the number of times they appear in GenBank. Transposon fragments and resistance genes are labeled; numbered boxes represent insertion sequences; different shapes represent other common MRR components

the movement of resistance genes to new locations in different mobile entities, freed from constraints of the original ME (O'Brien 2002). Insertion into an MRR also allows movement of genes by homologous recombination between common MRR components, which may be important if the original ME is damaged (O'Brien 2002). Cassettes can move by homologous recombination or in common cassettes (Partridge et al. 2002; Gestal et al. 2005; Toleman et al. 2006).

MRR appear to be modular structures that are subject to combinatorial evolution (Baquero 2004; Martinez et al. 2007; Walsh 2006). The same structures, composed of more than one ME-resistance gene combination or fragments thereof, may be found in many different MRR, often with precisely the same boundaries between them (Fig. 12.3). Events mediated by homologous recombination and by ME are both important in the remodeling of MRR and the relative importance of each is determined by opportunity. It seems logical that early developments are largely ME-driven and that further changes will likely occur mostly through recombinatory processes. The recurring themes in MRR suggest that although the processes involved in their assembly and evolution are complex, they are not random and should be predictable (Baquero 2004; Martinez et al. 2007; Walsh 2006).

12.3.5 The Arrival and Spread of New Members of the Gene Pool

If the early development of MRR is largely ME-driven, then new genes arriving in the gene pool will tend to initially disseminate as the ME-resistance gene complex (assuming they have secured a passage on an efficient vehicle such as a plasmid or bacterial strain). Two examples of newly arrived genes allow us to examine this.

1. $bla_{\text{OXA-23}}$ encodes a serine carbapenemase responsible for the highly carbapenem-resistant phenotype observed in outbreaks of *Acinetobacter baumannii* in this country (Playford et al. 2007; Valenzuela et al. 2007). This phenotype had not been seen before these outbreaks and, when examined, the gene proved to be in a composite transposon flanked by ISAba*1*, which others have also identified and shown to be mobile (Mugnier et al. 2009). Locally, the resistance gene was found in multiple different clones (Valenzuela et al. 2007), so that it appears that

mobilization of the resistance gene-complex may be the crucial element, whether transmitted as the composite transposon itself or as part of a larger structure.

2. Carbapenem resistance in Australian *Enterobacteriaceae* was recently identified as being due to a metallo-β-lactamase gene (*bla*$_{\mathrm{IMP\text{-}4}}$) found in an array of gene cassettes in an integron within a ~20 kb MRR (Espedido et al. 2008). This gene was also new to Australia and was found on a large number of closely related plasmids in Sydney, but in a different genetic context on apparently different plasmids in isolates from Melbourne, and in different arrays in an isolate from China (Espedido et al. 2008) and in local *P. aeruginosa* strains.

These findings support the idea that new arrivals in the local gene pool/s, whether as a composite transposon (e.g., ISAba1|*bla*$_{\mathrm{OXA\text{-}23}}$|ISAba1) or as a new gene cassette (*bla*$_{\mathrm{IMP\text{-}4}}$), can be expected in different contexts but with a fairly recognizable ME-resistance gene complex still intact. Conversely, genes that are already well established in the microflora should be found in a more mosaic structure with the initial ME-resistance gene complex less evident or at least apparently less important as a source of variation in the genetic context/MRR. Only one completely sequenced plasmid carrying the globally disseminated extended-spectrum β-lactamase (ESBL) gene *bla*$_{\mathrm{CTX\text{-}M\text{-}15}}$ is available in the GenBank database (Boyd et al. 2004) but we have recently characterized several related MRR carrying *bla*$_{\mathrm{CTX\text{-}M\text{-}15}}$ in more than a dozen plasmids from Sydney. Here, we see evidence of frequent IS-mediated rearrangements/deletions and although there is also evidence of movement of *bla*$_{\mathrm{CTX\text{-}M\text{-}15}}$ as part of smaller structures, we found only one example of apparent movement of the original ME-resistance gene complex, while the majority of structural variation is apparently the result of recombinatory processes (unpublished data).

12.3.6 Comparative Analysis of Multiresistance Regions

Patterns and predictable relationships in MRR will only become apparent after systematic comparison of many carefully selected examples. Until now most MRR have been analyzed as part of completely sequenced resistance plasmids, but sequencing more plasmids is an inefficient way to characterize large numbers of MRR. Modern high-throughput sequencing approaches using small (75–500 bp) reads are likely to be complicated by repeated elements found in MRR. These elements are often too long to sequence through in one reaction, producing mixed sequences from outwardly directed primers and making the assembly of final sequences more problematic (Frost et al. 2005).

Comparative analyses of MRR can be efficiently performed by a process of mapping based on the known likelihood of finding given components adjacent to one another. The very fact that this approach is efficient and successful (e.g., for *bla*$_{\mathrm{CTX\text{-}M\text{-}15}}$ contexts, above) is in itself evidence of the mosaic nature of these MRR. Apart from within resistance genes themselves (where minor mutations can lead to changes in phenotype), and across boundaries between ME or their fragments which are informative of transposition events and/or the original source of the DNA fragment, sequencing is generally unnecessary to obtain the most useful information about MRR. It is much more important to know which components are present and how they are organized. The challenge is to find efficient ways to map MRR and to collate the resulting data in a format amenable to analysis.

It may therefore be that the most efficient approach to an analysis of MRR is by a probabilistic method. Available data are dominated by many examples of single resistance genes with minimal flanking sequences, but there are a few whole plasmid sequences available. We favor an approach that combines gene-specific data (obtained by hybridization and/or sequencing) with the direct mapping of genetic contextual relationships (Fig. 12.4).

The MRR are a mosaic set of interrelated structures within a mobile gene pool and they evolve through the incorporation or substitution of different features, including the substitution of genes determining key phenotypes, as well as by the gain or loss of large genetic regions. It is therefore difficult to conceive of an MRR as a unitary structure using the usual paradigms of infectious diseases, in which we consider a pathogen with unique and relatively consistent defining characteristics. However, the elements that dictate the antibiotic resistance phenotype and the nature and efficiency of its spread throughout the gene pool have sufficient common features and predictable relationships to enable the modeling of the evolution of MRR and their epidemic potential. The necessity is unarguable and the most useful epidemiological analyses probably need to include:

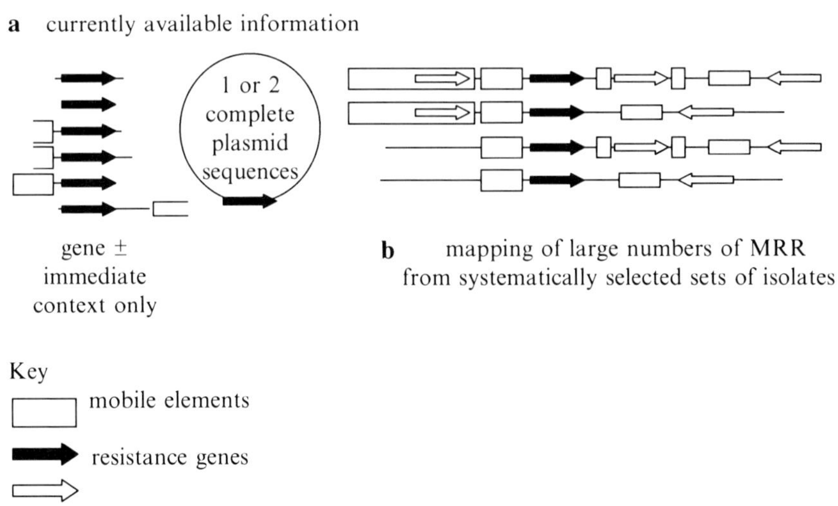

Fig. 12.4 An approach to the mapping and bioinformatics of the multi-resistance regions. Currently available information (**a**) is less informative than (**b**) complementary mapping and bioinformatics of multi-resistance regions (MRR). *Black and white arrows*, resistance genes; *white boxes*, different mobile elements

1. Appropriately representative sampling/surveillance of bacterial population/s
2. Efficient high-throughput hybridization methods for the recognition of important components (e.g., components of ME-resistance gene complexes)
3. An efficient bioinformatics tool to direct PCR mapping by identifying the most likely context of those components found in (2)
4. Determination of the DNA sequences of specific regions of interest - these logically include particular resistance genes (in which altered genotype may alter phenotype), and the signatures of recombination and transposition events found at the boundaries of discrete genetic elements within the mosaic structure of an MRR.

The third component of this is the most problematic. Analysis of the sequences from sources such as GenBank, which are required to make predictions about MRR structure, is often hampered by incorrect or incomplete annotations (Frost et al. 2005) that focus on open reading frames (ORF) rather than on informative boundaries between components. Many resistance genes are incorrectly/incompletely annotated or even missed and essentially identical ORF are given different names. Manual re-annotation of large complex sequences is time consuming, but these data can be managed by automated methods, which use context-specific grammars, rather than sequence alignments (Partridge et al. 2009). This allows a probabilistic approach to the mapping of MRR and complex mosaic sequences, which is discussed in detail in the following chapter.

12.3.7 Conjugative Plasmids: The Need for a New Metagenomics Strategy

In addition to MRR, the large self-mobilizing (conjugative) antibiotic-resistance plasmids, which so often carry them, are crucially important in the transmission dynamics of antibiotic resistance in the *Enterobacteriaceae*. However, very little data are available on the character and content of these plasmids. This may relate to the fact that it is difficult to obtain DNA sequences from them, as explained above for the MRR, or it may simply be that we have not focused sufficient attention on them. They are clearly the most important and efficient genetic vectors in the *Enterobacteriaceae*, and are responsible for the spread of most of the currently troublesome transmissible antibiotic resistance phenotypes.

Opportunities for homologous recombination increase with the extent of DNA relatedness between plasmids, but phylogenetic relatedness between large conjugative plasmids is difficult to define. By convention, we group plasmids according to the compatibility of their replication strategies, on the premise that plasmids with sufficiently similar replication strategies interfere and/or compete with each other in the same host bacterium and are therefore deemed to have "incompatible" replicons. The core functions which define this incompatibility relate to plasmid maintenance and replication processes. The associated genes are presumed to be more conserved and have therefore been targeted to develop successful typing systems (Carattoli et al. 2005), which assign conjugative plasmids to incompatibility (Inc) groups.

Surveys of *E. coli* show that such an approach identifies around three quarters of the conjugative plasmids, and often reveals multiple replicons (Zong et al. 2008). Multiple replicons may exist in a single conjugative plasmid, and multiple conjugative plasmids with different replicons may co-exist in a single cell. Distinguishing between these scenarios is difficult and labor-intensive, but an understanding of the makeup and genetic context of the replication regions may allow recognition of unique structures characteristic of particular plasmids or plasmid types, using modern genomic tools.

For example, most of the publicly available fully sequenced IncF plasmids appear to have multiple replicons, but whether the activity of a given replicon varies with the host bacteria is not well studied. These plasmids tend to be large (85–185 kb) and some have highly mosaic structures, including truncated replicon copies (Perichon et al. 2008). Detailed analysis of IncF plasmids carrying the $bla_{\mathrm{CTM\text{-}15}}$ gene in our own region reveals multiple different IncF replicons and even nonIncF replicons within these plasmids (unpublished data). Most of these replicons appear to be entire and may be fully functional.

The replicative strategy is a primary determinant of host range, and the presence of a potent "addiction" system may establish a plasmid as a permanent feature within a bacterial population. One can easily envisage such a plasmid becoming dominant, by virtue of an MRR, which has a highly protective phenotype or which provides an easy access point for other ME-resistance gene complexes. Strong and/or ongoing selection pressure (e.g., antibiotic exposure in a nosocomial environment) would be predicted to favor plasmids with effective addiction systems, broad host-range replicons, and increasingly versatile MRR. The epidemiology and evolution of such populations has not been systematically tested.

A practical approach to testing such hypotheses might reasonably focus on those factors, which determine the transmissible phenotypes, and those factors, which determine host range. Biological constraints upon sequence variation within replication-specific targets used for typing are poorly defined but may allow metagenomics approaches to discover related and as yet unrecognized sequences from short reads (<500 bp). Combined with a mapping approach to MRR as described above, this should allow development of a metagenomics profiling strategy. Much needs to be better defined, including the extent of divergence of different components within addiction systems (e.g., homologues and paralogues of the *sok/hok/mok*-like antisense RNA systems) (Gerdes and Wagner 2007), and their coassociations identified.

Multilocus sequence typing (Maiden 2006) is a useful phylogenetic tool for some bacterial genomes. It has been adapted to IncI1 (Garcia-Fernandez et al. 2008) and IncHI1 plasmids (Phan et al. 2009) but it is unknown whether this approach will break down in highly mosaic plasmids. Candidate targets would need to be tested for their informative capacity in a large plasmid population, in the way that candidate alleles are tested for MLST of bacterial genomes.

12.4 Concluding Remarks

There are a few key questions for those who seek to apply modern genomic tools to the management of major clinical problems such as the spread of antibiotic resistance in medically important bacteria. Firstly, how diverse is the gene pool in a given region for a given resistance phenotype? Are there predictable relationships between highly selectable elements such as antibiotic resistance genes? How fixed are these and how responsive are they to antibiotic selection once established?

There might be many dozens of transferable genes that could explain, say, gentamicin resistance, but how many are likely to be responsible in a given region? One might reasonably predict that the gene pool would be limited to those which are most successful locally, whether this is true globally or not. Local success will be serendipitous in part, but may also relate to factors that facilitate transmission. Such factors might include intrinsic mobility (such as association with a highly active transposon) and/or the ability to recombine into a plasmid with a high mobility and a broad host-range.

Secondly, how stable is this picture? Is genetic flux very rapid – do the subsets of genes determining the key phenotypes of today bear any relationship to those of tomorrow? Thirdly, how much does this vary from place to place and how predictable are these differences? It is possible to recognize genes with high-level epidemic potential. For example, a gene on a broad host-range conjugative plasmid in an MRR with extensive regions of homology to MRR already widely established in compatible plasmids is likely to be successful if there are strong selection pressures and few alternative sources of the resistance phenotype. At the other extreme, a gene in a disrupted ME-resistance gene complex in a small MRR with few common elements on a nonconjugative plasmid or on the chromosome is much less likely to spread other than clonally with the host strain. This is an oversimplification to illustrate the point, but it is clear that understanding these epidemiological differences is essential, and that this understanding can be advanced by the intelligent application of genomic tools. Well-integrated local and regional sampling is needed to answer these important questions.

Informative surveillance of mobile genetic material with a broad host range may require only representative sampling of the *Enterobacteriaceae* – readily achieved in the course of routine nosocomial Infection Control surveillance. The predictive power of such sampling is yet to be properly tested in clinical studies, but anecdotal evidence suggests that foreknowledge of the presence of a mobile highly-resistant MRR is useful (Thomas et al. 2005), as we accept to be the case for surveillance of organisms such as methicillin-resistant *S. aureus*. Applications of such knowledge in highly multiplexed PCR systems appears to be quite feasible, and the use of high throughput sequencing to identify signature regions within undifferentiated samples of the microflora may prove to be efficient.

Finally, there are a number of important unknowns about the evolution of the major resistance vehicles. Is there an optimal size for a resistance plasmid? Does the plasmid host-range evolve with natural selection, or is it more advantageous to have high compatibility with multiple plasmids so as to optimize genetic exchange opportunities? Do plasmids and/or MRR commonly cause resistance outbreaks by "infecting" bacterial populations? Do incoming MRR "infect" populations of local resident conjugative plasmids, with subsequent natural selection defining the final epidemiology, or do MRR and plasmids coevolve? Does this vary with the MRR and/or plasmid? If so, what are the characteristics which influence this? Are the MRR-plasmid relationships as stable over time and place as resistance gene-ME associations seem to be?

"Outbreaks" of plasmids (Espedido et al. 2005) and/or ME-resistance gene complexes (Valenzuela et al. 2007) obviously do occur in bacteria. It may be more usual, however, for a successful bacterial clone to be made more successful by the acquisition of a resistance plasmid and thus dominate the transmission of that resistance (e.g., *E. coli* ST131 carrying $bla_{\text{CTX-M-15}}$) (Nicolas-Chanoine et al. 2008). In any case, it is clear that when dealing with an outbreak of antibiotic resistant Gram-negative bacteria, epidemiological studies need to carefully consider those elements of the genetic context that determine the transmission and spread of the resistance trait itself. The multiple antibiotic resistance regions of Gram-negative eubacteria, and the movement of their components within the metagenome, provide a model of prokaryotic evolution, which should be able to be successfully approached by a combination of modern genomics and bioinformatics.

References

Babic M, Hujer AM, Bonomo RA (2006) What's new in antibiotic resistance? Focus on β-lactamases. Drug Resist Update 9:142–156

Baquero F (2004) From pieces to patterns: evolutionary engineering in bacterial pathogens. Nat Rev Microbiol 2:510–518

Bassler BL (2002) Small talk. Cell-to-cell communication in bacteria. Cell 109:421–424

Boyd DA, Tyler S, Christianson S, McGeer A et al (2004) Complete nucleotide sequence of a 92-kilobase plasmid harboring the CTX-M-15 extended-spectrum β-lactamase involved in an outbreak in long-term-care facilities in Toronto, Canada. Antimicrob Agents Chemother 48:3758–3764

Carattoli A, Bertini A, Villa L et al (2005) Identification of plasmids by PCR-based replicon typing. J Microbiol Methods 63:219–228

Chandler M, Mahillon J (2002) Insertion sequences revisited. In: Craig NL, Craigie R, Gellert M, and Lambowitz AM (eds) Mobile DNA II. ASM Press, Washington, D.C. pp 305–366

Chung EJ, Lim HK, Kim JC, Choi GJ et al (2008) Forest soil metagenome gene cluster involved in antifungal activity expression in *Escherichia coli*. Appl Environ Microbiol 74:723–730

Espedido B, Iredell J, Thomas L, Zelynski A (2005) Wide dissemination of a carbapenemase plasmid among gram-negative bacteria: implications of the variable phenotype. J Clin Microbiol 43:4918–4919

Espedido BA, Partridge SR, Iredell JR (2008) bla(IMP-4) in different genetic contexts in Enterobacteriaceae isolates from Australia. Antimicrob Agents Chemother 52:2984–2987

Faruque SM, Sack DA, Sack RB, Colwell RR, Takeda Y, Nair GB (2003) Inaugural article: emergence and evolution of *Vibrio cholerae* O139. Proc Natl Acad Sci USA 100:1304–1309

Finnan S, Morrissey JP, O'Gara F, Boyd EF (2004) Genome Diversity of *Pseudomonas aeruginosa* Isolates from cystic fibrosis patients and the hospital environment. J Clin Microbiol 42:5783–5792

Frost LS, Leplae R, Summers AO, Toussaint A (2005) Mobile genetic elements: the agents of open source evolution. Nat Rev Microbiol 3:722–732

Garcia-Fernandez A, Chiaretto G, Bertini A, Villa L, Fortini D, Ricci A, Carattoli A (2008) Multilocus sequence typing of IncI1 plasmids carrying extended-spectrum β-lactamases in *Escherichia coli* and *Salmonella* of human and animal origin. J Antimicrob Chemother 61:1229–1233

Gerdes K, Wagner EG (2007) RNA antitoxins. Curr Opin Microbiol 10:117–124

Gestal AM, Stokes HW, Partridge SR, Hall RM (2005) Recombination between the *dfrA12*-orfF-*aadA2* cassette array and an *aadA1* gene cassette creates a hybrid cassette, *aadA8b*. Antimicrob Agents Chemother 49:4771–4774

Gram H (1884) Über die isolierte Färbung der Schizomyceten in Schnitt- und Trockenpräparaten. Fortschritte der Medizin 2:185–189

Grindley NDF (2002) The movement of Tn*3*-like elements: transposition and cointegrate resolution. In: Craig NL, Craigie R, Gellert M, Lambowitz AM (eds) Mobile DNA II. ASM Press, Washington, D.C., pp 272–302

Hall LM, Henderson-Begg SK (2006) Hypermutable bacteria isolated from humans – a critical analysis. Microbiol 152:2505–2514

Hall RM, Collis CM, Kim M-J, Partridge SR et al (1999) Mobile gene cassettes and integrons in evolution. Ann NY Acad Sci 870:68–80

Holmes AJ, Gillings MR, Nield BS, Mabbutt BC et al (2003) The gene cassette metagenome is a basic resource for bacterial genome evolution. Environ Microbiol 5:383–394

Karaolis DKR, Somara S, Maneval DR, Jr et al (1999) A bacteriophage encoding a pathogenicity island, a type-IV pilus and a phage receptor in cholera bacteria. Nature 399:375–379

Li M, Diep BA, Villaruz AE et al (2009) Evolution of virulence in epidemic community-associated methicillin-resistant *Staphylococcus aureus*. Proc Natl Acad Sci USA 106:5883–5888

Liebert CA, Hall RM, Summers AO (1999) Transposon Tn21, flagship of the floating genome. Microbiol Mol Biol Rev 63:507–522

Maiden MC (2006) Multilocus sequence typing of bacteria. Annu Rev Microbiol 60:561–588

Martinez JL, Baquero F, Andersson DI (2007) Predicting antibiotic resistance. Nat Rev Microbiol 5:958–965

McGowan JE Jr (2006) Resistance in nonfermenting gram-negative bacteria: multidrug resistance to the maximum. Am J Med 119:S29–S36; discussion S62–S70

Minakhina S, Kholodii G, Mindlin S, Yurieva O, Nikiforov V (1999) Tn*5053* family transposons are *res* site hunters sensing plasmidal *res* sites occupied by cognate resolvases. Mol Microbiol 33:1059–1068

Mugnier PD, Poirel L, Nordmann P (2009) Functional analysis of insertion sequence ISAba1, responsible for genomic plasticity of *Acinetobacter baumannii*. J Bacteriol 191:2414–2418

Nicolas-Chanoine MH, Blanco J, Leflon-Guibout V et al (2008) Intercontinental emergence of *Escherichia coli* clone O25:H4-ST131 producing CTX-M-15. J Antimicrob Chemother 61:273–281

O'Brien TF (2002) Emergence, spread, and environmental effect of antimicrobial resistance: how use of an antimicrobial anywhere can increase resistance to any antimicrobial anywhere else. Clin Infect Dis 34:S78–S84

Partridge SR, Brown HJ, Hall RM (2002) Characterization and movement of the class 1 integron known as Tn*2521* and Tn*1405*. Antimicrob Agents Chemother 46:1288–1294

Partridge SR, Tsafnat G, Coiera E, Iredell JR (2009) Gene cassettes and cassette arrays in mobile resistance integrons. FEMS Microbiol Rev: 33(4):757–784

Perichon B, Bogaerts P, Lambert T et al (2008) Sequence of conjugative plasmid pIP1206 mediating resistance to aminoglycosides by 16S rRNA methylation and to hydrophilic fluoroquinolones by efflux. Antimicrob Agents Chemother 52:2581–2592

Phan MD, Kidgell C, Nair S et al (2009) Variation in *Salmonella enterica* serovar typhi IncHI1 plasmids during the global spread of resistant typhoid fever. Antimicrob Agents Chemother 53:716–727

Playford EG, Craig JC, Iredell JR (2007) Carbapenem-resistant *Acinetobacter baumannii* in intensive care unit patients: risk factors for acquisition, infection and their consequences. J Hosp Infect 65:204–211

Poirel L, Lartigue MF, Decousser JW, Nordmann P (2005) IS*Ecp1B*-mediated transposition of *bla*$_{\text{CTX-M}}$ in *Escherichia coli*. Antimicrob Agents Chemother 49:447–450

Reik R, Spilker T, Lipuma JJ (2005) Distribution of *Burkholderia cepacia* complex species among isolates recovered from persons with or without cystic fibrosis. J Clin Microbiol 43:2926–2928

Robinson A, Guilfoyle AP, Sureshan V et al (2008) Structural genomics of the bacterial mobile metagenome: an overview. Methods Mol Biol 426:589–595

Rossolini GM, D'Andrea MM, Mugnaioli C (2008) The spread of CTX-M-type extended-spectrum beta-lactamases. Clin Microbiol Infect 14(Suppl 1):33–41

Stroeher UH, Jedani KE, Dredge BK et al (1995) Genetic rearrangements in the rfb regions of *Vibrio cholerae* O1 and O139. Proc Natl Acad Sci USA 92:10374–10378

Thomas L, Espedido B, Watson S, Iredell J (2005) Forewarned is forearmed: antibiotic resistance gene surveillance in critical care. J Hosp Infect 60:291–293

Toleman MA, Bennett PM, Walsh TR (2006) IS*CR* elements: novel gene-capturing systems of the 21st century? Microbiol Mol Biol Rev 70:296–316

Valenzuela JK, Thomas L, Partridge SR, van der Reijden T, Dijkshoorn L, Iredell J (2007) Horizontal gene transfer in a polyclonal outbreak of carbapenem-resistant *Acinetobacter baumannii*. J Clin Microbiol 45:453–460

Waldor MK, Mekalanos JJ (1996) Lysogenic conversion by a filamentous phage encoding cholera toxin. Science 272:1910–1914

Walsh TR (2006) Combinatorial genetic evolution of multiresistance. Curr Opin Microbiol 9:476–482

Wood D, Smyth A (2006) Antibiotic strategies for eradicating *Pseudomonas aeruginosa* in people with cystic fibrosis. Cochrane Database Syst Rev 1:CD004197

Zhu J, Miller MB, Vance RE et al (2002) Quorum-sensing regulators control virulence gene expression in *Vibrio cholerae*. Proc Natl Acad Sci USA 99:129–134

Zong Z, Partridge SR, Thomas L, Iredell JR (2008) Dominance of blaCTX-M within an Australian extended-spectrum beta-lactamase gene pool. Antimicrob Agents Chemother 52:4198–4202

Chapter 13
Computational Grammars for Interrogation of Genomes

Jaron Schaeffer, Afra Held, and Guy Tsafnat

13.1 Introduction

Antibiotic resistance in bacteria is a growing health problem of major significance in both the developing and the developed world. The limited development of new antibiotics over the last three decades and the emergence of many new multi-drug resistant organisms have severely decreased our ability to treat bacterial infections (see Furuya and Lowy 2006 and Finch 2004 for an introduction and some history on antibiotic resistance). Mobile genetic structures are widely considered responsible for the emergence of bacterial strands resistant to multiple antibiotics due to their capacity to aggregate the multiple resistance genes (Levy and Marshall 2004; Furuya and Lowy 2006; Frost et al. 2005). Such genetic structures (called mobile genetic elements; MGEs) enable the multiple antibiotic resistance (R) genes to aggregate and be transmitted. MGEs typically consist of integration sites, transposase and/or integrase genes, and one or more R genes (see Box 13.1 for brief description, Frost et al. 2005 and Chap. 12 for a more comprehensive introduction). MGEs are usually arranged in semi-stable structures when in transit between DNA molecules in a cell. Once integrated into conjugative plasmids, the MGEs can horizontally transfer to other cells, even those of a different species (Lewin 2007; Bennett 2008).

Antibiotic chemotherapies aim to cure infections without unnecessarily increasing the prevalence of R genes in the population through excessive selection pressure. The ability to automatically recognize and explain the genetics underpinning mobility is a requirement for the detection of co-mobility of R genes. This co-mobility, in turn, is needed to inform the clinicians of the possible consequences of prescribing a drug if resistance to it is associated with resistance to another drug: namely, that such a prescription policy can lead to an increase in resistance to two or more antibiotics instead of one.

J. Schaeffer (✉)
Centre for Health Informatics, University of New South Wales, Sydney, NWS, Australia

V. Sintchenko (ed.), *Infectious Disease Informatics*,
DOI 10.1007/978-1-4419-1327-2_13,

Box 13.1 List of common mobile genetic elements (MGEs) associated with antibiotic resistance. For a more detailed introduction to MGEs, refer to Frost et al. (2005)

Gene cassettes: consist of a gene, often conferring resistance to one or more antibiotic agents, and a characteristic recombination site (Stokes and Hall 1989). This recombination site can interact with a recombination site present in integrons, resulting in the insertion of the corresponding gene cassette into the integron (Stokes et al. 1997). Repeated activation of this mechanism often leads to large cassette arrays that confer resistances to several antibiotics. A recent survey of GenBank found 132 unique resistance cassettes (Partridge et al. 2009).

Insertion sequences: insertion sequences are genes that code for a transposase protein. This protein can interact with inverted repeats on either side of the gene, leading to transposition of the gene. The insertion of the sequence at a new location duplicates up to nine bases on either end of the insertion sequence.

Transposons: transposons are similar to insertion sequences but usually larger. Normally, they contain at least one resistance gene, but may include an integron that in turn holds several resistance gene cassettes. Similarly to insertion sequences, between two and nine bases are copied in the transposition process at either end of the mobile unit.

Integrons: transposons in which the transposase gene is no longer functional are called integrons. Integrons can still move between molecules if an appropriate transposase protein exists in the cell (Stokes and Hall 1989). Typically, the protein would have been transcribed from a transposon present in the same cell but not necessarily on the same molecule as the integron. Integrons still maintain their gene cassette capture mechanism.

Composite transposons: Composite transposons consist of two similar insertion sequences that occur relatively close to each other. An error in the insertion sequence transposition mechanism can transpose both insertion sequences and the material between them. The DNA between the insertion sequences may potentially contain gene cassettes. The difference between a composite transposon and two independent insertion sequences is apparent from the direct repeats on either end.

13.2 Automatic Annotation of Bacterial DNA

MGE recognition is a DNA annotation problem and as such can potentially make use of established methods in this field. *DNA annotation* refers to marking up regions within a DNA sequence with a name and often an associated function. The function could be a protein product, an interaction with a protein, the phenotype associated with the DNA region, etc. Given the time and effort involved in manually annotating large genomes and the advances in sequencing throughput over the past two decades, it is not surprising that early research attention focused on the automation of these standard tasks.

Gene prediction refers to identifying the protein-coding and non-protein-coding genes in DNA. Most gene prediction tools rely on the identification of an open reading frame (Wheeler et al. 2003) followed by a machine-learning step to reduce the number of false positives (Delcher et al. 2007; Larsen and Krogh 2003). State-of-the-art systems have a prediction accuracy of 95–97% for coding regions (Overbeek et al. 2007). Intrinsic gene predictors such as GLIMMER (Delcher et al. 2007) and GeneMark (Besemer and Borodovsky 2005) rely on the statistical properties of the underlying sequence, typically modelled using hidden Markov models (HMM) (Rabiner and Juang 1986). Extrinsic approaches such as REGANOR (Linke et al. 2006) or CRITICA (Badger and Olsen 1999) employ homology using BLAST (Altschul et al. 1990) searches to identify the known genes (Bohnebeck et al. 2008; Partridge et al. 2009).

After gene-prediction, the next step in annotating a genome usually involves the identification of *gene products* or *gene function*. In particular, gene function prediction identifies the protein that is synthesized as a result of transcribing a gene. Automatic protein annotation is frequently done using BLAST and FASTA (Pearson and Lipman 1988), which rely on sequence similarity to predict homology with known (and ideally trusted) annotations from public databases such as GenBank (Benson et al. 2007), The Kyoto Encyclopaedia of Genes and Genomes (Kanehisa and Goto 2000) and the Gene Ontology (The Gene Ontology Consortium 2000; Ashburner et al. 2000). Results are usually refined and filtered either manually with the aid of annotation tools such as Artemis (Rutherford et al. 2000) or using some automatic tools that validate the annotations from multiple sources such as MAGPIE (Gaasterland and Sensen 1996) and GenDB (Meyer et al. 2003).

Further analysis can involve the establishment of protein families, i.e. proteins that perform the same or similar function or are homologous (phylogeny), or the comparison to established protein databases such as SwissProt (Wu et al. 2006), a manually curated protein database with a focus on high annotation quality and low redundancy, and TrEMBL (Bairoch and Apweiler 1999), which provides automatically derived annotations based on SwissProt.

13.3 Computational Grammars

A computational grammar (or *formal grammar*) consists of a set of symbols (sometimes called *tokens*; e.g. words in a natural language) and rules (sometimes called *productions*) that describe how symbols can be assembled into legal sentences. Grammar rules have two parts: the left hand side (LHS) of the rule represents some complex structure (such as a phrase or sentence). The right hand side (RHS) of the rule captures an ordered set of tokens that describe how the LHS can be assembled.

[1] **Sentence** → Noun VerbPhrase
[2] **VerbPhrase** → Verb NounPhrase
[3] **NounPhrase** → Noun
[4] **NounPhrase** → Determiner Noun
[5] **Noun** → Michael
[6] **Noun** → door
[7] **Verb** → opens
[8] **Determiner** → the

Fig. 13.1 Example grammars to describe a small subset of the English language. The first rule specifies that a sentence is only complete if it consists of a noun followed by a verb phrase; the second rule states that a noun phrase consists of a determiner followed by a noun or simply a noun. A verb phrase is defined as a verb followed by a noun phrase

The simple grammar shown in Fig. 13.1 consists of eight rules. The first rule's LHS is the single token 'sentence'. The RHS consists of two tokens ("Noun" and "VerbPhrase"). The rule defines the pattern of "Noun immediately followed by VerbPhrase" to be a Sentence. The second rules defines the pattern "Verb followed by a NounPhrase" to be a VerbPhrase. The third and fourth rules define the NounPhrase to be either a Noun or "Determiner immediately followed by Noun". The rest of the rules identify the part of speech (semantic type) of words (symbols).

Grammars can provide a readable and compact representation of even arbitrarily complex languages. For a comprehensive introduction, see Hopcroft et al. (2006).

The act of using such grammar rules to interpret a text is called *parsing* and a computer program that carries out the task is called a *parser* (or sometimes an interpreter). Various algorithms allow a parser to recognize patterns of RHS tokens in a text and label them with the LHS tokens. Thus, the parser "decodes" the semantics, or meaning, of a text, as prescribed in the grammar.

The output of a parser is a *parse tree* – a data structure in which every final or leaf-node corresponds to a symbol (word) from the text, and every internal node (non-leaf) corresponds to a semantic construct found in the LHS of a grammar rule. The parse tree represents the sentence structure and thus captures the relationships between different parts of speech, sentences, and words.

Figure 13.2 shows an example parse tree that results from applying the grammar in Fig. 13.1 to the sentence given in the previous example. The connections between the symbols indicate their relationship. The parse tree shows, for example, that the word "the" is a determiner for the word "door" (and not for "opens") and further that the pattern "the door" is called NounPhrase.

Computational grammars provide a basis for many aspects of computer science. Automatically understanding and translating programming and natural languages (Jurafsky and Martin 2008) as well as many aspects of theoretical computer science and computability theory are all strongly tied to formal grammar theory (van Leeuwen and Leeuwen 1994). The automatic extraction of grammars from example texts is a major research strand in machine learning and is particularly applicable to molecular biology, where the rules that govern self-assembly are not well understood.

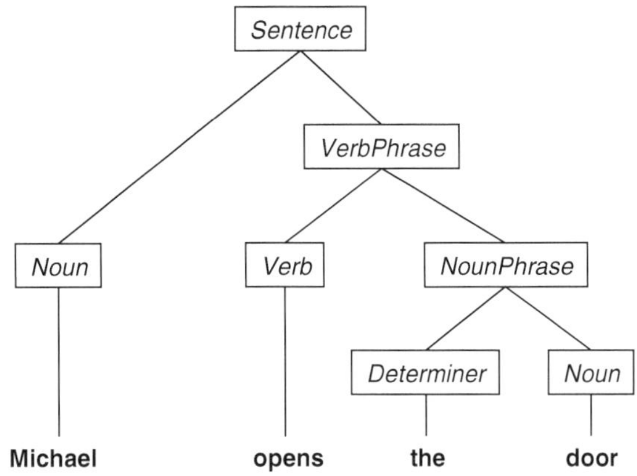

Fig. 13.2 Parse tree for the input sentence "Michael opens the door". The parse tree shows which structures have been found and how they form a sentence according to the rules from Fig. 13.1

In particular, HMM have been used to identify genes in prokaryotes (Delcher et al. 2007; Gheorghe and Mitrana 2004) and intron/exon junctions in eukaryotes (Burge and Karlin 1997).

The use of grammars in molecular biology is not limited to gene identification. It spans a diverse range of domains, and is especially useful where BLAST searches are not effective, such as when finding promoters (Leung et al. 2001) or identifying the mRNA secondary structure (Rivas and Eddy 2000). More examples can be found in a review by Searls (2002). In almost all applications, grammars of macromolecules represent arrangements of nucleic acids that correspond to patterns of interest. However, it is quite possible for grammars to operate on tokens that consist of entire genes and the other entities recognized in the DNA sequence, effectively modeling larger-than-gene structures such as MGEs.

13.4 Annotating Biological Structure Using Grammar Models

Theoretical knowledge about genetic structures in DNA and their relationship to each other can be expressed as constructs in a grammar, and a parser can then identify these structures in DNA sequences. For example, specific structures like genes can represent basic words, and the grammar can assemble such words into higher-order structures. The resulting parse tree shows the recognized annotations as a hierarchy of nested structures.

Figure 13.3 gives an example grammar that identifies the cassette arrays and their parts. Figure 13.4 shows the parse tree from this grammar as it is applied to a particular DNA sequence. A similarity based tokenizer identified the symbols 5'-CS, 3'-CS and aadA3. The parser identified those symbols as the start, end and middle of the array respectively, and that together they form an array.

[1] **CassetteArray** → CassArrStart CassArrMid CassArrEnd
[2] **CassArrStart** → 5'-CS
[3] **CassArrEnd** → 3'-CS
[4] **CassArrMid** → Cassette
[5] **CassArrMid** → CassArrmid Cassette
[6] **Cassette** → aadA3

Fig. 13.3 Example grammar that defines the parts of a cassette array. The elements 5'-CS and 3'-CS define the start and end of the array respectively and a cassette its middle. Only one cassette is given in this example for brevity, but any number of cassettes can be accumulated in a CassArrMid structure (rule 5) and many more rules of the form of rule 6 can be added

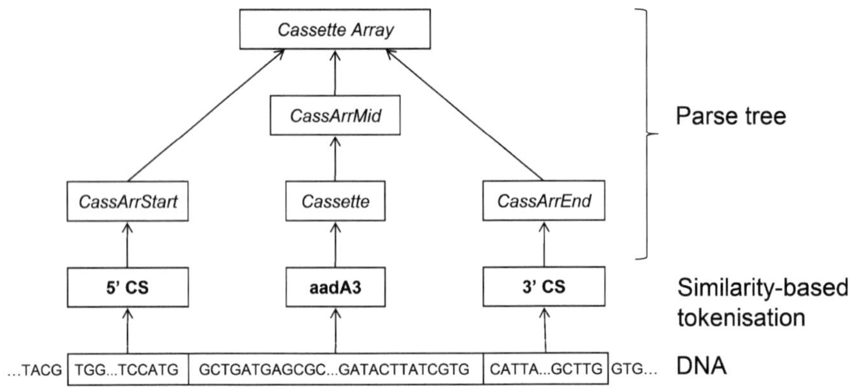

Fig. 13.4 A grammar for genetic structure annotation. The input DNA (*bottom*) is tokenized, and the tokens are interpreted by a parser following grammar rules similar to the ones in Fig. 13.1. The resulting parse tree offers a clear annotation of the structures found in the input DNA and the grammar rules that lead to their recognition. In this example, a 5′-CS token, an *aadA3* cassette and a 3′-CS have been interpreted by the parser as a complete cassette array

When designing a grammar for biological structure annotation, a number of design choices need to be considered, some of which will be elaborated on in this section.

13.4.1 DNA Tokenization

If a sequence of DNA or some other macromolecule is to be parsed, the basic unit of input needs to be defined. In the simplest example, a single nucleotide or amino acid can be used, and a trivial tokenizer would simply read the input sequence one letter at a time. This is an obvious and proven choice for grammars of relatively short DNA structures such as gene promoters or protein domains used by most approaches reviewed by Gheorghe and Mitrana (2004) and Searls (2002). However, due to the small symbol set, it is difficult to account for long structures, as the grammar rules become overly complex.

In analogy with the natural language, a grammar for recognizing larger-than-gene structures that uses the four DNA bases as its basic units of input is similar to a grammar for recognizing sentences from letters instead of words. A better choice may be to group functionally related nucleotides to form higher-level patterns for parsing steps. These patterns (often referred to as *features*) can, for example, cover the MGEs or other larger-than-gene structures typically consisting of a gene and associated protein interaction sites, various conserved sequences and direct and inverted repeats.

Performing such annotations manually is tedious and requires an understanding of the underlying biological process. Automatic motif finding methods (Sandve and Drablos 2006; Pavesi et al. 2004) such as classical consensus patterns (Brazma et al. 1998; Smith et al. 1990), pattern-inference methods from positive examples (Aiyar 2000; Schuler et al. 1991) or nucleotide-level grammar inference approaches (Muggleton et al. 2001) can assist the DNA tokenization by using the automatically inferred motifs as features.

13.4.2 Grammar Class and Parsing Algorithm

The type of grammar to be used depends on the complexity of the structures to be analyzed, is closely coupled to the choice of the parsing algorithm and may constrain the form of the grammar rules.

Context-free vs. *context-sensitive grammars:* Context free grammars always arrive at the same interpretation of a pattern, independent of the broader context within which it occurs. In contrast, context-sensitive grammars are able to distinguish between the instances of a symbol based on its context within the other symbols (Searls 2002): For example, the following rules could help analyzing whether the word "books" occurs as an object or a predicate within an already partially annotated sentence:

Subject **books** *Object* → *Subject* **Predicate** *Object*
Subject Predicate **books** → *Subject Predicate* **Object**

Here, "books" is recognized as a predicate only if surrounded by a subject and object and as an object only if preceded by a subject and a predicate. Parsing of these grammars, however, may require an exponential number of steps and is therefore only practical for short sequences, however various modifications exist to context free grammars that confer them with some context-sensitivity aspects without the big computational overheads (Grune and Jacobs 2008). Another advantage of context-sensitive grammars is that biological structures are usually defined in terms that allow them to be directly expressed as context sensitive grammars without introducing any artificial semantic constructs.

Deterministic vs. non-deterministic grammars: Deterministic grammars will always generate the same unambiguous parse tree for the same input, an appropriate property for many applications. Non-deterministic grammars, on the other hand, produce multiple parse trees; stochastic grammars, a subclass of the latter, have probabilities associated with each grammar rule (Baldi and Brunak 2001). Stochastic grammars are often appropriate for modeling non-determinism in biology and have mostly been applied as HMMs – a constrained form of stochastic grammars (Rabiner and Juang 1986; Ewens and Grant 2005; Koski 2001; Baldi and Brunak 2001). In particular, when multiple biological reading of DNA are possible, for example when multiple ORF shifts encode for different proteins which are both expressed, it would be possible to produce the alternative parses using a non-deterministic parser to show the various genes.

Bottom-up vs. top-down parsing: Bottom-up parsing refers to parsing techniques that build the parse tree from the leaves towards the root by successively combining input symbols (Grune and Jacobs 2008; Aho et al. 1986). Top-down parsers, in contrast, start with a root and try to find parse trees that match the input symbols (Dale et al. 2000). Top-down parsers are suitable for the recognition of patterns when the root (top node) of the parse tree is known, for example when annotating whole molecules. Bottom-up parsers are suitable when the structures to be annotated might be truncated, for example when annotating segments of DNA that include only a part of a structure.

13.4.3 Grammar Derivation

Grammars can be derived manually, automatically, or by a combination of both. The way a grammar is derived depends on the training data and the theories available in the biological domain that is to be modeled. In *automatic grammar derivation*, a grammar inference algorithm identifies the rules of a grammar using machine learning methods (for an introduction to machine learning, refer to Mitchell (1997). This is typically done by examining a set of annotated sequences and extracting frequent sequences of symbols, hypothesizing that frequent occurrences imply an evolutionary selection of the sequence and hence significance of the pattern.

Machine learning algorithms not only require a training set (ideally partitioned into positive and negative examples of the patterns to be inferred) for accurate results, they also require that the training set must be the representative of the sequences that will later be annotated by the parser. Consequently, the existence and size of such a corpus are crucial for its applicability and impact. Many automatic grammar inference methods have been developed for recognizing specific patterns in biological sequences from a training set (Rissanen et al. 2008; Baldi and Brunak 2001).

Manual derivation, on the other hand, requires that a well-defined theory exists in the biological domain. This theory can then be manually represented as a grammar. Grammar rules that encode how the symbols are mapped into semantic constructs are defined to reflect the biological concepts and processes. This typically involves a significant amount of computer programming; however, for context-free grammars, parser generators such as YACC (Johnson 1978) can facilitate the software development. A more complete description of this process is given in Grune and Jacobs (2008). A good example for manual derivation is the grammar presented in Searls (1988).

13.4.4 Validation of Grammatical Models

Once developed, grammar needs to be tested to ensure that they are accurate models of the biological processes that they are intended to represent. The ease with which a grammar can be evaluated depends on the availability of annotated sequences. If available, these can be used as a test set to measure the accuracy of annotations produced by the parser. Standard evaluation measures require that each symbol annotated be classified as either correctly (true positive) or incorrectly annotated (false positive), and that gaps in the annotation be classified as correctly left unannotated (true negative) or missed (false negative). Using these values, the quality of the grammar is reported as sensitivity, specificity and as F-score (van Rijsbergen 1979; Mitchell 1997).

If no test set exists for comparison, a panel of independent human experts can be asked to identify the semantic construct that each symbol belongs to. Agreement of the automatic annotations with those produced by the experts can then be used to show whether the grammar can annotate the sequences at least as well as the experts. Inter-annotator agreement within the panel is measured using a nominal agreement measure such as Fleiss' κ (Fleiss 1971). This measure is compared with the measure of the expert annotators and the grammar. Significantly lower κ value in the second measure indicates that the grammar created worse annotations than the experts.

13.5 Case Study: A Grammar Model for Cassette Array Modeling and Interrogation

In this section, we present an implementation of an automatic annotator for R gene cassette arrays (Box 13.1). The annotator uses BLAST to annotate the features of interest such as gene cassettes and conserved sequences that mark the start and end of several types of cassette arrays (Sect. 13.3). A context-sensitive grammar was derived manually (Sect. 13.3) to recognize the cassette arrays based on these features. The grammar model was used to conduct a major survey of antibiotics resistance cassettes (Partridge et al. 2009) and to discover two new gene cassettes (Tsafnat et al. 2009).

13.5.1 DNA Tokenization

Tokenization of the raw bacterial DNA was carried out with the help of a feature database (FDB). This database had been manually curated and comprised of 276 features including 214 gene cassettes, conserved sequences marking the start and end of the arrays and non-cassette sequences that are found in cassette arrays (but do not seem to interfere with the expression of R genes in the array). Instances of these features were found using BLAST (Altschul et al. 1990) and marked up in the bacterial DNA sequences. A default identity match of 97% was used and manually adjusted for the disambiguation of the features. Annotations gaps in the DNA were tagged using a special token denoted as λ. In a second run, these λ tokens were matched against the FDB using BLAST to find instances of truncated features. Full and partial feature instances as well as the remaining λ tokens were stored in a database and made available for processing by the parser.

13.5.2 Cassette Array Grammar

Gene cassette arrays are made up of an initial conserved sequence, a middle part (a sequence of gene cassettes) and a terminating conserved sequence (Hall and Collins 1998). Tsafnat with co-workers (2009) derived a 21-rule grammar that accurately annotated cassette arrays based on the input tokens described in Sect. 13.3. In cassette arrays, it is relatively common to encounter short DNA sequences (usually less than 300 base pairs) that are not gene cassettes, do not encode genes but do allow gene cassettes downstream from them to be expressed. Recognition of a sequence as such a *non-cassette insertion* depends on the context in which the sequence occurs and thus requires context-sensitive grammar rules.

The grammar was derived manually in collaboration with an expert and contained seven context-sensitive rules that facilitated the recognition of the array structure as compared to using context-free rules only. Rule application was performed by a deterministic parser in a bottom-up manner, gradually summarizing the sequence features to form the cassette array structure.

The grammar's ability to identify arrays was compared with three experts and achieved a very high agreement score of 94.8%, where agreement was measured using Fleiss' κ (Fleiss 1971). Two putative new gene cassettes (*qacK* and *dfrB7*) were discovered by investigating the length and context of annotation gaps (Tsafnat et al. 2009).

13.6 Interrogation of Annotated Structures

The annotation of thousands of sequences is of limited use without tools to retrieve, interrogate and visualize them. We borrow ideas from the discipline of information retrieval to present systems specific to genetic sequences and MGEs.

Most common information retrieval systems are retrospective search engines that work in two stages: an *indexing* stage that prepares the source data for quick search (Sect. 13.4.1), and a *retrieval* stage (Sect. 13.4.2) in which a search query is evaluated using the index, and the relevant components that satisfy the query are returned (Manning et al. 2008; van Rijsbergen 1979). Search indices are used in virtually any database and allow for efficient on-line query operations on large and complex amounts of data (Elmasri and Navathe 2007).

13.6.1 Indexing Hierarchical Genetic Structures

Given a set of bacterial DNA sequence data from chromosomes and plasmids, a grammar model can be used to obtain parse trees. One way of organizing these data is to transform the parse trees into an XML representation and persistently store them in a database such as Sedna (Fomichev et al. 2006). A query represented in a specific language, e.g. for searching for antibiotics resistance genes in MGEs, is translated to XQuery, a general query language for XML databases (Chamberlin et al. 2001). This XQuery program is then run against the XML database for result retrieval.

13.6.2 A Query Language for Structure Annotations

The tree representation of genetic structure annotations allows for a number of queries not possible using "flat" annotations only. In the context of MGE, it may, for example, be interesting to know if a certain resistance gene co-occurs with another in a unit of mobility, such as a gene cassette array or an integron. For this purpose, this query language provides, among the other things, an **in** operator of the form

X **in** *Y*,

restricting the results for a given query to structures where the predicate *X* (e.g. a gene) occurs inside the structure *Y* (e.g. a transposon). In addition, the query language interprets logical operators such as **and** and **or** as well as parentheses to distinguish between (*A* **and** *B*) **or** *C* and *A* **and** (*B* **or** *C*).

To execute a query, it is translated into the XQuery language and executed against the XML database of saved structure annotations. For example, the query

(*qacE* **and** *aadA3*) **in** *CassetteArray*

will return accessions containing cassette arrays in which both *qacE* and *aadA3* gene cassettes occur. This query is translated into XQuery

```
for $acc in doc("annotations.xml")//accession
where
$acc//feature[name eq "CassArray"]//feature[name eq "qacE"]
```

```
and
$acc//feature[name eq "CassArray"]//feature[name eq
"aadA3"]
order by $acc
```

This sample XQuery program is applied to an XML database (Fomichev et al. 2006) and returns all accessions from the XML annotation database annotations. xml that feature cassette arrays in which the resistance genes *qacE* and *aadA3* co-occur. Similarly, we can add a wildcard operator * that matches any structure. For example, the query

(*qacE* **and** *aadA3*) **in** *

will be satisfied by sequences that have any structure that contains both a *qacE* element and an *aadA3* element.

The **resists** operator allows querying for sequences based on some predicted phenotypes. This wildcard operator matches all names of genes that are known to confer resistance to an antibiotic. For example,

resists(*carbapenems*)

will match any element that confers resistance to carbapenems. Finally, nested searches allow for more powerful queries searching for potential co-mobility of arbitrary structures:

((qacE **in** *CassetteArray*) **and** **resists**(*carbapenems*)) **in** *Integron*

will match any sequence containing an integron that includes a cassette array carrying both the gene *qacE* and any gene conferring resistance to carbapenems.

13.6.3 Structure Visualization

A number of programs exist for DNA annotation visualization. Artemis (Rutherford et al. 2000) is a popular visualization tool that provides a number of features mainly centered on prokaryotic gene transcription and overlays of analysis results originating from the external tools. GO Bar (Lee et al. 2005) and DynGO (Liu et al. 2005) both visualize genes on the basis of the Gene Ontology (The Gene Ontology Consortium 2000) and Osprey (Breitkreutz et al. 2003) visualizes gene regulation networks. Other methods focus on the generic visualization of hierarchical information. For example, tree maps (Johnson and Shneiderman 1991) are better suited to display the large hierarchical structures than are the standard tree graphs, due to a better space utilization.

However, none of these tools meet the requirement of visualizing larger-than-gene genetic structures, such as the ability to incrementally explore complex hierarchical structures or direct links to relevant public databases. Generic parse tree visualization programs, on the other hand, do not offer explanations of the underlying

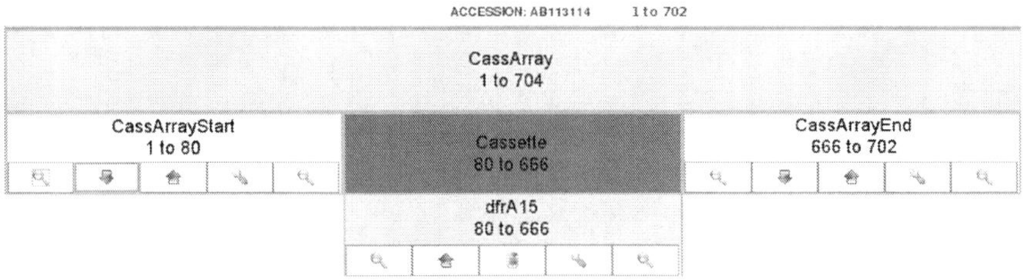

Fig. 13.5 Graphical tool to visualize structure annotations. The screenshot shows GenBank accession AB113114, where a cassette array consisting of three tokens has been found: a *CassArrayStart* token (1...80), a *Cassette* token (80...666) and a *CassArrayEnd* token (666...702). The arrow buttons allow expanding annotation details for each token. For example, pressing the down arrow under the *Cassette* token trigged expansion of the *dfrA15* token. This mechanism enables showing and hiding annotation parts as required and allows the visualisation of complex parse trees in a clearly arranged way. Other buttons allow viewing the token associated DNA as well as executing local sequence alignments against it

biology. The second and third authors implemented a DNA grammar visualization tool that addresses some of these shortcomings (Fig. 13.5), intended as a proof-of-concept implementation of what better visualization of complex hierarchical annotations in the biological domain could look like.

Navigation buttons allow structures (e.g. a CassArray token) to be expanded into their constituent parts, which in turn can be further explored through the same mechanism. Leaf nodes are accompanied by information on the feature that they match and links to knowledge repositories such as GenBank and GO. All nodes allow the DNA sequence part covered by them to be displayed. The tool (Held and Tsafnat 2007) is intended to complement the existing DNA visualization tools, thus only focusing on improving on the aspects mentioned above.

13.7 Conclusion

Generic annotation methods are insufficient for the recognition of MGEs, as they do not account for the variation in the genes accommodated. We suggested that computational grammars can be used for recognizing MGEs; by parsing an input DNA sequence using such a grammar, instances of structures of interest can be made explicit in the input. We presented some tools that can be used for the annotation of a batch of genetic sequences as well as for search through the annotations, and the visualization of individual ones.

We discussed in some detail the available strategies for tokenization – the recognition of genetic "words" - and those for putting them together into MGE "sentences". One way to tokenize DNA is to use a manually curated feature database containing annotations considered important for the specific genetic domain investigated. This is analogous to use a dictionary to find word patterns in a text without punctuation or words. However, this is only applicable if prior knowledge about the domain (i.e. the dictionary) is available. Context-sensitive grammars allow biological structures to be represented in similar terms to the way they are described in the literature. Once higher-order structures have been annotated in a sequence, these may be queried using a domain-specific query language, and individual structures can be visualized. This allows for the implementation of search operators specific to a certain research or clinical question. One example presented in the case-study was the implementation of search operators useful for the collection of evidence for the co-mobility of the resistance genes in MGEs.

Acknowledgements This research was supported by a Capacity Building Grant from New South Wales Health.

References

Aho AV, Sethi R, Ullman JD (1986) Compilers: principles, techniques, and tools. Addison-Wesley, Reading MA

Aiyar A (2000) The use of CLUSTAL W and CLUSTAL X for multiple sequence alignment. Methods Mol Biol 132:221–241

Altschul SF, Gish W, Miller W, Myers EW, Lipman DJ (1990) Basic local alignment search tool. J Mol Biol 215:403–410

Ashburner M, Ball C, Blake J, Botstein D et al (2000) Gene ontology: tool for the unification of biology. Nat Genet 25(1):25–29

Badger JH, Olsen GJ (1999) CRITICA: coding region identification tool invoking comparative analysis. Mol Biol Evol 16:512–524

Bairoch A, Apweiler R (1999) The SWISS-PROT protein sequence data bank and its supplement TrEMBL in 1999. Nucleic Acids Res 27:49–54

Baldi P, Brunak S (2001) Bioinformatics: the machine learning approach. MIT Press, Boston MA

Bennett P (2008) Plasmid encoded antibiotic resistance: acquisition and transfer of antibiotic resistance genes in bacteria. Br J Pharmacol 153:347–357

Benson DA, Karsch-Mizrachi I, Lipman DJ, Ostell J, Wheeler DL (2007) Genbank. Nucleic Acids Res 35:D21–D25

Besemer J, Borodovsky M (2005) GeneMark: web software for gene finding in prokaryotes, eukaryotes and viruses. Nucleic Acids Res 33:451–545

Bohnebeck U, Lombardot T, Kottmann R, Glöckner FO (2008) MetaMine – a tool to detect and analyse gene patterns in their environmental context. BMC Bioinform 9:459

Brazma A, Jonassen I, Eidhammer I, Gilbert D (1998) Approaches to the automatic discovery of patterns in biosequences. J Comp Biol 5(2):277–304

Breitkreutz B, Stark C, Tyers M (2003) Osprey: a network visualization system. Gen Biol 4:R22

Burge C, Karlin S (1997) Prediction of complete gene structures in human genomic DNA. J Mol Biol 268:78–94

Chamberlin D, Clark J, Florescu D, Robie J, Simeon J, Stefanescu M (2001) XQuery 1.0: An XML query language. W3C Working Draft, vol 7

Dale R, Moisl H, Somers H (2000) Handbook of natural language processing. CRC Press, London

Delcher AL, Bratke KA, Powers EC, Salzberg SL (2007) Identifying bacterial genes and endosymbiont DNA with Glimmer. Bioinform 23(6):673–673

Elmasri R, Navathe SB (2007) Fundamentals of database systems, 5th ed. Addison-Wesley, Reading, MA

Ewens WJ, Grant GR (2005) Statistical methods in bioinformatics: An introduction. Springer, Heidelberg

Finch RG (2004) Antibiotic resistance: a view from the prescriber. Nat Rev Microbiol 2:989–994

Fleiss JL (1971) Measuring nominal scale agreement among many raters. Psychol Bull 76:378–382

Fomichev A, Grinev M, Kuznetsov S (2006) Sedna: A native XML DBMS. LNCS 3831:272–281

Frost LS, Leplae R, Summers AO, Toussaint A (2005) Mobile genetic elements: the agents of open source evolution. Nat Rev Microbiol 3:722–732

Furuya E, Lowy F (2006) Antimicrobial-resistant bacteria in the community setting. Nat Rev Microbiol 4:36–45

Gaasterland T, Sensen CW (1996) MAGPIE: automated genome interpretation. Trends Genet 12(2):76–78

Gheorghe M, Mitrana V (2004) A formal language-based approach in biology. Comp Funct Genom 5:91–94

Grune D, Jacobs CJH (2008) Parsing techniques: a practical guide. Prentice Hall, Englewood Cliffs, NJ

Hall RM, Collis CM (1998) Antibiotic resistance in gram-negative bacteria: the role of gene cassettes and integrons. Drug Resist Updat 1:109–119

Held A, Tsafnat G (2007) ArrayVisual: an on-line visualization tool for DNA sequences annotated using grammars, http://www2.chi.unsw.edu.au:8080/VisualArray/start.html

Hopcroft JE, Motwani R, Ullman JD (2006) Introduction to automata theory, languages, and computation. Addison-Wesley, Reading, MA

Johnson B, Shneiderman B (1991) Tree-maps: a space-filling approach to the visualization of hierarchical information structures. Proc IEEE Conf Vis 1991:284–291

Johnson SC (1978) YACC-yet another compiler-compiler. Bell Laboratories

Jurafsky D, Martin JH (2008) Speech and language processing. Prentice Hall, New York

Kanehisa M, Goto S (2000) KEGG: Kyoto encyclopedia of genes and genomes. Nucleic Acids Res 28(1):27–30

Koski T (2001) Hidden Markov models for bioinformatics. Springer, Heidelberg

Larsen TS, Krogh A (2003) EasyGene – a prokaryotic gene finder that ranks ORFs by statistical significance. BMC Bioinform 4:21

Lee J, Katari G, Sachidanandam R (2005) GObar: a gene ontology based analysis and visualization tool for gene sets. BMC Bioinformat 6:189

Leung S, Mellish C, Robertson D (2001) Basic gene grammars and DNA-ChartParser for language processing of *Escherichia coli* promoter DNA sequences. Bioinform 17(3):226–236

Levy SB, Marshall B (2004) Antibacterial resistance worldwide: causes, challenges and responses. Nat Med 10:122–129

Lewin B (2007) Genes IX. Jones and Bartlett, Sudbury, MA

Linke B, McHardy A, Neuweger H, Krause L, Meyer F (2006) REGANOR: a gene prediction server for prokaryotic genomes and a database of high quality gene predictions for prokaryotes. Appl Bioinform 5:193–198

Liu H, Hu Z, Hu CH (2005) DynGO: a tool for visualizing and mining of gene ontology and its associations. BMC Bioinform 6:201

Manning CD, Raghavan P, Schuetze H (2008) Introduction to information retrieval. Cambridge University Press Cambridge, MA

Meyer F, Goesmann A, McHardy AC, Bartels D, Bekel T, Clausen J, Kalinowski J, Linke B, Rupp O, Giegerich R, Pühler A (2003) GenDB – an open source genome annotation system for prokaryote genomes. Nucleic Acids Res 31(8):2187–2195

Mitchell T (1997) Machine learning. McGraw-Hill, Columbus OH

Muggleton SH, Bryant CH, Srinivasan A, Whittaker A et al (2001) Are grammatical representations useful for learning from biological sequence data? A case study. J Comp Biol 8(5):493–521

Overbeek R, Bartels D, Vonstein V, Meyer F (2007) Annotation of bacterial and archaeal genomes: Improving accuracy and consistency. Chem Rev 107:3431–3447

Partridge SR, Tsafnat G, Coiera E, Iredell J (2009) Gene cassettes and cassette arrays in mobile resistance integrons. FEMS Microbiol Rev 33(4):757–784

Pavesi G, Mauri G, Pesole G (2004) In silico representation and discovery of transcription factor binding sites. Brief Bioinform 5(3):217–36

Pearson WR, Lipman DJ (1988) Improved tools for biological sequence comparison. Proc Natl Acad Sci U S A 85:2444–2448

Rabiner L, Juang B (1986) An introduction to hidden Markov models. Proc IEEE 77(2):257–286

Rissanen J, Grünwald P, Heikkonen J, Myllymäki P, Roos T, Rousu J (ed) (2008) Information theoretic methods for bioinformatics. Hindawi Publishing Corporation, Cairo, Egypt

Rivas E, Eddy SR (2000) The language of RNA: a formal grammar that includes pseudoknots. Bioinform 16:334–340

Rutherford K, Parkhill J, Crook J, Horsnell T et al (2000) Artemis: sequence visualization and annotation. Bioinform 16:944–945

Sandve GK, Drablos F (2006) A survey of motif discovery methods in an integrated framework. Biol Direct 1:11

Schuler GD, Alschult SF, Lipman DJ (1991) A workbench for multiple alignment construction and analysis. Struct Funct Genet 9:180–190

Searls DB (1988) Representing genetic information with formal grammars. Proc 7th Natl Conf Artif Intell, pp 386–391

Searls DB (2002) The language of genes. Nature 420:211–217

Smith HO, Annau TM, Chadrasegaran S (1990) Finding sequence motifs in groups of functionally related proteins. Proc Natl Acad Sci U S A 87:826–830

Stokes H, O'Gorman D, Recchia G, Parsekhian M, Hall R (1997) Structure and function of 59-base element recombination sites associated with mobile gene cassettes. Mol Microbiol 26:731–745

Stokes HW, Hall RM (1989) A novel family of potentially mobile DNA elements encoding site-specific gene-integration functions: integrons. Mol Microbiology 3:1669–1683

The Gene Ontology Consortium (2000) Gene Ontology: tool for the unification of biology. Nat Genet 25:25–29

Tsafnat G, Coiera E, Partridge SR, Schaeffer J, Iredell J (2009) Context-driven discovery of gene cassettes in mobile integrons using a computational grammar. BMC Bioinform 10:281

Van Leeuwen J, Leeuwen J (1994) Handbook of theoretical computer science. MIT Press, Cambridge, MA

Van Rijsbergen CJ (1979) Information retrieval. Butterworth-Heinemann, Newton, MA

Wheeler DL, Church DM, Federhen S, Lash AE et al (2003) Database resources of the National Center for Biotechnology. Nucleic Acids Res 31(1):28–33

Wu CH, Apweiler R, Bairoch A, Natale DA et al (2006) The Universal Protein Resource (UniProt): an expanding universe of protein information. Nucleic Acids Res 34:187–191

Chapter 14
In silico Discovery of Chemotherapeutic Agents

Lyn-Marie Birkholtz, Peter Burger, Samia Aci, Hélène Valadié, Ana Lucia da Costa, Loraine Brillet, Tjaart de Beer, Fourie Joubert, Gordon Wells, Vincent Breton, Sylvaine Roy, Abraham Louw, and Eric Maréchal

14.1 Introduction

In silico approaches to accelerate the identification, selection, and validation of novel targets ("target discovery") and of corresponding ligands ("drug discovery") follow basic principles that are briefly listed below:

1. The first principle is the *in silico* filtering of electronically stored information based on appropriate representations of knowledge. For example, a protein of an infectious agent might be a target if it is involved in a vital process. An *in silico* filter might be applied to identify such vital proteins. A drug candidate should comply with Lipinski's rule of five (Lipinski et al. 1997), which concerns its molecular mass, constituting atoms and solvent solubility. Incorporation of these attributes in an *in silico* filter allows the selection of drug-like compounds. The combination of filters is made possible using Boolean logical principles.
2. The second principle is the execution of *in silico* experiments or simulations. For example, in the target discovery process, the real effect of the impairment of a target might be simulated based on a modeled biological system in which a graphical representation of the targeted biological process is combined with a dynamic model that allows the prediction of the response of the system when some of its components are functionally altered. In the drug discovery process, simulation is even more critical since the filtering of information is clearly not sufficient. The modeled docking of small compound structures inside a cavity at the surface of a protein target is at the heart of most *in silico* strategies.

E. Maréchal (✉)
Institut de Recherches en Technologies et Sciences park le Vivant, CEA Grenoble, France

V. Sintchenko (ed.), *Infectious Disease Informatics*,
DOI 10.1007/978-1-4419-1327-2_14,

These general basic principles are clearly constrained by specific features, which are dictated by the biological uniqueness of each pathogen. It is not the purpose of this chapter to give a complete description of these specificities. Rather this chapter introduces one of the most complicated cases of the *in silico* drug discovery process – the search for novel compounds that might prevent or cure malaria infections.

Malaria is a life-threatening disease affecting approximately 500 million people worldwide (World Malaria Report 2005). Five species of malaria parasites can infect humans via mosquito transmission: *Plasmodium falciparum* (the species that causes the greatest incidence of illness and death), *P. vivax*, *P. ovale*, *P. malariae*, and *P. knowlesii*. These parasites belong to the Apicomplexa phylum, which contains thousands of other parasitic protists of medical and veterinary importance (Adl et al. 2005). Malaria was eradicated from temperate regions following concerted preventative sanitary actions that included insecticide-spraying campaigns and systematic treatments with the most cost-effective drugs, i.e., quinine and chloroquine (Desowitz 1992; Utzinger et al. 2002; Baldwin 2003). However, the prophylactic programs based on insecticide and drug treatments failed to control malaria in subtropical areas (Nchinda 1998). Resistance to chloroquine spread rapidly (Ridley 1998, 2002). Current efforts focus therefore on chemoprophylaxis using artemisinin, an antiplasmodial molecule from *Artemisia annua*, and its derivatives, which can be manufactured efficiently and cheaply. However, plans for the extensive use of artemisinin might be compromised by the emergence of the parasitic resistance that it will almost certainly trigger (Jambou et al. 2005; Towie 2006; Afonso et al. 2006). Given the small number of available drugs and the resistance they have already induced, the discovery of new targets and of new drugs remains a key priority.

A major landmark in the history of malaria was the launch of a collaborative genomic sequencing program in 1996 (for a review see Birkholtz et al. 2006), leading to the release of the complete genome of the 3D7 strain of *P. falciparum* in 2002 (Gardner et al. 2002). This unprecedented effort to sequence the genomes of eukaryotic pathogens was a technical challenge due to the extreme compositional bias of *Plasmodium* DNA (>80% A + T in *P. falciparum*), which accounted for the instability of these genomic fragments in bacteria (Gardner 1999; Carucci et al. 2004; Hall et al. 2004) and complicated the assembly of contigs (Hall et al. 2004). However, among eukaryotes, the *Plasmodium* genus is currently the best documented at the genomic sequence level, with well-established syntenic relations. At the level of the Apicomplexa phylum, additional complete genomes of *Cryptosporidium*, *Theileria*, and *Toxoplasma* have been also sequenced and annotated (Aurrecoechea et al. 2009).

All *Plasmodium* molecular data have been collected and organized in the PlasmoDB public database as early as sequencing outputs were made available (http://www.plasmodb.org; Aurrecoechea et al. 2009). The architecture of the relational database was designed following biologically relevant relationships, i.e., the "gene to mRNA to protein" dogma, using the Genomics Unified Schema (Kissinger et al. 2002), and ensures that gene loci are linked to annotation using the Gene Ontology standards (Aurrecoechea et al. 2009). Predictions of protein domains, post-translational modifications, subcellular targeting sequences, etc. are included. Furthermore, PlasmoDB is currently the only site where molecular data are (1) clustered based on sequence comparisons, (2) linked to generic schemes designed to view metabolic pathways, and (3) linked to X-omic functional information (transcriptome, proteome, interactome). This allows any biologist to exploit the integrated data with basic or combined queries, and it is therefore the first resource designed to allow the mining of biological knowledge in order to accelerate the development of new therapeutic strategies. PlasmoDB operates inside EuPathDB, a master Web portal for eukaryotic pathogens' genomes (Aurrecoechea et al. 2009). PlasmoDB was designed to allow the identification of novel target candidates following *in silico* approaches and the combining of "filters" inspired by our knowledge of the infection. Since the target discovery process cannot be disconnected from the downstream drug discovery process, it is briefly summarized in the first part of this chapter. Identifying novel drugs that might impair the function of these targets is not trivial. Actually, there are far more target candidates than derived drug candidates in the literature, since the latter process is not based on the simple filtering of biological knowledge appropriately stored in an electronic database. Therefore, there is a need for *in silico* experiments, which are mainly based on the simulation of the docking of small molecules in cavities at the surface of the protein targets. These interactions are believed to alter the target protein's structure in such a way that it becomes nonfunctional. *In silico* strategies to discover novel drug candidates will therefore be detailed in the second part of this chapter. Lastly, since *in silico* docking requires a large computational capacity, the deployment of grid infrastructures to expand the scale of these strategies will be described.

14.2 In Silico Identification and Selection of Chemotherapeutic Target Candidates

14.2.1 Target Discovery Overlapping with In Silico Drug Discovery

A target is a broad concept that qualifies a biological entity and/or a biological phenomenon at which part of a therapy is aimed. It follows that a target can be defined as a phenotype (e.g., symptoms of a disease), a biological process (e.g., a vital metabolic pathway in a pathogen), a subcellular organelle, a protein (e.g., a vital protein of a pathogen), and a protein domain (e.g., a pocket at the surface of a protein into which an active drug can dock; numerous targets can be defined on a given protein structure). It also follows that a target cannot be defined independently of the type of intervention one considers implementing, particularly when one intends to introduce a novel chemotherapeutic strategy. Intervention depends on the level of knowledge of the disease and relies on the availability of methods to design and introduce exogenous chemicals (biological extracts and natural substances, drugs purified from biological material or obtained by synthetic chemistry). As a result, in reviewing the strategies for target discovery, characterizations, validations, etc., the target should be defined. Here, targets will be discussed following their understanding in allopathic medicine, as molecular entities (genes, proteins, protein domains) or biological phenomena (molecular functions, pathways, phenotypes) organized in causal schemes. One is the DNA → RNA → protein ↔ function/phenotype simplified scheme; the other is the functional scheme in which the target plays a role like a metabolic scheme or a gene regulatory network.

In the last decade, access to complete genomic sequences of human and pathogens has brought the hope that target genes would be rationally identified, allowing the design of new cures. For diseases caused by parasites such as malaria, biological knowledge of all partners involved in the parasitic relationship (here *Plasmodium* parasites, *Anopheles* mosquito vectors, and the human host) is therefore a prerequisite to explore rational and creative ways of fighting the disease.

The introduction of a therapeutic treatment, which might be inspired by the characterization of a target, is a high-risk, lengthy, and expensive process. A robust, automated screening assay has to be developed to allow the detection of molecules with appropriate bioactivity (e.g., antagonists binding to a receptor, inhibitors of an enzymatic activity, molecules impairing a phenotype, etc.). From thousands to hundreds of thousands of molecules can thus be screened. From our experience in automated screening, and according to the cost of reagents and consumables, the total cost of such screening ranges from 15,000 to over 100,000 Euros. Risk of failure within the screening process is high, depending on the properties of the target, the quality of the screening assay, and the appropriate molecular diversity of the chemolibrary. More than 70% of the drug discovery and development projects fail (Frantz 2007). Lindsay (2005) reports that, whereas the number of targets identified by modern biological methods increases, yet clinical medicine declines. Any *in silico* analyses that could help select and characterize target candidates earlier in the drug discovery process, as well as help to discard risky targets, are therefore particularly valuable.

Thus, the objective of a target discovery project is not simply to end with a list of genes, but to advance and assist the subsequent therapeutic development processes, which could entail the search for novel bioactive molecules that might act as successful drugs as well as the development of new vaccines or the design of gene therapy strategies. Given the cost of drug/vaccine development, the feasibility of advancing the target discovery process is therefore a major additional criterion for the *in silico* target discovery.

14.2.2 Filters Combined with Boolean Logic

The acceleration of target discovery to combat diseases by employing appropriate databanks of genes' structures and the function they harbor is one of the most expected benefits of genomic sequencing projects. Considering the malaria example, the genomes of all partners, i.e., Human (International Human Sequencing Consortium 2001; Venter et al. 2001), *P. falciparum* (Gardner et al. 2002), and the main malaria vector *A. gambiae* (Holt et al. 2002) have been fully sequenced. The biology of each organism is a field of research by itself, and respective knowledge cannot be necessarily organized following the same principles, for historic and scientific reasons. Current genomic databases, however, have been organized using standard electronic formats allowing data exchanges, interoperability, and comparisons.

The search for new targets in genomic databases is mostly achieved by combinations of filters rationally designed given the knowledge of the disease. These filters are either basic (e.g., "List of genes that are vital for the infectious organism") or more sophisticated (e.g., "List of genes that are vital for the infectious organism"/"with no homologue in the human host in order to lower the toxicity risk of the treatment"/"with little allelic variation in order to lower the risk of resistance spreading"/"with genetic expression in the infectious stage that complies with the sought treatment, for instance the blood circulating stage"/"etc."). These combinations of filters are basic Boolean operations (AND, OR, NOT) applied to genomic and postgenomic databases. Outputs are lists of candidate target genes, the quality of which depends on the accuracy of the underlying scientific reasoning and the quality of the input data. *In silico* drug discovery is therefore a field of research where innovation lies in (1) the improvement of the quality of genomic data and postgenomic information, (2) the improvement of tools used to analyze and process genomic and postgenomic information, and (3) the rational combination of filters.

14.3 Case of Malaria *In Silico* Target Discovery

14.3.1 Targets Are Somewhere in Genomic and Postgenomic Databases

Considering malaria with a practitioner's eye, the organisms involved in the three–partner relationship fall in the following categories: (1) the patient, a human affected with malaria, often referred to as the host, (2) the vector, a mosquito also hosting the parasite and also affected, but seen as a contaminating intermediate from one human patient to another, and (3) the parasite, considered as the malaria causative agent, although the human and the mosquito also play their roles in the complete cycle. Currently, human DNA sequences, deduced genes, and corresponding annotations have been carefully organized as a source of data and information for a large variety of basic scientific questions, ranging from development and physiology to Mammal evolution, population genetics, etc. Some of these studies, in the field of medical sciences, aim at accelerating the design of new therapeutic strategies, but there is no data organization scheme that fulfils all of these purposes. It has now been acknowledged that organization schemes shall either be designed to address a focused research topic with a high level of precision, reliability, and quality, or shall be generalist. To our knowledge, little has been developed in the field of human genomics specifically focusing on malaria. Concerning generalist Internet databases, the Ensembl resource (Hubbard et al. 2007) provides access to the human genome, with comprehensive and integrated sources of annotation of other chordates, strain variation data, and ortholog/paralog annotations based on gene trees.

By contrast, *Anopheles* genomics is solely motivated by the fight against malaria, although mosquitoes happen to be insects and therefore interesting model organisms that could be compared with *Drosophila*. The first complete genomic sequence of *A. gambiae* was released a month before that of *P. falciparum* (Holt et al. 2002) and has been recently updated (Sharakova et al. 2007). The VectorBase (Lawson et al. 2007) provides internet access to *A. gambiae* genomic data and postgenomic information, together with those of other insect vectors of human pathogens but disconnected from any integrated genomic information on nonvector insects such as the fruit fly (FlyBase, Crosby et al. 2007) or honey bee (BeeBase, Elsik et al. 2007). The Ensembl resource mentioned earlier (Hubbard et al. 2007) allows access to both *Anopheles* and *Drosophila*, and since it is also a repository for human genomics, it represents a useful source of data for mosquito-specific target searches.

In the case of *P. falciparum*, PlasmoDB is currently the only database where molecular data have been designed to specifically address the fight against this infection. Genes have been tentatively clustered based on their homology with sequences of other organisms, allowing the search of *Plasmodium*-specific sequences. This search is linked to schemes designed to view metabolic pathways and X-omic functional information (transcriptome, proteome, interactome). Any biologist can exploit these integrated data with basic or combined Boolean queries (Coppel 2001; Kissinger et al. 2002; Bahl et al. 2003; Carucci 2005, Aurrecoechea et al. 2009; Saidani et al. 2009).

14.3.2 Translating Working Hypotheses into Boolean Searches

Boolean comparisons of genomic and postgenomic tables are key in the *in silico* target discovery process, and databases that allow comparative genomic approaches are therefore important instruments. The major working hypotheses underlying the search for novel antimalarial interventions should focus on the three-partner relationship:

- In the Human ↔ *Anopheles* relationship, targets on the mosquito side (mainly, targets for insecticides, i.e., vital processes and genes) are mainly sought (reviewed by Toure et al. 2004). On the human side, no molecular target has been currently described to repel or block *Anopheles* bites, in spite of clues that skin substances are involved in mosquito attraction (Schreck et al. 1990).

- In the Human ↔ *Plasmodium* relationship, targets on the *Plasmodium* side (mainly targets for chemotherapies and vaccines) are mainly sought. Human resistance is also studied (e.g., Hernandez-Valladares et al. 2004; Cunha-Rodrigues et al. 2006), with possible treatments based on immune response enhancers.
- In the *Anopheles* ↔ *Plasmodium* relationship, targets on both the *Anopheles* and *Plasmodium* sides are sought. The objective is the selection or transgenic production of mosquitoes that resist *Plasmodium* infection and have a better population fitness than the wild-type (Christophides 2005). This strategy is based on the hypothesis that transgenic mosquitoes might be released and that a fertilization barrier might also be designed so that *Plasmodium*-carrying mosquitoes are eradicated.

Among all these working hypotheses, the major effort is the search for *Plasmodium* chemotherapeutic targets that comply with practical constraints, i.e., the environmental risks accompanying insecticide spraying and the ethical concerns relating to transgenic organisms. Antimalarial chemotherapy is principally based on specific vital components and processes in the blood stages of the parasite (requiring therefore information on stage-specific gene expression). Parasite features currently investigated as promising targets for future drugs are very heterogeneous: Jana and Paliwal (2007) list membrane dynamics (lipid biosynthesis, membrane transporters), protein turnover (protein synthetic machinery and proteases), plant-like metabolism (shikimate pathway, isoprenoid biosynthesis), redox systems, organelles (mitochondria, apicoplast), nucleic acid metabolism (purine and pyrimidine pathways), etc. This heterogeneity reflects the diversity and creativity of biological researches.

To use these selection criteria for an *in silico* filtering, a translation into a formal language should occur. For instance, the assessment that a *Plasmodium*-specific organelle such as the apicoplast is a target for intervention (for review, Bisanz et al. 2008) can be translated into database queries based on a large variety of criteria:

- The apicoplast is an intracellular organelle that is limited by membranes, containing proteins encoded by both the apicoplast DNA, and by the nuclear chromosomes. Targets can therefore be sought in a list of genes encoded by the apicoplast DNA and nuclear genes whose products are directed to the apicoplast.
- The apicoplast is involved in specific metabolic processes (e.g., fatty acid, glycerolipid, and isoprenoid metabolism, etc.). Targets can therefore be sought among the genes whose products play a role in these metabolic pathways, including metabolite transporters.
- The apicoplast is an organelle, in which biogenesis (e.g., apicoplast DNA replication, transcription/translation, posttranslational modifications including methionine peptide deformylation; nuclear-encoded protein import), development, and division (organelle division machinery) depends on a series of genes that can also be examined for their potential as drug targets (Wiesner and Seeber 2005; Waller and McFadden 2005; Bisanz et al. 2008). Since the apicoplast

derives from an ancestral alga engulfed by a series of endosymbiotic events, the search for new drug targets in *Plasmodium* can also be expanded to all *Plasmodium* genes that were inherited from the ancestral algae and are therefore phylogenetically related to plants. All of these basic queries can be combined with tools made available by Internet resources such as PlasmoDB (Coppel 2001; Kissinger et al. 2002; Bahl et al. 2003; Carucci 2005; Aurrecoechea et al. 2009), or by local software suites.

14.3.3 In Silico Target Discovery Tools

Some of the main criteria used in the *in silico* identification of putative drug targets may include selecting the aspect of the parasite's biology to be interfered with; finding proteins or protein orthologs with sequence, functional and structural properties of interest; determining the level of conservation with host orthologs, which may affect cross-reactivity; defining the classes of compounds that the proteins interact with; analyzing the druggability of the protein active site and validating the protein as a suitable target or choosing targets that have been clinically validated in other species. The amount and quality of the target candidates one can predict based on *in silico* filtering depends on one's access to the relevant genomic and metagenomic information. Either one has experimental evidence regarding the structure and function of genes, or one has to rely on predictive tools. Obviously, given the millions of gene sequences in public databases, experimental investigations will not be carried out and the results of *in silico* mining will rely upon bioinformatics methods. Following are a list factors in the search of apicoplast targets as an example:

Prediction of gene structure will depend on the tools used to detect the genes (see GeneDB). In the case of *Plasmodium*, the strong A + T bias, the noncanonical structure of promoter regions and gene splicing sites, etc. imply that a comparison of the open reading frame predictions with sequenced ESTs and manual examination is required. Precise gene structures have recently been updated (Aurrecoechea et al. 2009). Prediction of nuclear genes coding for apicoplast proteins will depend on the accuracy of the tools developed to predict protein targeting, i.e., PATS (Zuegge et al. 2001) or PlasmoAP (Foth et al. 2003). The prediction of genes coding for enzymes involved in apicoplast metabolic reactions or transporters will depend on the quality of the automatic annotation of genes, i.e., the functional inference for each open reading frame (at the date of writing, > 60% of the *Plasmodium* genes have no functional annotation; Birkholtz et al. 2006), and the quality of the metabolic representations, i.e., KEGG (Kanehisa et al. 2006), MetaCyc (Caspi et al. 2006), and the Malaria Parasite Metabolic Pathways (Ginsburg 2006, 2009). Part of the functional annotation derives from the prediction of gene homologies and molecular phylogenies and depends on the performance of sequence comparison methods, statistics, and molecular phylogenetic reconstruction methods (Bastien et al. 2004a; 2004b; 2005; 2007). Moreover, part of the functional inference can derive from nonhomology-based annotation transfers, resulting in the creation of GO-databases, i.e., the *Plasmodium* OPI Databases (Zhou et al. 2008) or PlasmoDRAFT, which contain annotated predictions based on guilt-by-association methods using postgenomic data including those from the transcriptome, proteome, and interactome (Bréhélin et al. 2008). Prediction of nonprotein apicoplast targets depends on the availability of resources for nonprotein objects. Such resources do not exist.

In contrast to this focused search on the structure and function of the cell (and its components), global analyses based on postgenomic data provide invaluable additional clues. Thus, experimental and *in silico* interactome data have revealed unique protein–protein networks within the malaria parasite with controlling nodes indicating a "rich-club phenomenon" of interconnectivity (Birkholtz et al. 2008b). Computational prioritization of drug targets is utilized in PlasmoCyc, which contains an integrated pathway/genome database and has resulted in the identification of 216 chokepoint enzymes (Yeh et al. 2004). More recently, the *P. falciparum* metabolic pathways have been used to identify additional 22 potential new targets using *in silico* knock-out approaches (Fatumo et al. 2009).

PlasmoDB currently provides the most complete resource to assist with selection of malaria target candidates, but improvement of existing tools and development of new analytical tools are still needed to increase the quality of molecular and functional annotation, to allow biological knowledge representation and to improve the integration and mining of genomic/postgenomic molecular and functional data (Birkholtz et al. 2006). A real challenge is to move further from user-friendly but rigid Web portals to flexible and creativity-oriented knowledge accesses (with user-designed workflows, implying a strong interoperability).

14.3.4 Toward Druggable Plasmodium Genome

Despite the small number of genes that have been annotated in the *Plasmodium* genomes, tens of drug targets have already been deduced from *in silico* analyses. With the three-dimensional structures of previously characterized biological targets, classifiers based on machine learning methods can be developed by docking known drugs. Any protein whose structure has been assessed by crystallography or predicted by structure modeling (left of Fig. 14.1) might be subjected to this classifier in order to predict whether they share some of the key properties of drug targets. The structure of investigated proteins can be compared to databases of protein structures (Charette et al. 2006) (Fig. 14.1b). Alternatively, the structure of investigated proteins can be analyzed to detect surface cavities (e.g., Laurie and Jackson 2006; Nayal and Honig 2006) that can then be compared to databases of drug/ligand-binding cavities (Fig. 14.1c). Processing genes following this workflow allows the definition of a "druggable" genome. This is a longer term goal in the case of malaria *in silico* target discovery, due to the few *Plasmodium* protein structures currently available, as well as to the difficulties associated with the structural modeling of *Plasmodium* proteins (Birkholtz et al. 2006; de Beer et al. 2009).

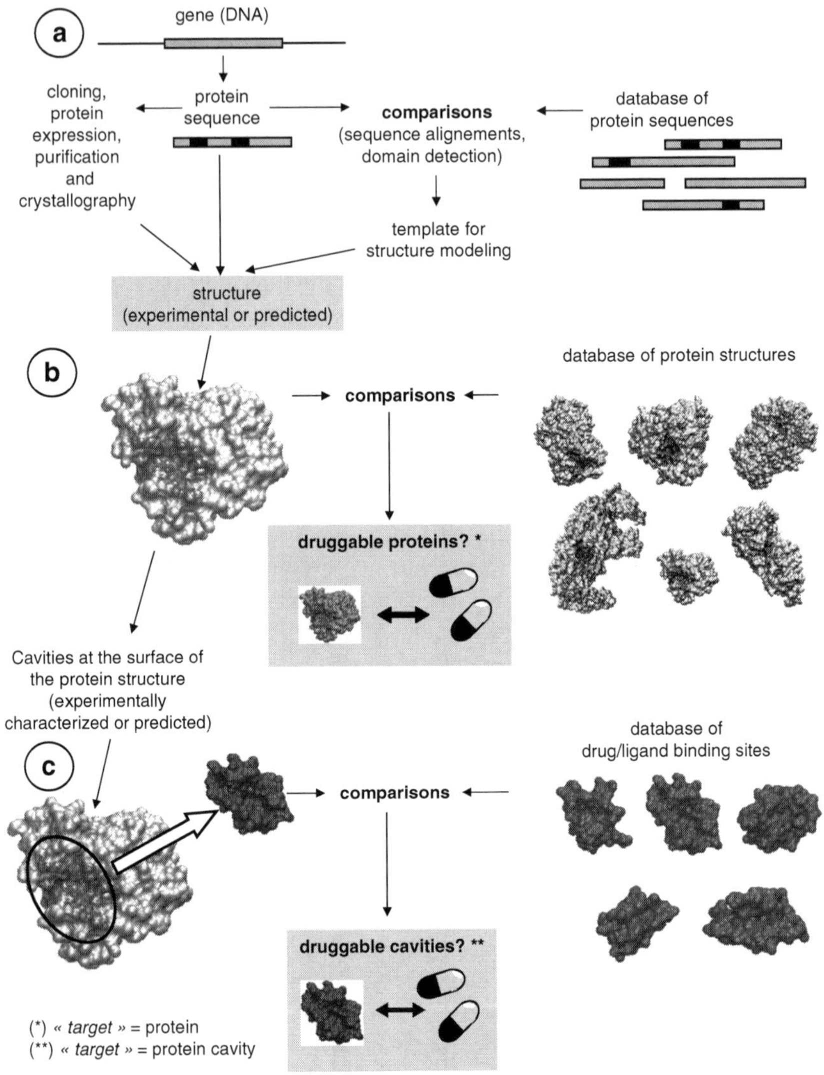

Fig. 14.1 *In silico* determination of the druggable genome. This scheme, adapted from Saidani et al. 2009, shows the *in silico* analyses (comparisons and docking) that can be performed in order to predict whether a protein sequence (A), protein structure (B), or protein surface cavity (C) share features with known protein targets (or known ligand/drug binding sites). The systematic investigation allows the definition of a set of predicted "druggable" genes

Progress toward the *Plasmodium* druggable genome and the linkage of genomic data and postgenomic information with chemical knowledge of drugs and drug-like small molecules represent future challenges for *in silico* target discovery. This information provides the basis for drug target databases including the TDR Targets Database (http://tdrtargets.org).

14.4 Strategies to Identify and Select Drug Candidates

14.4.1 In Silico and In Vitro Drug Discovery

There has been a steady decline in the number of new molecular entities entering clinical development and reaching the market over the past 10–15 years. This is due to high levels of drug attrition, mainly attributed to unanticipated efficacy and toxicity problems (Bhogal and Balls 2008). Reasons for this situation seem to reside in the extensive use of High-Throughput Screening (HTS) against ambiguous or single targets, which in effect reduces the biological context by separating the target from other cellular proteins and processes that might impact its function (Hellerstein 2008). Another contributing issue is the lack of diversity in existing chemical libraries (Lipkus et al. 2008). The phenotypic robustness of biological systems often reduces the effectiveness of a single-target compound (Hopkins 2008). Cell-based high content screening (HCS) circumvents this problem, since it allows the detection of small molecules acting in the cellular context (Muskavitch et al. 2008), but it leaves the question of the actual target unresolved. Disease-relevant *in silico* screens are therefore considered as advanced methods to be introduced as early as possible into the drug discovery process (Kassel 2004; Hall 2006; Lang et al. 2006; de Beer et al. 2009).

14.4.2 Structure-Based Drug Discovery

In silico structure-based drug design can be classified into receptor-based design and ligand-based design (Fig. 14.2) as follows:

- Receptor-based drug design exploits the three-dimensional structural description of a macromolecular drug target to predict the *in silico* binding of hypothetical ligands. These hypothetical ligands can be obtained from the *in silico* virtual screening of compound libraries against the target, receptor-based pharmacophore design, modification of a ligand known to bind to the target, and fragment-based inhibitor design (scaffold structures or de novo design) (Fig. 14.2).
- Ligand-based drug design aims to predict the effects of new compounds based on the properties of compounds previously known to affect the target. This may be pursued in the absence of a target structure.

Both design avenues are highly integrated, iterative, and knowledge-based and all substrategies should be investigated. The knowledge available on both the structure and inhibitors of a specific target determines the approach to be followed. Identified compounds are scored and ranked based on their physiochemical interactions with the target structure and the best scoring compounds are biochemically tested for inhibitory activity. Promising lead compounds are then verified by solving the

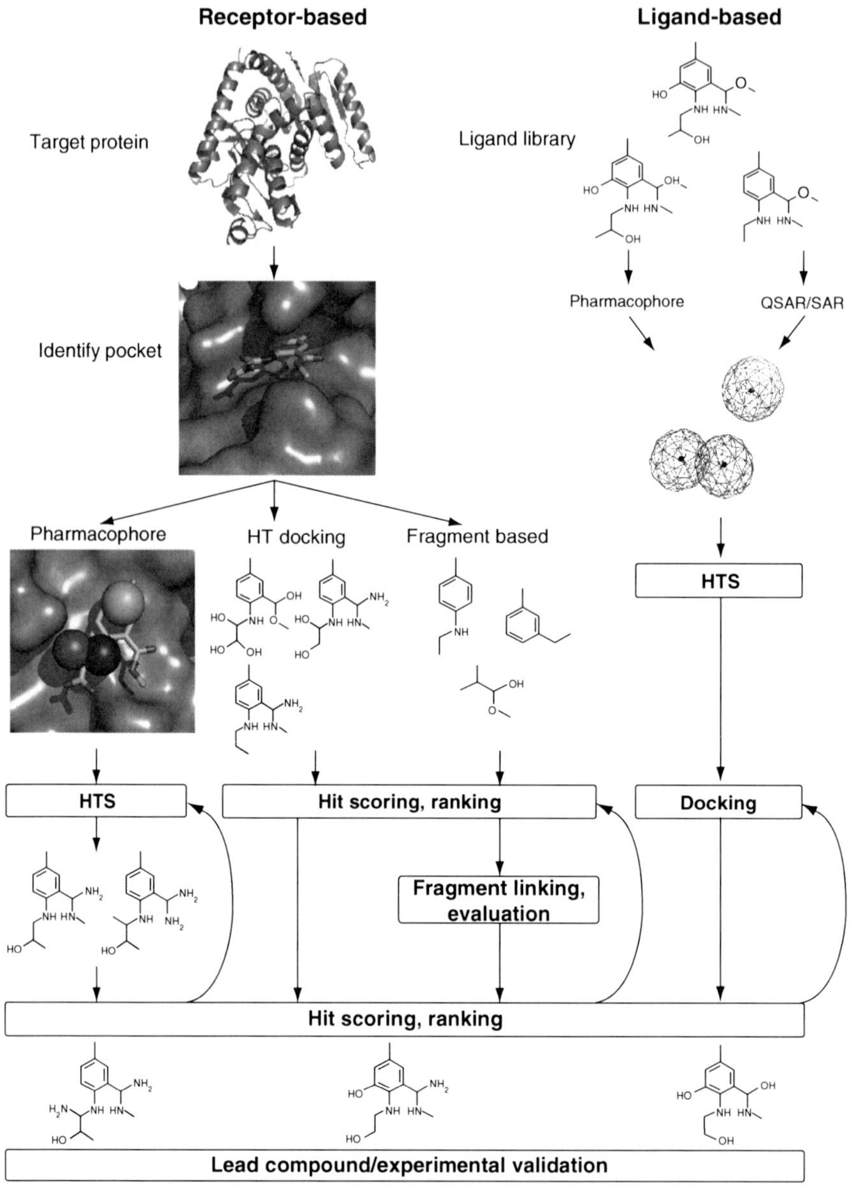

Fig. 14.2 *In silico* drug design pipelines. Parallel and integrative strategies include receptor-based design and ligand-based design

structure of the target-lead complexes to confirm predictions. This is followed by the *in silico* optimization of the lead compound and iterative testing.

Receptor-based drug design starts by describing the three-dimensional structure of the protein target. Subsequently, inhibitors are sought either by:

1. Docking ligands inside cavities within the structure surface and selecting those predicted to bind with the highest affinity (virtual HTS)
2. Deriving a pharmacophore model from the binding site and using it to screen for ligands or
3. Docking small molecular fragments within the binding site with a possibly weak affinity for assembly into a high-affinity ligand that occupies the entire cavity

Plasmodial proteins are notoriously difficult to express in heterologous systems, and the structures have proven difficult to solve experimentally. Some confounding characteristics of proteins from *P. falciparum* include large protein sizes, greater protein disorder, high pI, low complexity of parasite-specific inserted regions (Mehlin et al. 2006), and a marked A + T bias of the *P. falciparum* genome. These factors additionally contribute toward the low crystallization efficiencies of Plasmodial proteins. In the Protein Data Bank (PDB, http://www.pdb.org), only 118 entries correspond to structures of Plasmodial proteins, excluding sequences with high-level (>90%) identity. In contrast, querying the PDB for nonredundant (<90% identity) human protein entries reveals more than 4,500 structures.

Even though the number of Plasmodial protein structures is still sparse, there has been a notable increase in the corresponding protein structures between 2005 and 2008. This is due largely to the advent of structural genomics programs including the Structural Genomics Consortium (SGC, http://sgc.utoronto.ca) and the Structural Genomics of Pathogenic Protozoa (SGPP, http://www.sgpp.org). The SGC reported 25 distinct Plasmodial protein crystal structures from five species. The SGPP consortium has solved 16 Plasmodial proteins.

As an alternative to crystal structure resolution, many groups have resorted to homology modeling to predict the three-dimensional structure of a target. Successful homology modeling depends on the alignment of the target sequence with template structures. This is especially critical in the case of malaria, since *P. falciparum* proteins often contain long inserts that, along with low sequence similarity, make alignments problematic. Not surprisingly, proteins with long inserts appear to be avoided for modeling, and the problem of obtaining reliable alignments in their case is seldom discussed. A number of techniques can be used to circumvent this problem (for a recent review see de Beer et al. 2009). Once an alignment has been optimized, a series of models can be built to identify problem areas within the alignment. Despite the difficulties with the homology modeling of Plasmodial proteins, there have been some notable successes with a diversity of applications.

An example of a protein model used for *in silico* drug discovery is the DHFR (dihydrofolate reductase) domain of the bifunctional protein DHFR-TS domain (dihydrofolate reductase-thymidylate synthase). The effectiveness of existing drugs such as cycloguanil and pyrimethamine, which target the DHFR domain, has been reduced due to drug resistance. Hence, DHFR has been a popular target for homology modeling efforts (e.g., Toyoda et al. 1997; McKie et al. 1998; Lemcke and Christensen 1999; Rastelli et al. 2000; Santos-Filho et al. 2001; Delfino et al. 2002), which allowed the identification of new inhibitors in the nano- and micromolar range (Toyoda et al. 1997; McKie et al. 1998), the rationalization of the antifolate resistance mechanisms (McKie et al. 1998; Lemcke and Christensen 1999; Rastelli et al. 2000; Delfino et al. 2002), and the ability for the drug WR99210 to inhibit both pyrimethamine and cycloguanil-resistant mutants (Rastelli et al. 2000). A number of new inhibitors were also successfully designed, and the quality of the alignment used for modeling and dockings was subsequently confirmed with the crystal structure of the complete bifunctional enzyme (Yuvaniyama et al. 2003).

Gutiérrez-de-Terán et al. (2006) demonstrated the advantages of using multiple structures, including a homology model and a low-resolution crystal structure on *P. falciparum* plasmepsin IV as an example. The structural quality indicators for the homology model were better and more robust when calculating binding energies for an inhibitor series. Further improvements in predicting binding were gained by using a combined model employing both structures, as well as by using molecular dynamics to increase sampling quality. Other noteworthy examples have been discussed by de Beer et al. (2009). After a reliable structure for a *Plasmodium* drug target has been obtained, whether through modeling, X-ray crystallography, or NMR, it has to be extensively analyzed before the lead discovery process can be commenced. Protein quality assessment is needed to identify the limitations of the target structure to be used. The most reliable structures are those from X-ray and NMR, although one should be mindful of inaccuracies inherent in some crystal structures. Most deposited structures assume some isotropic variation of atomic positions and do not fully capture the dynamic and anisotropic nature of protein crystals (DePristo et al. 2004; Davis et al. 2008). It is essential that the dynamic nature of the target, which implies the use of multiple structures from crystallography or NMR, should be taken into account. This can be further supplemented with various *in silico* methods such as molecular dynamics or Monte Carlo sampling. It is generally believed that homology models with a > 50% sequence similarity can be reliably and independently used (Hillish et al. 2004).

Receptor-based virtual HTS involves the screening of large libraries of ligands by computational methods to simulate and evaluate the strength of the docking of ligand inside a cavity of a protein. The main issues of *in silico* docking experiments include the following, which, from our practical experience, should be examined carefully.

The selection of the screening method: according to Sousa et al. (2006), the five most popular algorithms are AutoDock (Huey et al. 2007), GOLD (Jones et al. 1997), FlexX (Rarey et al. 1996), DOCK (Ewing et al. 2001), and ICM (Abagyan et al. 1994). A recent program, Glide (Friesner et al. 2004) is also more and more associated with successful work in the literature. Among these top six, only AutoDock and DOCK are freely available for academic users. These software applications are based on different algorithmic approaches (detailed by Höltje et al. 2008) such as Incremental construction methods (FlexX, DOCK), Genetic Algorithms (GOLD, MolDock, Psi-Dock), the Tabu search that can be combined with Genetic Algorithms, Simulated annealing, and Monte-Carlo simulations (Glide), Shape-Fitting methods (FT-Dock), or miscellaneous other approaches.

Each structure-based screening tool addresses two correlated issues. The first issue is the ability to predict the best "pose," i.e., the best ligand-bound conformation and orientation in the target site (the *docking processing*), and the second one is the capacity to rank ligands and their poses by correctly evaluating their binding affinity with the target (the *scoring processing*). Since both tasks are complex and imply exponential computing time, each type of processing is generally based on different approximations or heuristics, especially in the context of High Throughput Docking (HTD).

Ideally, the docking processing should take into account the flexibility of both the protein and the ligand in order to massively and correlatively explore the two conformational spaces. Nowadays, most docking algorithms handle ligand flexibility using various approaches (Sousa et al. 2006; Höltje et al. 2008), which can be, for instance, the storage of multiple conformations in a database, the incremental construction of a ligand that was previously divided into fragments (like in FlexX), or the modification of specific dihedral angles during the genetic operations (mutations, cross-over) stage (in genetic algorithm like Gold). Incorporating target flexibility is a more complex challenge that is however essential in HTD, especially when using proteins models with low resolution, which can be the case for malaria proteins. According to Cavasotto and Singh (2008), the various observed movements that can modify docking results could be classified into three categories: side-chain movements, loop/backbone motions, and domain motions. The development of docking methods that address the flexibility of the protein is recent, thanks to improvements in the computer capacities. Cavasotto and Singh (2008) give a detailed review of the existing methods and their respective advantages and drawbacks. These authors emphasize that most of the current methods cannot handle some important protein motions like long loop or large backbone movements in a context of HTD because of extreme computational time requirements. This is why some current projects (like Docking@Grid, http://dockinggrid.gforge.inria.fr/) aim to use the computer Grid power to address this issue.

The scoring processing tries to give an accurate evaluation of the binding affinity between a ligand pose and the target site. It takes place during the docking stage, first to optimize the placement of a ligand and then to rank all the putative hits. Some robust methods (like free energy perturbations; Miyamoto and Kollman 1993) compute a reliable binding free energy, but they are so expensive in terms of computation time that they cannot be used for HTD. The conventionally used scoring functions are much faster; they are mainly grouped into three categories: empirical scoring ones, force-field-based ones, and knowledge-based ones (Höltje and al. 2008). Their main drawback is that the ranking based on these functions is not always reliable, even if they help to identify a restrained list of ligands with a higher percentage of hits than a random selection (Verkhivker et al. 2000). This can be a major issue, especially when the final goal is to give to the experimental biologists an extremely reduced subset of chemical compounds having a high probability to be active; a critical step for malaria targets, which are difficult to study *in vitro*. Filtering strategies to postprocess docking outputs have therefore been developed. Filtering by consensus based on different docking approaches

(Paul and Rognan 2002) or on different scoring functions (Bissantz et al. 2000; Teramoto and Fukunishi 2008), filtering by chemical diversity, and filtering by interaction fingerprint-based scoring (Heteny et al. 2003; Marcou and Rognan 2007) are some of the methods that are detailed by Höltje et al. (2008). They can be applied to detect false positives to improve the true positive rate. However, most of these approaches also have their constraints, which must be assessed before the approach is chosen in the context of the studied target. Consensus approaches or interaction fingerprint methods, for instance, require a large amount of information about the target (several X-ray structures, many known actives). Moreover, interaction fingerprints (built on a training set of known actives) present the risk of discarding new binding modes, and so perhaps of rejecting more original ligands. In conclusion, the choice of a docking program, of the scoring functions, and of the postfiltering strategies for a specific target requires a great deal of preparative work. Methodologies and algorithm evaluations abound in the literature but independent comparative studies are rare. It has been shown that some docking tools and some docking/scoring combinations are more robust than others (Bissantz and al. 2000; Kellenberger et al. 2004). Evidence also emphasized that the best possible option to design a strategy is to test a systematic combination of docking/scoring parameters on a reduced dataset (about 1,000 compounds) containing a few known ligands, and then to select for the full library a protocol that best discriminates true hits from random ligands. Likewise, the docking of known inhibitors has been used against wild-type and quadruple resistant mutant forms of *P. falciparum* DHFR (Fogel et al. 2008) to define a common interaction pattern between inhibitors and the different forms of the protein that describes selection criteria for further screening strategies.

The selection of the compounds before docking is extremely important. Several commercial and public chemical libraries are available for screening. Some of the major efforts to generate chemical databases include the ZINC database (Irwin and Schoichet 2005), the National Cancer Institute (NCI, http://cactus.nci.nih.gov), PubChem (http://pubchem.ncbi.nlm.nih.gov), the French National Chemical Library (http://chimiotheque-nationale.enscm.fr/), the Super Drug DataBase (Goede et al. 2005), the Drug Bank (Wishart et al. 2009), and the SuperNatural database (Dunkel et al. 2006). These databases are not all freely available for downloading and screening but are available online for similarity searches. Irwin and Schoichet (2005) suggested that the "gold standard" for docking databases in academia is the commercially available ones (e.g., the Available Chemical Database or ACD, http://www.mdli.com; the ACD screening compound set, http://www.ccdc.cam.ac.uk; the Cambridge Structural Database or CSD, http://www.ccdc.cam.ac.uk; and the ChemNavigator database, http://www.chemnavigator.com). These databases are a few of the most popular ones used in virtual screening and contain from a few hundred thousand up to ten million compounds. An important problem that should be handled is that the user is left with the decisions on the protonation states, charges, and tautomeric forms and the removal of salts (Irwin and Schoichet 2005). The ZINC database, containing over 8 million purchasable compounds, is the first database where all of these aspects have been addressed by the curators (Irwin and Schoichet 2005) and provides subsets such as lead-like, drug-like, fragment-like, Verneralis-filtered, etc., which have been prefiltered using specific criteria such as Lipinski's rule-of-five (http://zinc.docking.org). More generally, since HTD is CPU greedy and since postdocking analyses can be very demanding in human time, it is therefore strongly advised to first filter the starting compound library (Höltje et al. 2008; Dubois et al. 2008). The design of sublibraries (from a large library or from several libraries) is highly recommended and should take into consideration the size, chemical diversity, or specific properties of the chemical compounds according to the pursued goal. Different strategies and available software for the preparation of collections of compounds for virtual screening are described by Dubois et al. (2008).

Eventually, the preparation of the target needs meticulous attention. Usually, docking software explains how to prepare the protein file before submitting it to the screening processing. If the target structure has been resolved by X-ray crystallography, the structural information is derived from the Brookhaven Protein Data Bank. The user must first check the structure file; for instance, it is usual that some residues of the target structure are not resolved by X-ray crystallography due to a variety of reasons. If residues within or near the binding site are missing, they must be inserted by other methods. Usually, PDB files contain water molecules; these must be removed if they are not essential. Hydrogen atoms must be added, with particular care taken for the histidine residues, since they can have different protonated states depending on their local environment. Moreover, if the protein is considered as a rigid molecule during the docking, it is important to optimise the intermolecular (protein–ligand) and intramolecular (protein–protein) interactions by adjusting the torsions of the polar hydrogens (in the residues serine, theronine, and lysine), as well as the torsions of the residues that are the hydrogen bond donors or acceptors in the binding site. The software then usually makes some file format transformation (for instance from PDB format to Mol2 format) and conducts some specific preprocessing. These instructions are intended to answer some questions, which imply an in-depth knowledge of the protein. The latter point seems to be obvious but is difficult for recently discovered targets or for targets that are difficult to study in vitro, which is a common case for malaria targets. Questions that need to be answered are as follows: how large should the volume for docking around the active site be? Does the docking site contain molecules of water? which of these are necessary and should be conserved? if a cofactor binding site is overlapping the binding site of the ligand, should the cofactor be kept? if the site contains metallic ions, how should they be considered in the model? Some of these questions have been explored by Höltje et al. (2008), and most require extensive experimental data about the biological system (conditions of crystallisation, in vitro binding assays, mutagenesis results, etc.) to be provided before any answer can be offered.

Receptor-based pharmacophore approaches use resolved protein structures to derive pharmacophore features and, subsequently, pharmacophore models. These models are a set of structural features in a molecule, which are recognized at a receptor site and are responsible for the bioactivity of the molecule. This approach works particularly well when using structures resolved in complex with ligands, as the conformational changes associated with ligand binding and protein–ligand interactions can be inferred from the complexes. From these structures, a negative image of the active site can be constructed, which complements the interactions between the receptor and ligand as described by the pharmacophore model. These models are subsequently used to screen chemical libraries to find compounds matching the desired features.

Receptor-based HTS and pharmacophore approaches are complementary. Hits identified during virtual HTS need to be filtered and ranked using docking techniques, and only the best scoring compounds are then tested in vitro. The advantage of a pharmacophore-based method lies in the ability to generate a divergent set of compounds consisting of different scaffold structures. The derivation of the correct geometric orientation of the pharmacophore provides directionality during the search for ligands and the identification of novel features (Dror et al. 2004). The parameters describing protein movement in receptor-based pharmacophore strategies were developed to incorporate the inherent flexibility of protein structures in the drug design process and to reduce the entropic penalties that occur upon ligand binding to a target structure (Carlson et al. 2000). This led to a remarkable improvement in results compared to rigid pharmacophore models (Meagher and Carlson 2004). Because of the difficulty in obtaining three-dimensional structures for *Plasmodium* proteins, very few receptor-based pharmacophore studies have been performed (see de Beer et al. 2009).

Fragment-based drug design relies upon a library of smaller (<200–300 Da) but more diverse ligands than those used for HTS, which are docked into the cavities of a protein. The highest scoring hits are then used in subsequent steps of the rational drug design process. With structural insights, these fragments can be optimized quickly to a lead compound stage (Hesterkamp and Whittaker 2008), although linking the smaller ligands together in a complete and active compound can be a challenge (Villar 2007). However, the resulting molecules are likely to have better ligand efficiency than classical HTS-derived molecules (Erlanson 2006). Applications of this approach in the malaria field are limited but its future application may yield new classes of drug candidates (see de Beer et al. 2009).

The lack of a target three-dimensional structure does not preclude the use of an *in silico* approach to design novel drug candidates. With access to a set of structurally divergent compounds that bind in the active site, various *in silico* ligand-based drug design approaches can be followed (Fig. 14.2). All methodologies in this approach aim to reduce the chemical search space and may include similarity searching, substructure searching, as well as structure–activity relationship (SAR) or quantitative structure-activity relationship (QSAR) and ligand-based pharmacophores. These methods are usually tightly integrated since this approach is based on the assumption that molecules with similar physiochemical properties exert a similar biological activity (Sheridan and Kearsley 2002).

14.4.3 Target Similarity Searching, Substructure Searching, and QSAR

Similarity searching, substructure searching, and QSAR have been widely used to explore the chemical space of known inhibitors in the absence of a target structure. These techniques make use of molecular fingerprints that encode fragment-type descriptors that indicate the presence or absence of particular chemical features. Substructure searches can be defined as searches performed on complete structures to identify other compounds containing a specific query substructure. Maximum common substructure approaches are often preferred since they are more flexible than traditional similarity searching, which only considers global similarities between structures (Cao et al. 2008). Furthermore, similarity searching complements substructure searching since it often returns alternative structures. The predominant use of these methods is currently in the design of specific libraries used in virtual HTS (Gillet 2008). However, these methods can also be used to filter databases and to design custom libraries to be screened *in silico*. The use of similarity and substructure searching has become readily accessible by projects such as PubChem and DrugBank, and it is foreseen that it will play an increasingly important role in the drug discovery pipeline for malaria.

If a set of structurally divergent compounds with known inhibitory activities is available, a QSAR can be determined and used to statistically predict the inhibitory potential of new compounds. QSAR includes various levels of information that are captured in 2D-QSAR, 3D-QSAR, or 4D-QSAR models. Several QSAR studies have been performed on malaria with different levels of success (Marrero-Ponce et al. 2005; Dheyongera et al. 2005; Flipo et al. 2007; Fatorusso et al. 2008; Mahmoudi et al. 2008; Xie et al. 2006).

If a set of compounds that have sufficient structural diversity and act against a specific target is available, pharmacophore features can be extracted and used in the generation of ligand-based pharmacophore models. These models can then be screened against chemical databases to identify new lead compounds (Güner et al. 2004). The pharmacophore models can also be used to identify novel inhibitors with a wide diversity of backbones (scaffold-hopping) and of different chemotypes, which still have a similar biological activity (Sun 2008). As with the receptor-based pharmacophore approach, the advantage lies in the models' ability to generate a diverse set of compounds (Dror et al. 2004). The use of ligand-based pharmacophore approaches has clearly evolved as an important technique in the fight against malaria (de Beer et al. 2009).

14.5 Grid Infrastructures for In Silico Drug Discovery

Although virtual HTS is mainly achieved through clusters of computers physically connected to one another to screen compound sets against the target, powerful grid-computing strategies achieved by recent advances in the network linking of computers are increasingly being applied to HTS. Grid computing is an exciting new technology offering rapid computation, large-scale data storage, and flexible collaboration by harnessing together the power of a large number of commodity computers or clusters of other basic machines distributed worldwide and linked via a high-speed network (Andrade et al. 2007).

Several grid infrastructures with different sizes are available: the regional Auvergrid (http://www.auvergrid.fr), the French Grid 5000 (https://www.grid5000.fr), the E-science grid for Europe and Latin America (EELA, http://www.eu-eela.org), Enabling Grids for E-sciencE (EGEE, http://www.eu-egee.org/), EUChinaGrid (http://www.euchinagrid.org), EUMedGrid (http://www.eumedgrid.org), TWGrid (http://www.twgrid.org), North Carolina BioGrid (http://www.ncbiogrid.org), the Canadian BioGrid (http://www.cbr.nrc.ca), the Asia Pacific BioGrid (http://www.apbionet.org/grid), and the Cancer Biomedical Informatics Grid (Covitz et al. 2003). These grids focus on different problems ranging from genetic linkage analysis (Andrade et al. 2007) to molecular docking (Kasam et al. 2007) and metabolic pathway modeling (Kimura et al. 2004).

The malaria parasite presents various challenges, which can benefit from a grid-based approach. They include searching the *Plasmodium* genome and proteome for new drug targets, the identification of single nucleotide polymorphisms (SNPs) on human as well as *Plasmodium* genomes relating to drug sensitivity, drug resistance mechanism elucidations, and the epidemiological monitoring of outbreaks. Of these, drug discovery against malaria is a well-identified area of relevance for the grid paradigm. Various projects were initiated to use grids for the large-scale docking of ligands in target proteins to assist in the discovery of new drugs against malaria. For example, WISDOM-1 (World-wide In Silico Docking On Malaria) used EGEE, the largest multidisciplinary grid infrastructure in the world, to screen a filtered ZINC library against two *P. falciparum* plasmepsin proteins (plasmepsin II and IV) with FlexX (Kuntz et al. 1982) and Autodock (Goodsell and 1990). Around one million compounds were docked into each of the five PDB structures (1lee, 1lf2, 1lf3,1ls5): in total, 41 million dockings were achieved in 6 weeks (the equivalent of 80 years of CPU power). To further address parameters such as protein flexibility, an automatic procedure of refinement by molecular dynamics is applied on the best 5,000 docked compounds using the Amber software suite (Ferrari et al. 2007). In the course of the WISDOM-I program, previously characterized inhibitors, as well as novel promising groups of guanidino-based compounds, were selected *in silico* and are currently being investigated further (Kasam et al. 2007). Following the successful WISDOM-I round on both computational and biological sides, several teams in academic institutions worldwide proposed targets implicated in this disease, leading to the second assault or WISDOM-II program. Four different Plasmodial proteins (glutathione-S transferase, tubulin, and DHFR from both *P. vivax* and *P. falciparum*) were targeted (Salzemann et al. 2007). EGEE, Auvergrid, EELA, EUChinaGrid, and EUMedGrid were used to dock the whole ZINC database, representing 4.3 million compounds, into ligand sites defined at the surface of the four selected proteins, with the FlexX method. Comparisons of the predicted docking poses with the structures of compounds cocrystallized with the target were performed to evaluate the docking parameters.

During the 76-day duration of the project, nearly 140 million dockings were performed at a rate of almost 80,000 dockings per hour (equivalent to 413 years on a single PC). The outcome of these applications needs to be experimentally validated but illustrates the power of virtual HTS in substantially reducing search time as well as in providing a coarse filtering of large libraries. Libraries can be further reduced using more accurate docking, or can be screened using more stringent approaches. The use of grids as an initial screening tool will contribute significantly to the fight against malaria as more grids that can be applied to the search for new compounds become available.

14.6 Conclusions

In silico drug discovery has entered its mature stage. One indicator for this is the increasing number of successful drug discovery projects by virtual screening in the pharmaceutical industry. Since a 7- to 10-year period separates the initial phases of a drug discovery project and its communication, when successful, essentially by the publication of a patent, no statistics are available. Based on 2008 surveys of pharmacological research and development strategies, virtual models can reduce drug development costs by at least 25% and drug development time by up to 50%.

In this chapter, we have illustrated how *in silico* approaches can be creative and diverse, and how they rely on the quality of the biological expertise and its translation into accurate knowledge representations. There is a necessary connection with in vitro and in vivo approaches. *In silico* methods are particularly important in developing new treatments against infectious diseases, including virulent viruses, bacteria or eukaryotes, as the pathogens are difficult to handle in a laboratory environment, and since the number of target candidates is too high for the *in vitro* screening capacities. Future prospects include the modeling of biological systems to push the *in silico* tool a step further and computational testing of the responses to the drug candidates, in both the pathogen and the patient.

References

Abagyan R, Totrov M, Kuznetsov D (1994) ICM: a new method for protein modeling and design: applications to docking and structure prediction from the distorted native conformation. J Comput Chem 15:488–506

Adl SM, Simpson AG, Farmer MA et al (2005) The new higher level classification of eukaryotes with emphasis on the taxonomy of protists. J Eukaryot Microbiol 52:399–451

Afonso A, Hunt P, Cheesman S et al (2006) Malaria parasites can develop stable resistance to artemisinin but lack mutations in candidate genes *atp6* (encoding the sarcoplasmic and endoplasmic reticulum Ca2 + ATPase), *tctp*, *mdr1*, and *cg10*. Antimicrob Agents Chemother 50:480–489

Andrade J, Andersen M, Sillén A et al (2007) The use of grid computing to drive data-intensive genetic research. Eur J Hum Genet 15:694–702

Aurrecoechea C, Brestelli J, Brunk BP et al (2009) PlasmoDB: a functional genomic database for malaria parasites. Nucleic Acids Res 37:D539–D543
Bahl A, Brunk B, Crabtree J et al (2003) PlasmoDB: the Plasmodium genome resource. A database integrating experimental and computational data. Nucleic Acids Res 31:212–215
Baldwin PC (2003) How night air became good air, 1776–1930. Environ Hist 8(3):36
Bastien O, Aude JC, Roy S et al (2004a) Fundamentals of massive automatic pairwise alignments of protein sequences: theoretical significance of Z-value statistics. Bioinformatics 20:534–537
Bastien O, Ortet P, Roy S et al (2005) A configuration space of homologous proteins conserving mutual information and allowing a phylogeny inference based on pairwise Z-score probabilities. BMC Bioinform 6:49
Bastien O, Ortet P, Roy S et al (2007) The configuration space of homologous proteins: a theoretical and practical framework to reduce the diversity of the protein sequence space after massive all-by-all sequence comparisons. Future Generation Comput Syst 23:410–427
Bhogal N, Balls M (2008) Translation of new technologies: from basic research to drug discovery and development. Curr Drug Discov Technol 5:250–262
Birkholtz LM, Bastien O, Wells G et al (2006). Integration and mining of malaria molecular, functional and pharmacological data: how far are we from a chemogenomic knowledge space? Malar J 5:110
Birkholtz LM, Blatch G, Coetzer TL et al (2008a) Heterologous expression of plasmodial proteins for structural studies and functional annotation. Malar J 7:197
Birkholtz L, van Brummelen AC, Clark K et al (2008b) Exploring functional genomics for drug target and therapeutics discovery in Plasmodia. Acta Trop 105:113–123
Bisanz C, Botté C, Saïdani N et al (2008) Structure, function and biogenesis of the secondary plastid of apicomplexan parasites. In: Schoefs B (ed) Current research in plant cell compartments. Research Signpost, India, pp 393–423
Bissantz C, Folkers G, Rognan D (2000) Protein-based virtual screening of chemical databases. I. Evaluation of different docking/scoring combinations. J Med Chem 43:4759–4767
Cao Y, Jiang T, Girke T (2008) A maximum common substructure-based algorithm for searching and predicting drug-like compounds. Bioinformatics 24:i366–i374
Carlson HA, Masukawa KM, Rubins K et al (2000) Developing a dynamic pharmacophore model for HIV-1 integrase. J Med Chem 43:2100–2114
Carucci DJ (2005) Advances in malaria genomics since MIM Arusha, 2002. Acta Trop 95:260–264
Carucci DJ, Goodwin PM, Gottlieb M et al (2004) The *Plasmodium falciparum* genome project. In: Waters AP, Janse CJ (eds) Malaria parasites: genome and molecular biology. Caister Academic, England, pp 1–6
Caspi R, Foerster H, Fulcher CA et al (2006) MetaCyc: a multiorganism database of metabolic pathways and enzymes. Nucleic Acids Res 34:D511–D516
Cavasotto CN, Singh N (2008) Docking and high throughput docking: successes and the challenge of protein flexibility. Curr Comput Aided Drug Des 4:221–234
Charette BD, Macdonald RG, Wetzel S et al (2006) Protein structure similarity clustering: dynamic treatment of PDB structures facilitates clustering. Angew Chem Int Ed Engl 45:7766–7770
Christophides GK (2005) Transgenic mosquitoes and malaria transmission. Cell Microbiol 7:325–333
Coppel RL (2001) Bioinformatics and the malaria genome: facilitating access and exploitation of sequence information. Mol Biochem Parasitol 118:139–145
Covitz PA, Hartel F, Schaefer C et al (2003) caCORE: a common infrastructure for cancer informatics. Bioinformatics 19:2404–2412
Crosby MA, Goodman JL, Strelets VB et al (2007) FlyBase: genomes by the dozen. Nucleic Acids Res 3:D486–D491
Cunha-Rodrigues M, Prudencio M, Mota MM et al (2006) Antimalarial drugs – host targets (re) visited. Biotechnol J 1:321–332
Davis AM, St-Gallay, SA, Kleywegt, GJ et al (2008) Limitations and lessons in the use of X-ray structural information in drug design. Drug Discov Today 13:831–841
de Beer TAP, Wells GA, Burge PB et al (2009) Antimalarial drug discovery: in silico structural biology and rational drug design. Infect Disord Drug Targets 9:304–318

Delfino RT, Santos-Filho OA, Figueroa-Villar JD (2002) Molecular modeling of wild-type and antifolate resistant mutant *Plasmodium falciparum* DHFR. Biophys Chem 98:287–300

DePristo MA, de Bakker PI, Blundell TL (2004) Heterogeneity and inaccuracy in protein structures solved by X-ray crystallography. Structure 12:831–838

Desowitz RS (1992) Malaria: from quinine to the vaccine. Hosp Pract 27:209–214, 217–224, 229–232

Dheyongera JP, Geldenhuys WJ, Dekker TG et al (2005) Antimalarial activity of thioacridone compounds related to the acronycine alkaloid. Bioorg Med Chem 13:1653–1659

Dror O, Shulman-Peleg A, Nussinov R et al (2004) Predicting molecular interactions in silico. I. A guide to pharmacophore identification and its applications to drug design. Curr Med Chem 11:71–90

Dubois J, Bourg S, Vrain C et al (2008) Collections of compounds – how to deal with them? Curr Comput Aided Drug Des 4:156–168

Dunkel M, Fullbeck M, Neumann S (2006) SuperNatural: a searchable database of available natural compounds. Nucleic Acids Res 34:D678–D683

Elsik CG, Mackey AJ, Reese JT et al (2007) Creating a honey bee consensus gene set. Genome Biol 8:R13

Ewing TJ, Makino S, Skillman AG et al (2001) DOCK 4.0: search strategies for automated molecular docking of flexible molecule databases. J Comput Aided Mol Des 15:411–428

Fatumo S, Plaimas K, Mallm JP et al (2009) Estimating novel potential drug targets of *Plasmodium falciparum* by analysing the metabolic network of knock-out strains in silico. Infect Genet Evol 9:351–358

Ferrari AM, Degliesposti G, Sgobba M et al (2007) Validation of an automated procedure for the prediction of relative free energies of binding on a set of aldose reductase inhibitors. Bioorg Med Chem 15:7865–7877

Flipo M, Beghyn T, Leroux V (2007) Novel selective inhibitors of the zinc plasmodial aminopeptidase PfA-M1 as potential antimalarial agents. J Med Chem 50:1322–1334

Fogel GB, Cheung M, Pittman E et al (2008) In silico screening against wild-type and mutant *Plasmodium falciparum* dihydrofolate reductase. J Mol Graph Model 26:1145–1152

Foth BJ, Ralph SA, Tonkin CJ et al (2003) Dissecting apicoplast targeting in the malaria parasite *Plasmodium falciparum*. Science 299:705–708

Frantz S (2007) Pharma faces major challenges after a year of failures and heated battles. Nat Rev Drug Discov 6:5–7

Friesner RA, Banks JL, Murphy RB et al (2004) Glide: a new approach for rapid, accurate docking and scoring. I. Method and assessment of docking accuracy. J Med Chem 47:1739–1749

Gardner MJ (1999) The genome of the malaria parasite. Curr Opin Genet Dev 9:704–708

Gardner MJ, Hall N, Fung E et al (2002) Genome sequence of the human malaria parasite *Plasmodium falciparum*. Nature 419:498–511

Gillet VJ (2008) New directions in library design and analysis. Curr Opin Chem Biol 12:372–378

Ginsburg H (2006) Progress in in silico functional genomics: the malaria metabolic pathways database. Trends Parasitol 22:238–240

Ginsburg H (2009) Caveat emptor: limitations of the automated reconstruction of metabolic pathways in Plasmodium. Trends Parasitol 25:37–43

Goede A, Dunkel M, Mester N et al (2005) SuperDrug: a conformational drug database. Bioinformatics 21:1751–1753

Goodsell DS, Olson AJ (1990) Automated docking of substartes to proteins by simulated annealing. Prot Struct Funct Genet 8:195–202

Güner O, Clement O, Kurogi Y (2004) Pharmacophore modeling and three dimensional database searching for drug design using catalyst: recent advances. Curr Med Chem 11:2991–3005

Gutiérrez-de-Terán H, Nervall M, Dunn BM et al (2006) Computational analysis of plasmepsin IV bound to an allophenylnorstatine inhibitor. FEBS Lett 580:5910–5916

Hall SE (2006) Chemoproteomics-driven drug discovery: addressing high attrition rates. Drug Discov Today 11:495–502

Hellerstein MK (2008) A critique of the molecular target-based drug discovery paradigm based on principles of metabolic control: advantages of pathway-based discovery. Metab Eng 10:1–9

Hernandez-Valladares M, Rihet P, ole-MoiYoi OK et al (2004) Mapping of a new quantitative trait locus for resistance to malaria in mice by a comparative mapping approach with human Chromosome 5q31-q33. Immunogenetics 56:115–117

Hesterkamp T, Whittaker M (2008) Fragment-based activity space: smaller is better. Curr Opin Chem Biol 1:260–268

Höltje HD, Sipple W, Rognan D et al. (2008) Molecular modeling: basic principles and applications, 3rd edn. Wiley-VCH, Weihem

Holt RA, Subramanian GM, Halpern A et al (2002) The genome sequence of the malaria mosquito *Anopheles gambiae*. Science 298:129–149

Hubbard TJ, Aken BL, Beal K et al (2007) Ensembl 2007. Nucleic Acids Res 35:D610–D617

Huey R, Morris GM, Olson AJ et al (2007) A semiempirical free energy force field with charge-based desolvation. J Comput Chem 28:1145–1152

Irwin JJ, Schoichet BK (2005) ZINC-a free database of commercially available compounds for virtual screening. J Chem Inf Model 45:177–182

Jambou R, Legrand E, Niang M et al (2005) Resistance of *Plasmodium falciparum* field isolates to in-vitro artemether and point mutations of the SERCA-type PfATPase6. Lancet 366:1960–1963

Jana S, Paliwal J (2007) Novel molecular targets for antimalarial chemotherapy. Int J Antimicrob Agents 30:4–10

Jones G, Willett P, Glen RC (1997) Development and validation of a genetic algorithm for flexible docking. J Mol Biol 267:727–748

Kanehisa M, Goto S, Hattori M et al (2006) From genomics to chemical genomics: new developments in KEGG. Nucleic Acids Res 34:D354–D357

Kasam V, Salzemann J, Breton V et al (2007) Proceedings of the Fifth IEEE workshop on challenges of large applications in distributed environments

Kassel DB (2004) Applications of high-throughput ADME in drug discovery. Curr Opin Chem Biol 8:339–345

Kellenberger E, Rodrigo J, Muller P et al (2004) Comparative evaluation of eight docking tools for docking and virtual screening accuracy. Proteins 57:225–242

Kimura S, Kawasaki T, Hatakeyama M (2004) OBIYagns: a grid-based biochemical simulator with a parameter estimator. Bioinformatics 20:1646–1648

Kissinger JC, Brunk BP, Crabtree J et al (2002) The Plasmodium genome database. Nature 419:490–492

Kuntz ID, Blaney JM, Oatley SJ et al (1982) A geometric approach to macromolecule–ligand interactions. J Mol Biol 161:269–288

Lang P, Yeow K, Nichols A (2006) Cellular imaging in drug discovery. Nat Rev Drug Discov 5:343–356

Laurie AT, Jackson RM (2006) Methods for the prediction of protein–ligand binding sites for structure-based drug design and virtual ligand screening. Curr Protein Pept Sci 7:395–406

Lawson D, Arensburger P, Atkinson P et al (2007) VectorBase: a home for invertebrate vectors of human pathogens. Nucleic Acids Res 35:D503–D505

Lemcke T, Christensen IT (1999) Towards an understanding of drug resistance in malaria: three-dimensional structure of *Plasmodium falciparum* dihydrofolate reductase by homology building. Bioorg Med Chem 7:1003–1011

Lindsay MA (2005) Finding new drug targets in the 21st century. Drug Discov Today 10:1683–1687

Lipinski CA, Lombardo F, Dominy BW et al (1997) Experimental and computational approaches to estimate solubility and permeability in drug discovery and development settings. Adv Drug Deliv Rev 23:3–25

Lipkus AH, Yuan Q, Lucas KA et al (2008) Structural diversity of organic chemistry. A scaffold analysis of the CAS Registry. J Org Chem 73:4443–4451

Mahmoudi N, Garcia-Domenech R, Galvez J et al (2008) New active drugs against liver stages of Plasmodium predicted by molecular topology. Antimicrob Agents Chemother 52:1215–1220

Marcou G, Rognan D (2007) Optimizing fragment and scaffold docking by use of molecular interaction fingerprints. J Chem Inf Model 47:195–207

Marrero-Ponce Y, Iyarreta-Veitía M, Montero-Torres A et al (2005) Ligand-based virtual screening and in silico design of new antimalarial compounds using nonstochastic and stochastic total and atom-type quadratic maps. J Chem Inf Model 45:1082–1100

McKie JH, Douglas KT, Chan C et al (1998) Rational drug design approach for overcoming drug resistance: application to pyrimethamine resistance in malaria. J Med Chem 41:1367–1370

Mehlin C, Boni E, Buckner FS et al (2006) Heterologous expression of proteins from *Plasmodium falciparum*: results from 1000 genes. Mol Biochem Parasitol 148:144–160

Miyamoto S, Kollman PA (1993) Absolute and relative binding free energy calculations of the interaction of biotin and its analogs with streptavidin using molecular dynamics/free energy perturbation approaches. Proteins 16:226–245

Muskavitch MA, Barteneva N, Gubbels MJ (2008) Chemogenomics and parasitology: small molecules and cell-based assays to study infectious processes. Comb Chem High Throughput Screen 11:624–646

Nayal M, Honig B (2006) On the nature of cavities on protein surfaces: application to the identification of drug-binding sites. Proteins 63:892–906

Nchinda T (1998) Malaria: a reemerging disease in Africa. Emerg Infect Dis 4:398–403

Paul N, Rognan D (2002) ConsDock: a new program for the consensus analysis of protein-ligand interactions. Proteins 47:521–533

Rarey M, Kramer B, Lengauer T et al (1996) A fast flexible docking method using an incremental construction algorithm. J Mol Biol 261:470–489

Rastelli G, Sirawaraporn W, Sompornpisut P et al (2000) Interaction of pyrimethamine, cycloguanil, WR99210 and their analogues with *Plasmodium falciparum* dihydrofolate reductase: structural basis of antifolate resistance. Bioorg Med Chem 8:1117–1128

Ridley RG (1998) Malaria: dissecting chloroquine resistance. Curr Biol 8:R346–R349

Ridley RG (2002) Medical need, scientific opportunity and the drive of antimalarial drugs. Nature 415:686–693

Saidani N, Grando D, Valadie H et al (2009) Potential and limits of *in silico* target discovery – case study of the search for new antimalarial chemotherapeutic targets. Infect Genet Evol 9:359–367

Salzemann J, Kasam V, Jacq N et al (2007) Grid enabled high throughput virtual screening against four different targets implicated in malaria. Proceedings of HealthGrid conference 2007, Studies in Health Technology and Informatics, 126:47–54

Santos-Filho OA, de Alencastro RB, Figueroa-Villar JD (2001) Homology modeling of wild type and pyrimethamine/cycloguanil-cross resistant mutant type *Plasmodium falciparum* dihydrofolate reductase. A model for antimalarial chemotherapy resistance. Biophys Chem 91(3):305–317

Schreck CE, Kline DL, Carlson DA (1990) Mosquito attraction to substances from the skin of different humans. J Am Mosq Control Assoc 6:406–410

Sheridan RP, Kearsley SK (2002) Why do we need so many chemical similarity search methods? Drug Discov Today 7:903–911

Sousa SF, Fernandes PA, Ramos MJ (2006) Protein-ligand docking: current status and future challenges. Proteins 65:15–26

Teramoto R, Fukunishi H (2008) Structure-based virtual screening with supervised consensus scoring: evaluation of pose prediction and enrichment factors. J Chem Inf Model 48:747–754

Towie N (2006) Malaria breakthrough raises spectre of drug resistance. Nature 440:852–853

Toyoda T, Brobey RK, Sano G (1997) Lead discovery of inhibitors of the dihydrofolate reductase domain of *Plasmodium falciparum* dihydrofolate reductase-thymidylate synthase. Biochem Biophys Res Commun 235:515–519

Utzinger J, Tanner M, Kammen DM et al (2002) Integrated program is key to malarial control. Nature 419:431

Venter JC, Adams MD, Myers EW et al (2001) The sequence of the human genome. Science 291:1304–1351

Verkhivker GM, Bouzida D, Gehlhaar DK et al (2000) Deciphering common failures in molecular docking of ligand–protein complexes. J Comput Aided Mol Des 14:731–751

Villar HO (2007) Computational medicinal chemistry. Curr Top Med Chem 7:1509–1513

Waller RF, McFadden GI (2005) The apicoplast: a review of the derived plastid of apicomplexan parasites. Curr Issues Mol Biol 7:57–79

Wiesner J, Seeber F (2005) The plastid-derived organelle of protozoan human parasites as a target of established and emerging drugs. Expert Opin Ther Targets 9:23–44

Wishart DS, Knox C, Guo AC et al (2009) HMDB: a knowledgebase for the human metabolome. Nucleic Acids Res 36:D901–D906

World Malaria Report (2005) World Health Organization, WHO/UNICEF, Geneva

Xie A, Sivaprakasam P, Doerksen R (2006) 3D-QSAR analysis of antimalarial farnesyltransferase inhibitors based on a 2,5-diaminobenzophenone scaffold. Bioorg Med Chem 14:7311–7323

Yeh I, Hanekamp T, Tsoka S (2004) Computational analysis of *Plasmodium falciparum* metabolism: organizing genomic information to facilitate drug discovery. Genome Res 14:917–924

Yuvaniyama J, Chitnumsub P, Kamchonwongpaisan S (2003) Insights into antifolate resistance from malarial DHFR-TS structures. Nat Struct Biol 10:357–365

Zhou Y, Ramachandran V, Kumar KA et al (2008) Evidence-based annotation of the malaria parasite's genome using comparative expression profiling. PLoS ONE 3:e1570

Zuegge J, Ralph S, Schmuker M, et al (2001) Deciphering apicoplast targeting signals – feature extraction from nuclear-encoded precursors of *Plasmodium falciparum* apicoplast proteins. Gene 280:19–26

Chapter 15
Informatics for Healthcare Epidemiology

Bala Hota

15.1 Introduction

Healthcare epidemiology is a sub-discipline of epidemiology, with the practical focus on surveillance, prevention, and control of adverse events in healthcare. Healthcare epidemiology has gained prominence in an era of performance improvement in healthcare; this is so because the study of healthcare epidemiology is the study of the factors and outcomes of healthcare and the quality of healthcare being provided to patients within hospitals. Information technology has the potential to be transformative in the practice of healthcare, but its applications are in their infancy. Healthcare informatics is a tool to enable greater efficiency and broadened scope of performance measurement; as will be discussed, greater maturity in standards adoption and tool creation is needed to permit maximal use of healthcare information technology (HIT) to improve patient safety.

15.2 Performance Measurement and Healthcare Associated Infections

Since the 1999 Institute of Medicine Report, "To err is human" (Kohn et al. 2000), medical errors have been appreciated to be a significant cause of patient morbidity and improvements in patient safety and health care quality have become major areas of study. Public disclosure of performance information has been promoted as an important method to attain these goals (Lansky 2002). To this end, consumers of healthcare increasingly desire readily available information about the performance of hospitals and providers in healthcare delivery. Efforts to measure the performance

B. Hota
Rush University Medical Center, Chicago, IL, USA

V. Sintchenko (ed.), *Infectious Disease Informatics*,
DOI 10.1007/978-1-4419-1327-2_15, © Springer Science+Business Media, LLC 2010

of care delivery may take the form of process measures (i.e. measurement of adherence to guidelines in processes of care), or outcome measures (i.e. measurement of the occurrence of events) (McKibben et al. 2005). Healthcare epidemiology is critical in performance improvement efforts, as it provides foundational knowledge that can inform interventions.

Healthcare associated infections (HAIs) are a potentially rich area for performance improvement. The United States' Centers for Disease Control and Prevention (CDC) have estimated that annually, up to 10% of hospitalized patients in the U.S.A. (i.e. 2,000,000 patients) develop an HAI, leading to 100,000 deaths a year (Weinstein 1998). HAIs are increasingly considered controllable through surveillance, proper infection control practice, and bundling of effective preventive practices (Pronovost et al. 2006; Bleasdale et al. 2007; Vernon et al. 2006; Evans 2005; Haley et al. 1985). Examples of strategies to reduce HAI rates are the bundling of effective insertion, monitoring, and removal practices for central-line associated bloodstream infection (CLABSI) rates; the use of chlorhexidine bathing on CLABSI rates; bundling semi-recumbent positioning, subglottic suctioning, and monitoring for ventilator-associated pneumonia (VAP) rates; and the appropriate insertion and removal of urinary catheters, with monitoring and feedback of urinary tract infection (UTI) rates to healthcare personnel for reduction of UTI rates (Yokoe et al. 2008).

The Centers of Disease Control and Prevention (CDC) have conducted surveillance of HAIs since the 1970s. Infection control practitioners have by and large conducted this surveillance through the manual collection of data. Case definitions for infection were developed and recognized internationally through guidelines from the National Nosocomial Infections Surveillance (NNIS) Definitions (Garner et al. 1988). In the 1970s, infection control programs were implemented voluntarily to implement NNIS criteria and measure HAI rates. In 1985, the CDC published the results of the SENIC project, documenting reductions in HAI rates following the efforts of infection control programs (Haley et al. 1985). Mandatory reporting of HAI rates, public disclosure of HAI rates, and changes in reimbursement strategies have more recently been used to promote reductions in the burden of HAIs (Yokoe and Classen 2008). These initiatives have increased the demand for high-quality measures of healthcare-associated events. To allow the workforce of infection preventionists to focus on education and infection reduction efforts, the automation of HAI surveillance through the use of HIT is critical.

15.3 Electronic Health Records

HIT uses technological innovation to transform information into knowledge. Making use of large quantities of data, exposing these data in organized ways, and revealing underlying meanings of data are among the several key aspects of information technology in healthcare. Information technology is best considered a tool, not a solution, for quality improvement. Typically, while implementing IT tools in healthcare, hospitals will first identify problematic business processes before being able to use HIT to improve care. Therefore, the development and implementation of HIT solutions will only be successful after following a problems and requirements discovery process in which current processes and needed changes are identified, measured, and synthesized (Hota et al. 2008).

The substrate for information technology in healthcare is the availability of data for use with analytic and decision support software. These data can be categorized based on the type of data being stored, and the destination of the data once stored. Categories of data types generated by clinical care include clinical (e.g. laboratory, microbiology, pathology, radiology, and pharmacy), finance (e.g. discharge diagnosis codes, utilization of health care services), and administrative data (e.g. admission, discharge and registration data, bed location, demographic and address data). Since finance functions play a major role, as part of mature electronic record systems, data warehouses and applications of finance and administrative data sets to healthcare epidemiology are usually more advanced and standardized than the use of patient level laboratory or microbiology data, even though the latter are potentially richer and more accurate sources of data.

Clinical databases can also be categorized based on the capacity for sharing, or the target audience for data. Most clinical and administrative data are stored in an electronic medical record, or EMR. The focus of the EMR is patient and hospital centered data storage and retrieval. EMR data may be standardized within vendor solutions, but may not be designed with the goal of interoperability between centers. Data may lack interoperability: terminologies may be restricted to local code sets and may be difficult to export generally. The solutions for the restrictions imposed by the EMR model are the electronic health record (EHR) and personally health record (PHR). For the former, an emphasis is placed on sharing data between institutions. Given that many individuals seek care at multiple centers, such health information exchange would permit a medical record that can move with the patient between centers, thus reducing redundancy in tests ordered, and knowledge gaps based on inaccessible data. For the latter, a patient specific record of all medical visits and care is created. This record is then controlled and managed by a patient, and maintained over the patient's lifetime of care. For the field of healthcare epidemiology, access to regional data about HAIs and multidrug resistant organisms (MROs) acquired by patients could generate more accurate rates of MROs infections and enhance efforts to reduce the acquisition and transmission of MROs. For example, one regional health information exchange found that 10% of patients with methicillin-resistant *S. aureus* (MRSA) colonization or infection were shared between multiple centers – suggesting that information exchange could improve the identification and isolation of MRSA colonized individuals on admission (Kho et al. 2008).

Barriers to the use of information technology in healthcare epidemiology and infection control exist (Kilbridge and Classen 2008). These barriers include technical issues and non-technical issues. Technical issues are related to the lack of interoperability of data between centers; due to the evolution occurring in HIT, there are many systems for storing data. Many hospitals may have legacy systems or purpose-built homegrown systems, and data are frequently trapped in unique silos that cannot be scaled more broadly. As a result, data cannot be shared between centers, cannot have rules for event detection applied, and are semantically distinct to each center. The use of standards is an important step in overcoming this barrier.

Non-technical barriers to the use of healthcare IT in healthcare epidemiology relate to diffusion of capability and knowledge about the value of healthcare IT to patient safety. Many centers do not have EMRs, and may not see the value of implementing EMRs. Furthermore, the staff to support the application of EMR data to HIT may be lacking. Finally, data within EMRs may not be sufficient to deploy HAI detection algorithms; collection of these data may require changes in the business practice of documentation by healthcare personnel. A lack of appreciation of the application of these data may hamper efforts to change practices in hospitals. For example, documentation of device use electronically (e.g. central venous catheters, endotracheal tubes, urinary catheters) is essential for HAI detection, but is infrequently done. A change in this documentation step from paper to electronic methods may be resisted because of a lack of understanding of the value of these data. The recognition of these barriers by developers of policy may lead to opportunities to improve use of HIT to enhance patient safety (Kilbridge and Classen 2008).

15.4 Building Databases for Healthcare Infection Control

15.4.1 Standards in Healthcare Informatics

The application of information technology to healthcare epidemiology and performance measurement is in its infancy, largely due to segmentation of data stores between centers, which limits data sharing. As a result, multicenter research that demonstrates the value of HIT in patient safety and healthcare epidemiology may face many barriers and have limited quality; however, recent work suggests that positive trends are emerging (de Keizer and Ammenwerth 2008). Many hospitals have EMR with electronic data captured which is inaccessible for standardized reporting requirements (Kilbridge and Classen 2008); individual hospitals may have administrative data, finance data, and clinical data in electronic formats with unique database structures and semantic differences in representation of clinical values. Furthermore, even with the use of single vendors, implementations of EMR may vary slightly between centers, making data inoperable. The two problems of heterogeneity in system architecture and semantic knowledge seriously limit the development of solutions to conduct surveillance for healthcare events of interest.

To reduce the impact of silos of clinical information, the use of standards in HIT has been promoted (Overhage et al. 2001; National Electronic Disease Surveillance System (NEDSS) 2001; Khan et al. 2006; Wurtz and Cameron 2005). Standards exist or are being developed for the representation of clinical information, for documentation, for the messaging of data, and for security between parties sharing data.

The importance of standards is that they can permit the interchange of data between centers and investigators with minimal data manipulation, ensuring that standard algorithmic detection rules can be implemented at multiple sites. In the context of healthcare epidemiology, this ensures that performance measurement and interhospital comparison is a feasible goal.

A robust set of standards for data interoperability currently exists. The representation of clinical information has been specified by standards such as Systematized Nomenclature of Medicine (SNOMED – College of American Pathologists) and Logical Observation Identifiers Names and Codes (LOINC – Regenstrief Institute) for test names, test results, specimen sources, and test methods; these standard vocabularies have also been implemented as part of natural language processing techniques to help apply semantic frameworks to free text reports, e.g. radiology reports. Standards for data messaging have been specified as part of the health level seven (HL7) frameworks (Holena and Blobel 1997). Specifications for data transmission are available at http://www.hl7.org. These specifications create a set of rules that permit the sharing of data in understandable formats between centers. The clinical document architecture (CDA) and continuity of care record (CCR) formats specify requirements to be used in the arrangement and transmission of documents, and are versions of XML (Dolin et al. 2006; Ferranti et al. 2006).

Naming vocabularies are essential standards for data interoperability. Figure 15.1 illustrates the role of naming vocabularies in healthcare epidemiology analysis. Two vocabularies warrant special mention as they have emerged as core components of a public health information network in the United States: SNOMED and LOINC. These two vocabularies have been endorsed by the United States Department of Health and Human Services and CDC for the representation of concepts in electronic databases (Wurtz and Cameron 2005). Laboratory data sent to the National Healthcare Safety Network are suggested to be represented using these vocabularies (Edwards et al. 2008). LOINC coding is used to represent test names, and has varying levels of specificity. In a LOINC code, the test type, the specimen used, and the method of test can all be supported. SNOMED codes are used to represent coded results (e.g. organism names, susceptibility interpretations, and serologic interpretations). Using these standards, most information about a test can be standardized semantically (Wurtz and Cameron 2005).

Data messaging standards provide a framework for data transfer (Holena and Blobel 1997). Figure 15.2 shows examples of the HL7 version 2.5 and 3.0 standards. HL7 standards have several features of note. First, messages are event-driven, meaning that typically messages are generated and transferred at the time of data creation or updates. Second, HL7 messages are hierarchical and intent to preserve the relationships of data from multiple databases. Third, different kinds of clinical data are represented by individual HL7 message types. In HL7 version 2.x, delimiters [e.g. the pipe symbol (|), caret (^), and ampersand (&)] are used to denote unique fields in HL7 messages, and the type of HL7 message being sent is found at the message header (i.e. the MSH component) of HL7 messages. The HL7 3.0 and above use XML to markup messages. The clinical care record and CDAs extend the use of XML to allow the inclusion of narrative text and provider

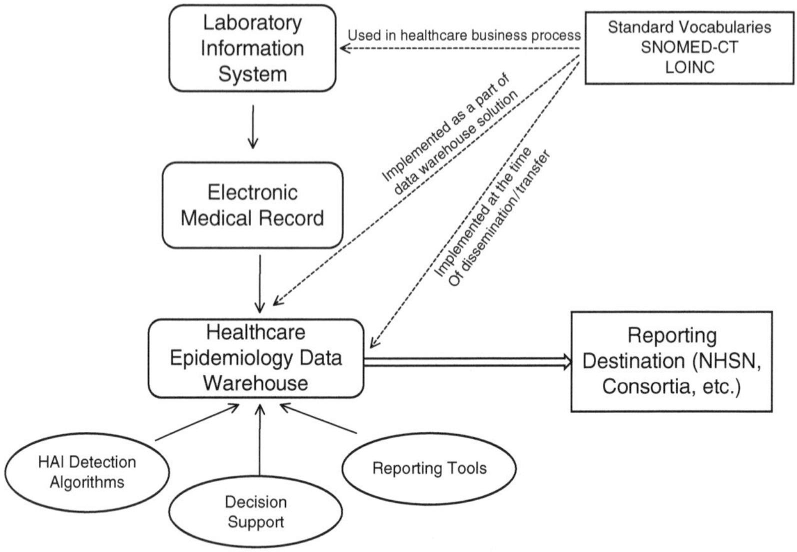

Fig. 15.1 Potential application points of standard vocabularies as a part of a healthcare epidemiology data warehousing solution

documentation in messages sent between systems (Dolin et al. 2006); (Ferranti et al. 2006). Using natural language processing and pattern matching, naming vocabularies can be applied to these narrative sections.

15.4.2 Data Auditing and Validation

Healthcare epidemiology concerns itself with the surveillance of healthcare events. Data obtained for surveillance is often displayed as charts and graphs for use in examining trends and detecting excess disease (e.g. outbreaks of infection or MROs). For a reporting solution to provide these data effectively, ongoing and systematic data auditing and validation is essential to allow confidence in the reports generated by the system. A problem common to clinical data warehouses is that captured data are either in a "free text" format that is not usable or use codes unique to the institution or laboratory information system which, though useful for within hospital trend analysis, do not scale well beyond single center measurement or require the local reinterpretation of measurement algorithms (Hota et al. 2008). Therefore, strategies for transforming data to ensure reliable and accurate measures are critical. Some examples of these strategies are the use of standard vocabularies as part of the business process of care and in local data stores; the mapping of local data to standard vocabularies for use in reporting and detection algorithms; systematic

mrn	ssn	last_name	first_name	dob	sex	race	race_desc
9999999	*9999999*	*CLINTON*	*BILL*	*8/16/49*	*M*	*W*	*WHITE*
9999999	9999999	CLINTON	BILL	8/16/49	M	W	WHITE
9999999	9999999	CLINTON	BILL	8/16/49	M	W	WHITE

ORDER_NO	street1	city	state	zip	obsdtime	obsenddtime	specrcvdtime	source_no	source_name
99999999	*1600 Pennsylvania Avenue NW*	*Washington*	*DC*	*20500*	*2/23/06*	*2/26/06*	*2/23/06*	*BLDV*	*BLOOD VENOUS*
99999999	1600 Pennsylvania Avenue NW	Washington	DC	20500	2/23/06	2/26/06	2/23/06	BLDV	BLOOD VENOUS
99999999	1600 Pennsylvania Avenue NW	Washington	DC	20500	2/23/06	2/26/06	2/23/06	BLDV	BLOOD VENOUS

loinc_no	loinc_name	snomed_no	snomed_result	obxobsdtime
11475-1	*Microorganism identified : PrId : Pt : xxx : Nom : Culture*	*L-25116*	*Streptococcus pneumoniae*	*2/23/06*
18895-3	CEFTRIAXONE : SUSC : PT : ISLT : ORDQN :	G-D450	Sensitive	2/23/06
18964-7	PENICILLIN : SUSC : PT : ISLT : ORDQN :	G-D450	Sensitive	2/23/06

↓

```
FHS|^~\&||^9999999^FEIIN|||20060227085000 || 20060227085000
BHS|^~\&||^9999999^FEIN|||20060227085000|| Stroger_20060227085000
MSH|^~\&||^9999999^FEIN|||20060227085000|| ORU^R01|20060227085001-0001|P|2.3.1
PID|1||9999999^^^^MR^&9999999&FEIN~9999999^^^^SS^SSA&Social Security Administration||CLINTON^BILL^^^^^L||19490816|M||
    W^WHITE^HL70005|1600 Pennsylvania Avenue NW^^Washington^DC^20500^ USA^H^^|| ^^PH^^^202^456-1111^^
OBR|1||99999999_1|11475-1^Microorganism identified : PrId : Pt : xxx : Nom :
    Culture^LN|||200602231434|200602261011||||||200602231614|BLDV^BLOOD VENOUS^HL70070|^^^^^^|^^^^^||||||||F|11475-
    1&Microorganism identified : PrId : Pt : xxx : Nom : Culture&In^^L-25116&Streptococcus
    pneumoniae&SNM|||||||||||||||||^^9999999^^^FEIN^||^WPN^PH^^^312^8644587^^|^^^^^^^^
OBX|1|CE|11475-1^Microorganism identified : PrId : Pt : xxx : Nom : Culture^LN|1|L-25116^Streptococcus
    pneumoniae^SNM||||||F|||200602231434|
OBX|2|CE|18895-3^CEFTRIAXONE : SUSC : PT : ISLT : ORDQN : ^LN|2|G-D450^Sensitive^SNM||||||F|||200602231434|
OBX|3|CE|18964-7^PENICILLIN : SUSC : PT : ISLT : ORDQN : ^LN|3|G-D450^Sensitive^SNM||||||F|||200602231434|
BTS|1|
FTS|1|
```

Fig. 15.2 Example of the conversion of table data to an HL7 2.3.1 message. This shows an ORU, or laboratory result, message type. Patient identifying information is in the PID segment, and laboratory test and result information are in the OBR and OBX segment. FHS, BHS, and MSH segments are headers for the message

auditing of individual records for accuracy in captured data; and the report based review of records to assess outlier events for erroneous data collection. Usually, discharge diagnoses are already coded with standard vocabularies, either as ICD-9 or 10 coding or diagnosis related groups (DRGs). In other databases, the use of standard vocabularies could be added as part of the business process of care: replacing local codes in laboratory information systems with a standard set of SNOMED and/or LOINC codes to represent test results and test names is one example. If this is not achievable, an alternative is to use translation, or mapping, tables to relate local terms to standard codes. Tools to achieve this effort exist: examples are WHONET (Stelling and O'Brien 1997; O'Brien and Stelling 1996; WHO|WHONET Software 2009), software from the World Health Organization available to standardize microbiology data and map microorganism names to SNOMED codes, or RELMA (McDonald et al. 2003); (RELMA – LOINC 2009), a tool to translate test names to LOINC values. Furthermore, WHONET uses a standard database schema, interacts with a variety of EHR and laboratory information systems, and also has an associated application, called BacLink, which can convert local data to standard SNOMED nomenclature (Stelling and O'Brien 1997; O'Brien and Stelling 1996; WHO|WHONET Software 2009).

Systematic auditing can be achieved through the regular random or sequential sampling of specific patient records with a manual review of medical charts to assess accuracy. As an example, one approach might be to obtain the first ten records in a month, or 1% of records in a quarter, and review these to assess completeness and accuracy of laboratory information, microbiology, pharmacy, or diagnosis codes. Finally, reports targeting outlier data measures could be of value to detect data quality issues. For example, a histogram of counts of diagnosis codes could be viewed to detect suspect codes. Also, this approach, when combined with auditing, can greatly assist validation. A report of patients with common conditions could be generated to prompt a sampling approach to assess accuracy. Patients with multidrug resistant organisms identified could populate a line list, which could then be prospectively reviewed for accuracy. Ad hoc reports, prior to dissemination, also provide an opportunity to validate electronic data. In general, a validation cycle should be incorporated to ensure reports properly represent the clinical or administrative question prompting a report (Wisniewski et al. 2003).

15.5 Information Systems for Healthcare Epidemiology

15.5.1 Use of Hit for Measurement

The use of EMRs in healthcare settings allows for the capture of clinical, finance, and administrative data in a single location, which can then be used to target areas that require improvement. Measures that document problems and interventions that target problem areas may rely on the electronic data for detection. IT is limited in its ability to improve patient safety, however, and should be considered as simply a

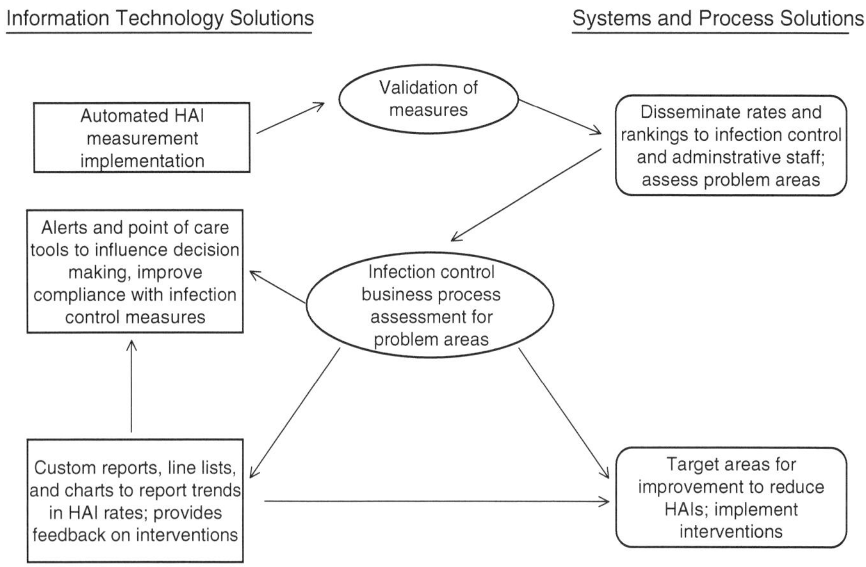

Fig. 15.3 The role of HIT in implementing an HAI measurement and reduction plan. Following business process assessment, HIT can be used to provide feedback of information as well as start decision support based interventions

tool to change behaviors. Figure 15.3 contrasts the pathway from measure development to decision support, and the role of HIT in these efforts.

Measures of HAI or adverse event rates are a good starting point for understanding problem areas. Once developed and validated, these measures can shine a light on inadequate processes of care in a hospital, and prompt a search for smaller components that require an assessment. For example, high CLABSI rates may indicate a breakdown in multiple areas of healthcare: improper line insertion practice, failure to remove central lines, changes in device utilization, poor dressing care, changes in comorbidities of patients admitted to the hospital, and changes in hand hygiene or environmental cleaning practice (Yokoe et al. 2008; Marschall et al. 2008). With such multifactorial causes of HAIs, HAI rates are the markers of many root causes. Therefore, designing an IT based solution to high HAI rates may be inappropriate, but may give an indication that changes in the business process of care are required.

If, on the other hand, an assessment of high HAI rates indicates that a process measure amenable to an IT intervention exists, then decision support tools may become essential in an improvement effort. The advantages of decision support are that it can be implemented at the point of care, it can be automatically triggered at the time of a decision being made, and it can (when effective) change behavior in a way that intervention teams tasked with changing behavior may not be able to do. Some examples of successful electronic HAI measures, decision support and alerting tools, and ordering systems to enhance care are outlined in Sect. 15.5.2.

15.5.2 Monitoring Infection Control Interventions

Many measures of HAI exist; vendors of EMR software may provide customized indicators of disease states unique to their implementations (Young and Stevenson 2008). These measures are often "black box" methodologies in which the exact algorithms used for measurement are proprietary and opaque to users. The future of HAI surveillance is the use of automated HAI measures that are interoperable between systems, produce results that are reliable within centers, and generate rates that can be used for between hospital comparisons or benchmarking. The use of vocabulary and messaging standards within EHRs may permit centralized processing of data for HAI measurement; for example, all positive and negative blood cultures may be sent to a regulatory or quality measurement agency or consortium, and BSI rates produced from the interoperable data that have been sent. Alternatively, the use of vocabulary standards at the database level may allow for the decentralized application of standard algorithms to detect and measure HAI rates. Finally, for each measure, the appropriate case-mix adjustment should be studied and applied; in many instances, these measures still require assessment, especially in the context of the increasing availability of electronic data (Harris and McGregor 2008; McGregor et al. 2006).

Research has been conducted to attempt the measurement of HAIs. Some measures have been studied using administrative and diagnosis coding data. Examples of efforts in this domain include work on SSI surveillance (Huang et al. 2007; Yokoe et al. 2004), the surveillance of *Clostridium difficile* (Dubberke et al. 2006), and the benchmarking of HAI rates (Julian et al. 2006; Sherman et al. 2006; Stevenson et al. 2008). Although administrative data is an easily obtained data source and is already standardized between institutions, concerns have been raised regarding the predictive value of administrative data for the detection of HAIs (Sherman et al. 2006; Stevenson et al. 2008). In contrast, laboratory reports, although a potentially richer source of data, often lack standardization, limiting general integration. The successful use of laboratory data for surveillance has been demonstrated for CLABSI surveillance (Trick et al. 2004). Efforts to conduct surveillance for ventilator associated pneumonia have been more complex, and algorithms likely will require electronic documentation of both laboratory results, radiological studies, and respiratory therapy care (Klompas et al. 2008a, b); if these datasets are available, however, a strategy that follows changes in ventilator settings, laboratory values, and sputum culture results holds promise (Table 15.1).

15.5.3 Decision Support

Moving beyond simply measuring HAI rates, detection algorithms to implement infection control interventions have been shown to accurately detect at-risk patients. Many centers have the facilities to identify patients with prior isolation of MROs from clinical or surveillance cultures. These patients may then be collated on admission on a "line-list" or daily report which can be utilized to recommend

Table 15.1 Examples of successful electronic approaches to HAI detection, with associated performance characteristics

Data source and measure	Numerator	Denominator	Performance characteristics
Administrative data			
Surgical site infection (SSI) (Yokoe et al. 2004)	Patients with ICD-9 CM codes suggesting SSI or with readmission ≤ 60 days AND antimicrobial use ≥ 9 days from procedure	Total number of patients with studied procedures	Sensitivity 79–97%; positive predictive value 20–42%
Clostridium difficile (Dubberke et al. 2006)	Patients with ICD-9 codes documenting *C. difficile* infection	Total number of admissions	Kappa 0.72; sensitivity 78%; specificity 99.7%
Laboratory/microbiology			
Central-line associated blood-stream infections (CLABSI) (Trick et al. 2004)	Patients with electronically applied NNIS criteria for CLABSI	Central-line days	Sensitivity 81%; specificity 90%; positive predictive value 81%; kappa 0.73
Hybrid			
Ventilator-associated pneumonia (Klompas et al. 2008a, b)	Patients with change in ventilator settings that persist ≥ 48 h AND fever OR WBC > 12,000 OR WBC < 4,000 cells/mm^3 AND sputum gram stain with ≥ 25 PMNs per HPF AND New radiographic infiltrate for ≥ 72 h	Ventilator days	Positive predictive value 100%

candidates for isolation to infection preventionists (Wisniewski et al. 2003; Evans et al. 2004). Furthermore, this list can be used to implement decision support, with automated alerts or ordering of contact isolation. Investigators at one center in the United States have created an index to identify patients at high risk of carriage of one MROs (methicillin-resistant *Staphylococcus aureus* (MRSA)) by targeting those not only with prior MRSA carriage, but also those with higher risk of carriage, based on the presence of longer durations of length of stay, older age, prior antimicrobial use, or use of hemodialysis (Evans et al. 2008). High-risk patients based on these criteria were fivefold more likely to be colonized with MRSA as compared with low risk patients, when patients were assessed with PCR. Automated alerts targeting those likely to be colonized with multidrug resistant organisms require further assessment, but may represent a novel method of detecting patients at risk for colonization with MROs, and help targeting the use of resources in the setting of increasing burdens of healthcare surveillance.

The most experience in process improvement using HIT has been gained with the implementation of computerized provider order entry and decision support systems. HIT has been used for many years to measure the adverse effects and increased costs of medication errors; mature systems exist for the provider entry of medications to eliminate the need for the handwriting of medication orders and corresponding opportunities for error. Antimicrobial use measurement, though complex, can help systems understand the utilization of antimicrobials and can help inform risk assessments for unit, hospital, and regional risks for MDRO acquisition (Fridkin et al. 1999; Monnet et al. 1998; Cunha 2002; Rogues et al. 2007; Charbonneau et al. 2006; Muller et al. 2006). A recent review highlights many accomplishments that have occurred in the domain of HIT and decision support tools for improvement in the prescribing of medications, with reductions in duplicate or redundant medication use, appropriate dosing, and allergy detection (Kuperman et al. 2007).

15.6 Reporting Tools

A major feature of healthcare epidemiology is the time-dependent nature of data being measured. Users of the data value graphical representations of collected information. Several methods of data presentation are typically used to display data and allow an understanding of rates, thus transforming information to knowledge. Interrupted time series, in which rates are graphed over time and critical events create interruptions in the series, are the predominant method of data presentation (Shardell et al. 2007). Changes in slope or intercept are noted on the graphs at the start and during interventions, as compared with a pre-intervention period. Rising in use are statistical process control (SPC) charts, which plot counts, rates, or time to events on the vertical-axis as compared with time on the horizontal axis (Walberg et al. 2008; Morton et al. 2001; Kahn et al. 1996). SPC charts have a well-established literature and history of use in the field of industrial management. A common feature among SPC charts is the graphing of a center line, or mean, value of a process, and warning lines, typically three standard deviations from the mean. Alerts are issued when trend data or points are found above or below the warning lines. A third method of data presentation is funnel plots, which graph rates as a function of denominator size (Spiegelhalter 2005). The underlying premise of these graphs is that rates with small denominators are inherently unstable, are more likely to show variability, and are more likely to regress to the mean from one period to another. These graphs have utility in comparing measures from centers or units of unequal size in a way that rank ordering rates do not; in addition, they provide alert confidence bands that change based on denominator size (Fig. 15.4).

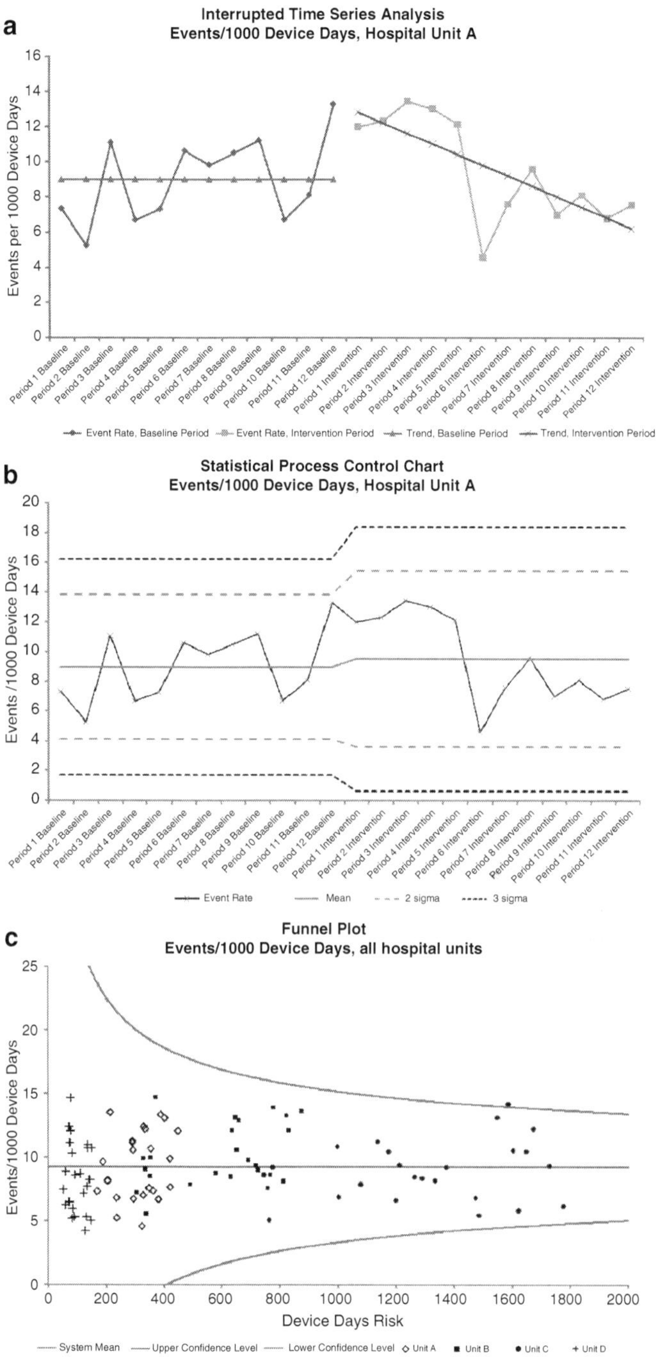

Fig. 15.4 Graphical displays of automated HAI measures. Data are for a hypothetical HAI measure obtained in four hospital units. Unit A is a 12 bed unit, unit B a 24 bed unit, unit C a 48 bed unit, and unit D a six bed unit. (**a**) Interrupted time series analysis; (**b**) Statistical process control chart; (**c**) Funnel plot of rates vs. device days at risk

The value of automated surveillance is that graphs can be produced and give feedback to infection preventionists, institutions, and regulatory agencies much faster than conventional manual measurement methods. If reliable, automated, measures of HAIs can be produced using HIT, then the possibility of a real-time assessment of the quality of healthcare can be achieved. As stantards adoption increases, such measurement need not be limited to technologically advanced centers, but can be instituted for a region. Regional differences in MROs epidemiology and risk adjustment based on case-mix could be used to generate near real-time measurement of HAI rates (see Chap. 17 for examples).

15.7 Concluding Remarks

Although the field of HIT has much to offer in terms of enhancing the measurement of HAIs and enabling automated surveillance of healthcare-associated events, continued innovation is essential. Areas of need include the adoption and validation of current manual surveillance processes to a more automated approach; the creation of new, reliable measures between centers to allow comparisons; the use of measures to permit global process assessment and improvement; the real-time reporting of measures to create feedback loops of improvement; and the implementation and dissemination of decision support tools that can be added to clinical workflows and can enhance patient care.

HIT alone does not represent a solution to healthcare quality issues, but it can be an effective tool to understand and improve care. The science of HIT implementation and assessment continues to improve, with more multicenter and methodologically sound research occurring. The future of HIT and healthcare epidemiology is likely to show better support of standards in EHR solutions, better interoperability, and more tools for clinical use to improve patient care.

References

Bleasdale SC, Trick We, Gonzalez IM et al (2007) Effectiveness of chlorhexidine bathing to reduce catheter-associated bloodstream infections in medical intensive care unit patients. Arch Intern Med 167(19):2073–2079

Charbonneau P, Parienti JJ, Thibon P et al (2006) Fluoroquinolone use and methicillin-resistant *Staphylococcus aureus* isolation rates in hospitalized patients: a quasi experimental study. Clin Infect Dis 42(6):778–784

Cunha BA (2002) Strategies to control antibiotic resistance. Semin Respir Infect 17(3):250–258

de Keizer NF, Ammenwerth E (2008) The quality of evidence in health informatics: how did the quality of healthcare IT evaluation publications develop from 1982 to 2005? Int J Med Inform 77(1):41–49

Dolin RH, Alschuler L, Boyer S et al (2006) HL7 clinical document architecture, Release 2. J Am Med Inform Assoc 13(1):30–39

Dubberke ER, Reske KA, McDonald LC et al (2006) ICD-9 codes and surveillance for *Clostridium difficile*-associated disease. Emerg Infect Dis 12(10):1576–1579

Edwards JR, Pollock DA, Kupronis BA et al (2008) Making use of electronic data: The National Healthcare Safety Network eSurveillance Initiative. Am J Infect Control 36:S21–S26

Evans B (2005) Best-practice protocols: VAP prevention. Nurs Manage 36(12):10–14

Evans RS, Lloyd JF, Abouzelof RH et al (2004) System-wide surveillance for clinical encounters by patients previously identified with MRSA and VRE. Stud Health Technol Inform 107(Pt 1):212–216

Evans RS, Wallace CJ, Lloyd JF et al (2008) Rapid identification of hospitalized patients at high risk for MRSA carriage. J Am Med Inform Assoc 15(4):506–512

Ferranti JM, Musser RC, Kawamoto K et al (2006) The clinical document architecture and the continuity of care record: a critical analysis. J Am Med Inform Assoc 13(3):245–252

Fridkin SK, Steward CD, Edwards JR et al (1999) Surveillance of antimicrobial use and antimicrobial resistance in United States hospitals: project ICARE phase 2. Project Intensive Care Antimicrobial Resistance Epidemiology (ICARE) hospitals. Clin Infect Dis 29(2):245–252

Garner, JS, Jarvis WR, Emori TG et al (1988) CDC definitions for nosocomial infections. Am J Infect Control 16(3):128–140

Haley RW, Culver DH, White JW et al (1985) The efficacy of infection surveillance and control programs in preventing nosocomial infections in US hospitals. Am J Epidemiol 121(2): 182–205

Harris AD, McGregor JC (2008) The importance of case-mix adjustment for infection rates and the need for more research. Infect Control Hosp Epidemiol 29(8):693–694

Holena M, Blobel B (1997) Healthcare information system approaches based on middleware concepts. Stud Health Technol Inform 45:178–185

Hota B, Jones RC, Schwartz DN (2008) Informatics and infectious diseases: What is the connection and efficacy of information technology tools for therapy and health care epidemiology? Am J Infect Control 36(3):S47–S56

Huang SS, Livingston JM, Rawson NS (2007) Developing algorithms for healthcare insurers to systematically monitor surgical site infection rates. BMC Med Res Methodol 7:20

Julian KG, Brumbach AM, Chicora MK et al (2006) First year of mandatory reporting of healthcare-associated infections, Pennsylvania: An infection control-chart abstractor collaboration. Infect Control Hosp Epidemiol 27(9):926–930

Kahn MG, Bailey TC, Steib SA et al (1996) Statistical process control methods for expert system performance monitoring. J Am Med Inform Assoc 3(4):258–269

Khan AN, Griffith SP, Moore C et al (2006) Standardizing laboratory data by mapping to LOINC. J Am Med Inform Assoc 13(3):353–355

Kho AN, Lemmon L, Commiskey M et al (2008) Use of a regional health information exchange to detect crossover of patients with MRSA between urban hospitals. J Am Med Inform Assoc 15(2):212–216

Kilbridge PM, Classen DC (2008) The informatics opportunities at the intersection of patient safety and clinical informatics. J Am Med Inform Assoc 15(4):397–407

Klompas M, Kleinman K, Platt R (2008a) Development of an algorithm for surveillance of ventilator-associated pneumonia with electronic data and comparison of algorithm results with clinician diagnoses. Infect Control Hosp Epidemiol 29(1):31–37

Klompas M, Kulldorff M, Platt R (2008b) Risk of misleading ventilator-associated pneumonia rates with use of standard clinical and microbiological criteria. Clin Infect Dis 46(9): 1443–1446

Kohn LT, Corrigan JM, Donaldson MS (2000) To err is human: Building a safer health system, Committee on quality of health care in America, Institute of Medicine report. National Academy of Press, Washington

Kuperman GJ, Bobb A, Payne TH et al (2007) Medication-related clinical decision support in computerized provider order entry systems: A review. J Am Med Inform Assoc 14(1):29–40

Lansky D (2002) Improving quality through public disclosure of performance information. Health Aff (Millwood) 21(4):52–62

Marschall J, Mermel LA, Classen D et al (2008) Strategies to prevent central line-associated bloodstream infections in acute care hospitals. Infect Control Hosp Epidemiol 29:S22–S30

McDonald CJ, Huff SM, Suico JG et al (2003) LOINC, a universal standard for identifying laboratory observations: a 5-year update. Clin Chem 49(4):624–633

McGregor JC, Perencevich EN, Furuno JP (2006) Comorbidity risk-adjustment measures were developed and validated for studies of antibiotic-resistant infections. J Clin Epidemiol 59(12):1266–1273

McKibben, L, Horan T, Tokars JI et al (2005) Guidance on public reporting of healthcare-associated infections: recommendations of the Healthcare Infection Control Practices Advisory Committee. Am J Infect Control 33(4):217–226

Monnet DL, Archibald LK, Phillips L et al (1998) Antimicrobial use and resistance in eight US hospitals: complexities of analysis and modeling. Intensive Care Antimicrobial Resistance Epidemiology Project and National Nosocomial Infections Surveillance System Hospitals. Infect Control Hosp Epidemiol 19(6):388–394

Morton AP, Whitby M, McLaws ML et al (2001) The application of statistical process control charts to the detection and monitoring of hospital-acquired infections. J Qual Clin Pract 21(4):112–117

Muller A, Mauny F, Talon D et al (2006) Effect of individual- and group-level antibiotic exposure on MRSA isolation: a multilevel analysis J Antimicrob Chemother 58(4):878–881

National Electronic Disease Surveillance System (NEDSS): A standards-based approach to connect public health and clinical medicine (2001) J Public Health Manag Pract 7(6):43–50

O'Brien TF, Stelling JM (1996) WHONET: removing obstacles to the full use of information about antimicrobial resistance. Diagn Microbiol Infect Dis 25(4):162–168

Overhage JM, Suico J, McDonald CJ (2001) Electronic laboratory reporting: barriers, solutions and findings. J Public Health Manag Pract 7(6):60–66

Pronovost P, Needham D, Berenholtz S et al (2006) An intervention to decrease catheter-related bloodstream infections in the ICU. N Engl J Med 355(26): 2725–2732

RELMA – LOINC. Available from: http://loinc.org/relma. Accessed 25 January 2009

Rogues AM, Dumartin C, Amadeo B et al (2007) Relationship between rates of antimicrobial consumption and the incidence of antimicrobial resistance in *Staphylococcus aureus* and *Pseudomonas aeruginosa* isolates from 47 French hospitals. Infect Control Hosp Epidemiol 28(12):1389–1395

Shardell M, Harris AD, El-Kamary SS et al (2007) Statistical analysis and application of quasi experiments to antimicrobial resistance intervention studies. Clin Infect Dis 45(7):901–907

Sherman ER, Heydon KH, St John KH et al (2006) Administrative data fail to accurately identify cases of healthcare-associated infection. Infect Control Hosp Epidemiol 27(4):332–337

Spiegelhalter DJ (2005) Problems in assessing rates of infection with methicillin resistant *Staphylococcus aureus*. Brit Med J 331(7523):1013–1015

Stelling JM, O'Brien TF (1997) Surveillance of antimicrobial resistance: The WHONET program. Clin Infect Dis 24:S157–S168

Stevenson KB, Khan Y, Dickman J et al (2008) Administrative coding data, compared with CDC/NHSN criteria, are poor indicators of health care-associated infections. Am J Infect Control 36(3):155–164

Trick WE, Zagorski BM, Tokars JI et al (2004) Computer algorithms to detect bloodstream infections. Emerg Infect Dis 10(9):1612–1620

Vernon MO, Hayden MK, Trick WE et al (2006) Chlorhexidine gluconate to cleanse patients in a medical intensive care unit: the effectiveness of source control to reduce the bioburden of vancomycin-resistant enterococci. Arch Intern Med 166(3):306–312

Walberg M, Froslie KF, Roislien J (2008) Local hospital perspective on a nationwide outbreak of *Pseudomonas aeruginosa* infection in Norway. Infect Control Hosp Epidemiol 29(7):635–641

Weinstein RA (1998) Nosocomial infection update. Emerg Infect Dis 4(3):416–420

WHO|WHONET Software. Available from: http://www.who.int/drugresistance/whonetsoftware/en/. Accessed 30 January 2009

Wisniewski MF, Kieszkowski P, Zagorski BM et al (2003) Development of a clinical data warehouse for hospital infection control. J Am Med Inform Assoc 10(5):454–462

Wurtz R, BJ Cameron (2005) Electronic laboratory reporting for the infectious diseases physician and clinical microbiologist. Clin Infect Dis 40(11):1638–1643

Yokoe DS, Classen D (2008) Improving patient safety through infection control: a new healthcare imperative. Infect Control Hosp Epidemiol 29:S3–S11

Yokoe DS, Noskin GA, Cunnigham SM et al (2004) Enhanced identification of postoperative infections among inpatients Emerg Infect Dis 10(11):1924–1930

Yokoe DS, Mermel LA, Anderson DJ et al (2008) A compendium of strategies to prevent health-care-associated infections in acute care hospitals. Infect Control Hosp Epidemiol 29:S12–S21

Young J, Stevenson KB (2008) Real-time surveillance and decision support : optimizing infection control and antibiotic choices at the point of care. Am J Infect Control 36(3):S67–S74

Chapter 16
Automated, High-throughput Surveillance Systems for Public Health

Ross Lazarus

16.1 Introduction

The emphasis in this chapter is on automated information systems capable of supporting high-throughput public health surveillance, adding public health value to large volumes of routinely collected health care data, at relatively low cost. Infectious disease is the main focus, but other important public health problems of current interest are addressed, as they share many challenges in terms of the information systems needed. In addition to access to complete and comprehensive EHR containing identifiable human data, these systems depend on reliable algorithms and decision rules that can be efficiently implemented and then objectively evaluated and iteratively refined. Examples of decision rules and production information systems are briefly reviewed to illustrate potential solutions to these challenges. Some of the material is abstract and generalized, but the practical and technical illustrations are necessarily concrete, being examples drawn from the author's collaborations in the United States, where reasonably complete EHR are relatively abundant and a number of automated, high-throughput systems of proven validity are operating (Lazarus et al. 2008).

Human EHR are the major focus of the work and systems described here, where the term EHR refers to reasonably complete data from an electronic medical record (EMR) system, or another electronic record of an individual's health care, including newer forms such as personally controlled health records (PCHR). Measurable public health benefits have been demonstrated for notifiable disease reporting systems using EHR. A desire to do something that might minimize the impact of an extremely improbable, but potentially high-impact, threat posed by biological terrorism or an emerging pandemic infection has motivated much recent activity, including work based on the hope that non-health related data sources, such as retail sales data, might be usefully repurposed for public health surveillance.

R. Lazarus
Channing Laboratory, Harvard Medical School, Boston, MA, USA

V. Sintchenko (ed.), *Infectious Disease Informatics*,
DOI 10.1007/978-1-4419-1327-2_16,

Additional benefit in terms of improved public health, arising from substantial and ongoing investment in systems using these non-health related data, remains to be demonstrated.

Public health activities are often aimed at preventing the occurrence or spread of illness, with a broad focus on whole communities or populations. This stands in stark contrast to investment in medical care in most developed countries, which is largely directed at diagnosis and treatment of existing pathology in individual patients. Historically, prevention has proved to be a very effective strategy to decrease the overall community burden, particularly for infectious diseases, such as tuberculosis or influenza. Enumerating and tracing of outbreaks is a core source of information for public health planning. Effective interventions often start with an index case and take actions that minimize the risk of further spread of infection, such as the delivery of effective curative treatment. The basic idea in public health surveillance is to routinely search for cases of illness that may have public health importance and to bring them to the attention of authorized public health officials, in a secure manner, so that they can efficiently intervene to prevent spread or other exposure. In addition to sources of data, appropriate governance and security systems, and appropriate automated case detection methods, this also requires some way of presenting the data to authorized users, in ways that help them to effectively manage public health resources. In the case of large-scale events, timeliness is of fundamental importance, as early response is likely to be far more efficient in most of the plausible scenarios.

Modern public health surveillance arose as a scientific response to unpredictable, explosive outbreaks of deadly diseases. It is hard to imagine the situation less than two centuries ago, when epidemic infections, including plague, influenza, and cholera, periodically ravaged Europe and Asia. Almost no effective control was implemented, because the causative organisms and their mechanisms of spread were not well understood. However, even in the absence of a formal scientific understanding of these diseases, some European authorities were able to effectively control the spread of plague to geographically isolated regions in the seventeenth and eighteenth centuries (Konstantinidou et al. 2009). More recently, John Snow combined surveillance, intervention, and evaluation to form what is generally recognized as the basis for modern population health science, or epidemiology (Paneth 2004). In a London cholera outbreak during 1854, he convinced local parish authorities to disable the mechanical Broad Street water pump. Subsequent surveillance data showed decreased cholera incidence rates among nearby residents. This was a defining moment in modern public health, although as Snow himself humbly pointed out, rates also fell in distant London communities, because the epidemic was already waning at the time of the intervention.

Surveillance in public health implies recording and reporting aggregate level details from individual cases of a disease over time. For example, effective interventions, such as curative or prophylactic antibiotic treatment, are often available to minimize risk of spread of disease from infected individuals. The goals of surveillance and analysis may include the following:

- Identifying, quantifying, and monitoring public health threats,
- Identifying opportunities to intervene efficiently, and
- Evaluating interventions.

16.2 Evaluation of Surveillance Systems

Measuring the performance of public health systems against pre-specified goals and other benchmarks is an integral part of good system design, so that competing methods can be objectively compared and existing systems can be iteratively improved (German et al. 2001). Surveillance systems can be modeled as disease detection systems or diagnostic tests applied to populations rather than to individuals. So when appropriate, independent "gold standard" data are available, surveillance system performance can be measured in the usual diagnostic test performance terms of validity, positive and negative predictive values, sensitivity, and specificity. However, as surveillance goals become less distinct, evaluation becomes increasingly challenging. For some potentially catastrophic events such a pandemic of avian influenza, there is no obvious way to externally validate the performance of any surveillance system until after an event has been observed.

The key ingredients of public health surveillance include partner organizations willing to serve as reliable sources of data, secure technologies to manage data and distribute queries and results, reliable statistical and other methods to identify opportunities for intervention, and objective measures of performance so impacts of implementations can be evaluated, compared, and improved. The specific objectives of a surveillance system are critical determinants of the kinds of data needed, the statistical or other methods used in evaluating the data, and the measurements needed to evaluate and improve performance.

16.3 Surveillance Goals

Public health surveillance systems can be divided into four broad classes, with decreasingly specific goals:

- *Notifiable disease surveillance*: Designed to monitor and control common and well known threats to public health – the oldest and best understood model. These systems usually rely on timely notifications from clinicians or laboratories about cases of communicable diseases.
- *Syndromic surveillance*: Generally aimed at the earliest possible detection of cyclical natural disease patterns, or potential deliberate bioterrorism. These systems usually monitor health care utilization patterns, in real time, and rely on detecting case features that are discernable before laboratory diagnoses are confirmed.

- *Adverse event surveillance*: Post-marketing surveillance for adverse events represents an important potential opportunity for EHR-based surveillance systems to help to overcome well-recognized deficiencies in the current regulatory processes for ensuring the safety of widely used therapeutic interventions, particularly of medications and vaccines.
- *Biosurveillance*: Attempting to borrow strength across all available and potentially informative sources of data in order to detect potentially important health perturbation, at the earliest possible time in the course of an event of public health importance.

Each of these activities can use EHR, but each requires very different decision rules, implementation, validation, evaluation, and presentation. In the sections that follow, technical aspects of each of these four broad public health surveillance goals are discussed in more detail, followed by sections in which some of the specific practical challenges to building and maintaining sustainable surveillance systems are discussed, including manual system challenges, sources of electronic data, security and protection of human subject data, statistical methods and visualization, technologies for data transfer, and governance for large-scale surveillance systems.

16.4 Notifiable Disease Surveillance

All functional public health jurisdictions operate some kind of case notification protocol for important conditions of public health significance. Some have automated notifiable disease reporting from electronic health data such as electronic laboratory records (ELR) (Klompas et al. 2007). With specific notifiable disease definitions, we can search electronic health data for matching criteria in terms of physician assigned diagnoses, laboratory test results, and other clinical observations (Lazarus et al. 2008). In general, notifiable disease surveillance involves relatively little statistical sophistication, as the goal is to identify all real cases and bring them to the attention of appropriate authorities in a secure and useful manner (Lazarus et al. 2008). Validation is important for new systems – it is expensive to conduct, but it is an essential component of good system design to ensure that public health resources are not diverted away by false alerts (Klompas et al. 2008a, b). The key issue for investment in public health activities is whether it efficiently leads to measurable improvements in the control of public health challenges.

16.4.1 Deficiencies in Existing Systems

In practice, most routine, planned public health case surveillance operates through practitioner initiated, manually transcribed case notification of designated diseases of potential public health importance. National US data from the CDC Morbidity and Mortality Weekly Reports (MMWR, http://www.cdc.gov/mmwr/) is a well-known example of this combined approach, where local and State records are centrally collated and made available for surveillance purposes. It is widely agreed that these systems suffer from under-enumeration and from incomplete data on reported cases, two deficiencies that automated systems can address directly. Automated ELR are increasingly being used to supplement manual reporting, but these also have substantial limitations that can be addressed using comprehensive EHR.

16.4.2 Challenges in Automated Disease Detection

When dealing with specific diseases, data streams, case detection algorithms, and the systems implementing all of these can be tested for validity and performance when we can calibrate against an independent source of case notifications. False negative and false positive alerts are both potentially very bad – the former leading to false reassurance and the latter to wasted investigational resources and other undesirable outcomes.

Distinguishing acute infection from chronic illness is useful in case notifications because they often trigger different kinds of intervention. For example, close contacts of a case of acute hepatitis B may benefit from immediate preventive intervention, and data on the changing epidemiology of the disease, particularly the impact of universal vaccination programs, have both planning and evaluation value (Klompas et al. 2008a, b). Surveillance using laboratory results alone is increasingly being used to supplement manual reporting systems, but cannot detect conditions requiring non-laboratory clinical findings, such as pelvic inflammatory disease, and cannot reliably distinguish between the acute, chronic, and resolved states of disorders such as viral hepatitis B or C.

Similarly, sensitivity and specificity will suffer if surveillance algorithms rely on diagnostic codes alone, because busy clinicians may not make ideal coding choices. In practice, clinicians tend to choose from a limited subset of codes when coding encounters using an EHR system. This behavior is often reinforced by the design of the EHR interface, where available code choices are deliberately restricted to make the interface less overwhelming for the user. Although distinct codes may exist in the full ICD-9 for some disorders, a clinician is likely to choose the same ICD-9 code to represent previous resolved illness, suspected current disease, confirmed acute disease, and current chronic disease, complicating the decision rules needed when these must be distinguished.

These challenges require careful design, testing, and implementation, but the problem is tractable. For example, an algorithm that takes into account the presence of elevated liver function tests, biochemical jaundice, a positive test for hepatitis B, and no prior ICD-9 codes or laboratory tests for hepatitis B can detect acute viral hepatitis B with sensitivity of 97.4% and specificity of 93.8% in the ESP system (Klompas et al. 2008a, b). This algorithm identified eight cases of acute hepatitis B without any false positives (Klompas et al. 2008a, b). Seven of the eight cases were novel, four were hitherto completely unknown to the health department, and three of the four previously reported cases had been misreported as chronic rather than acute cases. Surveillance of other notifiable diseases such as tuberculosis is also complicated by cases of culture-negative disease being missed by purely laboratory-based reporting. For tuberculosis, such false negative cases can be identified by finding associated prescriptions for pyrazinamide or other first-line antituberculous medications in EHRs and by finding the co-occurrence of ICD-9 codes for tuberculosis with pathology orders for diagnostic tests for tuberculosis (Calderwood et al. 2007). In prospective surveillance, the ESP algorithm identified and reported seven cases over an 18-month period, including two patients with culture-negative disease (Calderwood et al. 2007).

16.5 Syndromic Surveillance

As described above, automated systems can add value to EHR by detecting and reporting case notifications to local public health officials (Lazarus et al. 2008). These systems generally rely on automated decision rules applied regularly to all EHR data from a defined population and their performance and validity are well established (Klompas et al. 2008a, b). In notifiable disease reporting systems, the practitioners or the software "knows" what to look for and report to the public health official. However, for some potential public health threats, a practitioner initiated manual system, or even an EHR based automated disease identification and reporting system might not provide the timely detection needed for early intervention, because the initial effects are likely to be non-specific and will therefore not be picked up by a system designed to find cases of single, specific diseases (Lazarus et al. 2001, 2002).

This challenge arises when public health planners are asked to prepare for events of unknown, but extremely low probability, with potentially catastrophic consequences, such as mass exposure to a weaponized biological agent. In general, appropriate intervention is likely to be more effective earlier than later in most plausible scenarios, such as inhalational anthrax, since early appropriate antibiotic therapy is the only likely curative intervention. Fortunately there have not been any opportunities to test existing systems in real events, so it is very difficult to demonstrate any improvements in public health outcomes, although subjective reassurance might arguably be a sufficient goal.

16.5.1 Syndromes in Place of Specific Diseases

Surveillance for a known, specific contagious disease is reasonably tractable, using relatively simple decision rules, such as a laboratory report of a positive culture for the causative organism, and systems implementing these rules can be validated using independent case finding systems (Lazarus et al. 2001, 2002). Unfortunately, statistical power to reliably detect a very rare disease in a noisy EHR data stream may be very low, even with large sample sizes, and calibration is impossible without some real cases in the data stream (Kleinman et al. 2004). Defining decision rules for the reliable, early detection of a bioterrorism event such as the release of inhalational anthrax might be possible given appropriate definitions of "early" and "reliable", but the resulting rules cannot be externally validated until real events have been observed.

Many substantial practical problems must be overcome for the effective wide scale delivery of any of the few known plausible, large-scale biological terrorism agents, but although the risk of a large-scale event is vanishingly small, it is probably not zero (Haas 2002). These agents tend to produce specific patterns of early signs following exposure, in a limited number of broad categories. One of these patterns, such as *upper respiratory syndrome*, might be assigned to any individual with any one or more of dozens of carefully selected ICD-9 codes, chosen to reflect the presence of acute upper airways symptoms. Use of these coarse disease categories based on amalgamated ICD-9 codes may allow a large-scale event to be detected earlier because wide-scale exposure to any single agent is likely to cause an increase in only one of the syndrome categories. The basic idea is that if the release of an agent such as anthrax is effective and exposes a large number of individuals to an acute, inhalational route of infection, there will soon be an unusual rise in total periodic (e.g., hourly or daily) counts of "upper respiratory syndrome" cases. This idea is motivated by the observation that, clinically, inhalational anthrax tends to produce an early, influenza like prodrome. Given a method to obtain the summary data required for each data stream, syndromic surveillance can be framed in terms of looking for unexpected increases, localized in time and in space, for any of a dozen or so syndromes.

16.5.2 Choice of Syndromes and ICD Code Groupings

Syndromes are chosen to cover what most of the known, plausible, localized agents would be likely to produce. In the absence of a known attack, bioterrorism events have extremely low prior probability as the explanation for illness in a patient consulting a primary care or emergency room physician. For example, the initial clinical presentation of inhalational anthrax may not be reliably distinguished from a very severe instance of a common viral respiratory infection. Syndromic surveillance offers a way of being alerted to unusual numbers of respiratory infections that might be the first signal from a regional inhalational anthrax outbreak following an intentional or accidental exposure. In a real but as yet unrecognized event, a definitive diagnostic test (such as sputum culture for anthrax) is not likely to be initially ordered for the very first few cases, because in the absence of known exposure, anthrax is an extremely improbable explanation for the kinds of symptoms that are likely to be reported initially. More importantly, definitive laboratory test results may take considerable time to become available.

The range of organisms and toxins that are known to have high attack rates with effective, practicable dispersal methods suitable for large-scale bioterrorism is fairly narrow, and hence, it is possible to have a reasonable expectation of what to look for. These few agents, such as *Bacillis anthracis* (cause of anthrax), produce predictable symptom patterns as the infection progresses, so syndromic surveillance uses, for example, daily counts of broad groups of diagnostic codes in geographic regions. In the example of anthrax, upper respiratory coded events might be expected to enter the health data stream soonest after exposure. Daily counts by region for other broad syndromes are also of potential utility, including skin coded events, neurological events, gastrointestinal events, and so on. A public health surveillance system based on diagnoses grouped into broad syndromes will likely give as early a warning as we can get, if a substantial biological event should occur and the consequences should begin to enter the EHR stream feeding an automated syndromic surveillance system.

16.5.3 Early Detection and Alerting

Statistical methods are always required to make sense of syndromic surveillance data. We can count these syndrome events to help understand each day's count as it arrives, but an arbitrary count today (e.g., $n = 42$) for lower respiratory tract infection syndrome cases in a specific geographic area is just a number, and can only be interpreted using appropriate statistical methods and known historical count patterns (Kleinman et al. 2004). There are well known seasonal (e.g., winter lower respiratory infections), cultural (e.g., public holiday and weekend health care availability), geographic (e.g., sociodemographic differences in health care seeking behaviors), and other factors that confound these daily counts in addition to the usual daily random count variability.

16.5.4 Statistical Challenges

Any routine surveillance system evaluating syndrome counts in space and time will perform large numbers of statistical tests every day, so control of family-wise error is a fundamental concern, where a Type 1 statistical error or false alarm will have highly undesirable consequences. Conversely, ensuring that the model has appropriate statistical power to detect a true event is also challenging, because in the absence of real events to use for calibrating statistical inference and surveillance implementations, approximations using simulated data based on models incorporating subjective assumptions, such as the CDC simulated bioterrorism data streams (http://www.bt.cdc.gov/surveillance/ears/datasets.asp), are the only available option.

Paradoxically, we can never be certain that the assumptions are appropriate unless we are unfortunate enough to be able to record, study, and calibrate our systems in real events. Given that real events are exceedingly rare, parametric assumptions are hard to test, so permutation under the null hypothesis of no attack is a useful, if computationally expensive, way to estimate how improbable any given count of cases is in any given geographic area.

Once the counts for each syndrome, date, and region are stored in a database, we can model seasonal, geographic and other characteristics to check that nothing "unexpected" is going on from a statistical point of view. The right way to detect these events will depend on how the term "unexpected" is defined and there is a rich statistical literature on the challenge. There are non-parametric methods such as scan statistics that can be applied to syndromic counts in space and time, when we have no clear model of exactly what to expect to see. SaTScan is a statistical tool that analyzes spatial and temporal patterns to test for unexpected clustering under the null hypothesis of random distribution. It has been broadly applied to disease surveillance for applications ranging from infectious disease outbreak to cancer cluster detection and is freely available (http://www.satscan.org) (Kulldorff et al. 2005). SaTScan offers a method that makes few assumptions, and seems robust in a range of applications, providing empirical space and time case cluster probabilities adjusted appropriately for the large number of tests performed and providing an indication of exactly which region and period are of interest. Models incorporating more distributional assumptions may have greater statistical power for a given number of observations (Kleinman et al. 2004), but these models may be at risk of bias if those assumptions prove to be wrong, and in the absence of real examples, there is no empirical way to confirm their validity.

A manual notifiable disease system is unlikely to respond quickly enough to be useful for syndromic surveillance, as there is likely to be a substantial lag between multiple clinicians encountering new cases, and the manual report making its way through to being counted and distributed in an aggregate report. In addition most cases of acute illness are not routinely notified. Automated sources of data such as EHR from an ambulatory care practice or a hospital emergency room (ER) can be opportunistically repurposed for this purpose. For example, automated high volume but relatively non-specific "chief complaint" systems have been built, with sophisticated displays, such as AEGIS (Reis et al. 2007), allowing mapping of case volumes and statistics for chief complaints or syndromes by public health officials.

Given that an event can be reliably detected at an early stage, bringing this to the attention of appropriate public health and emergency officials remains a substantial challenge. The "last mile" problem in all of these systems has been addressed by investment in automated alert systems in Massachusetts (see below) and many other states of the USA. There is substantial political motivation and support for the implementation and maintenance of these expensive systems, even in the absence of high-risk threats, to satisfy a strongly felt desire to be doing something to address a variety of low-probability, but potentially very high-impact, events. The extent to which resources diverted to these activities improve public health practice and outcomes remains to be demonstrated.

16.6 Adverse Event Surveillance

Recent controversy about the adequacy of the post-regulatory monitoring of routinely administered vaccines and medications for particular subgroups of individuals indicates that improved public health "post marketing" surveillance is an urgent priority for improving the management of these risks at a whole population level. This is clearly a public health surveillance challenge and opportunity, because the whole population of medication exposed or vaccinated individuals must be considered in order to detect an elevated risk for some arbitrary adverse event associated with the exposure in a particular subgroup. EHR are likely to be one of the most valuable resources for automated systems to achieve these goals, and extremely large samples are required to ensure adequate statistical power to detect rare events, particularly when risk is only elevated among small subgroups of individuals.

16.6.1 Vaccine Adverse Event Surveillance

Although vaccination is effective in terms of preventing epidemics of potentially serious infectious illness, ensuring the safety of routine vaccination is a crucial activity, as new products are regularly being introduced and large numbers of healthy individuals are being exposed to them. No matter how well-intentioned, quantification of the risks involved is an important part of their ongoing evaluation. In general, risks are known to have a very low upper limit by the time the regulatory processes have been successfully completed. However, pre-marketing trials are of limited size and duration, and may have very low statistical power to detect extremely rare but serious adverse events. Large studies over long periods of time are needed to reliably detect very rare events. Aside from mandated regulatory animal experiments and pre-release human clinical trials, adverse events related to vaccination are generally collected and reported using relatively haphazard, incomplete manual systems.

16.6.2 Medication Adverse Event Surveillance

A similar situation pertains for widely used existing and newly introduced drugs. A number of recent, well publicized examples where there was a previously unrecognized increased risk of adverse events for patients using commonly prescribed medications (Graham et al. 2005) have demonstrated the importance of improving the current mechanisms for ongoing, active, very large-scale surveillance for unexpected and rare adverse events. Controlled clinical studies of the required sample sizes and duration are simply not practicable beyond the modest requirements of the regulatory process, but very large-scale observational studies may be possible in the form of post-marketing surveillance using multiple, federated collections of EHR and appropriate automated systems.

In all forms of adverse event detection, statistical and epidemiological methods play a key role, because the patients who are given a particular make of drug have characteristics that make some adverse outcomes more likely to occur. For example, patients with type II diabetes (T2D) are known to be at increased risk of cardiovascular disease, so if a drug used in the treatment of T2D is being evaluated for increased risk of adverse cardiovascular events, the biases associated with characteristics leading to the prescription of the drug, often termed "confounding by indication" must be controlled for reliable statistical inference about the risk associated with the drug itself. Specialized methods such as propensity scoring are required for reliable results, and for very rare events extremely large samples are needed over long periods.

16.7 Non-Specific Biosurveillance

There is great recent interest in surveillance for potential public health problems that may never have been previously seen. For example, how will we recognize when the first human cases of a pandemic outbreak of avian influenza have occurred? This turns out to be a far more difficult problem than notifiable disease reporting or the specific syndromic surveillance systems described above. Part of the challenge is that for efficient intervention in a large-scale event, timeliness is of the essence, so the earliest possible reliable detection and alerting are always sought, but this desire must balance the two possible kinds of highly undesirable errors – namely false positive reports inducing large-scale but useless response, and false negative reports inducing inappropriate reassurance that there is nothing unusual going on leading to lost opportunities for effective, early intervention.

Unlike notifiable or even rare potential bioterrorist vector disease cases, we do not have any well validated examples or reliable models of previously unseen, emerging, or unknown disease patterns, making it very hard to rationally design, and even harder to validate, reliable automated decision rules from existing EHR or non-health related data streams. In this situation, comparing competing models, each claiming to produce the earliest, most reliable signal, requires subjective judgment. Objective criteria about timeliness and reliability are available only under arbitrary assumptions using simulated null data generated under the null hypothesis of no events to establish Type 1 or false positive error rates, and "real" data simulated to represent real events of various kinds and scales, against a background of realistically noisy data. Simulating data to represent a large-scale public health event requires multiple subjective assumptions to be incorporated into both the null and real data models.

16.7.1 Non-Health Related Data Sources

It is widely hoped that the improved availability of non-health care related, unconventional data streams, such as orange juice (Fienberg and Shmueli 2005), facial tissue, or acetaminophen sale volumes from large retailers, might contain potentially important signals about health (e.g., https://www.rods.pitt.edu/site/), as health related issues might influence purchase patterns. Unfortunately these sales data are likely to be confounded by real, but far less informative variation related to advertising, availability, regional "specials", and other non-health related economic and market condition effects. Random and other non-health related variation might make it difficult or even impossible to extract any reliably health related signals from these data. Evidence that investment in gathering and processing these data can produce measurable public health benefit beyond some subjective measures of comfort is not yet available in any convincing form. Although improved so called "situational awareness" resulting from the substantial investment already made in non-health care related data streams and in non-specific surveillance methods may in itself be a worthy goal, there are many competing activities that can produce measurable improvements in public health given the same level of investment.

A privately funded web search term based system for identifying temporal and regional patterns of internet searches related to acute upper respiratory illness has been deployed at http://www.google.org/flutrends, demonstrating the potential utility of non-EHR data streams for surveillance. This system's main claim for benefit, in a high profile scientific publication, was a two weeks lead-time over existing CDC manual reporting systems for influenza like illness. It is disappointing that the authors (and reviewers) appear unaware that even greater lead times (up to 6 weeks) had been demonstrated in an EHR based automated system more than seven years previously (Lazarus et al. 2001). There is no evidence of substantial utility for less common but important non-respiratory public health threats, such as gastrointestinal illness. Again, despite enthusiastic and optimistic engagement, and the importance of novel public-private partnerships, the convincing demonstration of improved public health practice as a result of presumably substantial investment in processing data from this non-health care delivery data stream is not yet available.

16.7.2 The Challenge of Opportunity Cost

All public health activities, including newer biosurveillance systems have opportunity costs because resources devoted to those purposes become unavailable for other, potentially more efficient investment. Demonstrating return on investment in terms of improved public health may be an insurmountable methodological challenge unless goals are substantially sharpened. Although hundreds of millions of dollars have been invested, the literature on the application of non-EHR data based surveillance is relatively recent, and is notable in its optimism about their potential.

However, equally enthusiastic investment in rigorous evaluation of such surveillance in terms of measurable public health benefits is usually lacking. While process and activity are sometimes reported, even rudimentary validation against existing systems is generally absent, because without a source of known events to test the models, external criterion validation is not possible, and the performance and benefit of these expensive systems remain a subject of speculation, although substantial investment has been and continues to be diverted to these activities.

16.8 Finding and Harnessing Data

No matter what kind of surveillance is proposed, any useful surveillance system needs data. It is known that useful public health information can be gleaned from reliable and complete EHR, and it is clear that for post-marketing vaccine or medication adverse event surveillance, non-health care data streams are unlikely to have appropriate relevant exposure data to make them reliable or useful. The question of how useful information from retail (e.g., orange juice) sales or other non-health care data sources can be for large-scale outbreak detection remains to be answered.

For notifiable disease, syndromic surveillance, and adverse event reporting, data at the level of the individual patient are needed at some point in processing, even if such data do not allow the identification of a patient. However, only summary data may need to be distributed for effective public health planning, and for some kinds of broad scale intervention. Precise local regulatory requirements will vary by jurisdiction. As an example, some of the conditions currently prevailing in the United States are used to make the challenges more concrete, in terms of actually gaining access to individual patient electronic health data (EHR) or PCHR from one or more of the health delivery or insurance entities. Security, access control, and governance arrangements are crucial to securing collaboration and access to data needed for surveillance.

Public health information systems where potentially identifiable individual human data may be involved, such as those the CDC manages, require substantial levels of security and protection from unauthorized access. Integrating across multiple data resources often allows value to be added. Unfortunately, in the US system, individual applications have generally implemented their own security and authentication mechanisms, although more recently, Public Health Informatics Network Messaging System (http://www.cdc.gov/phin/activities/applications-services/phinms/) has been increasingly deployed for secure communication. The resulting, generally incompatible technologies currently deployed in secure public health systems are technically demanding to manage, and even more technically challenging to integrate across individual systems. Most of these applications have been very narrowly focused, and implemented as independent, isolated vertical information silos. They were deliberately made hard to access through specific and often proprietary access mechanisms, and were very hard to get data out of, because integration with other, independent systems was rarely a design goal. As a result, there are many large-scale public health systems in operation containing data that could be very useful, if they could be repurposed and combined with other data, but that are not currently easily integrated into any value-adding public health applications.

16.9 High Throughput Distributed Surveillance

There are many potential sources of EHR, such as health care payment and insurance processing, managed care organizations, and large group medical practices with electronic medical record systems, and there are many independent vendors and products available for implementing large-scale EHR systems. In order to obtain useful statistical power, public health surveillance requires consistent and complete data from very large numbers of individuals, particularly when rare adverse events are being sought. Currently, no single system or source covers any more than a fraction of the population of an entire state, so one of the challenges for practicable systems is to be able to combine data from multiple independent sources. This introduces many technical and administrative challenges. The most familiar information system model for doing this is to collect all the data in one single physical collection, and then process it using a single application. Unfortunately for identifiable EHR in the current US health system culture, this is simply not a sustainable option.

Experience in practical research projects over the last decade has clearly demonstrated that most large US data holders – termed covered entities under the U.S.A. HIPAA (http://hipaa.org) provisions – will not permit any individual level patient records to be moved outside their private networks, unless there is a statutory reporting requirement or some other special case. One potential solution is for the query or analysis to be sent to the data, in a distributed or federated information system model. Securing and administering these processes is not a trivial task, and some mechanism for ensuring that the analysis operates correctly at each EHR site, as well as for managing analysis queries and amalgamating all the individual query outputs, must be created and sustained.

Gaining data provider cooperation, in the absence of legal obligation, requires careful attention to specific restrictions on communication imposed as requirements for participation by the covered entities without whom there would not be any data to federate. The security requirements of the security and other corporate representatives of the entities volunteering to take part in data federation are very clear and very restrictive. Although there is a wide range of security considerations, there are two more or less uniform general principles – firstly, that no individual level data can be permitted to leave the private network and secondly, that only outgoing connections initiated from inside the private network are likely to be acceptable. These are the most important technical restrictions and they restrict the choice of available communication architectures markedly.

In essence, for a federated network to be attractive to potential data providers, the architecture is likely to be that queries are to be executed on data held in a common format, within the covered entity's private network, and that only summary data will generally be returned. The server running inside the private network will periodically contact (poll) the remote portal node, because it is unlikely that covered entities will permit unsolicited inbound connections to their servers. During each poll, it will pick up any waiting queries, and return completed results for amalgamation and presentation. This is the model that was demonstrated recently in the Distributed Research Network (DRN) project described below, and is the essence of a federated model that appears compatible with the requirements of the large-scale data providers who collaborated on that project.

In practice, there are many, far more detailed technical issues that will vary depending on the specifics of the implementation. These include the scripting and management of secure communications; specific task command syntax and local execution models; details of the shared data structures; details related to the actual analysis package; and the design and operation of the presentation layer at the central portal from where the queries originate, and where the individual entity results are amalgamated for presentation to the user. Although each of these issues is likely to depend on the specific implementation, the basic model of a successful large-scale voluntary system is likely to be distributed and federated.

16.10 Technical Aspects of Secure and Controlled Data Sharing

In the physical sciences, such as atmospheric science and astronomy, vast volumes of data are generated on a daily basis, and this is now routine for some life sciences such as genomics and genetics. It is widely understood that the value of all these data increases when they are made widely, conveniently, and securely available to as many authorized researchers as possible. In contrast with health services data flows, which have often been built with closed source, proprietary technologies in the past, there has been enormous investment in open source software, specifically designed to support the secure sharing of computational resources and large quantities of data, with very fine grained authentication, and high grade security.

16.10.1 The Globus Toolkit

One of the best known examples of an integrated set of technologies for the kinds of secure data transfers that characterize modern data rich shared computing resources and data is the Globus Toolkit, available from the Globus Consortium (http://globus.org). Globus is a very vigorous and diverse project. It can be characterized as a collection of fundamental, interoperable infrastructure components providing basic services such as authentication and access control, remote job execution, secure file transfer, and other basic communication protocols. Layered on top of these basic services are rapidly growing collections of specialized applications that re-use these infrastructure components to provide higher level applications. While there are important differences between the challenges faced in data sharing for public health surveillance, and those faced in the physical sciences, there are also many deep similarities. In particular, the need for secure, controlled federated query distribution in order to allow value to be added through the virtual integration of otherwise independent collections of individual health records enables far larger populations to be monitored than is possible with any single data source.

The National Center for Public Health Informatics (NCPHI) group has been working with the Globus Toolkit on a variety of public health projects in CDC priority areas. Details of their work are recorded at http://sites.google.com/site/phgrid/, where some of the important technical challenges associated with the sharing of potentially identifiable human data are described.

16.10.2 Internet Security

Potential data providers in the US such as health plans are covered entities working under HIPAA (http://www.hipaa.org) regulations. Security for the identifiable patient data they hold is a paramount concern. The primary issue is the threat vector represented by any incoming internet connections from the outside world to any servers on their private networks, so these are always handled on a case by case basis. Network security staff members are understandably very careful about how this is organized and will not permit any activity on their networks unless they are confident that the risks are appropriately managed.

One term widely used in talking about securing network traffic is the concept of a 'port' on a server. On the internet, each connected workstation or server is identified by a unique internet protocol "address", and most request packets addressed to an internet server are associated with a specific numeric address, or port, on that particular server where they are to be processed. A server may be executing a program or service that "listens" for incoming packets on one or more of those numeric ports. For example, a web server will typically "listen" on port 80 for unencrypted HTTP requests, and on port 443 for SSL encrypted Internet traffic. If an incoming request packet comes in from the internet, addressed to a particular server, with a numeric port label of 443, that server passes the contents of the request packet to the software running on that port – most likely an SSL enabled web server. As accepting a request and transferring it to a running program on a server represents an open communication channel, it requires careful control and substantial security expertise, because it is a major potential threat vector for unauthorized access to protected human data on the exposed systems and servers. Globus generally requires that a fully active grid node server responds to unsolicited incoming requests on literally hundreds of ports. For covered entities, the large number of incoming firewall exceptions required (see http://dev.globus.org/wiki/FirewallHowTo) exposes them to what they reasonably see as unacceptable administrative burden and risk.

16.11 Examples of Public Health Surveillance Systems

16.11.1 The National Bioterrorism Syndromic Surveillance Program

The CDC funded National Bioterrorism Syndromic Surveillance Program (NBSSP) was one of the earliest large-scale US distributed public health syndromic surveillance systems based on federated ambulatory and emergency care EHR data (Lazarus et al. 2001, 2002; Yih et al. 2004; Platt et al. 2003). Identifiable, patient-level information remained under the control of participating health care providers at all times in order to minimize the risk of inadvertent disclosure of protected health information. Software was distributed to process, display, and transfer summaries of that local data. Aggregate data were federated from six independent health plans and large group practices across five US states, covering approximately 25 million individuals. Each site extracted data in a uniform format from their local EHR system, with details of all encounters from the previous 24 h. Office visits or telephone encounters with diagnostic codes corresponding to syndromes of interest were counted by the distributed software. In order to minimize spurious correlation between repeat encounters for any given individual for the same episode of care, these were excluded within 6 weeks of a previously reported case for that individual and syndrome. De-identified daily counts of syndromes by zip code were transferred to the data center for statistical analysis, using the CDC PHIN-MS secure messaging software.

The system provided near real-time public health surveillance summaries by time and space for a range of aggregated syndromes in an effort to provide the earliest possible notification of possible disease outbreaks for participating public health officials. A variety of statistical methods were used to estimate the extent to which the number of cases seen each day was unexpected, based on seasonal and other known factors. Estimates were expressed in terms of how often a count of that particular magnitude might be seen, so a count expected at least once a month was far less interesting than one only anticipated every 100 years (Kleinman et al. 2004). These estimates were available to authorized users on a secured website maintained by the data center. When clusters of syndrome counts surpassed a statistical threshold (individually specified by each participating health department) an alert was automatically transmitted through the Massachusetts Health Alert Network for delivery to the appropriate public health officials.

16.11.2 The Electronic Medical Record Support for Public Health Project

The Electronic medical record Support for Public Health (ESP) project was built upon the experience of the NBSSP syndromic surveillance project but provided highly targeted notifiable disease surveillance and secure electronic case reporting to public health authorities, from EHR data. Traditional manual disease reporting systems are hampered by under-enumeration, delay, and incomplete data for reported cases. The automatic identification and reporting of these cases from EHR data overcomes many of these weaknesses in conventional surveillance. Collaborations between a large multispecialty group medical practice (Atrius Health, formerly known as Harvard Vanguard Medical Associates) and the Massachusetts Department of Public Health permitted the creation and deployment of a model system under the auspices of a CDC Center of Excellence in Public Health Informatics award. The rationale, development, architecture, algorithms, and surveillance results of this system have been reported in multiple publications (Klompas et al. 2007, 2008a, b; Lazarus et al. 2008). The source code and documentation are freely available at http://esphealth.org.

ESP is a generalizable model for secure public health information surveillance and reporting using EHR data (Lazarus et al. 2008). It is a standalone system, operating independently of the host EHR systems. This makes it easier to adjust for different source EHRs and to isolate computing burden from the host resources, to minimize impact on production functions. The ESP server is deployed inside the data center of the host practice or health information exchange, inaccessible behind the host organization's Internet firewall in order to minimize risk of the inadvertent exposure for sensitive clinical data. Once a case is identified, an electronic case report is securely transmitted as an HL7 document to local public health authorities. The case report includes patient demographics, contact information for the responsible clinician, patient symptoms, pertinent laboratory tests, prescribed treatments, and pregnancy status.

ESP is currently active in Atrius Health, a multispecialty, multisite practice with approximately 700 physicians that serve over 600,000 patients in Eastern Massachusetts. Clinical information on every patient encounter from the preceding 24 h is loaded each morning, and analyzed for cases of chlamydia, gonorrhea, pelvic inflammatory disease, acute hepatitis A, B, and C, active tuberculosis, and syphilis. Prior to transmission, cases can be reviewed by the practice's Infection Control personnel using an internal web-based case-management system limited to authorized users inside the host organization private network. Since deployment in January 2007, ESP has reported almost 2,500 notifiable disease cases.

Validation was undertaken by comparing the completeness, accuracy, and clinical detail of ESP case reports to existing public health reports for the same population (Klompas et al. 2008a, b). For the period of June 2006 through July 2007, 758 cases of chlamydia, 95 cases of gonorrhea, 20 cases of pelvic inflammatory disease, and four cases of acute hepatitis A were detected and reported (Klompas et al., 2008a, b). Specificity was measured using a manual chart review of all patients. The positive predictive value was 100% for chlamydia, gonorrhea, and hepatitis A and 95% for pelvic inflammatory disease. Archives of conventionally reported cases during this period served as an external validity criterion to quantify sensitivity to true disease. ESP detected 41% more cases than had been initially identified by conventional reporting including manual and automated ELR systems, and no additional cases of gonorrhea, pelvic inflammatory disease, or hepatitis A known to the health department were missed (Klompas et al. 2008a, b). One case of chlamydia was missed because of a coding error in the source EHR. Patient treatment information and pregnancy status for female patients were provided on all reports of chlamydia and gonorrhea whereas conventional surveillance only included treatment data on 88% of case reports and pregnancy status for 5% of female cases. The importance of improved case reporting completeness to public health authorities is that it enables far more efficient and effective intervention, as substantial effort is often wasted trying to communicate with busy clinicians in order to obtain these important case details.

16.11.3 The ESP Vaccine Adverse Event Reporting System

A new surveillance mechanism to automate the detection of adverse events following vaccination from EHR data, elicit clinician comments on potential events, and submit electronic case reports to the existing manual CDC and FDA vaccine adverse event reporting system (VAERS) is currently being integrated into the data analysis and external communication components of the ESP system. Existing EHR data flows from the notifiable disease detection algorithms described above are being repurposed as a model for other adverse event surveillance and reporting systems.

The ESP-VAERS event detection algorithm is predicated upon prospectively following patients who are given vaccines for novel diagnoses, abnormal laboratory values, elevated temperature, allergies, or new medication prescriptions for up to 42 days following the recorded administration of the vaccine. New ICD-9 codes and lab tests that arise during the risk period are compared to the patients' prior ICD-9s and test results in order to exclude pre-existing conditions or out of range values that were previously noted. The duration of the risk period, the threshold for abnormal lab values, and any pertinent ICD9 codes are tailored for each vaccine and potential adverse event – for example, fever is only sought for 72 h after vaccination whereas myocarditis is sought for 42 days. When a possible adverse event is identified, ESP delivers a message to the patient's primary care provider's secure in-basket. The message includes details of the purported adverse event and then invites the clinician to endorse or refute the case. If the clinician endorses the case, ESP submits an electronic case report directly to CDC and FDA's vaccine adverse event reporting system.

16.11.4 The Distributed Research Network

In response to the Institute of Medicine's call for improved post marketing surveillance for adverse events associated with medication and other routine health care interventions, there is substantial interest in the development of public health surveillance systems capable of supporting very large-scale analyses of comprehensive and reliable EHR data. No existing single EHR system is likely to provide enough complete health records to give sufficient statistical power to reliably detect very rare events sought in patients treated with particular medications or vaccines. The design and proof-of-concept demonstration of a national network capable of supporting this kind of research is a major technical and governance challenge, being addressed by the DRN collaborators, including academic researchers, and representatives from health care organizations willing to explore ways in which they can safely allow their very large collections of EHR to be repurposed.

A series of reports are being prepared and made available through the AHRQ web site as part of this project. The project itself will be completed soon, but some preliminary observations are summarized here because they provide useful insight into the likely future technical, administrative, and governance directions for large-scale adverse event surveillance systems.

The proof-of-concept demonstration conducted in February 2009 involved a distributed statistical query and response from participating covered entities. As planning progressed, the data providers were adamant that no individual level data would be permitted to leave their protected networks, even if the data contained none of the identifiable elements defined under the HIPAA regulations. They were willing to permit queries to return aggregate, summary data as long as the summaries from each participant were hidden from the user and amalgamated in an irreversible way before being presented, as there was substantial concern about the potential commercial value of individual institutional summary results for the types of research query being proposed.

Technically, the demonstration was based on Globus Toolkit security and communication infrastructure. As described above, Globus infrastructure assumes a large number of exposed server ports, and this default requirement was rejected by the participating data providers. Technical support for the demonstration was provided by the CDC NCPHI, who had already encountered this challenge, and had created and deployed a secure messaging service "wrapper" for Globus that required only a standard secure web browser port (443) to be opened for outbound connections at the host firewall to one specific external machine used as the query source and result destination. Permitting access only from one specific remote SSL secured machine is generally regarded as a highly secure, state of the art approach, particularly to a server on a special isolated network (usually termed a demilitarized zone or DMZ) inside the institutional firewall. Thus, the institutional technical staff is generally more comfortable about managing these specifically restricted access control rules, involving a single firewall port at most, accessed by one or only a few designated remote machines.

16.12 Concluding Remarks

EHR are now established as a valuable source of reliable data for public health surveillance. Security and control are two primary issues for large organizations responsible for managing identifiable EHR, and new, emerging grid based technologies offer substantial promise in terms of meeting these requirements. High throughput notifiable disease surveillance systems adding value to routinely collected EHR are practicable, and their performance can be directly and objectively validated against independent existing sources of similar data. Their effectiveness can be demonstrated, for example, through appropriate intervention in otherwise unrecognized cases. Syndromic surveillance system performance may be validated to some extent using data from known events such as seasonal respiratory illness, or known local outbreaks of food-borne illness, or simulated data. The comfort derived from knowing that something is being done may arguably be a worthwhile outcome, as we hope that the practical utility of many surveillance systems is never tested in a true outbreak.

Practicable automated EHR based adverse event detection systems are still in their infancy, but it would be expected that statistical findings should be replicable, in independent samples, to provide external validation. Using very large samples, they have a very high potential to prevent substantial death or injury before an excess adverse event risk is otherwise noticed. Although process and activity can be measured for generic biosurveillance systems, measuring performance in terms of public health outcomes, or the cost effectiveness of comparing competing methods, is not generally possible.

Notifiable disease surveillance usually involves reporting or counting specific positive laboratory tests and other clinical findings. Automated, high-throughput surveillance requires comprehensive and complete coverage of each individual, validated decision rules, and robust, efficient implementations to be useful for large volumes of data. Each decision rule must be able to determine when an event, such as a case of influenza like illness, or a case of syphilis, has been observed. The time, location, and demographic details of the case are then of potential use for presenting summaries of events and statistical inference to users.

Despite the abundance of statistical techniques and data for their testing, none of such methods appears to demonstrate outstanding performance over the wide range of potential kinds of aberrations or data patterns. The methods available in SaTScan (Kulldorff et al. 2005) involve relatively few assumptions and appear to calibrate well in real data, but there are many alternatives (Kleinman et al. 2004). It may be appropriate to offer multiple alternative statistical evaluations, based on different sets of assumptions, and therefore a human user can determine when more than one method shows a pattern suggestive of a low-probability event.

Explicit and rigorous evaluations are crucial design features for public health surveillance information systems. Competing systems should be compared on the basis of objective measures of effectiveness and cost efficiency, so that building these into the design at an early stage is highly desirable.

Acknowledgments The author acknowledges substantial contributions from Michael Klompas, Richard Platt, other members of the DACP/Channing Public Health Informatics group, and our many external collaborators, to the material presented here. This work was supported by grants from the Agency for Healthcare Research and Quality (HS 17045) and the CDC.

References

Calderwood M, Klompas M, Hou X et al (2007) Automated detection of tuberculosis using electronic medical record data. Adv Dis Surv 4:46

Fienberg SE, Shmueli G (2005) Statistical issues and challenges associated with rapid detection of bio-terrorist attacks. Stat Med 24:513–529

German R, Lee L, Horan J et al (2001) Guidelines Working Group Centers for Disease Control and Prevention (CDC). Updated guidelines for evaluating public health surveillance systems: Recommendations from the Guidelines Working Group. MMWR Recomm Rep 50(RR-13):1–35

Graham D, Campen D, Hui R et al (2005) Risk of acute myocardial infarction and sudden cardiac death in patients treated with cyclo-Oxygenase 2 selective and non-selective non-steroidal anti-inflammatory drugs: Nested case-control study. Lancet 365(9458):475–481

Haas CN (2002) On the risk of mortality to primates exposed to anthrax spores. Risk Anal 22:189–193

Kleinman K, Lazarus R, Platt R (2004) A generalized linear mixed models approach for detecting incident clusters of disease in small areas, with an application to biological terrorism. Am J Epidemiol 159:217–224

Klompas M, Lazarus R, Daniel J et al (2007) Electronic medical record support for public health (Esp): Automated detection and reporting of statutory notifiable diseases to public health authorities. Adv Dis Surv 3(3):1–5

Klompas M, Haney G, Church D, Lazarus R, Hou X, Platt R (2008a) Automated identification of acute hepatitis B using electronic medical record data to facilitate public health surveillance. PLoS ONE 3(7):e2626

Klompas M, Lazarus R, Platt R et al (2008b) Automated detection and reporting of notifiable diseases using electronic medical records versus passive surveillance – Massachusetts, June 2006–July 2007. MMWR Morb Mortal Wkly Rep 57:373–376

Konstantinidou K, Mantadakis E, Falagas M, Sardi T, Samonis G (2009) Venetian rule and control of plague epidemics on the Ionian Islands during 17th and 18th centuries. Emerg Infect Dis 15:39–43

MHR, Hartman J, Assunção RM, Mostashari F (2005) A space-time permutation scan statistic for the early detection of disease outbreaks. PLoS Med 2:216–224

Lazarus R, Kleinman KP, Dashevsky I, DeMaria A, Platt R (2001) Using automated medical records for rapid identification of illness syndromes (syndromic surveillance): The example of lower respiratory infection. BMC Public Health 1:1–9

Lazarus R, Kleinman K, Dashevsky I et al (2002) Use of automated ambulatory-care encounter records for detection of acute illness clusters, including potential bioterrorism events. Emerg Infect Dis 8:753–60

Lazarus R, Klompas M, Campion F et al (2009) Electronic support for public health: Validated case finding and reporting for notifiable diseases using electronic medical data. J Am Med Inform Assoc 16(1):18–24

Paneth N (2004) Assessing the contributions of John Snow to epidemiology: 150 years after removal of the Broad Street pump handle. Epidemiol 15:514–6

Platt R, Bocchino C, Harmon R et al (2003) Syndromic surveillance using minimum transfer of identifiable data: The National Bioterrorism Syndromic Surveillance Demonstration Project. J Urban Health 80:25–31

Reis BY, Kirby C, Hadden L et al (2007) Aegis: A robust and scalable real-time public health surveillance system. J Am Med Inform Assoc 14:581–588

Yih WK, Caldwell B, Harmon R et al (2004) National bioterrorism syndromic surveillance demonstration program. Morb Mortal Weekly Rep Suppl 53:S43–49

Chapter 17
Microbial Genotyping Systems for Infection Control

Matthew O'Sullivan

17.1 Introduction

It has long been recognized that admission to health care institutions is associated with a risk of acquiring infection (Best and Neuhauser 2004). Despite this and the institution of wide-ranging prevention measures, hospital acquired infections (HAI) are an increasing problem. While much of this can be explained by the changing demographics of the inpatient population, with an increase in the number of immunosuppressed and elderly patients who are undergoing more invasive procedures with indwelling prosthetic devices, there is also evidence of the emergence of more virulent nosocomial pathogens, which have evolved to thrive in the modern hospital environment. This evolution is characterized not only by an acquisition of resistance to a wide range of antibiotic and antiseptic agents, but also by other virulence mechanisms, which facilitate environmental persistence (Wagenvoort et al. 2000), and transmission from patient to patient (Casewell and Desai 1983; Papakyriacou et al. 2000; Phillips 1991).

HAI is a major cause of preventable healthcare-associated morbidity and mortality (see also Chap. 15). In 2005–2006, 31,639 hospital separations in Australia, or 0.5% of admissions, were coded as having the adverse event "infection following a procedure" (Australian Institute of Health and Welfare 2007). In the United States, it is estimated that there are 2 million nosocomial infections every year, resulting in 90,000 deaths and excess healthcare costs of approximately $5 billion dollars (Burke 2003). This occurs in spite of extensive infection control measures aimed at preventing both colonization and infection by nosocomial pathogens. Nosocomial acquisition of, and infection by, bacterial pathogens is increasing but remains under recognized. Current infection control measures concentrate only on a handful of multiresistant pathogens, and are often unsuccessful. New approaches to the identification and prevention of nosocomial infection are clearly required.

M. O'Sullivan (✉)
Centre for Infectious Diseases and Microbiology, Sydney West Area Health Service, Sydney, Australia

V. Sintchenko (ed.), *Infectious Disease Informatics*,
DOI 10.1007/978-1-4419-1327-2_17,

17.2 Hospital Infection Control Surveillance

An essential component of hospital infection control is surveillance for nosocomial infection. One function of surveillance is the reporting of specific infection rates for quality indicator purposes. A more valuable application of surveillance is the identification of clusters of infection, which may represent outbreaks of nosocomial transmission because of lapses in infection control precautions. Such clusters can then be investigated and measures instituted to terminate the outbreak.

A cluster is identified when the observed rate of infection is noted to be higher than the endemic, baseline rate. For a condition that is rare (i.e. the baseline rate is essentially zero), any two cases occurring contemporaneously may warrant investigation, and so a simple observation of laboratory notifications may be all that is necessary to identify clusters. Examples may vary from institution to institution, but could include vancomycin resistant *Staphylococcus aureus*.

More commonly, nosocomial outbreaks are due to organisms that also produce sporadic infection. These conditions thus have a measurable background incidence. Examples include methicillin-resistant *Staphylococcus aureus* infection or colonization, or *Clostridium difficile* diarrhea. In these examples, clusters of nosocomial transmission may be harder to discern because of randomly fluctuating background incidence. Here, the detection of clusters is greatly aided by statistical methods. Such methods identify a *statistically significant* increase in rate above the background incidence, which should prompt further investigation. A number of methods to perform these calculations have been described, including comparing the number of episodes in the time period to the long term mean, looking for a 100% increase compared to the previous time period or for a 50% increase compared to the mean of the previous three time periods (Hacek et al. 2004).

A particularly effective method for the statistical analysis of surveillance data is the use of process control charts (discussed in Chap. 15). Specifically, Shewhart and Exponentially Weighted Moving Average (EWMA) charts are particularly applicable to monitoring nosocomial infection events. These plot the incidence of the outcome of interest against time, with control limits for either the incidence (Shewhart) or its moving average (EWMA), which, if crossed, indicate an increase in the event rate beyond what would be expected by chance (Morton et al. 2001). Figure 17.1 gives an example of EWMA for MRSA incidence in an intensive care unit. Such methods have the ability to identify clusters of transmission or infection early and accurately (Wright et al. 2004). Separate charts may be generated for different wards or units in a hospital, allowing the detection of temporospatial clusters.

In an effort to improve interest in and compliance with infection control measures, the graphical presentation of incidence data to departmental staff, such as in the form of control charts, can highlight areas of concern and give positive feedback, which may itself be an effective intervention. This has been evaluated as an intervention to reduce nosocomial infection rates. To be effective,

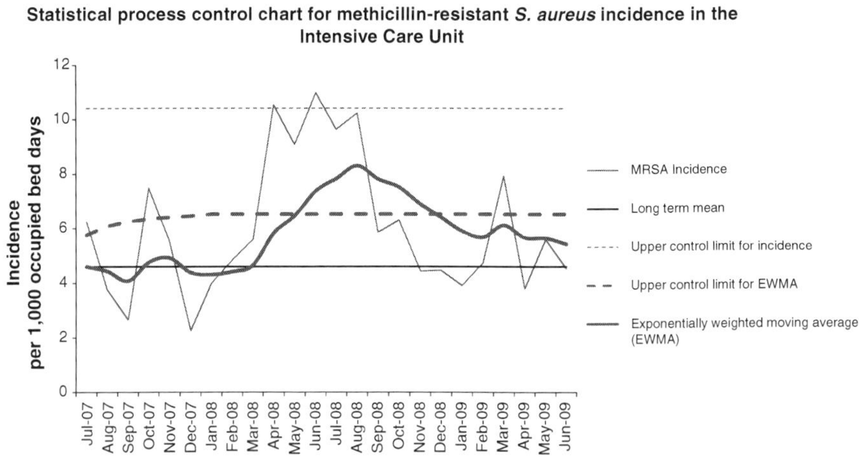

Fig. 17.1 Shewhart and exponentially weighted moving average (EWMA) control charts for MRSA incidence in and intensive care unit. Both the plot of incidence an EWMA cross their respective upper control limits in April/May 2008, indicating a statistically significant increase in incidence, which would alert staff to a possible cluster of transmission.

such strategies require an effective surveillance program with timely data entry and frequent feedback of the most current results. One study demonstrated a reduction in MRSA transmission with the introduction of such feedback (Curran et al. 2002).

17.3 Targeted Genotyping to Confirm Nosocomial Outbreaks

Traditionally, the molecular typing of organisms as an aid to infection control has been limited to investigating clusters identified by surveillance methods to see if the involved isolates are clonal. If clonality is established, then the clonal cluster is assumed to represent a true outbreak, and it is then investigated to identify infection control breaches and to institute measures to prevent further transmission. Often some cases can be excluded from the cluster when they are distinct from the clonal isolates, making the investigation of the outbreak more efficient (Hacek et al. 2004; Macfarlane et al. 1999; Peterson et al. 1993; Pfaller et al. 1991). If the clonality of a cluster is not established by molecular typing, then the cluster is usually presumed to be a "pseudo-outbreak," occurring by chance, with no further investigation being necessary (Hartstein et al. 1997; Imataki et al. 2006; Macfarlane et al. 1999). It is through these two factors - improving efficiency of outbreak investigation and the ability to identify "pseudo-outbreaks" - that molecular typing can improve the cost-effectiveness of nosocomial infection surveillance (Andrei and Zervos 2006).

The utility of molecular typing to confirm clusters depends on the background incidence of the organism in question. When the organism is not endemic and rarely isolated, then temporospatial clusters have a high probability of representing true outbreaks, and molecular typing may not be necessary. In fact, typing may be misleading in this circumstance if false negative results occur, leading to a true outbreak being mistakenly called a pseudo-outbreak.

Molecular typing can also be misleading, where the transmissible element is not the organism, but genetic material passing horizontally from one organism to another. Such genetic material will frequently encode antimicrobial resistance, but it is feasible that other virulence factors such as exotoxins could cause outbreaks in this way. An increasingly recognized example of this is the transmission of genetic material encoding carbapenemases conferring high-level resistance to a variety of gram-negative organisms. Molecular typing of the responsible organisms of such outbreaks will be misleading, since a single outbreak will be due to a variety of strains, or indeed a variety of species, all carrying the same genetic element (Peleg et al. 2005). This scenario is discussed further in Chap. 12.

When there is a low level but measurable background incidence of the organism in question, then targeted molecular typing becomes useful. This is commonly the case for organisms that may be community acquired, but may also spread in the hospital environment, or for organisms wherein long-term colonization may occur. Pimentel et al. were able to demonstrate, using molecular typing by pulsed field gel electrophoresis (PFGE), that what appeared to be a single large outbreak of multiresistant *Acinetobacter baumanii* was in fact two distinct outbreaks, one centered on a surgical unit and the other in ICU (Pimentel et al. 2005). Such information is valuable since it allows different infection control interventions targeted to each specific outbreak. In this case, the ICU outbreak was associated with contaminated respiratory ventilation equipment; in the surgical unit, complete ward closure and decontamination were required to terminate the outbreak. Mascini et al. describe the utilization of PFGE to determine whether cases of vancomycin resistant *Enterococcus* species (VRE) colonization were part of an evolving nosocomial epidemic or not; this allowed the targeting of infection control measures toward epidemic strains and led to successful termination of the outbreak (Mascini et al. 2006). Molecular typing with PFGE is also frequently used to confirm nosocomial outbreaks of methicillin-resistant *Staphylococcus aureus* (MRSA) (Imataki et al. 2006). Molecular typing can verify the termination of an MRSA outbreak and the effectiveness of infection control interventions (Hartstein et al. 1995).

When the organism in question is highly endemic in the hospital, molecular typing is still useful to confirm clusters, but the results must be interpreted with caution. This is because pseudo-outbreaks will occur more commonly, so the probability that a cluster represents a true nosocomial outbreak is lower. Yet, because of the high background incidence, there is a greater risk that the strains will appear clonally related by chance. In this case, the typing method utilized must have high discriminatory to avoid false positive results (Dziekan et al. 2000; Weber et al. 1997). A case in point is MRSA. In institutions with low baseline levels of MRSA colonization, targeted typing, even employing methods with a relatively low discriminatory power such as *spa* typing, has been successfully used to confirm outbreaks (Mellmann et al. 2006). However, many institutions have high endemic rates of MRSA colonization and infection, and increasing numbers of patients admitted to hospital from the community are already colonized with MRSA. Often these isolates may belong to only a restricted number of *spa* types. In this circumstance, a much more discriminatory typing method such as PFGE is required to distinguish true outbreaks from pseudo-outbreaks.

17.4 Universal Genotyping in Hospital Infection Control

Universal typing refers to the routine typing of all isolates as a primary part of cluster detection. This is in contrast to targeted typing, described earlier, that is only performed once a spatiotemporal cluster has been identified. Such an approach has found a place in the domain of community public health, wherein control programs for several organisms rely on typing for the initial identification of outbreaks. Clark and colleagues describe the success of the universal typing of *Mycobacterium tuberculosis* in identifying both laboratory contamination events and otherwise unknown transmission routes, which permitted more extensive case-finding and improved tuberculosis control (Clark et al. 2006). However, universal typing has only rarely been employed in hospital infection control. This is partly because, until recently, the available typing methods, such as PFGE, were expensive, slow, cumbersome, and low-throughput. With the development of PCR-based methods, rapid, high throughput typing is now possible using a variety of platforms. As the costs of these rapid methods continue to fall, universal typing with results available in real-time is becoming increasingly feasible. Some experts have argued against the use of universal typing in hospital infection control, arguing that it is likely to lead to the mis-identification of clonal clusters occurring by random chance (rather than representing true outbreaks) because of the imperfect discriminatory power of many molecular typing methods (Pfaller and Herwaldt 1997). This is a valid concern, but if the performance characteristics of the method being used are well defined and the results are interpreted appropriately, as discussed later, mis-identification should be able to be minimized.

For highly endemic organisms, outbreaks may be difficult to identify against the naturally fluctuating background incidence, so targeted molecular typing may be problematic. In this situation, cluster detection may be impossible without universal typing. This concept is illustrated with an example below (Figure 17.2).

It has been shown in short-duration studies of universal typing that frequent nosocomial transmission occurs in many organisms for which surveillance is not routinely performed. One such example examined nosocomial *Candida* sp. infection. In a retrospective study, Ásmundsdóttir et al. performed typing using PCR-fingerprinting

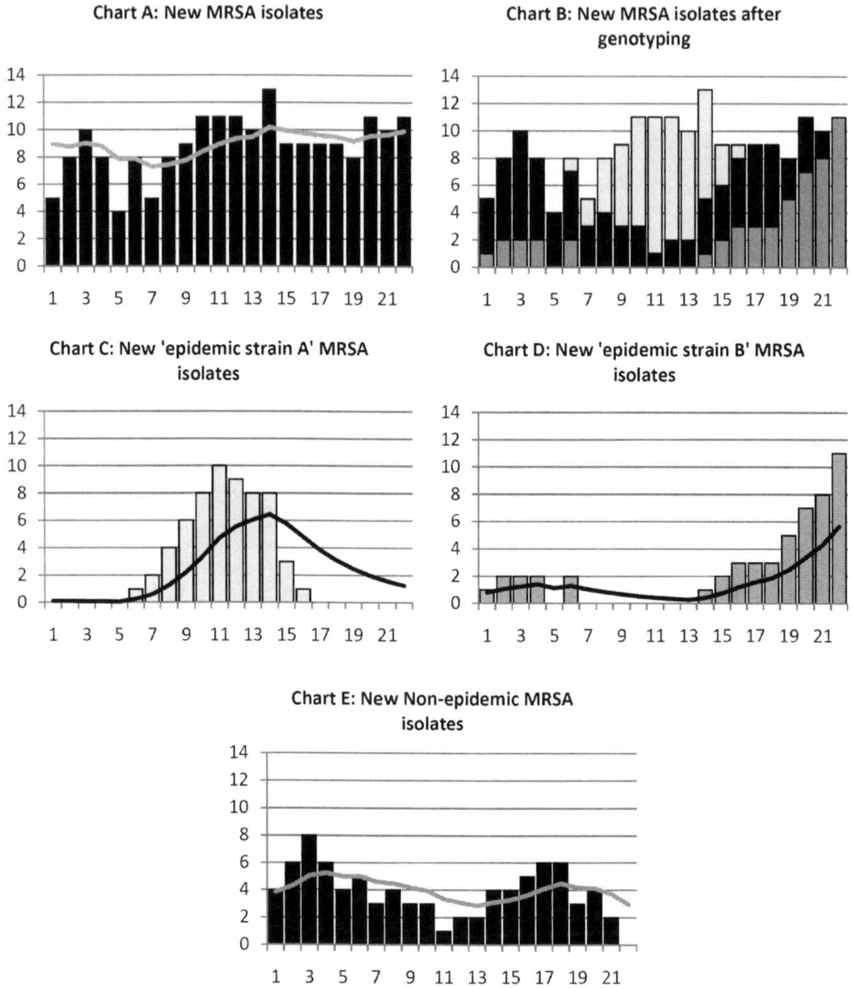

Fig. 17.2 Demonstration of utility of molecular typing for cluster detection using mock data from nosocomial surveillance of new MRSA acquisition in a hospital with high endemicity for MRSA. Horizontal axes – time in weeks. Vertical axes – count of new cases of MRSA. Solid line – Exponentially weighted moving average (EWMA). Chart A: surveillance without genotyping does not show any discernable clusters, and there is no obvious deviation of the EWMA. Chart B: graph of surveillance data incorporating genotyping information. Epidemic strain A is shown in light gray, epidemic strain B is shown in dark gray, and all other strains remain black. Two distinct temporal clusters are now discernable due to epidemic strains A and B, respectively. Charts C and D: surveillance data for epidemic strains A and B further highlighting the temporal clustering and demonstrating the deviation of the EWMA. Chart E: nonepidemic strains in the same period – no clustering is evident, and there is no deviation of the EWMA.

on all bloodstream isolates of *Candida* collected in Iceland over a 16-year period. Between 19 and 40% of isolates were suspected to have been acquired by nosocomial transmission on the basis of similar typing results and temporospatial occurrence (Asmundsdottir et al. 2008).

Other organisms for which frequent, unidentified transmission may occur in hospitals include, but are not limited to, methicillin susceptible *Staphylococcus aureus* (MSSA) (Chaves et al. 2005; Wilcox et al. 2000), *Staphylococcus epidermidis* (Muldrew et al. 2008), *Streptococcus pyogenes* (Ramage et al. 1996), *Streptococcus agalactiae* (Easmon et al. 1981; Kim et al. 2006), various gram negative bacilli (Almuneef et al. 2001; Prospero et al. 2006), and *Pneumocystis jiroveci* (Schmoldt et al. 2008).

Although not yet in wide routine use, several universal typing systems have been applied for hospital infection control. MRSA is one important organism for which traditional surveillance with targeted molecular typing fails to identify nosocomial outbreaks. This is largely because of MRSA's changing epidemiology, with the rising prevalence of this organism as a community pathogen (Otter and French 2006) and the high rates of endemicity of MRSA in many hospitals. In 1993, a study that used restriction enzyme analysis (REA) of plasmid DNA (REAP) typing found that the routine typing of all hospital MRSA isolates led to the identification of small clusters that would have gone unobserved if only traditional epidemiologic surveillance was used (Trilla et al. 1993).

Hacek et al. described universal typing by REA of genomic DNA with conventional electrophoresis in a tertiary facility for isolates of MRSA, VRE, and fluoroquinolone-resistant *Pseudomonas aeruginosa*. A 10% decrease in nosocomial infections was observed in the 24 months after the introduction of the system, compared with the 24 months prior, with an estimated cost saving of US$4,368,100 (Hacek et al. 1999). Mascini et al. used PFGE to distinguish epidemic from sporadic strains during a hospital outbreak of vancomycin resistant *Enterococcus faecium*, enabling infection control measures to be withheld for a large number of patients while still successfully terminating the outbreak (Mascini et al. 2006).

Microbial subpopulations differ in their virulence, which may manifest as an increased severity of a disease, or an enhanced ability to spread in the hospital environment and colonize the host. Genotyping data linked to clinical data have been able to establish the presence of hypervirulent clones in a number of microbial species. For example, some MRSA strains spread more easily and are more difficult to control than others (Amaral et al. 2005). In a study from the UK, colonization with one hypervirulent MRSA strain (sequence type [ST] 239) conferred a 4.5 times higher risk of intravenous-line associated BSI, compared with colonization with other MRSA strains (Edgeworth et al. 2007). Molecular typing using PFGE has demonstrated the emergence of the highly virulent USA300 clone of MRSA as a nosocomial pathogen in the United States (Patel et al. 2008).

Another clear example of molecular typing identifying a hypervirulent clone in a problematic nosocomial pathogen is *Clostridium difficile*. In 2003, it was reported that there had been an increasing number of cases and deaths from *C. difficile* diarrhea in Quebec, Canada, over the previous years. Subsequent molecular typing studies demonstrated that this increase was due the emergence of a clone, identified as ribotype 027 (Loo et al. 2005), which subsequently spread worldwide. Using PCR-ribotyping and multivariate analysis, Labbe et al. confirmed that this strain was more virulent, with a twofold increase in the risk of 30-day mortality compared with other ribotypes (Labbe et al. 2008).

Universal genotyping, when linked with clinical data (such as progression from colonization to infection, disease severity, complications and death) may be a powerful tool for the continuous monitoring for the emergence of new, hypervirulent strains of nosocomial pathogens. Such system, if had been in place in Quebec in 2002, could have identified the hypervirulent *C. difficile* ribotype 027 clone earlier, facilitating more timely, aggressive infection control measures that may have averted the subsequent world-wide epidemic.

17.5 Analysis of Genotyping Results

Traditionally, molecular typing results in infection control have been used to confirm or refute suspected outbreaks. In this scenario, an outbreak is considered confirmed if the strains are found to be indistinguishable (or closely related) by the typing method, and it is refuted if the strains are found to be unrelated by the typing method. However, for many organisms, such as MRSA, the population structure is very clonal, and indistinguishable results with even the most discriminatory typing method may not be sufficient to confirm an outbreak if the strain identified is one that is commonly circulating in the institution in question, or in the community in general. This has led to the criticism that universal typing may lead to the mis-identification of outbreaks (Pfaller and Herwaldt 1997). It follows, then, that a better way to utilize a typing result is to determine the *probability* that a set of indistinguishable strains represent an outbreak. This probability can be determined from both the surveillance data (using the magnitude of the increase in case frequency over the background rate) and the known molecular epidemiology of the organism in question (using the expected frequency of the particular strain type from the overall population).

This approach is analogous to the interpretation of any diagnostic test – wherein the likelihood ratio of a test result is applied to the pretest probability (in this case, the chance of an outbreak being present based on temporospatial surveillance data alone) to determine the posttest probability. In this case, the post-test probability is the probability that an outbreak has occurred, based on the combination of temporal, spatial, *and* genotyping data (Jaeschke et al. 1994). After this probability is determined, it can be decided whether further action is required, based on a certain threshold probability that would vary according to cluster frequency and available resources. Such an approach, previously advocated for MRSA *spa* typing (Harmsen et al. 2003), has been outlined in detail using nosocomial norovirus transmission as an example (Lopman et al. 2006).

17.6 Choosing Typing Method for Genotyping Systems

The choice of microbial genotyping method will be different for a particular organism in a given setting and will depend on the characteristics of the typing system, such as its discriminatory power, stability, ease of use, reproducibility, throughput, portability of results, and cost (Riley 2004). Rather than utilizing a single typing method, routine typing will often employ an initial, less discriminatory, typing method followed by more discriminatory methods for indistinguishable isolates. Selected methods referred to in this chapter are described in the Box 17.1.

Box 17.1 Typing methods

Pulsed field gel electrophoresis of restriction enzyme-digested genomic DNA (PFGE). This method consists of digesting DNA using a restriction enzyme that recognizes specific short DNA motifs and cleaves the DNA strand at that site. Variation between strains occurs because of mutations that create or remove restriction enzyme binding sites. Enzymes are chosen such that 10–20 DNA fragments are produced. These are then visualized using gel electrophoresis. No DNA amplification is employed. The technique is labor-intensive, low throughput, and the results do not lend themselves to digitization to establish libraries of strain types. However, it is a highly discriminatory method and remains the mainstay for the genotyping of many bacteria (Fig. 17.3a).

Rep-PCR. Amplification of bacterial DNA is performed using primers specific for an element that is found repeatedly interspersed throughout the genome. The direction of the primers is such that the intervening sequence, not the repetitive sequence, is the element amplified. Only when two repetitive elements are close together will a PCR product be produced, but with a good choice of target 10–20, PCR products can be obtained. These are then visualized by electrophoresis. This method has been commercialized in an automated high-throughput system (DiversiLab™), which also digitizes and analyses the results. It can be used for a wide variety of bacteria (Carretto et al. 2008) (Fig. 17.3b).

Multilocus variable number of tandem repeats analysis (MLVA). In this method, PCR is used to amplify a number of targets that vary in their size by virtue of short subunits that are repeated a variable number of times in a given isolate. The number of repeats at each locus is determined after the visualization of the amplification products with electrophoresis. The results can then be presented in a numerical format, rather than in a fingerprint format. With a good choice of targets, MLVA can be highly discriminatory. It is also high-throughput and reproducible (Fig. 17.3c).

(continued)

Box 17.1 (continued)

Sequence typing. This involves PCR amplification and sequencing of one or more targets that vary between isolates in their DNA sequence. Sequence typing is commonly used for the typing of viruses, often employing genes encoding surface proteins that are more variable and that broadly correlate with viral serotyping. *spa* typing is one form of sequence typing for *Staphylococcus aureus*. Sequence typing is robust and can be highly discriminatory if the correct targets are selected. With the greater availability, faster turnaround time, and lower cost of sequencing facilities, sequence typing is becoming increasingly used for hospital infection control applications. Multi-locus sequence typing (MLST) involves the sequencing of multiple targets. It is most commonly used to explore evolutionary relationships and population structure, so the targets used are genes that evolve slowly and so are less discriminatory (Fig. 17.3d).

17.7 Integrating Genotyping with Surveillance Systems

An ideal microbial genotyping system for infection control would integrate genotyping data with that from patient information systems, medical records, and laboratory information systems. This would then be analyzed to alert infection control practitioners of spatio-temporal-genotypic clusters of infection, which are suggestive of outbreaks and require further investigation and intervention. The addition of clinical outcome data from the medical record would allow continuous monitoring for the emergence of hypervirulent strains, which could prompt a higher level of infection control precautions. For such a system, the genotyping method used would ideally be inexpensive, rapid, and high throughput to allow universal genotyping when necessary. The method would so be highly discriminatory to reduce false positive cluster detection and to produce results that could easily be tracked in a database. The results would be expressed as a probability that isolates were related, as discussed earlier. The medical records would be electronic, and would automatically be screened for outcomes of interest, as discussed in detail in Chaps. 15, 16, and 20.

There are few publications describing such comprehensive genotype-based surveillance systems for hospital infection control, and none which link typing information to clinical outcome data. However, increasing progress is being made toward such systems. Mellman with colleagues successfully incorporated *spa* typing of MRSA with spatiotemporal epidemiologic data in a German tertiary care facility to automatically generate prospective "clonal alerts" that were found to identify clusters of MRSA transmission more reliably than in the case of clusters identified by frequency data alone or by the infection control professionals after a review of the microbial data and patient information (Mellmann et al. 2006). Fontana et al.

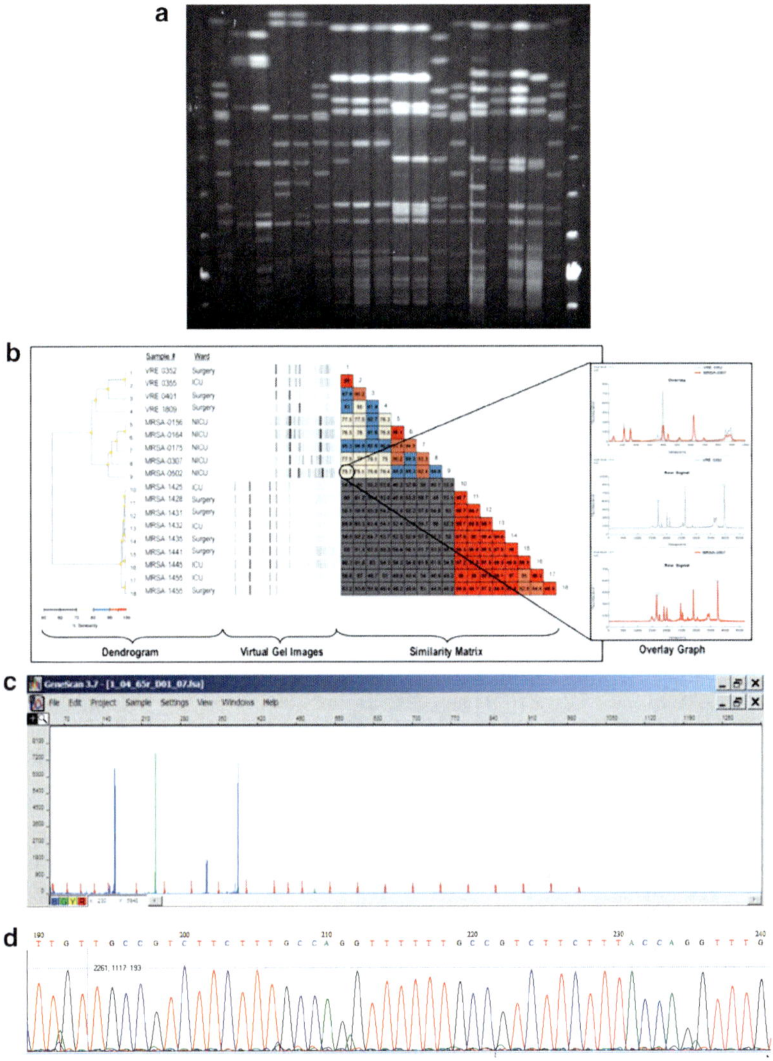

Fig. 17.3 Molecular typing methods. (**a**) Pulsed field gel electrophoresis (PFGE) of restriction-enzyme digested genomic DNA. (**b**) rep-PCR results and analysis using the automated DiversiLab™ system (figure courtesy of BioMerieux). (**c**) Multilocus variable number of tandem repeats analysis (MLVA) with product detection by capillary electrophoresis. (**d**) Sequence based typin

describe a system by which temporospatial clusters of nosocomial infections are identified in the laboratory by software (VIGI@ct), which takes data from the laboratory information system and the patient information system. Rapid targeted typing performed by either fluorescent AFLPs or the DiversiLab™ rep-PCR typing system is then conducted to confirm clusters. This was found to be useful to identify nosocomial clusters of a variety of pathogens (Fontana et al. 2007, 2008).

17.8 Conclusion

Molecular typing is a valuable tool for the identification of nosocomial infection outbreaks. Recent advances have produced rapid, discriminatory, high-throughput, and inexpensive typing methods that may permit the routine use of universal typing. This may prove to be invaluable for the surveillance of organisms with high endemicity, but typing results must be applied appropriately to avoid false conclusions. When combined with outcome information from electronic medical records, universal typing could facilitate continuous monitoring for the emergence of hypervirulent nosocomial pathogens.

References

Almuneef MA, Baltimore RS, Farrel PA et al. (2001) Molecular typing demonstrating transmission of gram-negative rods in a neonatal intensive care unit in the absence of a recognized epidemic. Clin Infect Dis 32:220–227

Amaral MM, Coelho LR, Flores RP et al. (2005) The predominant variant of the Brazilian epidemic clonal complex of methicillin-resistant *Staphylococcus aureus* has an enhanced ability to produce biofilm and to adhere to and invade airway epithelial cells. J Infect Dis 192:801–810

Andrei A, Zervos MJ (2006) The application of molecular techniques to the study of hospital infection. Arch Pathol Lab Med 130:662–668

Asmundsdottir LR, Erlendsdottir H, Haraldsson G et al. (2008) Molecular epidemiology of candidemia: evidence of clusters of smoldering nosocomial infections. Clin Infect Dis 47:e17–24

Australian Institute of Health and Welfare (2007) Australian hospital statistics 2005–06. Health services series no. 30. Cat. no. HSE 50. AIHW, Canberra

Best M, Neuhauser D (2004) Ignaz Semmelweis and the birth of infection control. Qual Safety Health Care 13:233–234

Burke JP (2003) Infection control – a problem for patient safety. N Engl J Med 348:651–656

Carretto E, Barbarini D, Farina C et al. (2008) Use of the DiversiLab semiautomated repetitive-sequence-based polymerase chain reaction for epidemiologic analysis on *Acinetobacter baumannii* isolates in different Italian hospitals. Diagn Microbiol Infect Dis 60:1–7

Casewell MW, Desai N (1983) Survival of multiply-resistant *Klebsiella aerogenes* and other Gram-negative bacilli on finger-tips. J Hosp Infect 4:350–360

Chaves F, Garcia-Martinez J, de Miguel S et al. (2005) Epidemiology and clonality of methicillin-resistant and methicillin-susceptible *Staphylococcus aureus* causing bacteremia in a tertiary-care hospital in Spain. Infect Control Hosp Epidemiol 26:150–156

Clark CM, Driver CR, Munsiff SS et al. (2006) Universal genotyping in tuberculosis control program, New York City, 2001–2003. Emerg Infect Dis 12:719–724

Curran ET, Benneyan JC, Hood J (2002) Controlling methicillin-resistant *Staphylococcus aureus*: a feedback approach using annotated statistical process control charts. Infect Control Hosp Epidemiol 23:13–18

Dziekan G, Hahn A, Thune K et al. (2000) Methicillin-resistant *Staphylococcus aureus* in a teaching hospital: investigation of nosocomial transmission using a matched case-control study. J Hosp Infect 46:263–270

Easmon CS, Hastings MJ, Clare AJ et al. (1981) Nosocomial transmission of group B streptococci. Br Med J (Clin Res Ed) 283:459–461

Edgeworth JD, Yadegarfar G, Pathak S et al. (2007) An outbreak in an intensive care unit of a strain of methicillin-resistant *Staphylococcus aureus* sequence type 239 associated with an increased rate of vascular access device-related bacteremia. Clin Infect Dis 44:493–501

Fontana C, Favaro M, Pistoia ES et al. (2007) The combined use of VIGI@ct (bioMerieux) and fluorescent amplified length fragment polymorphisms in the investigation of potential outbreaks. J Hosp Infect 66:262–268

Fontana C, Favaro M, Minelli S et al. (2008) *Acinetobacter baumannii* in intensive care unit: a novel system to study clonal relationship among the isolates. BMC Infect Dis 8:79

Hacek DM, Suriano T, Noskin GA et al. (1999) Medical and economic benefit of a comprehensive infection control program that includes routine determination of microbial clonality. Am J Clin Pathol 111:647–654

Hacek DM, Cordell RL, Noskin GA et al. (2004) Computer-assisted surveillance for detecting clonal outbreaks of nosocomial infection. J Clin Microbiol 42:1170–1175

Harmsen D, Claus H, Witte W et al. (2003) Typing of methicillin-resistant *Staphylococcus aureus* in a university hospital setting by using novel software for spa repeat determination and database management. J Clin Microbiol 41:5442–5448

Hartstein AI, Denny MA, Morthland VH et al. (1995) Control of methicillin-resistant *Staphylococcus aureus* in a hospital and an intensive care unit. Infect Control Hosp Epidemiol 16:405–411

Hartstein AI, LeMonte AM, Iwamoto PK (1997) DNA typing and control of methicillin-resistant *Staphylococcus aureus* at two affiliated hospitals. Infect Control Hosp Epidemiol 18:42–48

Imataki O, Makimoto A, Kato S et al. (2006) Coincidental outbreak of methicillin-resistant *Staphylococcus aureus* in a hematopoietic stem cell transplantation unit. Am J Hematol 81:664–669

Jaeschke R, Guyatt GH, Sackett DL (1994) Users' guides to the medical literature. III. How to use an article about a diagnostic test. B. What are the results and will they help me in caring for my patients? The Evidence-Based Medicine Working Group. J Am Med Assoc 271:703–707

Kim HJ, Kim SY, Seo WH et al. (2006) Outbreak of late-onset group B streptococcal infections in healthy newborn infants after discharge from a maternity hospital: a case report. J Korean Med Sci 21:347–350

Labbe AC, Poirier L, Maccannell D et al. (2008) *Clostridium difficile* infections in a Canadian tertiary care hospital before and during a regional epidemic associated with the BI/NAP1/027 strain. Antimicrob Agents Chemother 52:3180–3187

Loo VG, Poirier L, Miller MA et al. (2005) A predominantly clonal multi-institutional outbreak of *Clostridium difficile*-associated diarrhea with high morbidity and mortality. N Engl J Med 353:2442–2449

Lopman BA, Gallimore C, Gray JJ et al. (2006) Linking healthcare associated norovirus outbreaks: a molecular epidemiologic method for investigating transmission. BMC Infect Dis 6:108

Macfarlane L, Walker J, Borrow R et al. (1999) Improved recognition of MRSA case clusters by the application of molecular subtyping using pulsed-field gel electrophoresis. J Hosp Infect 41:29–37

Mascini EM, Troelstra A, Beitsma M et al. (2006) Genotyping and preemptive isolation to control an outbreak of vancomycin-resistant *Enterococcus faecium*. Clin Infect Dis 42:739–746

Mellmann A, Friedrich AW, Rosenkotter N et al. (2006) Automated DNA sequence-based early warning system for the detection of methicillin-resistant *Staphylococcus aureus* outbreaks. PLoS Med 3:e33

Morton AP, Whitby M, McLaws M-L et al. (2001) The application of statistical process control charts to the detection and monitoring of hospital-acquired infections. J Qual Clin Pract 21:112–117

Muldrew KL, Tang YW, Li H et al. (2008) Clonal dissemination of *Staphylococcus epidermidis* in an oncology ward. J Clin Microbiol 46:3391–3396

Otter JA, French GL (2006) Nosocomial transmission of community-associated methicillin-resistant *Staphylococcus aureus*: an emerging threat. Lancet Infect Dis 6:753–755

Papakyriacou H, Vaz D, Simor A et al. (2000) Molecular analysis of the accessory gene regulator (agr) locus and balance of virulence factor expression in epidemic methicillin-resistant *Staphylococcus aureus*. J Infect Dis 181:990–1000

Patel M, Waites KB, Hoesley CJ et al. (2008) Emergence of USA300 MRSA in a tertiary medical centre: implications for epidemiological studies. J Hosp Infect 68:208–213

Peleg AY, Franklin C, Bell JM et al. (2005) Dissemination of the metallo-beta-lactamase gene blaIMP-4 among gram-negative pathogens in a clinical setting in Australia. Clin Infect Dis 41:1549–1556

Peterson LR, Petzel RA, Clabots CR et al. (1993) Medical technologists using molecular epidemiology as part of the infection control team. Diagn Microbiol Infect Dis 16:303–311

Pfaller MA, Herwaldt LA (1997) The clinical microbiology laboratory and infection control: emerging pathogens, antimicrobial resistance, and new technology. Clin Infect Dis 25:858–870

Pfaller MA, Wakefield DS, Hollis R et al. (1991) The clinical microbiology laboratory as an aid in infection control. The application of molecular techniques in epidemiologic studies of methicillin-resistant *Staphylococcus aureus*. Diagn Microbiol Infect Dis 14:209–217

Phillips I (1991) Epidemic potential and pathogenicity in outbreaks of infection with EMRSA and EMREC. J Hosp Infect 18(Suppl A):197–201

Pimentel JD, Low J, Styles K et al. (2005) Control of an outbreak of multi-drug-resistant *Acinetobacter baumannii* in an intensive care unit and a surgical ward. J Hosp Infect 59:249–253

Prospero E, Barbadoro P, Savini S et al. (2006) Cluster of *Pseudomonas aeruginosa* catheter-related bloodstream infections traced to contaminated multidose heparinized saline solutions in a medical ward. Int J Hyg Environ Health 209:553–556

Ramage L, Green K, Pyskir D et al. (1996) An outbreak of fatal nosocomial infections due to group A streptococcus on a medical ward. Infect Control Hosp Epidemiol 17:429–431

Riley LW (2004) Principles and approaches. Molecular epidemiology of infectious diseases: principles and practices. ASM Press, Washington, DC

Schmoldt S, Schuhegger R, Wendler T et al. (2008) Molecular evidence of nosocomial *Pneumocystis jirovecii* transmission among 16 patients after kidney transplantation. J Clin Microbiol 46:966–971

Trilla A, Nettleman MD, Hollis RJ et al. (1993) Restriction endonuclease analysis of plasmid DNA from methicillin-resistant *Staphylococcus aureus*: clinical application over a three-year period. Infect Control Hosp Epidemiol 14:29–35

Wagenvoort JHT, Sluijsmans W, Penders RJR (2000) Better environmental survival of outbreak vs. sporadic MRSA isolates. J Hosp Infect 45:231–234

Weber S, Pfaller MA, Herwaldt LA (1997) Role of molecular epidemiology in infection control. Infect Dis Clin North Am 11:257–278

Wilcox MH, Fitzgerald P, Freeman J et al. (2000) A five year outbreak of methicillin-susceptible *Staphylococcus aureus* phage type 53,85 in a regional neonatal unit. Epidemiol Infect 124:37–45

Wright MO, Perencevich EN, Novak C et al. (2004) Preliminary assessment of an automated surveillance system for infection control. Infect Control Hosp Epidemiol 25:325–332

Chapter 18
Temporal and Spatial Clustering of Bacterial Genotypes

Blanca Gallego

18.1 Introduction

Infectious disease surveillance involves the monitoring of available infection-related data with the goal of detecting and, consequently, preventing and controlling outbreaks. Current methods of disease surveillance incorporate temporal, spatial, and multivariate information and make use of a wide range of statistical and machine learning algorithms for the classification, clustering and analysis of the data (Buckeridge et al. 2005; Shmueli and Fienberg 2005; Sonesson and Bock 2003; Song and Kulldorff 2003; Wagner et al. 2006). One important data source for the detection and monitoring of infectious disease outbreaks are collections of bacterial isolates. In particular, molecular characterization of pathogens from infected patients can be used as a biomarker of transmission and provides help with the identification of unsuspected transmission sites, reinfection and laboratory cross-contamination (McNabb et al. 2004; Tauxe 2006; Torpdahl et al. 2007).

Identifying patients that share the same pathogen genotype is often not enough to proceed with a public health investigation. The automated spatio-temporal clustering of pathogen genotypes can aid routine epidemiological surveillance by providing an operational definition of outbreak that adjusts to the local epidemiology of the disease as well as to the availability of public health resources (Gallego et al. 2009). In what follows, we provide a review of the detection of spatio-temporal clusters in biosurveillance and present a specific example of outbreak definitions using the spatio-temporal clustering of bacterial genotypes.

18.2 Detection of Spatio-Temporal Clusters

Given a set of events, a spatio-temporal cluster is loosely defined as a set of occurrences in a bounded space and time that are related to each other and are therefore unlikely to have occurred by chance. In a more rigorous statistical

B. Gallego
Centre for Health Informatics, University of New South Wales, Sydney, NSW, Australia

V. Sintchenko (ed.), *Infectious Disease Informatics*,
DOI 10.1007/978-1-4419-1327-2_18, © Springer Science+Business Media, LLC 2010

context, clustering represents the departure from a null hypothesis of spatio-temporal randomness. However, often some degree of clustering is regarded as belonging to a background spatio-temporal pattern, leading to a redefinition of clustering as the departure from this pattern. Within epidemiology, events usually represent disease counts and their relatedness may be due to human factors (e.g., infectious or genetic factors) or external factors (environmental, social) which give rise to spatio-temporal variations in risk. This chapter is concerned with the clustering of bacterial genotypes for the early detection of infectious disease outbreaks. We will start with a discussion on methods of prospective temporal surveillance followed by the introduction of the spatial dimension. Finally, we will discuss the methods' applicability to bacterial genotypes. A snapshot of the cluster detection methods most commonly used in biosurveillance systems can be found in Table 18.1.

18.2.1 Temporal Surveillance Methods

In its most basic form, temporal surveillance consists of the monitoring of a univariate time series with the goal of detecting an important change in the underlying process as early and accurately as possible. This requires information on the expected background activity or baseline, as well as a definition of the deviation from expected. When the signal is specific and sparse (e.g., laboratory test results), it is often monitored using simple nonstatistical rules. Less specific and noisy signals (e.g., syndromic data), require the use of statistical detection algorithms. In such statistical models, the quantity under surveillance is assumed to be a random variable that follows a given probability distribution. Under the null hypothesis of no outbreak, the baseline described by this distribution is usually either population-based (expected number of cases is proportional to population at risk) or expectation-based (expected number of cases is estimated from historical data). Deviation from expected is measured by an alarm function and an alarm limit is imposed to determine when the deviation is significant.

Many detection algorithms have been used in statistical surveillance. They differ on their baseline, alarm function and limit calculations depending on the application. Some of the algorithms most commonly used in public health biosurveillance systems include variations of the traditional statistical process control (SPC) methods such as: the likelihood ratio (LR) (Frisen and Demare 1991), the cumulative sum (CUSUM) method (Hawkins and Olwell 1998) and the exponentially weighted moving average (EWMA) method (Roberts 1959). A review and comparison of statistical surveillance algorithms can be found in (Brookmeyer and Stroup 2004; Buckeridge et al. 2005; Sonesson and Bock 2003). A description of their optimality properties is given in (Frisen 2003). Often the time series under analysis is pre-processed before the application of a detection algorithm with the goal of filtering background temporal structure and thus improving surveillance performance. Typical pre-processing methodologies include forecasting values using Poisson regression models (Williamson 1999), adaptive Kalman filters (Harvey 1993) and wavelet analysis (Shmueli 2005).

Table 18.1 Common methods of temporal and spatio-temporal surveillance

	Temporal surveillance		Spatio-temporal surveillance	
SPC methods	LR, CUSUM, EWMA, Shiryayev-Roberts	(Frisen and Demare 1991; Hawkins and Olwell 1998; Roberts 1959)	SPC + spatial clustering (e.g., spatial Scan Statistic)	(Neill and Moore 2004; Sonesson 2007)
Scan statistics methods	Temporal scan statistic (similar to an SPC method but with process control levels specified from a maximization conditional to the total number of events)		Spatio-temporal Scan Statistic	(Kulldorff 2001; Kulldorff et al. 2005)
Pre-processing methods	Regression, KF, wavelet analysis	(Harvey 1993; Shmueli 2005; Williamson 1999)		
Artificial intelligence methods	BCD, WSARE		BCD, WSARE, SVM	(Wong et al. 2002, 2003)

Note: *SPC* statistical process control; *CUSUM* cumulative sum; *EWMA* exponentially weighted moving average; *LR* likelihood ratio; *BCD* biosurveillance using a change-point detector; *WSARE* what's strange about recent events; *SVM* support vector machines

Current public health surveillance systems collect a wide range of data from a variety of sources such as emergency department visits, over-the-counter medication sales or reports of notifiable diseases. This data translates into multivariate time series often accompanied by additional co-variate information (e.g., spatial, demographic, clinical etc). Monitoring of multiple values can be performed by combining the outputs of the individual time series analysis or via the application of a multivariate algorithm. Both the SPC and scan statistics methodologies have been extended to account for covariate and multivariate information (Burkom and Murphy, 2007; Kulldorff et al. 2007). Other proposed methods for monitoring multiple temporal values make use of artificial intelligence techniques such as Bayesian networks and associations rule search and include the Biosurveillance using a change-point detector (BCD) algorithm (Wong et al. 2002) and the What's strange about recent events (WSARE) method (Wong et al. 2003). For a discussion on the performance of some of these multivariate detection systems, see (Buckeridge et al. 2005).

18.2.2 Spatio-Temporal Surveillance Methods

A large number of approaches have been proposed to search for spatial clustering in a set of data. These include heterogeneity methods (Potthoff and Whittinghill 1966), distance methods (Cuzick and Edwards 1990; Tango 1995), risk surface methods (Clayton and Kaldor 1987; Kelsall and Diggle 1998), moving window methods (Kulldorff 1997) and cluster modeling methods (Lawson and Denison 2002). An overview of the literature on spatial clustering in epidemiology can be found in (Elliot et al. 2000; Lawson 2001; LeSage et al. 2009). However, many of the proposed spatial clustering methods do not offer a formal indication of the location of clusters or their statistical significance, and are therefore not appropriate for infectious disease surveillance.

Here, we are interested in detection algorithms that extend the temporal surveillance methods to account for spatial variability. One possibility is to treat location as any other covariate such as age, and use the methodologies mentioned in the previous section. This approach, however, does not account for the special geographical properties of spatial information. Similarly, performing separate tests for each spatial point or region is generally inappropriate since events in each location are not likely to be independent. An approach that overcomes these drawbacks is to scan over all possible sets of regions using scan statistics. First proposed by Naus (1965) and further elaborated in the public health context by (Kulldorff 1997), the spatial scan statistics method looks for spatial regions where the probability of an incident case occurring is higher than outside. This model can be applied to temporal regions or extended to look for clusters in the spatio-temporal space (Kulldorff 2001; Kulldorff et al. 2005). It can also be combined with temporal SPC methods (Neill and Moore 2004; Sonesson 2007). The scan statistics methodology can, in theory, be applied for the scanning of any combination of multi-dimensional objects. In practice, however, the scanning algorithm is computationally very expensive and the search must be limited to a few degrees of freedom. For instance, Kuldorff's models search for contiguous clusters within circular (or at most elliptical) spatial regions. A faster spatial scan algorithm has been described (Neill and Moore 2004).

The task of the spatio-temporal detection of disease outbreaks remains a work in progress. Open questions include: analysis and recognition of spatio-temporal epidemic signatures, faster appropriate algorithms for prospective spatio-temporal detection and the evaluation of existing models for the surveillance of data with spatial information.

18.3 New Surveillance Data Types

In infectious disease surveillance, data types range from the very specific such as lab-test results and clinicians reports to the pre-diagnostic (such as emergency department chief complaints) and the pre-clinical (such as over-the-counter drug sales or school absentees reports). There is generally a trade-off between data quality and specificity and its timeliness and level of coverage. Syndromic data represents early warning signals characterized by large, noisy datasets. It is useful for the faster detection of large outbreaks but its associated public health action is often unclear. Higher diagnostic precision implies larger communication times and smaller coverage, since it is associated with patients undergoing diagnosis and maybe testing. It is more useful for the detection of moderate and small outbreaks and it can prompt well-defined public health action. Modern cheaper rapid pathogen genotyping techniques have allowed for highly specific epidemiological signals that can be generated very soon after a patient visit. This, combined with the implementation of electronic laboratory reporting, can improve the timeliness and completeness of the collection of diagnostic datasets (Overhage et al. 2001; Panackal et al. 2002).

Outbreak detection algorithms have different performance characteristics depending on the properties of their target datasets (Buckeridge 2007). Most of the statistical methods described so far have been designed for the analysis of syndromic data. These algorithms fail when applied to datasets that are sparse or have low signal-to-noise ratios, which is often the case for pathogen genotyping data. For example, Burkom and Murphy (2007) indicate the failure of existing temporal detection algorithms for sparse time series with low mean values, while Edgerton et al. discuss alternatives to the spatial scan statistics model when observed events are sparse in a large percentage of the spatial zones (Edgerton et al. 2007). Similarly, Gallego and colleagues find that the space-time permutation scan statistics fails to find statistically significant clusters when applied to sparse sets of *Salmonella typhimurium* isolates with identical genotypes (Gallego et al. 2009).

In contrast to the analysis of large syndromic datasets, very little has been said about algorithms suitable for the clustering of sparse spatio-temporal information in the context of public health biosurveillance. There is a need for the development of detection algorithms that make use of spatial, lower count, more specific biosurveillance data.

18.4 Infectious Disease Surveillance Using Genotype Clustering

18.4.1 Outbreak Definitions

Many definitions of the infectious disease outbreak have been coined, all of them subjective to the context in which they are applied. However, these definitions always contain two elements: there must be transmission of a pathogen and this transmission must be epidemic or unusual in nature. In recent years, pathogen genotyping profiles have proven to be good biomarkers of transmission and have been used to infer epidemiological links and to guide outbreak investigations. For example, in the United States, the tuberculosis genotyping and surveillance network has developed a standard for cluster investigations based on the IS*610* restriction fragment length polymorphism (RFLP) and spoligotype patterns of isolates (Crawford et al. 2002). For many types of infections, the epidemic nature of a set of same-genotype isolates needs to be assessed using spatio-temporal considerations. For example, in Denmark, an outbreak was defined as occurrence of at least five cases of foodborne infection detected within a 4-week period with respective *Salmonella typhimurium* (STM) isolates with indistinguishable MLVA (multiple-locus variable-number tandem-repeats analysis) profile (Torpdahl et al. 2007). Looking for spatio-temporal clusters adds value because it provides valuable information on transmission patterns and it lowers the threshold for investigating possible outbreaks. Ultimately, an operational definition of outbreak must include the criteria that controls public health action, both in terms of the severity, communicability and local epidemiology of the disease as well as in terms of the public health resources regarding investigative methods and options for effective prevention and control.

Recently, Gallego et al. introduced a scalable definition of outbreak that was based on the temporal and spatial clustering of molecular genotypes and could be tuned to accommodate the requirements and resources available for outbreak investigations (Gallego et al. 2009). Given a set of genotyped isolates from infected patients, each with an associated date (e.g., specimen collection date) and location (e.g., patient's residential address), the model clusters the isolates according to their genotype, and the temporal and spatial distance among them. In this way, a spatio-temporal cluster is defined using three parameters: the minimum number of same-genotype isolates, maximum time between consecutive isolates, and the maximum distance between spatially adjacent isolates (all locations in a genotype cluster are linked, forming a spanning tree, and two isolates are spatially adjacent when they are connected by an edge of the tree). These clustering parameters can be adjusted to the desired level of outbreak investigation. In what follows, this operational definition of outbreak is implemented using a dataset of STM isolates from patients in the state of New South Wales, Australia.

18.4.2 Clustering Cases of Foodborne Disease

The dataset used in this section consists of all confirmed (STM) isolates from patients referred to the New South Wales reference facility for enteric pathogens at the Centre for Infectious Diseases and Microbiology, Institute of Clinical Pathology and Medical Research (ICPMR) in Sydney between October 2006 and May 2008. Isolates were fingerprinted using multiple-locus variable-number tandem-repeats analysis (MLVA). MLVA (Lindstedt et al. 2004) detects short sequence repeats that vary in copy number in five regions or loci of the microbial genome. An MLVA profile consists of five numbers corresponding to the allele numbers assigned to each locus. MLVA has high discriminatory power within clonal species and has been found useful in epidemiological surveillance of salmonellosis (Chan et al. 2001; Torpdahl et al. 2007). Each isolate was also marked with an associated collection date and postcode of patient's address. The average turn-around time for MLVA genotyping was between 3 and 7 days after identification of STM.

The clustering algorithm described in (Gallego et al. 2009) was implemented using different clustering parameters. A genotyping cluster was defined as a maximal set of at least N isolates that share an identical MLVA type. A temporal cluster of parameter t was defined as a genotyping cluster for which the time difference between any two consecutive collection dates is at most t days. Similarly, a spatial cluster of parameter d was defined as a genotyping cluster for which the spatial difference between any two adjacent (refer to footnote) isolates is at most d kilometers. The distance between two isolates was estimated as the distance between the geographical centers of the corresponding patients' postcodes. Finally, a spatio-temporal cluster was defined as the combination of a temporal and a spatial cluster.

There were 1,464 isolates, displaying 345 different MLVA profiles. Most MLVA profiles (60.6%) appeared only once, while the most common MLVA profile 3–12–9–10–550 was found in 136 (9.3%) of the isolates. The average number of isolates per genotyping profile was 4.2. If we believe MLVA profiling to be a good biomarker of STM transmission, we can search for potential outbreaks of salmonellosis by looking for MLVA clusters. We start by deciding on the minimum number N of same-MLVA cases that will constitute an outbreak. The larger N is the fewer the number of potential outbreaks for our investigation. Here, the number of potential outbreaks decreased from 345 ($N = 1$) to 136 ($N = 2$), 43 ($N = 5$), and 25 ($N = 10$). The sharpest change took place between $N = 1$ and $N = 6$ (*see* Fig. 18.1). Also dependent on N are the spatio-temporal properties of the potential outbreaks. In this study, a genotyping cluster is characterized by: (a) size - number of isolates, (b) temporal duration - number of days from the first to the last collection dates, (c) surface area - sum of the areas of the patients' postcodes plus those of the enclosed

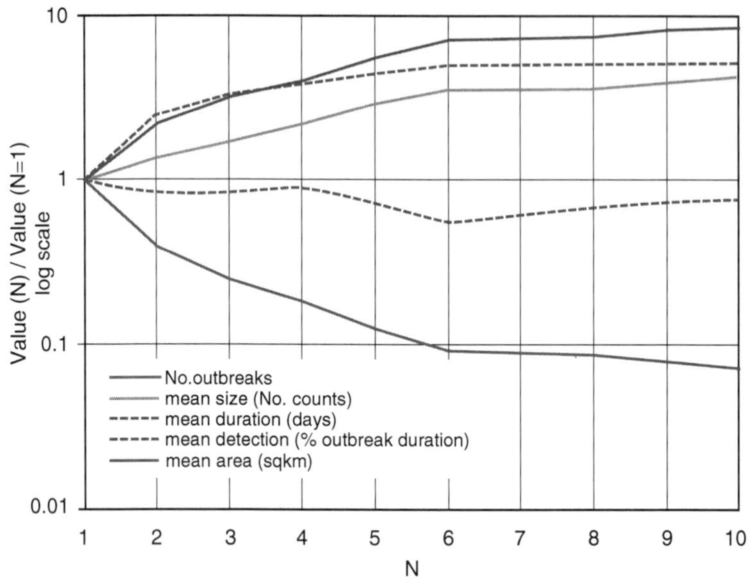

Fig. 18.1 Relationships between the size and duration of different outbreaks

postcodes, and (d) detection time - date at which the set of same genotype isolates that fulfill the appropriate spatio-temporal restrictions (if any) reaches N or more isolates. This time has been measured as a fraction of the duration of the outbreak and represents the time at which the cluster would have been detected prospectively. Figure 18.1 illustrates the properties that characterize the average cluster for different values of N. As expected, the mean size, duration, and area of the potential outbreaks increases with N. The capacity for early outbreak detection is best for $N = 6$ (35% of the outbreak duration).

Table 18.2 presents the sizes, temporal durations, surface areas and detection times of the 43 genotyping clusters, each containing at least five isolates ($N = 5$). The cluster durations ranged from 37 (MLVA 1–16–0–0–490) to 566 days (MLVA 3–9–9–12–523) and averaged 340 days. Their mean area was 6,861 km^2 representing 0.86% of total area of New South Wales. MLVA 3–12–10–12–523 occupied the largest area (29,911 km^2 or 3.73% of total area). Many of these clusters were confirmed by epidemiological investigations. Figure 18.2 shows a map of the spatio-temporal cluster characterized by MLVA 3–10–14–11–496, with a minimum number of isolates $N = 5$, a maximum temporal distance between consecutive cases $t = 2$ days, and a maximum spatial distance between adjacent cases $d = 10$ km. This cluster corresponds to a salmonellosis outbreak associated with contaminated pork in a Chinese bakery that took place in western Sydney at the end of March 2007.

Spatio-temporal clustering also has an effect in the number of potential outbreaks and their characteristics. Variation with temporal and spatial “distances” t and d in the number of spatio-temporal genotyping clusters with $N = 5$, as well as their

Table 18.2 MLVA clusters of *Salmonella typhimurium* in NSW, containing at least five cases between October 2006 and May 2008

Cluster	MLVA profile	Size (no. isolates)	Duration (days)	Detection (% duration)	Area (sq km)
1	3–12–9–10–550	136	102	21.6	28093.8
2	3–10–8–9–523	95	560	5.4	24021.4
3	3–12–11–10–523	79	505	2.2	10349.5
4	3–10–14–11–496	78	530	18.3	19324.1
5	3–11–10–8–523	71	409	16.1	15341.8
6	3–12–10–12–523	68	525	15.6	29911.0
7	3–9–8–12–523	51	557	5.7	9180.4
8	3–17–16–13–523	37	135	8.9	4018.4
9	3–9–7–12–523	34	554	4.3	11360.4
10	5–14–9–9–490	31	525	16.0	22421.4
11	3–9–9–12–523	22	566	14.8	15963.3
12	3–13–10–12–523	22	501	2.4	7995.8
13	3–11–7–12–523	21	561	0.4	684.2
14	3–14–11–9–523	18	397	2.5	3863.7
15	1–16–0–0–490	18	37	21.6	84.1
16	4–10–13–0–544	15	117	16.2	10185.6
17	3–14–8–13–523	13	228	41.7	3681.2
18	3–15–0–0–517	12	505	47.7	9967.9
19	3–9–7–13–523	11	449	12.9	606.3
20	4–16–13–0–517	11	399	73.4	10234.3
21	3–10–14–11–523	11	464	22.8	3762.2
22	3–12–9–12–523	10	472	74.2	9104.2
23	4–16–14–0–517	10	533	3.2	207.7
24	3–11–7–13–523	10	83	8.4	166.2
25	4–16–10–0–517	10	118	3.4	1010.9
26	3–12–12–12–523	9	83	53.0	49.7
27	3–12–10–10–523	9	532	75.9	645.9
28	3–13–11–12–523	8	423	51.5	3771.6
29	3–10–9–9–523	8	556	89.9	525.7
30	4–16–15–0–517	8	118	28.8	82.2
31	3–13–12–10–523	7	172	77.9	9268.4
32	3–13–8–12–523	6	345	97.7	4194.5
33	4–13–10–0–490	5	228	100.0	660.3
34	3–13–15–9–523	5	489	100.0	6044.9
35	3–13–11–10–523	5	377	100.0	716.5
36	3–12–12–11–523	5	46	100.0	9214.9
37	3–13–12–20–523	5	43	100.0	82.3
38	3–9–8–13–523	5	420	100.0	847.7
39	4–16–12–0–517	5	157	100.0	40.1
40	1–14–0–0–490	5	81	100.0	43.5
41	4–14–10–0–490	5	198	100.0	2449.4
42	3–25–13–12–523	5	202	100.0	2502.7
43	4–12–0–0–462	5	323	100.0	2355.0

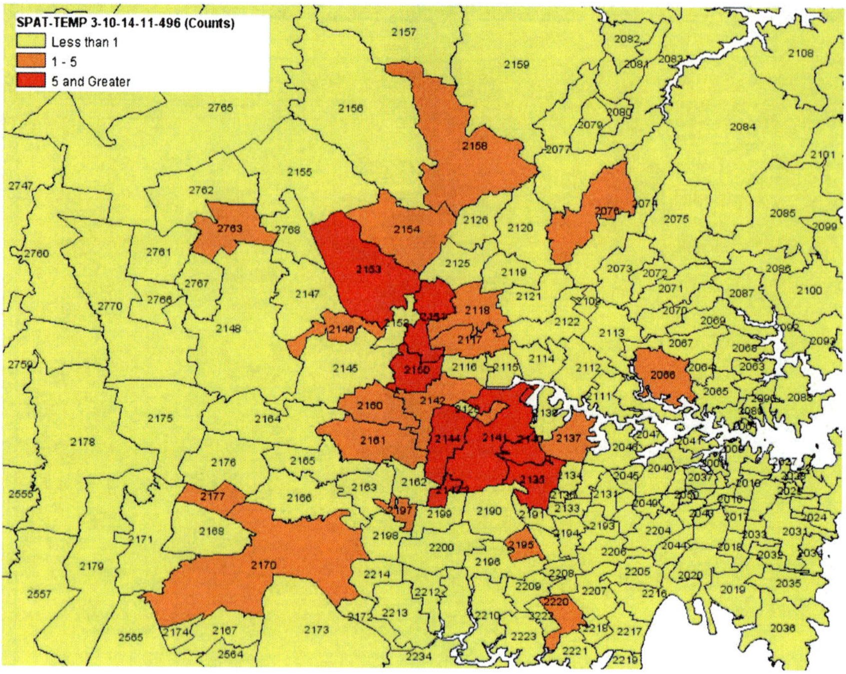

Fig. 18.2 Spatio-temporal cluster of *Salmonella typhimurium* corresponding to point-source outbreak in metropolitan Sydney

average size and detection time, are shown in Fig. 18.3. The number of potential outbreaks (Fig. 18.3a) generally decreases with increasing restrictions in time and space (that is, with smaller t and d). There is only 1 outbreak taking place in 1 day ($t = 0$) and one postcode ($d = 0$). The mean outbreak size (Fig. 18.3b) varies from 7 to 24.2 isolates and it is most sensitive to variations in t between 0 and 3 days and in d between 5 and 15 km. For large enough values of t and d, both the average duration and the average area of the clusters (not shown here) tend also to decrease with increasing temporal and spatial restrictions. Variations for small values of t and d are less straightforward. Mean duration has a small local maximum around $t = 3$ days and $d = 10$ km, while mean area has a local maximum at $t = 1$, 2 days and $d = 0$ km. The outbreak detection time (Fig. 18.3c) tends to increase with increasing restrictions in t and d. That is, clustering in space and time generally has the effect of decreasing the effectiveness of prospective surveillance.

18.5 Concluding Remarks

The operational definition of infectious disease outbreaks using the spatio-temporal clustering of bacterial genotypes integrates pathogen genotyping data into public health actions, thus aiding in the detection of small and medium size epidemics.

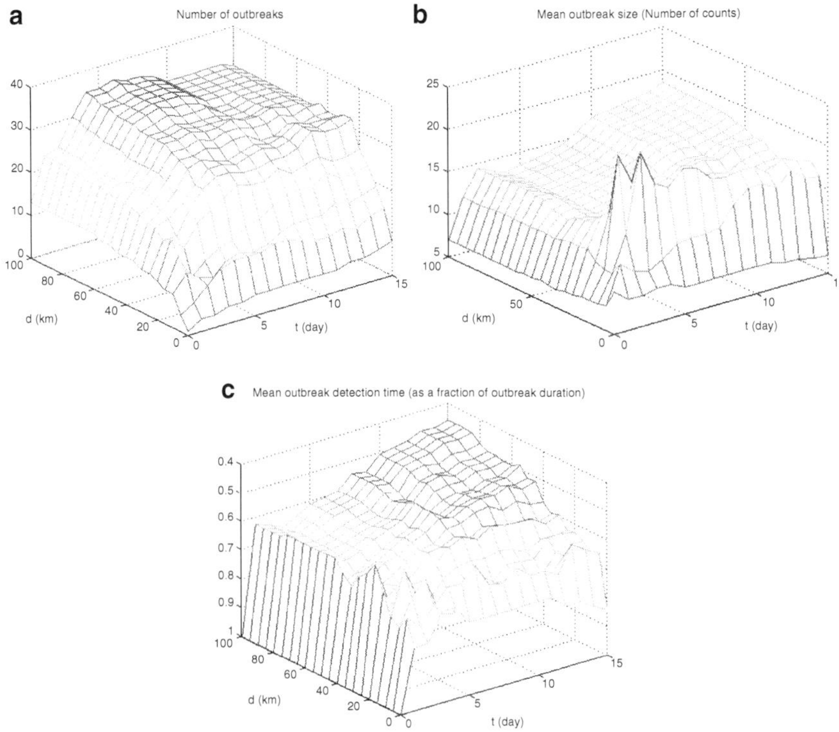

Fig. 18.3 Variations in the number of *Salmonella typhimurium* MLVA clusters (**a**), mean cluster size (**b**), and mean cluster detection (**c**) with changes in the temporal and spatial clustering parameters t and d

The timeliness of the response is limited by the turn-around time of the molecular genotyping techniques and the coverage is limited by the proportion of the infected population that undergoes testing. On the other hand, any spatio-temporal analysis of patients' isolates will also be restricted by the fact that associated locations and dates are generally just an estimate of the place and time of transmission.

A promising extension of this work consists of including other molecular profiling data relevant to the spatio-temporal clustering. Subtyping methods for the purpose of clonal discrimination cannot often identify the phenotypic characteristics associated with more rapidly changing regions of the pathogen's genome. The integration of genetic information regarding characteristics such as the pathogen's drug resistance profile or virulence would complement the biosurveillance task and assist in designing more effective public health interventions (Sintchenko et al. 2007).

Acknowledgments The author acknowledges substantial contributions from Qinning Wang, Gwendolyn L Gilbert, Vitali Sintchenko and Peter Howard of the Centre for Infectious Diseases and Microbiology, Institute of Clinical Pathology and Medical Research, Sydney West Area Health Service and The University of Sydney. This work was supported by the Australian Research Council.

References

Brookmeyer R, Stroup D (2004) Monitoring the health of populations: statistical methods for public health surveillance. Oxford University Press, Oxford

Buckeridge D (2007) Outbreak detection through automated surveillance: A review of the determinants of detection. J Biomed Inform 40:370–379

Buckeridge D, Burkom H et al. (2005) Algorithms for rapid outbreak detection: a research synthesis. J Biomed Inform 38:99–113

Burkom H, Murphy S (2007) Data classification for selection of temporal alerting methods for biosurveillance. In: Zeng D et al. (eds) Intelligence and Security Information: Biosurveillance. Lecture Notes in Computer Science 4506. Springer, pp. 59–70

Chan M-S, Maiden M et al. (2001) Database-driven multi locus sequence typing (MLST) of bacterial pathogens. Bioinformatics 17:1077–1083

Clayton D, Kaldor J (1987) Empirical Bayes estimates of age-standardised relative risks for use in disease mapping. Biometrics 43:671–681

Crawford J, Braden C et al. (2002) National Tuberculosis Genotyping and Surveillance Network: design and methods. Emerg Infect Dis 8:1192–1196

Cuzick J, Edwards R (1990) Spatial clustering for inhomogeneous populations. J R Stat Soc Series B 52:73–104

Edgerton J, Burkom H et al. (2007) Modifications to spatial scan statistics for estimated probabilities at fine-resolution in highly skewed spatial distributions. Adv Dis Surv 4:89

Elliot P, Cuzick J, et al. (2000) Geographical and environmental epidemiology: methods for Small area studies. Oxford University Press, Oxford

Frisen M (2003) Statistical surveillance: optimality and methods. Int Stat Rev 71:403–434

Frisen M, Demare J (1991) Optimal surveillance. Biometrika 78:271–290

Gallego B, Sintchenko V et al. (2009) Biosurveillance of emerging biothreats using scalable genotype clustering. J Biomed Inform 42:66–73

Harvey A (1993) Time series models. The MIT Press, Boston

Hawkins D, Olwell D (1998) Cumulative sum charts and charting for quality improvement. Springer, New York

Howard S, Burkom YE 2A, Feldman 1J, Lin1 A (2004) Role of data aggregation in biosurveillance detection strategies with applications from ESSENCE. MMWR Morb Mortal Wkly Rep 53(Suppl):67–73

Kelsall J, Diggle P (1998) Spatial variation in risk of disease: a nonparametric binary regression approach. J R Stat Soc Series C 47:559–573

Kulldorff M (1997) A spatial scan statistic. Commun Stat Theory Methods 26:1481–1496

Kulldorff M (2001) Prospective time-periodic geographical disease surveillance using a scan statistic. J R Stat Soc Series A 164:61–72

Kulldorff M, Heffernan R et al. (2005) A space-time permutation scan statistic for disease outbreak detection. PLoS Med 2:e59

Kulldorff M, Mostashari F et al. (2007) Multivariate scan statistics for disease surveillance. Stat Med 26:1824–1833

Lawson A (2001) Statistical methods in spatial epidemiology. Wiley, London

Lawson A, Denison D (2002) Spatial cluster modelling: an overview. *Spatial Cluster Modelling*. CRC Press, New York

LeSage J, Banerjee S et al. (2009) Spatial statistics: methods, models & computation. Comput Stat Data Anal 53:2781–2785

Lindstedt B, Vardund T et al. (2004) Multiple-locus variable-number tandem-repeats analysis of *Salmonella enterica* subsp. enterica serovar Typhimurium using PCR multiplexing and multicolor capillary electrophoresis. J Microbiol Methods 59:163–172

McNabb S, Kammerer J et al. (2004) Added epidemiologic value to tuberculosis prevention and control of the investigation of clustered genotypes of *Mycobacterium tuberculosis* isolates. Am J Epidemiol 160:589–597

Naus J (1965) The distribution of the size of the maximum cluster of points on a line. J Am Stat Assoc 60:532–538

Neill D, Moore A (2004) Rapid detection of significant spatial clusters. In: Proceedings of the 10th ACM SIGKDD Conference on Knowledge Discovery and Data Mining, pp 256–265

Overhage J, Suico J, et al. (2001) Electronic laboratory reporting: barriers, solutions and findings. J Public Health Manag Pract 7:60–66

Panackal A, Mikanatha N, et al. (2002) Automatic electronic laboratory-based reporting of notifiable infectious diseases at a large health system. Emerg Infect Dis 8:685–691

Potthoff R, Whittinghill M (1966) Testing for homogeneity. II. The Poisson distribution. Biometrika 53:183–190

Roberts SW (2000) Control chart tests based on geometric moving averages. Technometrics 42(1):97–101

Shmueli G (2005) Wavelet-based monitoring in modern biosurveillance. Robert H Smith School Research Paper No RHS-06–002 Available at SSRN: http://ssrncom/abstract = 902878

Shmueli G, Fienberg S (2005) Current and potential statistical methods for monitoring multiple data streams for biosurveillance. In: Alyson G, Wilson D, Olwell A (eds) Statistical methods in counterterrorism: game theory, modeling, syndromic surveillance, and biometric authentication. Springer, New York

Sintchenko V, Iredell JR, et al. (2007) Genomic profiling of pathogens for disease management and surveillance. Nat Microbiol Rev 5:464–470

Sonesson C (2007) A CUSUM framework for detection of space-time disease clusters using scan statistics. Stat Med 26:4770–4789

Sonesson C, Bock D (2003) A review and discussion of prospective statistical surveillance in public health. J R Stat Soc 166:5–21

Song C, Kulldorff M (2003) Power evaluation of disease clustering tests. Int J Health Geogr 2:1–8

Tango T (1995) A class of tests for detecting "general" and "focused" clustering of rare diseases. Stat Med 14:2323–2334

Tauxe RV (2006) Molecular subtyping and the transformation of public health. Foodborne Pathog Dis 3:4–8

Torpdahl M, Sorensen G et al. (2007) Tandem repeat analysis for surveillance of human *Salmonella typhimurium* Infections. Emerg Infect Dis 13:388–395

Wagner M, Moore A et al. (2006) Handbook of biosurveillance. Elsevier Academic Press

Williamson GD, Weatherby HG (1999) A monitoring system for detecting aberrations in public health surveillance reports. Stat Med 18:3283–3298

Wong W, Moore A et al. (2002) Rule-based anomaly pattern detection for detecting disease outbreaks. AAAI-02. Edmonton, Alberta, pp 217–223

Wong W, Moore A et al. (2003) What's strange about recent events? J Urban Health 80:66–75

Chapter 19
Infectious Disease Ontology

Lindsay Grey Cowell and Barry Smith

19.1 Vocabulary Resources for Biomedicine

Vocabulary resources have been used in biology and medicine at least since the time of Linnaeus, whose work on classification extended not only to organisms but also, in his *Genera morborum* (1763), to the classification of diseases. Linnaeus' work (and through it Aristotle's ideas on classification) continues to play an influential role in terminology and taxonomy work today.

Initially, vocabularies and terminologies existed in the form of printed dictionaries compiled for human use, and such resources continue to play an important role, for example, in education. The primary use of vocabulary resources of interest to us, however, is in fostering the presentation of biomedical and clinical data and information in ways that can support the use of computation in research. In this context, vocabulary resources have been developed for purposes of bibliographic search, coding of clinical and public health data, and database interoperability. For example:

- The Medical Subject Headings (MeSH) vocabulary (http://www.nlm.nih.gov/mesh/meshhome.html), first published in 1954, is used to support literature indexing and document retrieval for the MEDLINE database of biomedical literature.
- The International Classification of Diseases (ICD) (http://www.who.int/classifications/icd/en/), first published as the International List of Causes of Death in 1893, is the international standard for coding diagnostic information for health and vital records and is also commonly used for hospital billing purposes.
- SNOMED (http://www.snomed.org), first released in 1965, was initially developed to support documentation of pathology data and is projected to become a worldwide reference vocabulary for structured clinical documentation.

L.G. Cowell (✉)
Department of Biostatistics and Bioinformatics, Duke University Medical Center, Durham, NC, USA

V. Sintchenko (ed.), *Infectious Disease Informatics*,
DOI 10.1007/978-1-4419-1327-2_19, © Springer Science+Business Media, LLC 2010

- The Gene Ontology (GO) (http://www.geneontology.org/), created in 1998, is a vocabulary resource for the annotation of gene and gene-product data facilitating interoperability between a large number of diverse databases, especially in the domain of model organism research.

In the last decade, there has been an increasing need for biology and medical terminologies to support more sophisticated computational algorithms requiring high precision. This is a consequence of (a) tremendous increases in the volumes and types of data and information coming out of biomedical and clinical research, resulting in the need for computational assistance for the analysis and interpretation of these data, (b) pressure to implement electronic health records, and (c) increased interest in the possibilities of automated reasoning for biomedical research, clinical decision support, and biosurveillance.

In addition to the increased need for machine interpretable vocabulary resources, there is a growing need also for vocabularies to be interoperable across institutional and disciplinary boundaries. In both the biological and clinical domains, interoperability across subdisciplines is critical to advancing scientific understanding. The emergence of translational medicine as a new field and the push to use clinical data for research have increased the need for interoperability between the academic and clinical care domains. The formation of public data repositories and the movement of patients between health care systems both put additional requirements on vocabularies to be interoperable across institutions. Analogous requirements are also increasingly being felt in the domains of public health and disease and pathogen surveillance.

Unfortunately, existing biomedical and clinical vocabularies are in many ways incompatible because they were developed for a variety of different purposes and by multiple separate communities. They have different underlying semantics, employ different linguistic and logical structures, and manifest varying degrees of formal rigor (described in detail below). As a consequence, they are not interoperable and most do not support sophisticated computing of the sort that is becoming central to informatics-driven biomedical research. Increasing reliance on the computer processing of data and information and the requirement for cross-domain interoperation have highlighted the need for more structure and formal rigor in vocabulary resources. Because of their enhanced formal capabilities to support computing, interoperation, and reasoning, ontologies are being advanced as a new kind of terminology resource that can provide a necessary foundation for biomedical and clinical research in the future.

In what follows, we describe the different types of vocabulary resources available in the infectious disease domain, covering the spectrum of terminology-based representational artifacts from simple taxonomies, wordlists, glossaries, and loosely structured thesauri through data dictionaries to the more highly formalized "ontologies" now increasingly being applied in biomedical research. We will emphasize those features of formal ontologies that make them most useful for computational applications. We will then describe the various uses of ontologies in biomedical and clinical research, describe existing vocabulary resources relevant to infectious diseases, and conclude with some speculations concerning the potential uses of ontologies in the future.

19.2 Types of Vocabulary Resources

All vocabulary resources consist of terms; they differ in how these terms are presented and organized. Most importantly for our present purposes, vocabulary resources differ in whether terms are provided with definitions, in the types of relationships asserted between terms or the entities to which the terms refer, and in the degree of logical rigor underlying definitions and relations.

The simplest vocabulary resources are term lists (with or without definitions), containing no information about how the terms or the entities to which the terms refer are related to each other beyond what can be inferred from the terms themselves when considered linguistically. Examples include nomenclatures such as the Human Genome Organization (HUGO) Gene Nomenclature (http://www.hugo-international.org/committee_nomen.htm) and the Nomenclature for Factors of the Human Leukocyte Antigen (HLA) system (http://www.anthonynolan.org.uk/HIG/nomen/reports/homen/reports.html).

The majority of vocabulary resources, however, assert a simple term hierarchy or taxonomy in which the relationships between terms indicate that one term has a narrower meaning than another, or that one type of thing (e.g, dog) is classified as a subtype of another type of thing (e.g., animal). ICD and MeSH are examples of this type of resource. Vocabulary resources that assert a richer set of relations are less common. The best example is the Foundational Model of Anatomy (FMA) (http://sig.biostr.washington.edu/projects/fm/), which includes backbone hierarchies structured by means of taxonomic (*is_a*) and partonomic (*part_of*) relations and various formally defined spatial relations representing adjacency, connectedness, and relative position.

Many vocabulary resources, including many medical glossaries, have poor structural organization and provide at best definitions written in natural language for interpretation by human users. This means that they are poorly suited for computational purposes. Providing definitions based on a formal theory [such as (Rosse and Mejino 2003)] enhances the potential utility of a vocabulary resource for computation, but requires a non-trivial investment of resources, especially for the large vocabulary resources often found in the biomedical domain.

Similarly, there is great variability in the degree to which the relations used in the structure of vocabulary resources are formalized in a way that supports automatic reasoning. In MeSH, for example, relations are presented primarily in an implicit fashion through the relative position of terms in the MeSH hierarchy. Among vocabulary resources with explicitly asserted relations, the vast majority provides either no definition of the relations, or provides only natural language descriptions of the intended meaning of relational expressions. At the other, more formally rigorous, end of the spectrum are a growing number of vocabulary resources employing relations defined according to a formal theory, for example, within the context of the Semantic Web (Ruttenberg et al. 2007) and of the Open Biomedical Ontologies (OBO) Foundry Initiative (Smith et al. 2007).

Following what is increasingly becoming standard usage, we shall here employ the term "ontology" to refer to a vocabulary resource that is structured by means of relations between its terms and is logically formalized in the sense that the developers adhere to a logical theory in the definition of terms and relations, for example, as outlined in Rosse and Mejino (2003) and Smith et al. (2005). Vocabulary resources of this sort are standardly represented as graph-theoretical structures built up out of terms as the nodes of the graph and relations as edges (Bechhofer et al. 2004). While there are a variety of other meanings associated with the term "ontology," the usage here is consistent with that of large influential ontology developer and user groups, including the Gene Ontology Consortium (http://www.geneontology.org/), the W3C community (http://www.w3.org/), and the OWL Web Ontology Language community (http://www.w3.org/2004/OWL).

The different uses for which the different vocabulary resources have been built have determined to a large extent the degree and type of structure, level of detail, and logical formalism used in their construction. We argue, however, that even when the intended application does not require a highly structured and formalized vocabulary resource, there are benefits to be gained from developing the resource with a structured and formalized approach in ways that adhere to best practice guidelines. First, such an approach results in vocabulary resources that have fewer developer-introduced errors. Second, the resulting vocabulary resources can be subjected to automated error checking (Ceusters et al. 2004a, 2005; Smith et al. 2004). Third, structured and formalized resources are likely to be free of idiosyncratic features and are therefore more broadly applicable. Thus, the development of a structured and formalized vocabulary resources can facilitate their reusability and utility as biomedical research becomes increasingly reliant on computation (Yu 2006).

A simple illustration of the advantages already resulting from a greater formal organization of a vocabulary resource is how this organization makes possible a more complete and more focused retrieval of data. Without formal organization, searches against data catalogued on the basis of mere word lists are restricted to the use of string matches, which is highly ineffective especially in a domain like infectious diseases, where data are derived from many heterogeneous sources and nomenclature is poorly standardized. Formal organization means that, when collecting information about a given disease or pathogen, we can automatically extend our search to include corresponding subtypes or variants independently of how the latter are named. Another simple benefit of formal organization is the ability to ensure that the effects of changes to a classification are automatically propagated to all relevant parts of the classification.

The use of classifications that rest on a well-defined and reliably executed application of the subclass or subtype relations (called in what follows "*is_a*") is crucial to the realization of these benefits. Here, the test of reliability is conformity to the rule: if type *A* is classified as a subtype of *B*, then all instances of *A* (e.g., all cases of a given infectious disease) are also instances of *B*.

One consequence of conformity to this rule is that the *is_a* relation will be transitive (if we know that *A is_a B* and *B is_a C*, then we can infer also that *A is_a C*). For example, if we know that *Staphylococcus aureus is_a Staphylococcus* and *Staphylococcus is_a bacterium*, then we can infer that *Staphylococcus aureus is_a bacterium*.

Another consequence is that all instances of *A* will inherit the properties shared by all instances of *B*. For example, bacteria of the genus Staphylococcus are facultative anaerobes. If this is asserted in the ontology, along with *Staphylococcus aureus is_a Staphylococcus*, *Staphylococcus aureus* will inherit the property of being a facultative anaerobe. Inheritance is an important source of potential benefits from the use of vocabulary resources in automatic reasoning. Definition and use of the *is_a* relation are discussed in more detail below.

Terminological note. Where type *A* stands in an *is_a* relation to type *B* in a classificatory hierarchy, we shall also describe "*A*" as the child term and "*B*" as parent. Any given child can have sibling terms in the sense of terms that share a common parent. Further discussion of the different types of vocabulary resources can be found in Yu (2006), Bodenreider and Stevens (2006), Cimino and Zhu (2006), and Coonan (2004).

19.3 Features of Ontologies Needed to Support Informatics

For ontologies to support sophisticated computational algorithms with high precision, it is necessary that they be developed in accordance with certain principles of ontology development best practice. In particular, adherence to the following has been shown to enhance support for computation: (a) the use of Aristotelian definitions with a single mode of classification, (b) the use of single inheritance hierarchies, (c) the use of relations with formal, logical definitions based on a distinction between types and instances, and (d) writing definitions and ontology assertions as compositions of ontology terms and relations rather than as natural language.

The definition of types in an ontology serves an important purpose beyond describing the meaning of the term that refers to the type, and that is to specify the placement of the types in the ontology's inheritance hierarchy. This is accomplished through the use of Aristotelian definitions, the form of which is *A is_a B* which *C*, where *A* is the type being defined, *B* is its *genus* (parent or supertype), and *C* is the *differentia* (Rosse and Mejino 2003; Michael et al. 2001). It is the first part of the definition, *A is_a B*, that results in inheritance, as *A* will inherit all of the properties of *B*, including those properties *B* inherits from its parent. *B* may have many subtypes, and it is the differentia, *C*, that distinguishes *A* from the other subtypes of *B*. For example, in a hierarchy of disease types, one could define *infectious disease* as a *disease* that is caused by an infection.

In addition to the use of Aristotelian definitions, it is recommended that a single mode of classification be adopted for any given hierarchy, that is, all types within a single hierarchy should be differentiated based on the same type of criterion. It is further recommended that each type has only a single parent type. Hierarchies in which all types have only a single parent are referred to as single inheritance hierarchies, whereas hierarchies in which types can have more than one parent are referred to as multiple inheritance hierarchies. The problem with using multiple modes of classification and with allowing multiple inheritance is that the meaning of the *is_a* relation becomes uncertain, resulting in errors on the part of both maintainers and users of an ontology (Bodenreider et al. 2004) and the inability to use the hierarchy for automated reasoning. For example, in SNOMED, *is_a* has in some contexts the meaning "has cause" (e.g., *Tuberculosis of meninges is_a Mycobacteriosis*), while in others it means "has location" (e.g., *Tuberculosis of meninges is_a Disorder of meninges*). The use of *is_a* with multiple meanings is often referred to as "*is_a* overloading" (Guarino 1998). While in practice it can be difficult to avoid multiple inheritance, even within a single mode of classification, multiple modes of classification (and therefore multiple meanings for *is_a*) should be avoided by using the corresponding specific relations (e.g., *has_location*). The benefits are not only an ontology that has fewer errors, is easier to maintain, and can be used for automated reasoning, but also a reduced loss of information by using the more specific representation. Other considerations in the classification of biological entities are outlined in detail in (Michael et al. 2001; Bodenreider et al. 2004).

Successful inferencing over the relations asserted between ontology types relies on a single, logical definition for each relation with clearly specified implications. This is best accomplished by distinguishing between types (e.g., influenza infection) and instances (e.g., each of the individual cases of influenza infection), and defining the relations between types in terms of the relations between the corresponding instances (Smith et al. 2005). Thus, a type-level relation R will be defined in terms of the instance-level relation **R** by: $X\ R\ Y =_{\text{def}}$ for every instance x of X, there exists at least one instance y of Y such that $x\ \mathbf{R}\ y$, where uppercase indicates types (X, Y) and lowercase indicates instances (x, y). For example, *human has_part brain* means that every instance of *human* has as part of it some instance of *brain*. Defining the relations between types in terms of the relations between instances, and specifying that the type-level relation $X\ R\ Y$ holds when the instance-level relation $x\ \mathbf{R}\ y$ holds for *all* instances of X ensures that $X\ R\ Y$ holds universally. This, in turn, ensures transitivity, which can be used for automated reasoning: if $X\ R\ Y$ and $Y\ R_1\ Z$, then there is some relation R_2 such that $X\ R_2\ Z$. The distinction between types and instances corresponds to the distinction between A-boxes and T-boxes used in the Owl/Semantic Web community (Baader 2007).

In almost all natural-language-based vocabulary resources thus far, terms and definitions have been treated in effect as black boxes, so that their logical content is not accessible to computational tools. The GO, along with its sister ontologies in the OBO Foundry, has initiated an ambitious strategy to expose the compositional character of compound terms and definitions by conceiving them as cross-products of simpler terms, some of which are derived from other ontologies (Smith et al.

2007; Hill et al. 2002). For example, rather than defining *Tuberculosis of the meninges* with the natural language phrase "Tuberculosis of the meninges is a *Mycobacterium tuberculosis* infection in which the site of infection is the meninges," one can instead use formally defined relations between ontology terms to create structured phrases such as:

Tuberculosis of the meninges is_a Mycobacterium tuberculosis infection THAT *has_location meninges*

where *meninges* is a term in an anatomy ontology, such as the FMA, and *Mycobacterium tuberculosis infection* is a term in an ontology of infectious diseases, such as the IDO described below, and is itself defined as a cross-product. By this means, the potential for the ontology to support automatic reasoning and error checking is enhanced, and so also is its capacity to integrate data in the direction of enhanced semantic interoperability.

That the enhanced formalism and logical rigor of ontologies relative to other vocabulary resources brings significant benefits to applications is perhaps best evidenced by the relative numbers of citations for the GO, SNOMED, and the Unified Medical Language System (UMLs) in the PubMed database. As the name implies, the UMLS, initiated in 1986, is an attempt to provide a unified terminology system for the medical domain. The goal is two-fold: to make the many medically relevant vocabulary resources interoperable, and to create a single, broad coverage resource. The strategy used by the UMLS developers is to integrate the many existing medical terminologies by providing joint access to them through mappings between their terms. The UMLS includes the GO and SNOMED, as well as MeSH and ICD, among its source terminologies. Despite its short history and small domain relative to SNOMED and the UMLS, the GO has become the most cited vocabulary resource in PubMed, with over 450 citations per year (Bodenreider 2008). In contrast, the number of UMLS citations has remained constant over the last 10 years (Bodenreider 2008). From 2001 to 2007, among papers that cite the GO, SNOMED, the UMLS, the FMA, MeSH, the National Cancer Institute Thesaurus (NCIT), and the Logical Observation Identifiers, Names, and Codes (LOINC) vocabulary, the proportion of GO citations increased from about 5% to about 85%, while the proportion citing SNOMED decreased from about 20% to about 5% and the proportion citing the UMLS decreased from about 55% to about 5% (Bodenreider 2008).

As can be seen from the description of ontology uses below, the utility of ontologies in computational applications depends not just on adherence to development principles like those outlined above, but also on the breadth of the developer and user communities. When each community develops and uses its own ontology, many of the benefits of ontology are not realized. To address both of these issues, the Open Biomedical Ontologies (OBO) Foundry (http://obofoundry.org) (Smith et al. 2007) was initiated in 2006. The goals of the Foundry are to foster the pursuit of best practice in ontology development on the basis of an evolving set of design principles and to provide a foundation for the coordinated development of ontologies by large developer and user communities. Its ontologies are designed to represent in an interoperable fashion the biomedical reality from which data are sampled. Their development within the framework of a common top-level ontology, the

Basic Formal Ontology and the consistent employment of a constrained set of logically defined relations allows Foundry ontologies to be used together as modules of a larger system for computational applications.

There are currently some 35 member ontologies at varying stages of development in the OBO Foundry. There are OBO Foundry ontologies covering many of the domains relevant to infectious diseases, including proteins [the Protein Ontology (Natale et al. 2007)]; cells [the Cell Ontology (Bard et al. 2005)]; human anatomy [the FMA (Rosse and Mejino 2003)]; anatomy for important vector species [the Tick Gross Anatomy Ontology and the Mosquito Gross Anatomy ontology (http://www.anobase.org/)]; and biological processes, molecular functions, and cellular components [the Gene Ontology (http://www.geneontology.org)].

19.4 Uses of Ontologies in Informatics-Driven Research and Care

Vocabulary resources have a long history of use in clinical settings, primarily to support the coding of clinical data for health records, laboratory reports, and hospital billing, the coding of public health data for monitoring disease incidence and prevalence, and the coding of knowledge for clinical decision support systems. In basic biomedical research, the primary use of vocabulary resources has, until recently, been to support bibliographic searches and database integration. However, the logical rigor and formalism underlying biomedical ontologies has increased significantly in recent years, allowing biomedical ontologies to be applied for a larger variety of purposes.

For ontologies and the data annotated in their terms, we find a variety of different types of uses in biomedicine, outlined in Yu (2006), Bodenreider (2008), and Rubin et al. (2008), including terminology management; text-mining; integration, interoperability, and sharing of data; data interpretation and analysis; and knowledge reuse, reasoning, and decision support. Ontologies support terminology management in aligning independently developed terminologies with overlapping content (Rickard et al. 2004; Zhang and Bodenreider 2005). They also bring benefits in managing changes to terminologies by allowing flexible response to new scientific discoveries, as contrasted with the relative inflexibility of more traditional database approaches, where a database schema may need to be revised in its entirety when one aspect of classification changes.

Ontologies are increasingly used to add value to more traditional vocabulary resources, whose informal structure and lack of systematic definitions "is generally deemed to be inadequate with respect to the requirements of health care information systems that depend on clear communication of complex medical and biological information in a form that is usable by computers" (Yu 2006). Applying a formal structure to vocabulary resources allows enhanced opportunities for both manual and automatic error checking. Ontological methods are used to detect errors in definitions and to analyze the meanings of terms and represent those meanings formally (Ceusters et al. 2005; Smith et al. 2004; Pisanelli 2004). Additionally, ontological methods are used to detect errors in classification, such as the improper assignment of *is_a* relations arising through inadequate treatment of negation, or the improper assignment of part-whole relations resulting from an inconsistent use of terms in different parts of terminology (Ceusters et al. 2004, 2005).

In the area of text mining, vocabulary resources are used to facilitate the retrieval of information from biomedical literature [reviewed in (Bodenreider 2008; Spasic et al. 2005)]. The greatest success has come from the assignment of terms from vocabulary resources to individual documents within large collections, a process referred to as indexing. MeSH has long been used to index documents within the PubMed database (Nelson et al. 2001), and, more recently, ontologies have been used for this purpose, allowing text-mining algorithms to take advantage of the richer set of relations and their formal definitions (Ide et al. 2007; Muller et al. 2004; Doms and Schroeder 2005). The identification of documents that are relevant to a query within a collection (document retrieval) is greatly facilitated by utilizing the ontologies' structure. For example, the hierarchy of *is_a* relations can be used to expand a query to include parents or children of the original query term, significantly improving recall. *part_of* relations can be similarly used, retrieving documents that refer to fingers or palms in response to a query for documents that refer to hands.

After the identification of relevant documents, text-mining often progresses to information extraction, the identification within documents of statements about prespecified entities. Named entity recognition is the simplest approach in which a list of entities of interest is provided as input to the information extraction algorithm. The terms from ontologies can serve as an important source of term lists for named entity recognition, and the ontologies' structure can serve to improve information extraction just as document retrieval is improved.

Within the area of infectious disease research, ontology-supported text-mining is used to monitor news reports from all over the world so as to detect disease outbreaks, monitor the geographic distribution of diseases (BioCaster, http://biocaster.nii.ac.jp/; EpiSpider, http://www.epispider.org/), and predict candidate vaccine epitopes (Schonbach et al. 2004). Ontologies have also been developed to support text-mining about Dengue fever, specific Dengue virus serotypes (Rajapakse et al. 2008), and vaccine development and efficacy. The Vaccine Investigation and Online Information Network (VIOLIN, http://www.violinet.org) was established as a central repository for literature related to vaccine research and the data resulting from vaccine research. In addition to a variety of data analysis tools, VIOLIN provides several text-mining tools supported by its Vaccine Ontology (VO), as well as MeSH and the Textpresso Ontology (Muller et al. 2004).

Currently the most successful use of ontologies is to support integration, interoperability, and the sharing of data through data annotation. The best example is use of the GO for the creation of annotations by the curators of model organism databases (Blake et al. 2006; Cherry et al. 1997; Grumbling and Strelets 2006) and genome annotation centers (Camon et al. 2004). GO curators are striving to capture, in a form accessible to computational algorithms, information about the contributions of gene products to biological systems, as reported in the scientific literature. The annotation process unfolds in a series of steps (Blake et al. 2007). First, specific experiments, documented in the biomedical literature, are identified as relevant to the responsibilities of a given ontology curator. Second, the curator applies expert knowledge to the documentation of the results of each selected experiment. This process entails determining which entities (e.g., which proteins) are being studied in the experiment, the nature of the experiment itself, and (in the case of the Gene Ontology) the molecular functions, biological processes, and cellular components that the experiment identifies as being associated with that gene product. The curator then creates an annotation, which captures the appropriate relationships between the corresponding ontology types and the database entry for the gene product type. The annotated data then become accessible through the use of the associated Gene Ontology term as a search vehicle and becomes automatically combined with many other types of relevant and useful information as a result of the fact that the curators of many other types of data are using the same controlled vocabulary resource to annotate their data. Developing the ontology in tandem with the process of curation of data also provides a means of ensuring that the ontology is maintained in a way that keeps pace with the advance of science as recorded in the published literature and ensures that the vocabulary provides the resources needed to express the most recent scientific results.

The GO and other ontologies are used for annotation of genes and gene products in a variety of databases relevant to infectious disease research. In addition to the annotation of data for humans and for model organisms, such as mice, which are used to study the host immune response, ontologies are used to annotate data in:

- The ApiDB databases (http://eupathdb.org/eupathdb/), which include genomic and other data for *Cryptosporidium*, *Giardia*, *Plasmodium*, *Theileria*, *Toxoplasma*, and *Trichomonas* strains
- VectorBase (http://www.vectorbase.org), which includes genomic and other data for invertebrate vectors of human pathogens, including *Anopheles gambiae*, *Aedes aegypti*, *Ixodes scapularis*, *Pediculus humanus*, and *Culex quinquefasciatus*
- The Integrated Microbial Genomes System (http://img.jgi.doe.gov/), Microbes Online database (http://www.microbesonline.org/), the Pathogen-Host Interaction Data Integration and Analysis System (http://phidias.us), BioHealthBase (http://www.biohealthbase.org/), and the National Microbial Pathogen Data Resource (http://www.nmpdr.org/), among others (Medigue and Moszer 2007), together include annotations for the genomes of hundreds of bacterial and viral species, as well as a significant number of eukaryotic pathogen species
- Many of the databases listed at http://databases.biomedcentral.com/under the "infectious diseases" subject area

In addition to the annotation of genomic data, the use of ontologies and other vocabulary resources to annotate other types of data is also becoming common. For example, data in ArrayExpress (http://www.ebi.ac.uk/microarray-as/ae/) has been annotated with GO terms as well as terms from the Microarray Gene Expression Data (MGED) Ontology (Whetzel et al. 2006). Of particular interest in the study of infectious diseases is the use of MeSH to annotate human disease names to microarray data in the Gene Expression Omnibus (Butte and Chen 2006) and the use of GO and SNOMED to annotate pathways and integrate disease and pathway information (Chabalier et al. 2007).

Ontology annotations not only provide a basis for database interoperability, but also significantly enhance the interpretation of data from genome-wide and high-throughput experiments, as for example in Baranzini et al. (2009), Valouev et al. (2008), Kim et al. (2008), and Grinde et al. (2007). A variety of software tools have been developed to use ontologies and other vocabulary resources for the analysis and interpretation of microarray data, including Onto-Tools (Chabalier et al. 2007), GoMiner (Zeeberg et al. 2003), GOTree Machine (Zhang et al. 2004), MeSHer (Djebbari et al. 2005), and more recent tools (Bresell et al. 2006; Osborne et al. 2007).

Ontology annotations have formed the basis for new bioinformatics approaches for the analysis of such data (Osborne et al. 2007; Ochs et al. 2007). One such method for the analysis of microarray data is the CLASSIFI algorithm (Lee et al. 2006), which determines, for sets of genes clustered based on their expression levels, whether particular gene ontology terms are overrepresented within any set of genes. Ontologies have also been used to enhance clustering algorithms for microarray data by using the ontology annotations as a second cluster variable (Brameier and Wiuf 2007; Huang et al. 2006; Liu et al. 2004). In another study, proteins were clustered based on the similarity of their GO annotation profiles. The annotations for each protein were represented as a graph, and the graph similarity for pairs of proteins was used as the distance measure for clustering (Wolting et al. 2006). This method was applied to sets of proteins from two different protein array screens, and in both cases, proteins not identified in the original study were implicated to play a role in the biological process under study (Wolting et al. 2006). Finally, ontologies have been used to integrate text-mining approaches with microarray data analysis to facilitate disease gene identification (Tiffin et al. 2005).

An important benefit of ontologies is that they facilitate knowledge reuse. While knowledge-based systems that support applications such as decision support in health care are typically dependent on large amounts of current domain knowledge, the capture of such knowledge in computationally accessible information systems through data curation is an expensive and arduous process. In the domain of molecular biology, the widespread adoption of the Gene Ontology as a standard vocabulary has worked well, eliminating the need for developers of different information systems to expend resources capturing the same knowledge. In the clinical domain, however, knowledge capture has standardly been performed with the aid of locally developed database schemas and vocabulary resources, both structured to the specific application at hand. Such database schemas and vocabulary resources do not support the reuse or accumulation of data and often lose their validity within a short space of time. Increasingly, therefore, there is a move, illustrated by the caBIG endeavor, to foster the development of reusable resources for data capture in which, again, ontologies and ontology-related technologies are again playing an important role.

The use of ontologies to support automated reasoning is an active area of research and recent work, described below, has shown that the benefits of even primitive reasoning algorithms can be significant. These results have led to increased interest in developing vocabularies with sufficient formalism to support reasoning as well as in developing reasoning algorithms that make use of the types of information captured in ontologies. An important application area of automated reasoning is clinical decision support.

Query engines have been developed in such a way that the ontology itself is a directly query-able knowledge resource. For example, Emily (Detwiler et al. 2004) is a system used to query the FMA for structural relationships between anatomical entities. The FMA also serves as a source of anatomical knowledge in a reasoning application used to predict the consequences of penetrating injury (Rubin et al. 2006). The system is used to determine which organs are injured and whether vital structures, such as a coronary artery, are injured given particular projectile trajectories (Rubin et al. 2006). HyBrow is a system that uses ontologies and ontology annotations as sources of existing knowledge to test whether hypotheses are consistent with existing knowledge and data, to rank hypotheses by the amount of supporting evidence, and to test the implications of hypotheses (Racunas et al. 2004).

Clinical decision support systems (CDSS) are commonly used in the infectious diseases field for diagnostic assistance, guidance in the prescription of anti-infectives, biosurveillance, and vector control [Global Infectious Disease and Epidemiology Network, http://www.gideononline.com, and (Schurink et al. 2005; Thursky 2006; Sintchenko et al. 2008; Pestotnik 2005; Coleman et al. 2006; Buckeridge 2007; Buckeridge et al. 2005; Veenema and Toke 2006)]. Vocabulary resources, such as classifications of drug types, serve as a source of knowledge for CDSS. In most cases, however, simple terminology lists or term hierarchies are used, and when vocabulary resources with more complex relations are used, the resources are developed for the purposes of the specific application and do not have sufficient logical formalism to serve the purposes of broad interoperability. For example, the clinical vocabulary resource with the broadest scope, and which also has many ontology-like features, is SNOMED. A recent review of the literature found little evidence that SNOMED is being used for direct care purposes such as CDSS (Cornet and de Keizer 2008). The use of ontologies, as we have defined them, in CDSS is still a young field of research. One prominent example is the use of ontologies in the Dengue Decision Support System (http://www.rams-aid.org/) developed by the Risk Assessment and Management Solutions for Arthropod-borne and Infectious Diseases group at Colorado State University. There is a growing effort within the OBO Foundry community to develop ontologies with coverage of the clinical domain and to develop ontology-based reasoning algorithms, including those useful within CDSS.

19.5 Vocabulary Resources Relevant to the Field of Infectious Diseases

We provide a brief review of vocabulary resources that have content relevant to the infectious diseases domain, restricting ourselves primarily to those resources that are freely available, widely used, and likely to persist. For each resource, we describe its intended use and evaluate its adequacy and prospects for general use in infectious disease research and clinical care, taking account of the considerations outlined below.

The vocabulary resources relevant to this review can be divided into two broad groups: resources produced primarily as terminologies for use in the clinical domain, and resources developed in support of research in the basic biological sciences. In light of the increasing focus on translational medicine, we take it that the trajectory of clinical and biomedical sciences is toward an ever closer alignment of these two groups of resources, which have hitherto evolved almost entirely independently. Therefore, one focus of our evaluation has been to gauge the degree to which existing clinical and biomedical terminology resources can support this trajectory. The second focus is on evaluating the degree to which such resources support the increasing demand for more sophisticated information processing capabilities.

19.5.1 Medical Subject Headings Controlled Vocabulary

MeSH is a general-purpose vocabulary, initially developed for purposes of indexing and cataloging medical literature, now used to support many text- and literature-mining endeavors. Terms from the MeSH controlled vocabulary are used to annotate biomedical journal article citations and abstracts for the MedLine database. Query interfaces to MedLine, such as PubMed, use MeSH to support the retrieval of MedLine records in ways that supplement the use of simple string searches.

MeSH is a controlled vocabulary organized as a thesaurus consisting of sets of terms or "descriptors" in a hierarchical structure that permits searching at various levels of specificity. The relationship between terms in a hierarchy is not *is_a*; rather the terms appear in the MeSH term hierarchies on the basis of relatedness as assessed in terms of fields of study or research (a strategy designed to maximize the utility of MeSH as a literature indexing resource). For example, most of the content relevant to the infectious disease domain is found under one of descriptors *Anatomy, Organisms, Diseases* or *Biological Sciences*. Under *Biological Sciences*, one finds *Public Health*, under which one finds *Disease Outbreaks*, *Disease Reservoirs*, and *Disease Transmission*, along with terms such as *Consumer Product Safety* and *Equipment Reuse*. A natural language note is associated with each term.

MeSH is marked by a broad coverage of topics relevant not only to the domain of infectious diseases but also to microbiology and host immunity. Of all the vocabulary resources we have evaluated, MeSH has the broadest coverage across the entirety of the infectious disease/immunology domain. However, the terms are not linked to any relations, which limits the usefulness of the information contained in MeSH for many purposes. Despite its broad coverage of the subject matter, MeSH cannot be used as a computable vocabulary resource for infectious diseases, though it is highly useful in supporting a variety of string- and statistics-based forms of data and literature mining. Its utility in this respect has been enhanced by its recently completed alignment to the GO (Tveit et al. 2004).

19.5.2 International Classification of Diseases

ICD version 10 (ICD-10) is a member of a family of World Health Organization (WHO) international classifications designed to promote international comparability in the collection, processing, classification, and presentation of diagnostics in health epidemiology, health management, and mortality statistics. ICD-10 is a classification of diseases and other health problems developed for the purposes of compiling statistics of disease or causes of death. ICD-10 is used to record disease and other health problems on health and vital records such as death certificates. These records are subsequently used to compile national mortality and morbidity statistics by WHO member states. ICD-10 is also used for general epidemiological and health management purposes, such as monitoring the incidence and prevalence of diseases.

ICD-10 is organized as a term hierarchy in which terms are names of diseases and each term is associated with a code of up to six digits in length, indicating the term's placement in the hierarchy. Terms are defined primarily by their placement in the hierarchy along with statements of inclusion and exclusion. For example, *Tuberculosis* is defined by being a subclass of *Certain infectious and parasitic diseases*, along with the statements "Includes: infections due to *Mycobacterium tuberculosis* and *Mycobacterium bovis. Excludes:* congenital tuberculosis, pneumoconiosis associated with tuberculosis, sequelae of tuberculosis, silicotuberculosis."

ICD's coverage of the domain in terms of types of infectious diseases is broad, but information about other aspects of infectious disease is limited and thus the scope of ICD-10 is considered narrow. Because ICD provides a disease classification constructed primarily on the basis of anatomy, it has a relatively robust classification of pathological structures resulting from disease, such as carcinomas and neoplasm, whose classification follows the anatomical partition. For the infectious disease domain, however, a different approach would be needed. The ICD-10 classification of infectious disease is based on many different and inconsistently used classification criteria resulting in a disorganized hierarchy that is counter-intuitive, difficult to navigate, and difficult to construct queries for. Furthermore, there are no formal definitions for terms and no logical basis for the hierarchical structure used. Thus, ICD-10 could not sensibly be used to support either interoperability with other information resources or reasoning within the context of its own hierarchy.

19.5.3 The Systematized Nomenclature of Medicine – Clinical Terms

While Systematized Nomenclature of Medicine – Clinical Terms (SNOMED CT) is not a fully open source vocabulary resource, its broad scope and the long experience of its use and maintenance, combined with its presumptive status as an international master vocabulary for the coding of clinical information, mean that it is an especially important vocabulary resource for analysis and critical review.

The intended use for SNOMED CT is documentation and reporting of health care information throughout the health care process (medical history, illnesses, treatments, laboratory results, etc.) in software applications used for clinical data collection. The intention is that the processing of health care information recorded in SNOMED CT terms can be used to improve patient outcomes by providing health care providers with more easily accessible and complete information, as well as to conduct outcomes research, to evaluate the quality and cost of care, and to design effective treatment guidelines.

SNOMED CT comprises concepts, concept descriptions, and relationships. A concept is described as a clinical meaning. Concepts are defined by the relationships between them. The primary defining relationship is the *is_a* relation, but there are an additional 50 defining attribute relationships, such as *Finding_site* and *Associated_morphology*.

In general we find that SNOMED CT contains a large number of terms relevant to the infectious disease domain, but that these terms and their organization are biased toward capturing information about clinical observations and about patients in patient records. Terms and relations describing pathogens and the host immune responses to these pathogens are correspondingly lacking. The emphasis on clinical findings and their attributes is not surprising given SNOMED CT's intended use for the documentation and reporting of clinical data, but this does handicap SNOMED CT in terms of its usefulness for translational medicine. This handicap could be overcome if SNOMED CT were developed in accordance with a set of rigorously applied principles sufficient to allow its interoperation with vocabulary resources from the biological domain.

The logical formalism underlying SNOMED CT has been evaluated previously (Ceusters et al. 2004; Bodenreider et al. 2004). Our evaluation based on the infectious disease-relevant content is consistent with these previous evaluations. We observed problems with SNOMED CT's classification hierarchies resulting primarily from the use of multiple modes of classification and a lack of adherence to basic principles of sound classification. The result is the assertion of type-supertype relations that do not hold. For example, the SNOMED class *Infectious disease* is asserted to have subclass *Abrasion AND/OR friction burn with infection*, where neither an abrasion nor a friction burn is itself an infectious disease. Similarly, *Incomplete illegal abortion with genital tract or pelvic infection* is a subtype of *Infectious disease* in SNOMED CT, asserting that a type of abortion is an infectious disease.

As SNOMED becomes more widely used, and begins to serve as a platform to ensure cross-language interoperability of clinical data, it will become ever more urgent that SNOMED meets the highest standards of logical coherence. The SNOMED International Health Terminology Standards Development Organization has recognized many of the above problems and is taking steps to correct them.

19.5.4 The Disease Ontology

The Disease Ontology (DO) was developed for the annotation of patient DNA samples collected with the patients' associated healthcare information. Broader motivations for the creation of the DO were to provide a public domain vocabulary resource for use in data mining against medical records and in annotating model organism phenotype data using terms for human disease.

The DO is organized as a taxonomy of diseases with terms, taken over primarily from ICD, referring to types of diseases. The hierarchy is intended to reflect the *is_a* relation between disease types. Few terms are defined, but the definitions thus far included are natural language expressions, usually taken from MeSH, SNOMED CT, or the NCI thesaurus. The current DO hierarchy improves somewhat on ICD version 9, and plans for further improvements to the DO are based on a strategy of aligning DO to the SNOMED CT disease typology.

Despite the DO claim of organizing disease terms based on types using an *is_a* relation, the DO hierarchy is poorly organized, mixing not only types of infection with types of disease, but also mixing types based on anatomical location, properties of infection (e.g., latent), type of infectious agent, developmental stage, type of geographical area to which a disease is endemic, and properties of infectious agents (e.g., zoonotic). The mixing of modes of classification and the use of multiple inheritance results in the inheritance of properties that do not hold for a type. For example, *Tuberculosis* is a subtype of *Respiratory Tract Infections* in DO, but not all instances of tuberculosis infection are an infection of the respiratory tract. *Tuberculosis* is also a subtype of *Opportunistic Infections*, which is a subtype of *Virus Diseases*, but Tuberculosis is not a viral disease. The DO has a limited utility as a general vocabulary resource for the infectious disease domain due to its limited scope and its disorganized classification hierarchy containing false assertions. The DO developers are, however, aware of these problems, and have initiated efforts toward realizing the necessary reforms.

19.5.5 General Conclusions Concerning Clinical Vocabularies

The most common use of clinical vocabulary resources thus far is as dictionaries with the potential to support forms of computer-aided retrieval of information. Vocabularies such as SNOMED CT also have in a certain logical structure, which means that they may be able to support more advanced services, including data integration (e.g., the integration of public health data), patient status descriptions, providing codes for problem lists or drug adverse events, and support for text-mining (Bodenreider 2006). In addition, they can support certain kinds of reasoning. They are increasingly used in association with basic biology vocabulary resources as tools for clinical and translational research, which are reviewed next.

19.5.6 The Gene Ontology and OBO Foundry Ontologies

We focus here on ontologies within the OBO Foundry, as these ontologies are being developed with the intention of broad interoperability and of their joint use for computation. Although there are still gaps in the domain jointly covered by Foundry ontologies, there is steady progress towards broad coverage of the biomedical domain, including both basic biological and clinical entities.

To fully support informatics-driven infectious disease research, prevention, and treatment, vocabulary resources that cover physiologic and pathologic entities are needed, and within each of those categories, resources are needed that cover: objects, such as molecules and cells; qualities, functions, and roles of the objects; and processes. The domain of physiologic objects is already well covered within the OBO Foundry by ontologies such as the many anatomy ontologies, the Cell Ontology, the Protein Ontology, and the GO Cellular Component Ontology. In addition, the domains of physiologic processes and molecular functions are also well covered by the GO Biological Process Ontology and the GO Molecular Function Ontology.

However, there are important gaps in the current coverage of the infectious diseases domain by OBO Foundry ontologies. In particular: terms for population-level processes, such as the epidemiological spread of disease; terms for cellular functions, such as the presentation of antigen to naïve T cells; terms for pathological anatomical entities, such as granulomas, and pathological processes, such as hematogenous seeding; terms for roles, such as host, pathogen, vector, carrier, and reservoir; terms for qualities, such as immunocompromised and virulent; and terms for relevant clinical entities, such as clinical phenotypes. In addition, important information is not captured, even about the entities already represented in Foundry ontologies, due to the restricted set of relations currently used. There are, however, large consortia of individuals committed to the development of Foundry ontologies, including the development of a set of ontologies developed specifically for the coverage of the infectious diseases domain (described below). Thus, we anticipate good coverage of the relevant entities in the near future.

Previous evaluations of the GO's implementation and underlying formalism found flaws (Smith et al. 2004; Kohler et al. 2006; Smith and Kumar 2004), but the GO Consortium has responded by working to educate curators and make the necessary changes to the GO ontologies. For example, efforts are under way to create genus-differentiate definitions (Rosse and Mejino 2003) for all terms, to standardize naming conventions, to utilize rigorous definitions of the GO's two relations, *is_a* and *part_of* (Smith et al. 2005), and also to add further relations, including relations spanning GO's three constituent ontologies. Development of OBO Foundry ontologies, including revisions and expansion to the GO, adheres to a set of guidelines (http://www.obofoundry.org/crit.shtml) that include the features outlined above and are designed to maximize the long-term utility of Foundry ontologies, in particular for computational applications.

19.5.7 Inadequacy of Current Resources

The existing vocabulary resources in medicine, such as SNOMED-CT, and many of the other source terminologies collected by the UMLS are highly valuable for purposes of data retrieval. However, they were independently developed by separate specialist groups, and thus manifest a low degree of interoperation. They use different naming conventions, different modes of classification, different relations, and different formalisms. Moreover, each has its own independently derived technical implementation. The resulting vocabulary resources are therefore inadequate for purposes of computational and translational medicine; their representations are lacking in both the needed formal rigor and in their coverage of the relevant biological domains. They fall short as cross-domain applications requiring high precision because they employ uneven standards of rigor. Thus, any information resource created using terms from these terminologies contains insufficient formalism for the sorts of reasoning applications needed for future biomedical and clinical research and translational medicine. Furthermore, the representation of information about the immunobiology and pathogenesis of infectious diseases has thus far been neglected in these terminologies, and this is so even for SNOMED-CT, currently the medical terminology with the broadest coverage.

The medical vocabulary resources are also marked by a focus on billing, hospital management and liability issues, and hence by a centrality in their organization on findings, observations, and procedures, with associated epistemological problems. These factors hinder their interoperability with counterpart vocabulary resources developed in the basic biological sciences, where approaches to developing computable vocabulary resources have been developed and tested to a larger degree than in the clinical realm, primarily because the biological data are more highly structured and more readily accessible to researchers.

Biologically focused ontologies and terminologies accordingly employ a more rigorous formalism than do the medical terminologies. Even here, however, the biological content relevant to our purposes is lacking. Formal, computable representations of information about infectious diseases, immunology, and disease pathogenesis are thus still needed.

19.6 The Infectious Disease Ontology Consortium

The last five years have seen a surge of interest in biomedical ontology, yet broad coverage, computable vocabulary resources for the infectious diseases domain are lacking. This is resulting in both an urgent need for ontology development in this field and there is an opportunity for a coordinated, community-wide development effort producing broad interoperability across the disease-specific specialties and across the clinical care, public health, and biomedical research domains.

To provide the foundation for such a community-wide ontology development effort, we have established a methodology for the development of ontology modules that together cover the entire infectious disease domain (http://www.infectiousdiseaseontology.org). The methodology relies on the use of a general IDO that serves as a core for the development of domain-specific extensions (e.g., tuberculosis). This methodology offers many benefits. The core IDO ensures interoperability between the domain-specific extensions, while the modular approach allows for each module to be developed and maintained by researchers expert in that domain. The division of labor allows for rapid progress toward the needed set of ontologies, ensures the biological accuracy of the modules, and increases the likelihood of the broad adoption of the ontologies by the infectious disease research community.

IDO and its extensions are being built by relating terms from OBO Foundry ontologies using relations from the Foundry's relation ontology where possible, and creating new terms and relations as needed. There are many benefits from building IDO and its extensions from OBO Foundry ontologies. In addition to the formalism underlying Foundry ontologies subsequently ensuring their support for sophisticated computation both within and between ontologies, building from Foundry ontologies means extensive use of existing ontology resources, both eliminating redundant effort and providing a significant head-start to ontology development. By building on OBO Foundry ontologies, IDO and its extensions are automatically interoperable with other ontologies that also build from Foundry ontologies as well as with the large information resources, such as UniProt and others mentioned above, that use Foundry ontologies for their wide base of existing annotations. Finally, as OBO Foundry ontologies, and in particular GO, are widely used, the use of Foundry ontologies in constructing IDO and its extensions improves the chances that IDO and its extensions will be accepted by the biological ontology and database communities.

To facilitate participation in the development and use of the infectious disease ontologies, we have established an Infectious Disease Ontology Consortium. In addition to development of the core IDO, consortium members are developing extensions for malaria, dengue fever, *Staphylococcus aureus* bacteremia, tuberculosis, brucellosis, influenza, HIV, and infective endocarditis. The Vaccine Ontology described earlier is also being developed as an IDO extension.

The IDO extensions are being tested for interoperability and for their use in a variety of computational applications. In response to these tests, the ontologies are refined for continued improvement. For example, the Vaccine Ontology is being applied to text-mining within the VOLIN project; the *Staphylococcus aureus* bacteremia ontology is being applied to the prediction of disease genes; the influenza ontology is being applied to influenza surveillance within the context of the Centers for Excellence in Influenza Research and Surveillance program; and the Dengue fever ontology is being utilized with the Dengue Decision Support System (DDSS).

The DDSS project (http://www.rams-aid.org) is the most developed and best demonstrates the long-term potential of computing with ontologies. The goal of the

DDSS is to guide the implementation of locally appropriate Dengue and Dengue vector control programs. The DDSS makes use of the Mosquito Insecticide Resistance Ontology (http://www.obofoundry.org/), the Vector Surveillance Ontology, the Vector Control Ontology, and the Dengue ontology.

19.7 Conclusions

Here, we have described the various types of vocabulary resources used to support informatics. We have emphasized the formal features of ontologies that enhance their utility for informatics applications relative to other types of vocabulary resources. We have discussed the current uses of vocabulary resources with a particular focus on the use of ontologies in the domain of infectious diseases. We have included a brief review of existing vocabulary resources relevant to the infectious diseases domain and have found that they are lacking in terms of their support of computational applications and translational medicine. We have described the Infectious Disease Ontology suite of ontologies and now invite all interested parties to participate in the development, testing, and refinement of these ontologies.

Acknowledgments LGC's contributions were supported by a Career Award from the Burroughs-Wellcome Fund and NIAID grants R01 AI077706 and R01 AI068804. BS's contributions were funded in part through the NIH Roadmap for Medical Research grant to the National Center for Biomedical Ontology (1 U 54 HG004028). Initial development of the Infectious Disease Ontology as well as the Infectious Disease Ontology meetings were generously supported by the Burroughs-Wellcome Fund.

References

Baader F (2007) The description logic handbook: theory, implementation, and applications. Cambridge University Press, Cambridge

Baranzini SE, Wang J, Gibson RA, Galwey N, et al (2009) Genome-wide association analysis of susceptibility and clinical phenotype in multiple sclerosis. Hum Mol Genet 18:767–778

Bard J, Rhee SY, Ashburner M (2005) An ontology for cell types. Genome Biol 6:R21

Bechhofer S, van Harmelen F, Hendler J, Horrocks I, McGuinness DL et al (2004) OWL Web Ontology Language Reference

Blake JA, Eppig JT, Bult CJ, Kadin JA, Richardson JE (2006) The Mouse Genome Database (MGD): updates and enhancements. Nucleic Acids Res 34:D562–D567

Blake JA, Hill DP, Smith B (2007) Gene Ontology annotations: what they mean and where they come from, Vienna, pp 79–82

Bodenreider O (2006) Lexical, terminological and ontological resources for biological text mining. In: Ananiadou S, McNaught J (eds) Text mining for biology and biomedicine. Artech House, Norwood, MA, pp 43–66

Bodenreider O (2008) Biomedical ontologies in action: role in knowledge management, data integration and decision support. Yearb Med Inform 67–79

Bodenreider O, Stevens R (2006) Bio-ontologies: current trends and future directions. Brief Bioinform 7:256–274

Bodenreider O, Smith B, Kumar A, Burgun A (2004) Investigating subsumption in DL-based terminologies: a case study in Snomed-CT, KR-MED Proceedings 2004, pp 12–20

Brameier M, Wiuf C (2007) Co-clustering and visualization of gene expression data and gene ontology terms for *Saccharomyces cerevisiae* using self-organizing maps. J Biomed Inform 40:160–173

Bresell A, Servenius B, Persson B (2006) Ontology annotation treebrowser: an interactive tool where the complementarity of medical subject headings and gene ontology improves the interpretation of gene lists. Appl Bioinform 5:225–236

Buckeridge DL (2007) Outbreak detection through automated surveillance: a review of the determinants of detection. J Biomed Inform 40:370–379

Buckeridge DL, Burkom H, Campbell M, Hogan WR, Moore AW (2005) Algorithms for rapid outbreak detection: a research synthesis. J Biomed Inform 38:99–113

Butte AJ, Chen R (2006) Finding disease-related genomic experiments within an international repository: first steps in translational bioinformatics. AMIA Annu Symp Proc 106–110

Camon E, Magrane M, Barrell D, Lee V, Dimmer E et al (2004) The Gene Ontology Annotation (GOA) Database: sharing knowledge in Uniprot with Gene Ontology. Nucleic Acids Res 32:D262–D266

Ceusters W, Smith B, Kumar A, Dhaen C (2004a) Mistakes in medical ontologies: where do they come from and how can they be detected? In: Pisanelli D (ed) Ontologies in medicine. IOS, Amsterdam, pp 145–164

Ceusters W, Smith B, Kumar A, Dhaen C (2004b) Ontology-based error detection in SNOMED-CT. MedInfo 11:482–486

Ceusters W, Smith B, Goldberg L (2005) A terminological and ontological analysis of the NCI thesaurus. Methods Inform Med 44:498–507

Chabalier J, Mosser J, Burgun A (2007) Integrating biological pathways in disease ontologies. Stud Health Technol Inform 129:791–795

Cherry JM, Ball C, Weng S, Juvik G, Schmidt R et al (1997) Genetic and physical maps of *Saccharomyces cerevisiae*. Nature 387:67–73

Cimino JJ, Zhu X (2006) The practical impact of ontologies on biomedical informatics. Yearb Med Inform 124–135

Coleman M, Sharp B, Seocharan I, Hemingway J (2006) Developing an evidence-based decision support system for rational insecticide choice in the control of African malaria vectors. J Med Entomol 43:663–668

Coonan KM (2004) Medical informatics standards applicable to emergency department information systems: making sense of the jumble. Acad Emerg Med 11:1198–1205

Cornet R, de Keizer N (2008) Forty years of SNOMED: a literature review. BMC Med Inform Decis Mak 8 Suppl 1:S2

Detwiler LT, Chung E, Li A, Mejino JL Jr, Agoncillo A, et al (2004) A relation-centric query engine for the Foundational Model of Anatomy. Stud Health Technol Inform 107:341–345

Djebbari A, Karamycheva S, Howe E, Quackenbush J (2005) MeSHer: identifying biological concepts in microarray assays based on PubMed references and MeSH terms. Bioinformatics 21:3324–3326

Doms A, Schroeder M (2005) GoPubMed: exploring PubMed with the Gene Ontology. Nucleic Acids Res 33:W783–W786

Grinde B, Gayorfar M, Rinaldo CH (2007) Impact of a polyomavirus (BKV) infection on mRNA expression in human endothelial cells. Virus Res 123:86–94

Grumbling G, Strelets V (2006) FlyBase: anatomical data, images and queries. Nucleic Acids Res 34:D484–D488

Guarino N (1998) Some ontological principles for designing upper level lexical resources. In: Rubio AGN, Castro R, Tejada A (eds) Proc of First Int Conf Lang Res Eval, Granada, Spain, pp 527–534

Hill DP, Blake JA, Richardson JE, Ringwald M (2002) Extension and integration of the gene ontology (GO): combining GO vocabularies with external vocabularies. Genome Res 12:1982–1991

Huang D, Wei P, Pan W (2006) Combining gene annotations and gene expression data in model-based clustering: weighted method. OMICS 10:28–39

Ide NC, Loane RF, Demner-Fushman D (2007) Essie: a concept-based search engine for structured biomedical text. J Am Med Inform Assoc 14:253–263

Kim CH, Lillehoj HS, Hong YH, Keeler CL Jr (2008) Comparison of transcriptional changes associated with *E. acervulina* and *E. maxima* infections using cDNA microarray technology. Dev Biol 132:121–130

Kohler J, Munn K, Ruegg A, Skusa A, Smith B (2006) Quality control for terms and definitions in ontologies and taxonomies. BMC Bioinform 7:212

Lee JA, Sinkovits RS, Mock D, Rab EL, et al (2006) Components of the antigen processing and presentation pathway revealed by gene expression microarray analysis following B cell antigen receptor (BCR) stimulation. BMC Bioinform 7:237

Liu J, Wang W, Yang J (2004) Gene Ontology friendly biclustering of expression profiles. Proc IEEE Comput Syst Bioinform Conf 436–447

Medigue C, Moszer I (2007) Annotation, comparison and databases for hundreds of bacterial genomes. Res Microbiol 158:724–736

Michael J, Mejino JL Jr, Rosse C (2001) The role of definitions in biomedical concept representation. Proc AMIA Symp 463–467

Muller HM, Kenny EE, Sternberg PW (2004) Textpresso: an ontology-based information retrieval and extraction system for biological literature. PLoS Biol 2:e309

Natale DA, Arighi CN, Barker WC, Blake J, Chang TC et al (2007) Framework for a protein ontology. BMC Bioinform 8 Suppl 9:S1

Nelson SJ, Johnston D, Humphreys BL (2001) Relationships in medical subject headings. In: Bean CA, Green R (eds) Relationships in the organization of knowledge. Kluwer Academic, Dordrecht; sold and distributed in North, Central, and S. America by Kluwer Academic, pp ix, 232 p

Ochs MF, Peterson AJ, Kossenkov A, Bidaut G (2007) Incorporation of gene ontology annotations to enhance microarray data analysis. Methods Mol Biol 377:243–254

Osborne JD, Zhu LJ, Lin SM, Kibbe WA (2007) Interpreting microarray results with gene ontology and MeSH. Methods Mol Biol 377:223–242

Pestotnik SL (2005) Expert clinical decision support systems to enhance antimicrobial stewardship programs: insights from the society of infectious diseases pharmacists. Pharmacotherapy 25:1116–1125

Pisanelli D (2004) If ontology is the solution, what is the problem? In: Pisanelli D (ed) Ontologies in medicine. IOS, Amsterdam, pp 1–19

Racunas SA, Shah NH, Albert I, Fedoroff NV (2004) HyBrow: a prototype system for computer-aided hypothesis evaluation. Bioinformatics 20 Suppl 1:i257–i264

Rajapakse M, Kanagasabai R, Ang WT, Veeramani A, Schreiber MJ, et al. (2008) Ontology-centric integration and navigation of the dengue literature. J Biomed Inform 41:806–815

Rickard KL, Mejino JL Jr, Martin RF, Agoncillo AV, Rosse C (2004) Problems and solutions with integrating terminologies into evolving knowledge bases. MedInfo 11:420–424

Rosse C, Mejino JLV (2003) A reference ontology for bioinformatics: the foundational model of anatomy. J Biomed Inform 36:478–500

Rubin DL, Dameron O, Bashir Y, Grossman D, Dev P, et al (2006) Using ontologies linked with geometric models to reason about penetrating injuries. Artif Intell Med 37:167–176

Rubin DL, Shah NH, Noy NF (2008) Biomedical ontologies: a functional perspective. Brief Bioinform 9:75–90

Ruttenberg A, Clark T, Bug W, Samwald M, Bodenreider O et al (2007) Advancing translational research with the Semantic Web. BMC Bioinform 8 Suppl 3:S2

Schonbach C, Nagashima T, Konagaya A (2004) Textmining in support of knowledge discovery for vaccine development. Methods 34:488–495

Schurink CA, Lucas PJ, Hoepelman IM, Bonten MJ (2005) Computer-assisted decision support for the diagnosis and treatment of infectious diseases in intensive care units. Lancet Infect Dis 5:305–312

Sintchenko V, Coiera E, Gilbert GL (2008) Decision support systems for antibiotic prescribing. Curr Opin Infect Dis 21:573–579

Smith B, Kumar A (2004) On controlled vocabularies in bioinformatics: a case study in the Gene Ontology. BIOSILICO: Drug Discov Today 2:246–252

Smith B, Köhler J, Kumar A (2004) On the application of formal principles to life science data: a case study in the Gene Ontology. Data Integration in the Life Sciences (DILS). Springer, New York, pp 79–94

Smith B, Ceusters W, Klagges B, Kohler J, Kumar A et al (2005) Relations in biomedical ontologies. Genome Biol 6:R46

Smith B, Ashburner M, Rosse C, Bard J, Bug W et al (2007) The OBO Foundry: coordinated evolution of ontologies to support biomedical data integration. Nat Biotechnol 25:1251–1255

Spasic I, Ananiadou S, McNaught J, Kumar A (2005) Text mining and ontologies in biomedicine: making sense of raw text. Brief Bioinform 6:239–251

Thursky K (2006) Use of computerized decision support systems to improve antibiotic prescribing. Expert Rev Anti Infect Ther 4:491–507

Tiffin N, Kelso JF, Powell AR, Pan H, Bajic VB, et al (2005) Integration of text- and data-mining using ontologies successfully selects disease gene candidates. Nucleic Acids Res 33: 1544–1552

Tveit H, Mollestad T, Laegreid A (2004) The alignment of the medical subject headings to the Gene Ontology and its application in Gene annotation. Lecture Notes Comput Sci 3066:798–804

Valouev A, Johnson DS, Sundquist A, Medina C, et al (2008) Genome-wide analysis of transcription factor binding sites based on ChIP-Seq data. Nat Methods 5:829–834

Veenema TG, Toke J (2006) Early detection and surveillance for biopreparedness and emerging infectious diseases. Online J Issues Nurs 11:3

Whetzel PL, Parkinson H, Stoeckert CJ Jr (2006) Using ontologies to annotate microarray experiments. Methods Enzymol 411:325–339

Wolting C, McGlade CJ, Tritchler D (2006) Cluster analysis of protein array results via similarity of Gene Ontology annotation. BMC Bioinform 7:338

Yu AC (2006) Methods in biomedical ontology. J Biomed Inform 39:252–266

Zeeberg BR, Feng W, Wang G, Wang MD, et al (2003) GoMiner: a resource for biological interpretation of genomic and proteomic data. Genome Biol 4:R28

Zhang B, Schmoyer D, Kirov S, Snoddy J (2004) GOTree Machine (GOTM): a web-based platform for interpreting sets of interesting genes using Gene Ontology hierarchies. BMC Bioinform 5:16

Zhang S, Bodenreider O (2005) Alignment of multiple ontologies of anatomy: deriving indirect mappings from direct mappings to a reference. AMIA Annu Symp Proc 864–868

Chapter 20
Populations, Patients, Germs and Genes: Ethics Of Genomics and Informatics in Communicable Disease Control

Gwendolyn L. Gilbert and Michael Selgelid

20.1 Introduction

Infectious diseases are still among the major causes of morbidity and mortality worldwide. Current estimates are that each year – mainly in developing countries – 500 million people become ill and more than 1 million die from malaria; 2 million of the 33 million people living with human immunodeficiency virus (HIV) infection die of acquired immune deficiency syndrome (AIDS); and 1.7 million of the 14 million with active tuberculosis (TB) die from it (WHO 2000, 2007, 2009). Millions of children, particularly, die each year from respiratory and diarrheal diseases, the rates of which are largely determined by political, socioeconomic and environmental factors. Although there has been a progress in the control of vaccine preventable diseases in developing countries, vaccines for malaria, TB and HIV/AIDS remain elusive and increasing antimicrobial resistance makes treatment difficult, even when it is available.

In industrialized countries, food-borne, respiratory and healthcare associated infections (HAIs) cause significant excess morbidity, mortality and healthcare costs. In the USA each year, an estimated 1.7 million HAIs cause ~100,000 deaths; 76 million food-borne diseases lead to 5,000 deaths. Many of these infections and deaths could be prevented if evidence-based control measures were properly implemented (Mead et al. 1999; Klevens et al. 2007). Clearly "smarter" strategies are needed to control communicable diseases.

Modern technology has enabled large scale screening for human genomic markers of susceptibility or resistance to infection and comparative studies of microbial genomes and is providing new knowledge about relationships between humans and

G.L. Gilbert (✉)
Centre for Infectious Diseases and Microbiology, Institute of Clinical Pathology and Medical Research, Westmead Hospital, Sydney West Area Health Service and The University of Sydney, Sydney, New South Wales, Australia

V. Sintchenko (ed.), *Infectious Disease Informatics*,
DOI 10.1007/978-1-4419-1327-2_20, © Springer Science+Business Media, LLC 2010

disease-causing microbes. This knowledge will reveal new targets for vaccines, antimicrobial agents, diagnostics and disease surveillance, which can be exploited to improve disease prevention, control and management.

Because infectious (or communicable) diseases affect communities – rather than merely independent individuals – new strategies to control and prevent infection involve complex relationships within and between populations. The disproportionate burden of communicable diseases among the most disadvantaged populations provides a challenge for new technology to improve disease prevention and control where conventional strategies have failed.

20.2 Infectious Diseases Ethics

The emergence of the discipline of human bioethics in the 1950s and 1960s coincided with a prevalent (but, with hindsight, unwarranted and dangerous) belief that the problems of infectious diseases had been solved by sanitation, immunization and antibiotic therapy. The much-quoted pronouncement that "it is time to close the book on infectious disease" is usually attributed to former US Surgeon General William Stewart. Although there appears to be no evidence that he ever actually said this, "the sentiment was certainly widely shared" at the time (Sassetti and Rubin 2007). This widespread complacency remained largely unchallenged throughout most of the twentieth century. It was dispelled by the unfolding of HIV pandemic and the plethora of other emerging and re-emerging infectious diseases that followed (or in some cases preceded) it, but it had already contributed to the gross neglect of infectious diseases by bioethicists (Smith et al. 2004; Francis et al. 2005; Selgelid and Selgelid 2005). AIDS was a rare exception, but many of the ethical issues it raised – confidentiality, discrimination, patients' rights and sexual freedom – were not specifically related to its status as an infectious disease.

Belatedly, this neglect is now being addressed; infectious diseases have at last come to the attention of bioethicists. During the twenty-first century, public health ethics has become a rapidly growing sub-discipline of bioethics, and much of the public health ethics literature has focused on infectious disease in particular. In addition to AIDS, attention has especially focused on severe acquired respiratory syndrome (SARS), pandemic influenza planning and issues related to bioterrorism (Reid 2005; Thompson et al. 2006; Miller et al. 2007). There has also been debate about the ethics of issues such as: intellectual property rights, relating to antimicrobial agents and their implications for the access to essential treatment of infectious diseases (Gupta et al. 2005) and the relationship between marketing of antimicrobials and the emergence of antibiotic resistance (Selgelid 2007).

Although infectious diseases are no longer the most common cause of death worldwide, they are still major contributors to illness, loss of productivity and premature death in developing countries and among poor and disadvantaged people everywhere, despite the long history of successful prevention and control. Communicable diseases have implications far beyond their effects on individual sufferers and their immediate families. Because they can be rapidly fatal in previously healthy people and their spread is often unpredictable and indiscriminate, they can cause fear, panic, social disruption, political overreaction and victimization, out of proportion to the actual disease burden or risk (Smith et al. 2004). The explosive, but relatively short-lived, spread and high mortality of SARS in 2003 led to international socioeconomic repercussions affecting tourism, trade and international relations and costing billions of dollars. Disproportionate responses are often exacerbated by the florid language used by media and politicians, with analogies to terrorism or war ("flesh-eating"; "silent killer"; "superbug"; "plague"; "attack"; "struck down"). The fact that many communicable diseases are preventable or can be successfully treated can provoke recriminations against individuals or institutions, which are perceived to have failed.

Infectious disease ethics occupies a position between the individualistic perspective of conventional bioethics and the traditionally more collective approach of public health and incorporates elements of both. The former emphasizes the right of individuals to make decisions about their health, based on their own interests or preferences (autonomy), limited only by the potential of those decisions to harm others (the harm principle). The latter is based on utilitarian principles, whereby decisions are determined by the best overall outcomes (in terms of aggregate and/or average human well-being), even if some individuals may be disadvantaged as a result. Recently, Margaret Battin and colleagues have suggested a new approach – that ethical decision-making about infectious diseases should take place behind a Rawlsian "veil of ignorance", a concept developed as the basis for making fair decisions about distributive justice (Battin et al. 2009; Rawls 1971). They propose that patients with infectious diseases – and indeed anyone – can be seen, actually or potentially, as both a victim and a vector of infection (Battin et al. 2009). Behind the "veil of ignorance", the decision-maker does not know her actual status – victim and/or vector – but acknowledges that she could be, or could become, either.

Infection affects communities, not just individuals; everyone is both part of a human social network and host to billions of micro-organisms which can spread from person to person. Most of these microbes are benign or even essential to health, but a minority are potentially harmful to people carrying them or to others with whom they interact. Levels of susceptibility to infection vary between individuals, as determined by, *inter alia*, where and how they live, their age, underlying health, nutritional status, life-style choices and genetic makeup and the measures they take to protect themselves, such as immunization. No one can be reliably protected from infections due to respiratory viruses, food-borne bacteria or pathogens spread by mosquitoes. Like the patient with fever and cough or diarrhea, each of us is a potential victim and a potential vector. Ethical infectious disease policy will respect the interests of both patients with infections – who want care and protection, without discrimination – *and* of the rest of the community who seek protection from infection. The latter include not only apparently healthy individuals, some of whom are unwitting carriers of potentially dangerous pathogens, but also people at increased risk of infection, because of underlying disease or genetic predisposition.

In this chapter we explore how recent advances in microbial and/or human genomics and modern information technology can improve our understanding of communicable diseases and provide better strategies to manage, prevent and control them. We try to anticipate and suggest ways to meet the social and ethical challenges that will arise. Some ethical issues, such as those relating to research in developing countries or human genomics, are neither new nor specific to communicable diseases and have been debated at length. Others, which arise from application of new microbial diagnostics and pathogen profiling, enhanced communicable disease surveillance and informatics, have been explored less extensively, if at all. We examine issues such as informed consent, privacy and confidentiality, autonomy, resource allocation, quality control, compliance with evidence-based practice and disease surveillance, prevention and control, in these contexts, from behind a "veil of ignorance," by assuming that anyone could be victim or vector of infection.

20.3 Challenges in Infectious Diseases Genomics Research

20.3.1 Genetics and Disease Susceptibility

It is well established that susceptibility to infection varies between individuals and that a component of this variation is inherited (Cooke and Hill 2001). For example, malaria parasites are known to have contributed, over millennia, to the evolution of the human genome, by selecting gene mutations, such as those causing sickle cell disease and glucose-6-phosphate deficiency (G6PD) that enhance survival of heterozygous carriers living in malaria-endemic areas (Daily et al. 2008). Differences in susceptibility to malaria and TB have been recognized between different but closely related ethnic groups (Modiano et al. 1996); and large epidemiological, twin and genetic studies have provided insights into the hereditable proportions of susceptibility or resistance to a number of infectious diseases. There are well documented associations between certain human leukocyte antigen (HLA) genes and susceptibility to severe malaria, rapid progression of HIV infection to AIDS, development of overt TB disease or leprosy and hepatitis B carriage (Cooke and Hill 2001). However, HLA genes account for only a small component of genetic susceptibility to infection, which (like many other types of disease) is apparently determined by interactions between many different genes, acquired characteristics (e.g., nutrition, previous exposure) and environmental factors.

Sequencing of the human genome and advances in metagenomics have provided opportunities to search more broadly for genetic traits that contribute to infectious disease susceptibility and host–pathogen interactions that can be targeted by new vaccines or drugs. Genome-wide mapping and analysis of hundreds of polymorphic markers in family groups and matched case/control studies of diseases of interest are currently underway. The aim is to identify genomic regions linked to communicable disease risk. These studies are difficult, because the diseases most suited to this type of investigation are most common in the poorest countries with limited health (and research) infrastructure and whose residents are understandably wary of possible exploitation by researchers from rich countries (Cooke and Hill 2001).

20.3.2 The Malaria Genomic Epidemiology Network

The MalariaGEN project illustrates some of the ethical challenges involved in human genomics research. It was established in 2005, with joint funding from the Gates Foundation and the Wellcome Trust. Members of this network of independent investigators contribute to a central DNA repository and to databases of core phenotypic data. One of the goals of MalariaGEN is to determine why only a small proportion of children develop life-threatening malaria, in communities where all children are repeatedly infected with the malaria parasite, *Plasmodium falciparum*. Researchers are using the technique of genome-wide association (GWA) analysis, which involves mapping half a million or more single nucleotide polymorphisms (SNPs) in thousands of individuals – without the need for whole genome sequencing – to identify sequence variants that correlate with disease risk, using statistical inferences based on common patterns.

The study has the potential to benefit millions of children but involves the complex methodological, social and ethical challenges which are common to any clinical research in developing countries or human genomics research anywhere. The involvement of numerous independent investigators, in rich and poor countries, from disciplines as varied as clinical and community medicine to state-of-the-art genomics and bioinformatics, requires a balance between standardization and uniformity of practice, on the one hand, and the need for sensitivity to diverse cultural settings, on the other (The Malaria Genomic Epidemiology Network 2008).

Informed consent and privacy. Children with severe malaria often die within hours of the admission to hospital. This raises logistical issues of recruiting subjects, classifying clinical phenotypes correctly and collecting specimens for genetic studies, without compromising medical care in the resource-poor settings where most cases occur. Language and cultural barriers complicate effective communication with the parents of potential research participants. It can be difficult to convey the distinction between diagnosis and medical research. Unfamiliar concepts must be explained in the local language – perhaps through the use of metaphors drawn from local experience – but even then there may be misunderstandings. Guidelines for obtaining informed consent, without creating undue anxiety, are being developed and carefully evaluated by MalariaGEN researchers, in collaboration with local communities.

Actual and perceived protection of the anonymity of research participants and their communities is critical to the development of trust between researchers and participants. In the MalariaGEN project, local databases which contain both phenotypic and genotypic data are designed to comply with appropriate ethical guidelines to ensure data security. A data access committee oversees researchers' access to individual genomic data. Qualitative research is underway to identify the concerns, of community members and other stakeholders, about the collection and use of ethnicity data in relation to genomic epidemiology, which could result in stigmatization if misused. Although it is commonly claimed that the use of de-identified data cannot harm research subjects, this is not necessarily so; research findings can sometimes lead to the development of policies or behaviors that are harmful to (e.g., ethnic) groups of which the subject is a member. This kind of risk should be explained to parents of potential research subjects as part of the informed consent process. Guidelines are essential for the publication and release of ethnicity data to provide maximum scientific benefit while respecting and protecting the interests of participants and their communities.

Ownership of data and intellectual property. When many different research groups and parent institutions are involved, ownership of data and intellectual property is complex and potentially contentious. There is often institutional pressure on researchers to patent any discoveries with the potential for commercial development. The principle agreed by MalariaGEN is that intellectual property protection will be sought only if it will facilitate the translation of research results into affordable health benefits for the populations most in need. Any resulting financial gains will be returned to the participating communities.

20.3.3 The Human Microbiome Project

The Human Microbiome project (McGuire et al. 2008) is another multicenter program, which entails familiar ethical, legal, and social challenges in a novel setting. It is an investigation of the relationship between humans and microbial societies that inhabit all body surfaces and play a vital role in human health. It will establish a database of microbial DNA and RNA, based on sampling of 15–18 mucosal and skin sites from about 250 healthy individuals aged between 18 and 40 years of age, about half of whom will provide a follow-up set of samples within 12 months. Blood will be collected and stored for human genome and immune response investigation from a subset of around 10 participants. Extensive demographic and medical historical data will be collected.

Informed consent, respect for autonomy, and communication. Disclosure of the possible risks involved in providing samples for this project is difficult because of the current dearth of knowledge about the human microbiome and what future research questions may arise from linking microbial with human genomic data. As in other areas of research involving biobanking, there is controversy as to whether participants should be asked to give consent only for specific investigations already planned or blanket consent for future research. Almost by definition, blanket consent involves consent to research that neither subject nor researcher may, at the time it is given, be able to understand or predict. It has been argued, however, that requesting general consent is acceptable so long as participants are well informed about the uncertainties, and there is a strong governance structure to protect the privacy of participants and ensure that future research is consistent with their expectations (Caulfield et al. 2008). This would generally involve the appointment of an independent multidisciplinary monitoring body, including lay representatives, to promote public trust and ensure respect for participants' autonomy; therefore, blanket consent would be limited to future research approved by this body.

It is likely that analysis of the preliminary results of this project will identify characteristics of individual microbiomes, which could affect the health of the participant (e.g., risk of obesity or type 2 diabetes or changes due to medical interventions, such as antibiotic therapy). The point at which information, which could affect lifestyle or medical decisions, should be shared with participants or their physicians will be controversial. The study will almost certainly identify healthy individuals who are infected or colonized with potential pathogens that could cause future disease, under circumstances which are currently unpredictable and likely to vary between individuals. Should participants be told that they are potential victims or vectors if the level of risk is unknown? Researchers are unlikely to be qualified to manage potential clinical issues; at what stage should a medical practitioner be consulted, if at all?

The answers will depend on the validity and clinical significance of the findings and whether the participant has expressed a desire to know the results. For example, identifying nasal colonization with *Staphylococcus aureus* would require a different response from the discovery that the participant has asymptomatic genital infection with a sexually transmissible pathogen, which is a potential risk to others. If there were no apparent risk (e.g., of infection) to others, the participant's "desire to know" may be a key consideration. For this kind of research, discussion and negotiation on details regarding disclosure of findings to the subject and/or others should arguably become a more important part of the informed consent process.

Data confidentiality and security. Confidentiality of individual genomic and microbiomic data will compete with the need for researchers to share data and will depend on the extent to which data can be linked to individuals. For the human microbiome project, microbial DNA sequence data will be coded and released into publicly accessible databases, but clinical information and individual human DNA data will be coded and stored in controlled-access databases for later correlation with microbial data. Only aggregate human genomic data will be released into public databases. Whether, how, and by whom data are linked remain controversial because of the existing uncertainty about the extent to which microbial data can reveal individual identity and could be used to stigmatize individuals or groups. These are among the risks that will be discussed with participants when seeking informed consent.

Representativeness and justice. In most clinical research projects, subjects are selected and so not truly representative of the whole population. This means that the risks and potential benefits are not equally shared and the results may not be generalizable. The human microbiome project excludes children and older adults, to ensure that interpretation is not complicated by metabolic changes related to growth, puberty, or aging. However, subjects are chosen to include as many racial and ethnic groups as possible even though this could risk identifying false associations due to unrecognized confounding factors. While these problems are often unavoidable, they must be recognized and accounted for in the data analysis and conclusions.

20.4 Application of Pathogenomics and Informatics Research to Communicable Disease Diagnostics and Prevention

Over the past half-century or so, the natural histories of many human infectious diseases have changed, often fundamentally and often as a result of deliberate or unwitting human intervention. For example, immunization has (actually or almost) eliminated a few (smallpox, polio, and measles) and has controlled many other diseases (diphtheria, rubella, tetanus, hepatitis B). However, although vaccines are available, they have been less successful in controlling some diseases (e.g., pertussis, TB, influenza), and immunization remains elusive for many (e.g., most respiratory and diarrheal diseases, malaria, and HIV infection). Antimicrobial agents are available for the treatment of many types of infection but, with few exceptions, their efficacy has been compromised by the development of resistance in target pathogens. On the other hand, changes in land and water use, agriculture, animal husbandry, transportation, climate, or lifestyle, as well as increasing numbers of people who are immunocompromised because of AIDS or immunosuppressive drug therapy, have led to the emergence of new and opportunistic human pathogens which were once regarded – if they were recognized at all – as animal, rather than human, pathogens or as harmless commensals.

Recently, studies of microbial genomes have helped explain many of these phenomena at the molecular level and have led to changes in anthropocentric concepts of pathogens and commensals. In future, they will reveal new ways to protect humans from illness and death by identifying new targets for antimicrobial agents or vaccines. At least one genome (and often several) of all significant human pathogens has now been fully sequenced. Comparison of genomes of different strains of the same and related species can provide extensive information about microbial evolution and the relative importance of different types of genetic variation (e.g., mutation, insertion, deletion, duplication, recombination, or lateral transfer) and how they occur. We now know that many of the genes that determine virulence or antibiotic resistance are transferred on mobile genetic elements (plasmids, bacteriophages, transposons, pathogenicity islands) between different strains or species; this can dramatically amplify the effects of selection pressures (see Chap. 12 for details). These mobile elements can be exploited in the development of diagnostic and surveillance tools, but they also complicate the interpretation of test results and attempts to control disease transmission.

20.4.1 Diagnostics and Antibiotic Resistance: Ethical Implications

Increasingly sophisticated "smart" diagnostics, which are currently under development, will potentially allow more sensitive and specific pathogen detection and profiling (Sintchenko et al. 2007) which could significantly improve communicable disease diagnosis, management, and control. If their benefits are to be fully realized, the predictive values of new tests (i.e., the ability to predict whether or not the patient has the infection which the test is intended to diagnose) must be thoroughly evaluated, with reference to clinical outcomes not just through comparisons with existing diagnostic methods. Moreover, the evaluation should not end with their introduction into routine practice.

Currently, the microbiology laboratory's task is to identify a relevant pathogen in a clinical specimen and report it, with an antibiotic susceptibility profile, if appropriate. Conventional diagnostic methods are relatively slow, and the interpretation of results is often subjective. For example, whether or not a pathogen is identified and reported in a culture from a site with normal flora may depend on the skill and experience of the laboratory scientist. The interpretation of the result depends on clinical information, which is often not available to the scientist, and technical information which may not be available to the clinician – such as the type and quality of specimen, diagnostic method used, and the pathogen strain. The clinician's interpretation of the result will often determine the antibiotic choice, but if this is inappropriate, the outcome may be compromised (Khatib et al. 2006; Chapman et al. 2008).

In the near future (and to some extent already), multiplexed nucleic acid detection (NAD) systems, which target 10s, 100s, or even 1,000s of highly specific nucleic acid sequences, will identify, in virtually real-time, any of a large number of possible pathogens relevant to the site of the specimen or the clinical syndrome. At the same time, they will also determine whether the pathogen identified carries specific virulence determinants or antibiotic resistance genes and/or whether its profile is similar to those of pathogens isolated from other people (a cluster of infections) (see Sect. 20.4.2). New or unusual pathogens can be included in these systems at little or no extra cost, which will save time by identifying less common or less obvious pathogens sooner than is currently possible.

In a clinical research setting, the ability to study the prevalence and clinical associations of many different species or genetic markers simultaneously will provide new knowledge about the etiology, epidemiology, and pathogenesis of infectious disease syndromes and interactions between species. Multiplexed NAD systems will allow inclusion of species which are usually harmless commensals but occasionally are potential pathogens, copathogens or opportunists (Wang et al. 2008; Masue et al. 2007; Mckechnie et al. 2009). With appropriate analysis of clinical, epidemiological, and microbial data, this will help define their role and the circumstances, if any, in which they cause disease.

Properly designed clinical research studies (currently, a rarity in diagnostic microbiology) will clarify the circumstances in which the detection of virulence or antibiotic resistance markers in mixed flora is significant (Table 20.1). For example, genes that encode resistance to newer β-lactamase and carbapenem antibiotics or vancomycin are often carried in commensal gut flora, but can be transferred to virulent Gram negative bacilli (such as *Enterobacteriaceae*) or enterococci, respectively, under selection pressure from antibiotic therapy (Chapman et al. 2008; Iredell et al. 2006). Multiresistant *Enterobacteriaceae* or vancomycin resistant enterococci (VRE) are much more likely

Table 20.1 Examples of nucleic acid detection methods to detect virulent and/or antibiotic resistant pathogens and potential interpretation problems

Species	Gene marker(s)	Potential confounders	Indication for testing
Methicillin resistant *Staphylococcus aureus* (MRSA)	Methicillin resistance gene – *mec*A – or other SCC*mec*[a] genes; *S. aureus* species-specific gene – *nuc*	Other methicillin resistant staphylococci in same specimen (false positives); sequence variation in SCC*mec* (false negatives)	Screening for carriage (nares, groin etc); rapid identification in blood culture
Vancomycin resistant enterococci (VRE)	*van*A, *van*B, *vanB*2/3 (genes encoding vancomycin resistance)	Enteric anaerobes and streptococci carrying *vanB* (Ballard et al. 2005)	Screening – rectal swabs
Clostridium difficile	*tcd*A, *tcd*B or *tcd*C (toxin genes)	Not all *C.difficile* isolates contain relevant toxin genes	Diagnosis of diarrhea
Diarrhegenic *E. coli*	Shiga toxin – stx1, stx2+/–*eae*; +/–other toxin/virulence genes	Genes may be present but not expressed; sequence variants	Diagnosis of diarrhea
Various commentals and environmental bacterial species	Antibiotic resistance genes or transmissible elements	Genes may be carried, harmlessly by commensals	Screening to guide future therapy or infection control

[a] SCC*mec* is the staphylococcal cassette chromosome (a mobile genetic element), which carries the methicillin resistance gene *mecA*

than commensal or environmental bacteria to cause disease or spread to other patients and are more difficult to treat than their antibiotic susceptible counterparts.

However, even after careful evaluation, in a research setting, there are pitfalls in the translation of new diagnostic methods into practice. Although the interpretation of conventional microbiological results is often empirical and subjective, it is based on years of experience. Faster and more sensitive methods will provide more information, more timely and reproducible results and detection of a broader range of pathogens than conventional methods; they may uncover new infectious disease syndromes or identify previously unrecognized carriers. Confirmation that a new (and usually more expensive) assay will improve clinical outcomes requires ongoing prospective analysis of reliability and cost-effectiveness, which is difficult in a diagnostic laboratory setting. However, without it, the use of new assays could lead to unnecessary therapy or medicalization of "normal" conditions.

For example, screening patients for carriage of multiresistant organisms, such as methicillin resistant *Staphylococcus aureus* (MRSA) using rapid NAD methods, can improve hospital infection control by allowing more timely and appropriate isolation of patients and can guide appropriate antibiotic therapy. However, the sensitivity and specificity of some NAD methods differ from those of conventional methods, leading to potentially adverse consequences. Failure to identify some carriers (Thomas et al. 2008) will increase the risk of transmission to other patients. On the other hand if NAD assays identify more carriers than conventional methods, it can be difficult to distinguish increased sensitivity from false positive results. Either way, it will mean that more patients will be isolated, possibly unnecessarily (Humphreys 2008), which is costly, can adversely affect clinical care (Stelfox et al. 2003) and may cause unnecessary anxiety.

These uncertainties emphasize the importance of not only carefully evaluating the performance characteristics of a new test, but also of defining its purpose and clinical impact. Is it performed for the benefit of the patient on whom it is performed or for the benefit of other patients? While benefits to other patients may justify screening and isolation of patients who are colonized with multiresistant organisms, the degree of benefit, cost-effectiveness, and possible alternative strategies to achieve similar results must be assessed (Jeyaratnam et al. 2008; Wenzel et al. 2008; Buhlmann et al. 2008). A key question in public health (and infection control) ethics is: how great must the expected danger to public health (or to hospital patients) be to justify involuntary isolation of an individual who is a potential source of danger? Assuming that the appropriate metric of danger to public health is the "disability-adjusted life year" (DALY), for how many DALYs (x) would confinement of a person (e.g., a carrier of MRSA) for time t be justified, assuming that the free movement of that person could be expected, on average, to result in x or more DALYs?

The effects of changes in test turn-around times, reliability and predictive values, on patient care should be critically assessed, as new diagnostic methods are introduced. As part of this assessment, the point at which clinical research – with its ethical safeguards such as the informed consent of subjects – merges into routine practice will need to be defined. We need to develop standards for interpretation and reporting of the results of new diagnostic tests, in consultation with clinicians, to improve consistency. At present, introduction of new diagnostic and screening methods generally occurs independently in individual laboratories; test evaluation is often limited to comparison with existing methods and continued satisfactory performance in quality assurance programs, as required by accreditation bodies. Differences between methods used by different accredited laboratories suggest that some are "better" than others, but this information is not readily accessible to clinicians or patients and the criteria on which choices are based are often poorly defined. New tests are usually more expensive that conventional methods. It is usually assumed – and often true – that any increased costs are justified by better patient outcomes and savings elsewhere, but formal cost-effectiveness studies that are needed to confirm this are rarely done. Even if extensive evaluation demonstrates that a new method can improve patient outcomes and/or reduce costs, introduction of the test is often prevented or delayed because of the difficulty of transferring costs (and savings) between cost centers.

In summary: the widespread application of the new science of pathogenomics to infectious diseases diagnosis – with appropriate prospective evaluation of the clinical impact – should not only improve outcomes, but also provide a better understanding of many aspects of human infection and disease such as:

- The spectrum of diseases caused by known pathogens
- The possible infectious etiology of diseases of unknown cause
- The ecology of human microflora and factors that affect them
- The incidence and significance of colonization with different strains of known pathogens and of carriage, by commensals or opportunistic pathogens, of virulence or antibiotic resistance genes
- Potential interactions that may affect virulence, simultaneous carriage of combinations of pathogens, and/or commensal species
- The routes and mechanisms of transmission of pathogens between people and of genes between different microbial strains or species

20.4.2 Strain Typing for Pathogen Tracking

Surveillance is essential for disease control. It has been described as "the eyes of public health" (Fairchild et al. 2008). Although laboratory-confirmed cases of infectious diseases represent a small minority of notified cases (and an even smaller proportion of all cases), laboratory notification is more specific, reliable, and consistent than clinician notification. For many notifiable infectious diseases, simple species identification of the pathogen is inadequate and strain typing is required to monitor trends or to investigate outbreaks. However, until recently, the efficacy of surveillance has been limited by the fact that conventional strain typing methods are relatively slow, insensitive, and often performed only by specialized public health laboratories. Recent developments in microbial genomics have led to the development of faster and more discriminatory methods (see Chap. 2, 4, and 17), but their introduction has been limited and haphazard, in part because of inadequate recognition of the importance of improved strain typing methods, for disease control.

Delays of 2–3 weeks, in obtaining strain typing results mean that recognition of outbreaks is delayed and subsequent investigation of the cause is compromised. For example, in cases of food-borne disease, it may be impossible to identify a common food source, because victims cannot remember what they ate weeks before. Outbreaks involving large geographic areas, which are investigated in different jurisdictions and laboratories, may only be recognized after very large numbers of people have been affected, if at all.

In the early stages of the 2009 "swine flu" outbreak, there was no rapid strain typing method to distinguish the novel influenza H1N1 strain from other circulating H1N1 seasonal influenza A strains. This meant that many recent travelers to Mexico, where the outbreak began, or to the USA or Canada, where human-to-human transmission was reported early, were isolated for many days, awaiting results from the few reference laboratories able to identify the strain (initially, only after it was isolated in cell culture). However, sequences of several relevant antigen genes (hemagglutinin [H], neuraminidase [N], and polymerase [P]), from "swine flu" H1N1 strains isolated in different parts of the world, were published (http://www.ncbi.nlm.nih.gov/genomes/FLU/SwineFlu.html) within a very short period. This meant that culture-independent strain identification and typing methods soon became available to diagnostic laboratories around the world and played an important role in subsequent surveillance and control.

The availability of culture-independent diagnostic and strain typing systems for many pathogens of public health importance will make it possible for diagnostic laboratories to simultaneously identify relevant pathogenic species and their strain profiles, in a single assay, and report the results to public health authorities, within hours. Faster recognition and investigation of outbreaks will limit the number of cases and reduce the risk of new outbreaks. Some rapid strain typing methods are already available and in use. However, like diagnostic methods, new strain typing methods need to be carefully evaluated to ensure that their use translates into better public health outcomes. Unfortunately, the variety of different methods, the speed with which they are already being introduced, and limited funding for surveillance studies make prospective evaluation of risks, costs, and benefits, difficult. In addition, prospective evaluation will be impracticable without easy access to patient demographic, clinical, and outcome data and will be impracticable without sophisticated informatics tools to analyze these data.

20.5 Information Science and Technology for Patient Management and Communicable Disease Control

20.5.1 Health Information Systems

Rapid advances in medical science and therapeutics and increasing specialization have increased the demand for more accessible diagnostic, epidemiological, and therapeutic information, interpretive reporting of diagnostic test results, and clinical decision support systems. Electronic patient records (EPRs), networked with relevant clinical databases and information systems, are a logical response to these needs and are predicted to improve the quality, efficiency, safety. and reduce the cost of healthcare (Hillestad et al. 2005). They could also significantly improve disease surveillance and control (Friedman 2006; Chaudhry et al. 2006) and population health.

Linking clinical data from EPRs with microbiological results will enhance and personalize clinical decision support, e.g., for antibiotic prescribing (Sintchenko et al. 2008; Thursky et al. 2007). Linking clinical information with strain typing data will allow comparison of strains from different patients, in order to identify linked cases or outbreaks and to define their limits in space and time, much more rapidly than is currently possible (Gallego et al. 2009). Prospective surveillance of aggregated clinical, diagnostic, pathogen profile, and outcome data will help identify previously unrecognized risk factors or microbial strains which are associated with more severe disease or adverse outcomes. They will provide a basis for risk assessment tools to alert public health or infection control practitioners to the need for quarantine or investigation of contacts. The elements of an integrated clinical and public health information system may include:

- On-line laboratory test order entry and reporting systems
- Rapid, microbiological diagnosis and strain typing
- Access to components of individual EPRs, including demographics and relevant medical history (e.g., medical or environmental risk factors, presenting complaint, and laboratory test results)
- Data mining/analysis software that can identify and interpret epidemiological links
- Risk assessment and decision support systems to guide public health or infection control action
- Online prescribing and decision support to guide antibiotic therapy, if required, based on laboratory results and clinical history

20.5.2 Practical Application

Imagine this (future) scenario (only some components of which are currently plausible or - some would argue - even desirable):

- A patient presents with symptoms of an infectious disease; the doctor records the clinical findings in the EPR and orders diagnostic tests online.
- An informatics program with appropriate scanning software will scan the EPR for relevant demographic and medical risk factors and may prompt the doctor to seek additional information (e.g., about recent travel, diet, or contacts).
- The program will analyze the clinical data, provide a differential diagnosis and a list of appropriate laboratory tests, and recommend empirical antibiotic therapy,

if indicated, based on therapeutic guidelines, local susceptibility data and the medical history. (Artificial intelligence systems capable of making diagnostic and management decisions are still largely aspirational).

- The doctor will confirm, change, or override the laboratory test orders or prescription before transmitting them, electronically, to the laboratory and pharmacy, respectively.
- A pharmacy information system will establish that the drug dose is correct and will check for possible interactions with other current medications before the drug is dispensed and ready for the patient to collect, along with a personalized information sheet about precautions and potential adverse side-effects.
- The laboratory request form and a list of specimens required will be available when the patient arrives at the specimen collection center; specimens will be delivered to the laboratory and processed rapidly.
- If a relevant pathogen is identified, appropriate strain typing and/or antibiotic susceptibility testing will be performed. A personalized laboratory report, with interpretative information, will be generated and sent immediately or after review by a clinical microbiologist.
- The treating doctor's report may include a modified recommendation for treatment (e.g., a different antibiotic, based on the pathogen susceptibility or a recommendation to discontinue treatment); in some cases, a warning of potential complications (based on patient and pathogen profiles) will be added.
- If the infection is notifiable a second report will be sent, automatically, to the relevant public health authority. The strain profile will be compared with those of other strains in a database linked to similar laboratory databases within the same jurisdiction, country or, potentially, internationally.
- This analysis will identify outbreaks and monitor the geographic and temporal distribution of different strains in different populations, which may provide early warning of the emergence of new strains or detect potential vaccine failures. Spatial and temporal parameters for the detection of outbreaks due to the same strain will be modifiable to account for varying geographic areas or time periods from a few days to months or years.
- If an outbreak is identified, the report may also list other individuals infected with the same pathogen strain and any relevant medical or epidemiological risk factors (recorded in their EPRs) and suggest appropriate public health action or a possible common source or index case.

The use of integrated clinical and laboratory systems and informatics tools, linked to decision support systems, with continuous analysis and feedback of epidemiological, clinical, outcome and other data, could improve our understanding of disease epidemiology. It would enable assessment and improvement of the predictive accuracy (likelihood ratios) of diagnostic and pathogen profiling methods and the efficacy and cost-effectiveness of treatment and preventive interventions; it should improve clinical outcomes. Nevertheless, as with other novel health management systems, if there is inadequate validation or precautions against inappropriate use, it could lead to unnecessary anxiety, the stigmatization of infected patients, unwarranted infringement of liberty (if coercive public health restrictions are inappropriately applied) and increased healthcare costs.

20.6 Ethical Implication of Improvements in Biosurveillance

20.6.1 Electronic Patient Records

EPRs are computerized medical records, which allow storage, easy retrieval, searching, and sharing of different types of medical and non-medical data (including laboratory results). Many different EPR systems have been described but are still in limited use in hospitals and healthcare systems. Many potential benefits – including better medical care, reductions in medical errors and litigation, and significant cost savings – have been claimed, but, so far, there is limited hard evidence to support the claims. A report commissioned by the Rand Corporation, in 2005, suggested that the introduction of EPRs could save >$US 80 billion in healthcare costs in the USA (Hillestad et al. 2005). However, this has been recently disputed by physicians from Harvard Medical School hospitals – where EPRs have been in use for many years – who claimed that, despite some real benefits of EPRs, the projected cost-savings and quality improvement were exaggerated (Groopman and Jartzband 2009). They expressed concern about the potential use of EPRs to gather evidence about costs, which could be used to limit the use of expensive medical or surgical interventions, and warned against the introduction of expensive technology without rigorous evaluation and evidence.

There has been very little analysis of the potential improvement in disease surveillance by the use of EPR data (and, to our knowledge, none specifically related to communicable disease control). In paper-based medical systems, "privacy is protected by chaos" (Rothstein 2008), records are fragmented and often difficult to compile or locate. EPRs can facilitate the optimal use (mining, analysis, linkage) of data to improve health outcomes and save lives. To achieve this, EPRs would need to be universal (everyone has one), longitudinal (cradle – or womb – to grave) and networked with each other and with other information systems (Fairweather and Rogerson 2001); for example, in the USA, the Nationwide Health Information Network (NHIN) is being established to develop electronic formats that will make records of different types that are compatible and transportable across networks and across the country.

The characteristics which make EPRs most useful are also those that cause most public concern about the potential for inappropriate access and use. Patients will be reluctant to disclose intimate information, no matter what the potential public benefit, if they fear that it could be used to their disadvantage by government officials, employers or insurance companies. Safeguards based on sound ethical principles will be needed to protect privacy and to prevent harm or disadvantage to individuals while promoting public health and gaining optimal benefit from limited public health resources.

Despite increasing concern and legislation relating to the privacy of personal information (e.g., in Australia, the Federal Privacy Act, 1988 – http://www.austlii.edu.au/au/legis/cth/consol_act/pa1988108/), health information is generally treated as a separate category of personal information (e.g., New South Wales [NSW] Health Records and Information Privacy Act, 2002 – http://www.lawlink.nsw.gov.au/lawlink/privacynsw/ll_pnsw.nsf/pages/PNSW_03_hripact). If the use of health information for disease surveillance were to be expanded, there would be certain requirements for the protection of privacy, such as:

- Development of ethical standards for the development, implementation, evaluation and modification of bioinformatics software programs for the storage and analysis of patient data (Gotterbarn and Rogerson 2006)
- Publicly debated, transparent and binding software and hardware standards to protect privacy, confidentiality, integrity and security of data
- Clearly defined principles governing access to identified data for the purposes of disease surveillance or research, including by whom and under what circumstances access is allowed, how it will be monitored and under what circumstances the individual must either give consent or be informed that their record has been accessed

Breaches of privacy may be objective (i.e., resulting in fraud or denial of a service or of freedom) or subjective (i.e. resulting in second or third parties having access to intimate information, which may cause distress, without objective harm). These different consequences may need to be considered differently in assessing the risks associated with the use of EPRs. It has been suggested (Dyson 2008) that the best way to prevent breaches of privacy would be to allow individuals to control access to their own data. However, informed consent for the selective release of medical records (McKinney et al. 2005) would be difficult to obtain and is unlikely to be practicable if data are to be accessible in an emergency or for disease control purposes.

A number of standards exist already, including some designed to protect confidentiality of data transferred across national borders in compliance with international health-related applications e.g., International Organization for Standardization (ISO) 22857:2004 (Kalra and Ingram 2006).

20.6.2 Communicable Disease Notification and Surveillance

Even for communicable disease surveillance, some data can be de-identified and used to monitor trends in disease rates, to identify risk factors and to assess the effectiveness of public health interventions. However, communicable disease surveillance often requires individual patient identification to allow contact tracing, outbreak investigation and the implementation of appropriate control measures and to determine the outcomes. For example, under the NSW Public Health Act, 1991 (http://www.austlii.edu.au/au/legis/nsw/consol_act/pha1991126/), disclosure of certain data is allowed, but there are strict principles governing the collection, storage, access and use of information. In practice, there is little public opposition to the notification of identifiable, personal information to health authorities for communicable disease surveillance, which is accepted as necessary in the public interest. However, this may be, in part, because current communicable disease surveillance systems are generally slow, insensitive, and nonspecific. They are relatively ineffective in detecting, preventing, or interrupting disease outbreaks (Eng and Eng 2004) but also difficult for unauthorized individuals to access and use inappropriately. Thus, privacy is protected by "information friction" (Dyson 2008). The type of future networked EPRs and databases envisaged in the scenario above will be more effective than conventional systems, but potentially more at risk of abuse, with more serious consequences.

Protection of genetic or infectious diseases data is necessary to prevent objective breaches of privacy, such as harassment or stigmatization, which could lead to denial of insurance or jobs. However, improvements in disease control, based on efficient surveillance across large populations could not be achieved if large numbers of people refused to participate because of fear that results could be misused. Denmark has one of the most advanced EPR networks, which allows individuals to block information in their records. This option is reported to be rarely exercised but greatly valued (Rothstein 2008). At present, the disclosure of health information for public benefit is often regulated by laws that are so broad that, in practice, no limits are placed on their scope. EPR networks could, paradoxically, protect privacy more effectively, by allowing limits to be imposed on the scope of data that could be accessed. Scanning software could be programmed to select only information relevant to a specific purpose, using 'contextual access criteria' – software algorithms which specify that, for an enquiry of type X, only data A, B, and C are needed.

Networking of EPRs and other information systems raises new issues relating to informed consent. For effective disease surveillance, all patient records would need to be accessible to data scanning software. Limiting the data that can be accessed to what is relevant may be theoretically possible but defining, in advance, what is relevant may be difficult. Informed consent for individual investigations or routine surveillance would be impracticable. It will, therefore, be important that the implementation of electronic health data management systems and their use for disease surveillance be preceded and accompanied by adequate information, public debate, transparency and appropriate safeguards.

20.6.3 The Use of New Laboratory Data

Networking laboratory information systems. The use of laboratory data for electronic disease surveillance would require that laboratory results from different laboratories mean the same thing. While this may seem obvious, existing differences in result interpretation, predictive values of different methods and lack of consensus on optimal methods, mean that considerable harmonization of laboratory practices will be required. Different laboratory management structures, funding sources, referral patterns, and accountabilities between private and public laboratories or between primarily diagnostic and reference/public health laboratories will make this difficult, but not impossible.

Laboratory staff and directors are often reluctant to share details of tests numbers or methods, quality assurance programs are generally conducted anonymously, and accreditation authorities are required to maintain strict confidentiality, in relation to procedures (including any deficiencies) within individual laboratories. Clearly, issues of trust, commercial confidentiality and quality assurance will need to be addressed at the same time, as details of, laboratory testing methods and interpretation and compatibility of different types of information system.

The ability to generate personalized interpretive laboratory reports based on demographic and clinical data in the EPR would assist clinicians who are often unfamiliar with rapidly changing laboratory methods and their interpretation. The ability of the laboratory information system to rapidly identify a possible outbreak, by identifying clusters of microbial isolates with similar genetic profiles could significantly reduce the size and impact of communicable disease outbreaks. Personalized, targeted decision support can potentially reduce inappropriate antibiotic use, healthcare costs (Sintchenko et al. 2005), the emergence of drug resistance and adverse drug effects.

The use of laboratory data for clinical quality and safety. Laboratory information systems can be used by health authorities to monitor the quality of patient care in individual hospitals (Fairweather and Rogerson 2001) by gathering statistics about infections which develop after a patient's admission to hospital – such as *S. aureus* or specifically MRSA blood stream infections. This has benefits for both potential patients and the general public who arguably have a right to information about the quality of care in hospitals to which they may be admitted in future. In some countries, data related to HAIs are publicly reported, and the occurrence of cases judged to be preventable may incur penalties. For example, the Centers for Medicare and Medicaid Services (CMS) in the USA have recently announced that they will no longer reimburse healthcare facilities for costs related to certain HAIs that could have reasonably been prevented through the use of evidence-based guidelines (http://www.idsociety.org/newsArticle.aspx?id = 6,852).

Many health professionals and administrators are concerned about financial penalties for "preventable" infections and about possible misinterpretation of publicly reported HAI rates because of differences in case-mix and reporting systems (Stone et al. 2005) between hospitals. Some commentators fear that hospitals may refuse to care for high-risk patients who are more likely to develop infections. However, electronic reporting and data scanning software have the potential to analyze individual patient risk factors and adjust incidence data according to the differences in case mix between different types of hospital.

Like most other applications, the use of surveillance data for quality assurance has the potential to improve patient care and the performance and accountability of individual clinicians and healthcare organizations, but there is, also the potential for misuse, breaches of confidentiality and data security not only for patients, but also for professionals, who are usually very wary of any type of performance monitoring.

20.6.4 Surveillance Ethics: A New Paradigm

Advances in surveillance technologies raise the need for the development of frameworks and guidelines for surveillance ethics. Research ethics has traditionally been a central theme of bioethics discourse, for which monitoring guidelines and procedures are well established in health and research institutions, but the ethics of disease surveillance is a relatively unexplored area in need of debate. On the one hand, there are questions about the technical similarities and/or differences between surveillance and research and how they affect practice, if at all (Fairchild and Bayer 2004). In theory, these may be the questions of definition and semantics, but there are currently major differences, which may or may not be justified, in the way these two areas are perceived by practitioners and funding bodies. From an ethical perspective, the key question is whether there are *morally relevant* differences between research and surveillance such that the ethical requirements for the former should not also apply to the latter. According to research ethics, for example, informed consent is paramount and the interests of the individual are supposed to take priority over those of science or society (Declaration of Helsinki - available at: http://www.wma.net/e/policy/b3.htm). Given that research and surveillance are similar insofar as both aim to generate information to promote health outcomes, the crucial questions are whether, why and how much, if at all, ethical requirements for disease surveillance should be less stringent than those of biomedical research.

References

Ballard SA, Pertile KK, Lim M, Johnson PD et al (2005) Molecular characterization of vanB elements in naturally occurring gut anaerobes. Antimicrob Agents Chemother 49(5):1688–1694

Battin MP, Francis LP, Jacobson JA, Smith CB (2009) The multiple perspectives of the *patient as victim and vector* view. In: The patient ad victim and vector: Oxford University Press, Oxford, pp 93–109

Buhlmann M, Bogli-Stuber K, Droz S, Muhlemann K et al (2008) Rapid screening for carriage of methicillin-resistant *Staphylococcus aureus* by PCR and associated costs. J Clin Microbiol 46(7):2151–2154

Caulfield T, McGuire AL, Cho M, Buchanan JA et al (2008) Research ethics recommendations for whole-genome research: consensus statement. PLoS Biol 6(3):e73

Chapman S Jr, Iredell JR, Chapman S, Jr, Iredell JR (2008) Gram-negative sepsis in the intensive care unit: avoiding therapeutic failure. Curr Opin Infect Dis 21(6):604–609

Chaudhry B, Wang J, Wu S, Maglione M et al (2006) Systematic review: impact of health information technology on quality, efficiency, and costs of medical care. Ann Intern Med 144(10):742–752

Cooke GS, Hill AV (2001) Genetics of susceptibility to human infectious disease. Nat Rev Genet 2(12):967–977

Daily JP, Sabeti P, Daily JP, Sabeti P (2008) A malaria fingerprint in the human genome? [comment]. N Engl J Med 358(17):1855–1856

Dyson E (2008) Reflections on privacy 2.0. Sci Am 299(3):26–31

Eng TR, Eng TR (2004) Population health technologies: emerging innovations for the health of the public. Am J Prev Med 26(3):237–242

Fairchild AL, Bayer R (2004) Public health. Ethics and the conduct of public health surveillance. Science 303(5658):631–632

Fairchild AL, Bayer R, Colgrove J (2008) Privacy, democracy and the politics of disease surveillance. Public Health Ethics 1:30–38

Fairweather NB, Rogerson S (2001) A moral approach to electronic patient records. Med Inform Internet Med 26(3):219–234

Francis LP, Battin MP, Jacobson JA, Smith CB et al (2005) How infectious diseases got left out-and what this omission might have meant for bioethics. Bioethics 19(4):307–322

Friedman DJ (2006) Assessing the potential of national strategies for electronic health records for population health monitoring and research. Vital Health Stat 2 143:1–83

Gallego B, Sintchenko V, Wang Q, Hiley L et al (2009) Biosurveillance of emerging biothreats using scalable genotype clustering. J Biomed Inform 2(1):66–73

Gotterbarn D, Rogerson S (2006) Software design ethics for biomedicine. In: Nagl S (ed) Cancer bioinformatics. Wiley, London, UK, pp 213–231

Groopman J, Jartzband P (2009) Obama's $80 billion exaggeration. Wall St J March 11

Gupta R, Gabrielsen B, Ferguson SM, Gupta R et al (2005) Nature's medicines: traditional knowledge and intellectual property management. Case studies from the National Institutes of Health (NIH), USA. Curr Drug Discov Technol 2(4):203–219

Hillestad R, Bigelow J, Bower A, Girosi F et al (2005) Can electronic medical record systems transform health care? Potential health benefits, savings, and costs. Health Aff (Millwood) 24(5):1103–1117

Humphreys H (2008) Can we do better in controlling and preventing methicillin-resistant *Staphylococcus aureus* (MRSA) in the intensive care unit (ICU)? Eur J Clin Microbiol Infect Dis 27(6):409–413

Iredell J, Thomas L, Espedido B, Iredell J et al (2006) Beta-lactam resistance in the gram negatives: increasing complexity of conditional, composite and multiply resistant phenotypes. Pathology 38(6):498–506

Jeyaratnam D, Whitty CJ, Phillips K, Liu D et al (2008) Impact of rapid screening tests on acquisition of meticillin resistant *Staphylococcus aureus*: cluster randomised crossover trial.[see comment]. Br Med J 336(7650):927–930

Kalra D, Ingram D (2006) Ethical issues of electronic patient data and informatics in clinical trial settings. In: Nagl S (ed) Cancer bioinformatics. Wiley, London, UK, p 233–73

Khatib R, Saeed S, Sharma M, Riederer K et al (2006) Impact of initial antibiotic choice and delayed appropriate treatment on the outcome of *Staphylococcus aureus* bacteremia. Eur J Clin Microbiol Infect Dis 25(3):181–185

Klevens RM, Edwards JR, Richards CL, Jr, Horan TC, et al (2007) Estimating health care-associated infections and deaths in U.S. hospitals, 2002. Public Health Rep 122(2):160–166

Masue N, Deguchi T, Yokoi S, Yamada T, Ohkusu K et al (2007) System for simultaneous detection of 16 pathogens related to urethritis to diagnose mixed infection. Int J Urol 14(1):39–42

McGuire AL, Colgrove J, Whitney SN, Diaz CM et al (2008) Ethical, legal, and social considerations in conducting the Human Microbiome Project. Genome Res 18(12):1861–1864

McKechnie ML, Hillman R, Couldwell D, Kong F et al (2009) Simultaneous identification of 14 genital micro-organisms in urine using a multiplex PCR-based reverse line blot (mPCR-RLB) assay. J Clin Microbiol 47(6):1871–1877

McKinney PA, Jones S, Parslow R, Davey N et al (2005) A feasibility study of signed consent for the collection of patient identifiable information for a national paediatric clinical audit database. Br Med J 330(7496):877–879

Mead PS, Slutsker L, Dietz V, McCaig LF et al (1999) Food-related illness and death in the United States. Emerg Infect Dis 5:607–625

Miller S, Selgelid MJ, Miller S, Selgelid MJ (2007) Ethical and philosophical consideration of the dual-use dilemma in the biological sciences. Sci Eng Ethics 13(4):523–580

Modiano D, Petrarca V, Sirima BS, Nebie I et al (1996) Different response to *Plasmodium falciparum* malaria in west African sympatric ethnic groups. Proc Natl Acad Sci U S A 93(23):13206–13211

Rawls J (1971) A theory of justice. The Belknap Press of Harvard University Press, Cambridge, MA
Reid L (2005) Diminishing returns? Risk and the duty to care in the SARS epidemic. Bioethics 19(4):348–361
Rothstein MA (2008) Keeping your genes private. Sci Am 299(3):40–45
Sassetti CM, Rubin EJ (2007) The open book of infectious diseases. Nat Med 13:279–280
Selgelid MJ (2007) Ethics and drug resistance. Bioethics 21(4):218–229
Selgelid MJ, Selgelid MJ (2005) Ethics and infectious disease. Bioethics 19(3):272–289
Sintchenko V, Iredell JR, Gilbert GL, Coiera E (2005) Handheld computer-based decision support reduces patient length of stay and antibiotic prescribing in critical care. J Am Med Inform Assoc 12(4):398–402
Sintchenko V, Iredell JR, Gilbert GL, Sintchenko V, Iredell JR, Gilbert GL (2007) Pathogen profiling for disease management and surveillance. Nat Rev Microbiol 5(6):464–470
Sintchenko V, Coiera E, Gilbert GL, Sintchenko V, Coiera E, Gilbert GL (2008) Decision support systems for antibiotic prescribing. Curr Opin Infect Dis 21(6):573–579
Smith CB, Battin MP, Jacobson JA, Francis LP et al (2004) Are there characteristics of infectious diseases that raise special ethical issues? Develop World Bioeth 4(1):1–16
Stelfox HT, Bates DW, Redelmeier DA (2003) Safety of patients isolated for infection control. J Am Med Assoc 290(14):1899–1905
Stone PW, Braccia D, Larson E (2005) Systematic review of economic analyses of health care-associated infections. Am J Infect Control 33(9):501–509
The Malaria Genomic Epidemiology Network (2008) A global network for investigating the genomic epidemiology of malaria. Nature 456:732–737
Thomas L, van Hal S, O'Sullivan M, Kyme P et al (2008) Failure of the BD GeneOhm StaphS/R assay for identification of Australian methicillin-resistant *Staphylococcus aureus* strains: duplex assays as the "gold standard" in settings of unknown SCCmec epidemiology. J Clin Microbiol 46(12):4116–4117
Thompson AK, Faith K, Gibson JL, Upshur RE et al (2006) Pandemic influenza preparedness: an ethical framework to guide decision-making. BMC Med Ethics 7:E12
Thursky KA, Mahemoff M, Thursky KA, Mahemoff M (2007) User-centered design techniques for a computerised antibiotic decision support system in an intensive care unit. Int J Med Inform 76(10):760–768
Wang Y, Kong F, Yang Y, Gilbert GL (2008) A multiplex PCR-based reverse line blot hybridization (mPCR/RLB) assay for detection of bacterial respiratory pathogens in children with pneumonia. Pediatr Pulmonol 43(2):150–159
Wenzel RP, Bearman G, Edmond MB, Wenzel RP et al (2008) Screening for MRSA: a flawed hospital infection control intervention. Infect Control Hosp Epidemiol 29(11):1012–1018
WHO. Millenium Development Goals 6: combat HIV/AIDS, malaria and other diseases. Journal [serial on the Internet]. 2000 Date: Available from: http://www.who.int/topics/millennium_development_goals/diseases/en/index.html
WHO. Global summary of the AIDS epidemic. Journal [serial on the Internet]. 2007 Date: Available from: http://www.who.int/hiv/data/2008_global_summary_AIDS_ep.png
WHO. Global tuberculosis control – epidemiology, strategy, financing. Journal [serial on the Internet]. 2009 Date: Available from: http://www.who.int/tb/publications/global_report/2009/en/index.html

Glossary

Analysis workflow: The transformation of raw data into biological evidence by applying algorithms, tools and services in a certain order

Annotation: The routine process of assignment of functions to genes in a sequenced genome or the extraction of biological knowledge from raw nucleotide sequences

Antisense: Nucleic acid molecules that bind a complimentary strand of nucleic acid to modify gene expression

Assembly: Construction of longer sequences, such as contigs or genomes, from shorter sequences, such as sequence reads with or without prior knowledge on the order of the reads or reference to a closely related sequence

Bayes' rule: A mathematical identity [$Pr(x|y)=Pr(y|x)\ Pr(x)/Pr(y)$] that allows one to swap variables in a conditional probability expression

Bioinformatics: The application of molecular biology as an information science, especially the use of computational tools and algorithms in genomics research

Biomarker: A biological characteristic which is objectively measured and evaluated as an indicator of normal or pathological processes or host responses to a therapeutic intervention

Biosurveillance: A systematic process that monitors the environment for pathogenic bacteria, viruses and other biothreats. Disease surveillance systematically collects and analyzes this data for the purpose of detecting cases and outbreaks of disease

BLAST (basic logical alignment and search tool): A computer program for finding sequences in databases that have identity to a query sequence

Browser: Interface to the Web that permits the display of Web pages and other applications

Clade: A group of organisms that shares a common ancestor to the exclusion of the other considered taxa

Cladistics: A school of thought that emphasizes reconstructing evolutionary relationships solely through recognizing clades by a set of specific criteria for inference

Clone: Clone can be identified using molecular epidemiological methods. Strains belong to a clonal cluster if they share at least five out of seven housekeeping genes according to multilocus sequence typing

Controlled vocabulary: A set of terms used in a database to describe a particular biological object or process. Use of these terms avoids confusion when describing the same type of biological object or process in different databases

Core genome: The set of genes found in all members of a single species

Data: Any and all complex data entities from observations, experiments, simulations, models and higher order assemblies, along with the associated documentation needed to describe and interpret them

Data integration: The process of combining disparate data and providing a unified view of these data

Data mining: Automatically searching large volumes of data for patterns or associations

Data warehouse: An information infrastructure that enables researchers and clinicians to access and analyze detailed data and trends. Created by collecting databases and linking them using common data elements

De novo gene prediction: An approach to gene prediction in which the only inputs are genome sequences; no evidence derived from RNA is used

DNA sequencing: Biochemical methods for determining the order of the nucleotide bases, adenine, guanine, cytosine and thymine, in a DNA oligonucleotide

Electronic laboratory reporting (ELR): The automated reporting of notifiable disease data via a secure, electronic connection by laboratories to state and local health departments or public health authorities

Electronic medical record (EMR): Computer-based patient medical record

Epitope: The regions of an antigen that bind to antigen-specific membrane receptors on lymphocytes

Exon: DNA or mRNA sequences that include a series of codons carrying information for a part of the amino acid sequence of a protein.

Free text: Data that has no particular structure other than normal grammar; may show substantial variation between records

Functional genomics: Exploration of the function of genes and other parts of the genome

Gene cassettes: Gene cassettes consist of a gene, often conferring resistance to one or more antibiotic agents, and a characteristic recombination site, which can interact with a recombination site present in integrons, resulting in the insertion of the corresponding gene cassette into the integron

Genome: The complete set of genetic information in an organism. In bacteria, this includes the chromosome(s) and extrachromosomal genetic information, e.g., plasmids

Genome-level characters: Features of a genome or its products other than the linear sequences of nucleotides or amino acids that can be assessed for phylogenetic analysis

Genomics: The study of the entire genome of an organism; structural genomics includes whole-genome sequencing, whereas functional genomics aims to determine the functions of all genes

Genotype: The entire genetic constitution of an organism or the genetic composition at a specific gene locus or set of loci

Geographic information system (GIS): A computer system designed to allow users to collect, manage and analyze large volumes of spatially referenced information and associated attribute data

Grid: A fully distributed, dynamically reconfigurable, scalable and autonomous infrastructure to provide location-independent, secure and efficient access to a coordinated set of services encapsulating and virtualizing resources

Informed consent: A legal term referring to a situation where a person can be said to have given his or her consent based upon an appreciation and understanding of the facts and implications of an action

Health Level 7 (HL7): A health data interchange standard designed to facilitate the transfer of health data resident on different and disparate computer systems in a health care setting

Homoplasy: A pattern of character states that supports an alternative to the true, accepted or most parsimonious evolutionary tree that is generally caused by evolutionary changes

Horizontal gene transfer: Any process in which an organism transfers genetic material to another cell that is not its offspring. This process is in contrast to more common vertical gene transfer, which occurs when genetic information is passed from parent to offspring

Hospital-acquired infection (HAI): An infection that is associated with a stay in a hospital. An infection is considered nosocomial or hospital-acquired if it occurs 48 h or more after a hospital admission

Infectome: System of networks of interacting host and pathogen's genes, proteins and metabolites involved in a process of infection and disease

Information retrieval: An electronic process that selects documents from a collection based on a user's query

Insertion sequences: Insertion sequences are genes that code for a transposase protein. This protein can interact with inverted repeats on either side of the gene, leading to transposition of the gene

International Classification of Diseases (ICD): A standard vocabulary for diseases, health status, types of patient visits to doctors and other health providers, and external cases of injuries

Intron: Portions of a gene between the coding exons that are also transcribed, but are enzymatically removed from the mRNA before its translation into a protein

Knowledge base: A repository for the knowledge used by a knowledge system

Knowledge-based system: A computer system that represents and uses knowledge to carry out a task

Metagenomics: The high-throughput study of sequences from multiple genomes recovered from samples that contain mixed microbial populations

Metadata: Data about data; may be regarded as a subset of data which adds relevance and purpose to data and enables the identification of similar data in different data collections

MHC – major histocompatibility complex: A large genomic region or gene family which plays an important role in the immune system

Microbiome: Collective system of genomes of all microbial flora of the human

Middleware: A software stack composed of security, resource management, data access and other services and applications, users and resource providers to operate effectively

Natural language processing: Automated methods for converting free-text data into computer-readable format

Network: Series of points or nodes interconnected by edges, edges can have direction or different weights

Next-generation sequencing: Novel approaches to DNA sequencing that dispense with the need to create libraries of clone sequences in bacteria and holds the promise of providing faster and cheaper sequencing

Notifiable disease: A disease that by public health law must be reported to some jurisdictions, typically a local public health authority, by laboratories, hospitals, or individual clinicians

Ontology: The systematic description of a given phenomenon, which often includes a controlled vocabulary and relationships, captures nuances in meaning and enables knowledge sharing and reuse. Typically, ontology defines data entities, data attributes, relations and possible functions and operations

Outbreak detection: A process or set of processes that detects the existence of an outbreak

Pan-genome: The set of all genes found in members of a single species

Ontology: A formal description of set of entities within a body of knowledge and the relationships between those entities, used to reason about the entities. Usually is represented as hierarchical, and often richly interconnected, set of objects, concepts and other entities that embody knowledge about the field

Orthologs: Homologous genes in two or more organisms that are related only by lineage splitting and not by gene duplication

Parsing: A segmentation of a string of letters together with a labeling of the segments

PCR or polymerase chain reaction: A method for amplifying a specific region of DNA fragment using enzyme DNA polymerase and short primer sequences to delimit the amplified region. The repetitive cycle of reactions results in the exponential production of new DNA molecules

Phenetics: Phylogenetic reconstruction based on measures of overall similarity

Pharmacophore: Functional group linked to a molecular scaffold that is necessary to ensure the optimal supramolecular interactions with a specific biological target and to trigger (or block) its biological response

Polymorphism: Genetic variability across a population of the same species. Within a particular gene, there may be single or multiple base changes which may not affect an individual microorganism or may cause a biological change. They are distributed in a population in different frequencies depending on when they occurred and their biological effects

Quantitative structure–activity relationship (QSAR): Rules deduced from the comparison of active and inactive analogues of a drug, correlating features of the drug scaffold (such as the presence of specific categories of pharmacophores) with their effects on the target

Quorum sensing: The communication and coordination of bacteria through signaling molecules

Scaffold: A central structure of a molecule, on which one can add or remove functional groups

Single nucleotide polymorphism (SNP): Sites in the genome where individual organisms differ in their DNA sequence, often by a single base, usually with very low population frequencies

Species: For the purpose of sequence analysis, a population of biologically alike microorganisms that share nearly the same genetic makeup

Standard vocabulary: Systems of names that are assigned to concepts or entities that can create order within databases

System biology: Integrative discipline that seeks to explain the properties and behavior of complex biological systems in terms of their components and their interactions

Systematized Nomenclature of Medicine (SNOMED): A standard vocabulary system for medical databases; contains more than 144,000 terms and is available in at least two languages. Developed by the College of American Pathologists

TCR – T-cell receptor: A molecule found on the surface or T cells which is essential for recognizing antigens bound to major histocompatibility complex molecules

Transposons: Transposons are similar to insertion sequences but usually larger. Normally, they contain at least one resistance gene, but may include an integron that in turn holds several resistance gene cassettes

Terminal branch: The part of an evolutionary tree that leads only to the taxon considered (not internode branches)

Universal genetic code: A misnomer based on an earlier, incorrect belief that all genomes share the same code for specifying amino acids from triplets of nucleotides

Virulence factor: A protein or a gene that is required for a pathogen to cause disease

Whole-genome shotgun sequencing: An approach to determine the sequence of a genome in which the genome is broken into numerous small fragments. These fragments are then assembled en masse. The individual sequences are assembled into larger sequences (known as contigs) that correspond to substantial portions of the genome

Index